AAPS Introductions in the Pharmaceutical Sciences

Volume 15

Series Editor

Claudio Salomon, National University of Rosario
Rosario, Argentina

Founding Editor

Robin Zavod, Chicago College of Pharmacy
Midwestern University
Downers Grove, IL, USA

The *AAPS Introductions in the Pharmaceutical Sciences* book series is designed to support pharmaceutical scientists at the point of knowledge transition. Springer and the American Association of Pharmaceutical Scientists (AAPS) have partnered again to produce a second series that juxtaposes the *AAPS Advances in the Pharmaceutical Sciences* series. Whether shifting between positions, business models, research project objectives, or at a crossroad in professional development, scientists need to retool to meet the needs of the new scientific challenges ahead of them. These educational pivot points require the learner to develop new vocabulary in order to effectively communicate across disciplines, appreciate historical evolution within the knowledge area with the aim of appreciating the current limitations and potential for growth, learn new skills and evaluation metrics so that project planning and subsequent evolution are evidence-based, as well as to simply "dust the rust off" content learned in previous educational or employment settings, or utilized during former scientific explorations. The *Introductions* book series will meet these needs and serve as a quick and easy-to-digest resource for contemporary science.

Kenneth Alexander • Perry Romanowski
Editors

Cosmetic Science

Formulation and Technologies

Springer

aaps®

Editors
Kenneth Alexander
Pharmacy
The University of Toledo College of Pharmacy and Pharmaceutical Sciences - Professor Emeritus
Grand Rapids, OH, USA

Perry Romanowski
Vice President
Element 44 Inc
Chicago, IL, USA

ISSN 2522-834X ISSN 2522-8358 (electronic)
AAPS Introductions in the Pharmaceutical Sciences
ISBN 978-3-032-03595-0 ISBN 978-3-032-03596-7 (eBook)
https://doi.org/10.1007/978-3-032-03596-7

This Springer imprint is published by the registered company Springer Nature Switzerland AG
The registered company address is: Gewerbestrasse 11, 6330 Cham, Switzerland

This Cosmetic Science Volume is dedicated to:

My wife Dolores who was always there to encourage me to proceed with this endeavor.

My co-editor who was there to provide the morale for both of us to keep us on track.

As the co-editor, I am honored to have contributed to this work alongside my esteemed colleague. This volume reflects our shared passion for advancing cosmetic science and supporting the professionals who drive this industry forward. I would also like to express my gratitude to my wife, Shannon, whose unwavering support, and encouragement made this endeavor possible.

Finally, to all the men and women who work in the Cosmetic and Personal Care Industries. It is they who daily and diligently pursue the formulation, production, testing, packaging, distribution, and sales of all the products the people of the world use in their lives to increase their beauty, comfort, and convenience.

It is hoped that this volume increases their knowledge and peaks their enthusiasm for their industry and in which they work as well as their educational pursuits.

Preface

Cosmetic science brings together chemistry, biology, material science, and human behavior to support the development of personal care products. From the early use of scented oils and plant extracts to today's complex formulations, the field continues to adapt and respond to changing needs, resources, and expectations. Science plays a key role in how these products are made, how they function, and how they're received by the people who use them.

This volume, *Current Perspectives in Cosmetic Science*, brings together recent thinking, technical practices, and real-world concerns across the personal care sector. It provides a balanced mix of historical background, formulation science, and forward-looking insights. The goal is to offer useful information to anyone involved in the creation, evaluation, or regulation of cosmetic and personal care products.

The Scope of This Volume

This book begins with a review of history, tracing how fragrance and personal care have developed over time. From ancient techniques to modern formulations, the early chapters provide context for understanding how science and industry have shaped today's products.

The middle sections take a closer look at fragrance creation, raw materials, and formulation approaches. Topics include:

- The role of perfumers and how fragrance is structured
- Natural and synthetic ingredients, including their roles and challenges
- Differences between product formats such as fine fragrance, rinse-off, and leave-on formulas
- Practical concerns such as phase separation, stability, and odor control
- Evaluation methods for performance, aesthetics, and sensory appeal

The final chapter explores developments in ingredient delivery, shifts in consumer priorities, and regulatory updates. These themes are likely to influence future research, product design, and market direction.

Why This Matters

Consumer interest in ingredients, sourcing, safety, and environmental impact has grown. These shifts are influencing how companies design, produce, and communicate about their products. Cosmetic science sits at the center of this conversation, where product claims meet real-world results.

Regulations are part of the picture too. As new rules and standards emerge around the world, those working in product development and testing must stay informed and flexible. This includes being able to explain choices clearly and support claims with evidence.

Collaboration between different fields—chemistry, toxicology, dermatology, sensory science, and data analysis—has always been part of this work. That continues to be true as cosmetic science responds to new questions and expectations.

Who This Book Is For

This volume is intended for a wide range of readers, including:

- Formulators working on ingredient choice, product design, or stability
- Scientists and researchers focused on function, testing, or performance
- Regulatory teams supporting claims, documentation, and compliance
- Educators and students looking for a grounded, up-to-date resource

Whether you're starting out or have spent years in the lab, this book is meant to support your work and offer new perspectives on familiar topics.

Acknowledgments and Final Thoughts

Progress in this field depends on the people who show up every day to solve problems, test ideas, and make better products. The authors, reviewers, and contributors to this volume reflect that effort. Their willingness to share knowledge helps move the field forward.

We hope this book serves as a reliable and thought-provoking resource. Whether used in the lab, classroom, or office, we believe it will provide practical value to those working in cosmetic and personal care science.

Grand Rapids, OH, USA Kenneth Alexander
Chicago, IL, USA Perry Romanowski

Contents

Editors and Contributors

About the Editors

Kenneth Alexander is Professor Emeritus at the University of Toledo College of Pharmacy who after 44 years of teaching retired in 2016. During his tenure at the University of Toledo, he developed the MS Industrial Pharmacy program, which graduated approximately 140 students from 1988 through 2016. He also developed the BSPS Pharmaceutics program in 2002 as well as the BSPS Cosmetic Science program in 2012. He is the author of more than 80 publications, 6 book chapters, 100 posters, and 1 book. He personally mentored more than 106 MS students through their thesis and had more than 80 of these students go on to pursue their PhD at other institutions. The remainder pursued employment in pharmaceutical and chemical industries in the USA or returned to their country to prosper in their pharmaceutical industry. He is currently a private consultant for Formulation and Product Development as well as Pharmaceutical CE courses for Pharmacists and Pharmacy Technicians.

Perry Romanowski has spent the past 30 years researching and developing products in the personal care and cosmetic industry. His primary focus has been on formulating hair-related products. He is currently vice president of Element 44 Inc, which specializes in science education and is responsible for the consumer-focused podcast, TheBeautyBrains.com, and the cosmetic science website, ChemistsCorner.com. Previously, he worked for Alberto Culver, serving as a senior project leader for innovation. He has been an active member of the Society of Cosmetic Chemists (SCC) and is past president. Currently, he is the Education Committee chair for the IFSCC. He earned his BS in Chemistry from DePaul University and his MS in Biochemistry. He has written and edited numerous articles and books. He also teaches the SCC's Beginning Cosmetic Chemistry continuing education class. Additionally, he provides consulting services for cosmetic formulating, testing, and Internet solutions. He can be reached through his website, ChemistsCorner.com.

Contributors

Tasnem H. Abdelfattah Allam is in the last year of the Bachelor of Pharmacy program at Ajman University, UAE. She is recognized for consistently excelling academically and being among the top students in her class. She has received recognition for exceptional academic accomplishments and has set herself apart through her leadership abilities, guidance of younger students, and involvement in joint research initiatives.

Pooja G. Achar is a research scholar at the Department of Pharmacology, JSS College of Pharmacy, JSS Academy of Higher Education & Research (JSS AHER), Mysuru, Karnataka, India. She is a dedicated researcher in the field of pharmacology, specializing in drug development, molecular pharmacology, and translational medicine. As a part of JSS AHER, her research focuses on exploring novel therapeutic approaches and evaluating pharmacokinetics and pharmacodynamics to improve drug efficacy and safety.

Hillary Belzer is the founder and curator of The Makeup Museum, a virtual museum she established in 2008. Dedicated to the preservation, research and exhibition of makeup from antiquity through the twenty-first century, The Makeup Museum is a progressive, unique space that explores makeup artistry and packaging, culture, and history from all eras. With a bachelor's degree in art history and a master's degree in cultural studies, she ensures The Makeup Museum is a leader in conceptualizing new ways of thinking about makeup and its cultural and artistic significance. Her specialties include the intersection of fine art and makeup, '90s cosmetics, and little-known makeup histories.

Sai H. S. Boddu, PhD, is a Professor of Pharmaceutics in the College of Pharmacy and Health Sciences, Ajman University (AU), UAE. He has over 20 years of pharmaceutical experience in both academic and industrial settings. He has developed several pharmaceutical formulations and novel nanocarrier systems for Rx and OTC drugs. He holds 2 patents and has published more than 120 research and review articles in various peer-reviewed journals. He is a co-Editor of the book titled *Excipient Applications in Formulation Design and Drug Delivery*, published by Springer. He serves as an Associate Editor of *Cancer Reports* and Scientific Advisor on the *Journal of Pharmaceutical Sciences*. He is an active member of the American Association of Pharmaceutical Scientists (AAPS).

Kathy Cruz's education includes a BS degree in Biochemistry from SUNY Stony Brook and an MA degree in Cosmetic Science from Fairleigh Dickinson University. With over 25 years of cosmetic product development experience, she spent most of her career working on the many cosmetic brands of Avon Products, Inc. and COTY. She now works at Kobo Products, Inc., where she is the Director of Kobo's US Application laboratory. New to the raw ingredient part of the industry, she thoroughly enjoys her new role as she gets to support many cosmetic and personal care companies with their product development launches.

Andrea Echeverry, PhD is a lifelong biologist. She graduated with a BS in Microbiology and Biotechnology with a minor in Aerobiology from Florida Atlantic University. During her PhD studies in Microbiology & Immunology at the University of Miami, she developed the first neonatal mouse model of intestinal infection using fully pathogenic human-isolated strains of *Yersinia enterocolitica*. Despite having made many discoveries and after a decade in academic research, she decided to become independent to pursue her own scientific interests in gut and skin microbiomes, where she could apply her knowledge in microbial pathogenesis and chronic low-level inflammation. This led her to become a hair loss specialist as this field allowed her to study microbiome interactions in the smallest human organ, the hair follicle.

She is an expert in scalp micropigmentation, hair transplant scar camouflaging, and biocompatible hair implantation. With these techniques, she improves cosmetically, people suffering from hair loss or hair thinning due to gut and skin dysbiosis and environmental factors. As an advocate for those struggling with scalp health issues, she's now a practitioner of Trichology in Florida, Georgia, and Puerto Rico. She is particularly passionate about integrating permanent cosmetics like scalp micropigmentation with trichology principles to help people realize their ideal cosmetic goals. Her work has allowed her to collaborate with dermatologists and hair transplant surgeons to promote the importance of the skin and scalp microbiome, which she believes is often overlooked in the pursuit of aesthetic perfection. She has acted as an advisor and speaker in skin and scalp microbiome health, contributing her unique perspective and knowledge to the broader community. She is the previous past President and Treasurer for the Association of Professional Trichologists, a nonprofit organization dedicated to creating unity in the hair loss industry.

B. R. Ganavi is a research scholar in the Department of Pharmacology, JSS College of Pharmacy, JSS Academy of Higher Education & Research (JSS AHER), Mysuru, Karnataka, India. She is a dedicated researcher specializing in pharmacological studies with a focus on drug discovery and development. As a key contributor to the Department of Pharmacology at JSS College of Pharmacy, she has been actively involved in exploring novel therapeutic strategies, conducting preclinical and clinical studies, and advancing research in targeted drug delivery systems.

Hosahalli Veerbhadrappa Gangadharappa's, PhD, research expertise spans diverse areas of pharmaceutical sciences, including pharmaceutics, pharmaceutical nanotechnology, graphene-based drug delivery systems, and silk fibroin-based drug delivery systems. His innovative work focuses on developing novel drug delivery platforms that address critical challenges in the pharmaceutical field, particularly in targeted and controlled drug delivery. Beyond his academic and research accomplishments, he is known for his commitment to mentoring young scientists and fostering innovation. His extensive knowledge and dedication to advancing pharmaceutical sciences have earned him a reputation as a prominent figure in the field. His contributions have played a pivotal role in shaping the future of drug delivery systems and nanotechnology applications.

Allison Turquoise Kent-Gunn Garibay is a leader in cosmetic packaging, specializing in sustainability, consumer insights, and innovative design solutions. As a Cosmetic Packaging Sales Director, she collaborates with brands to develop packaging strategies that align with their values and meet evolving consumer demands. By leveraging her robust social media presence of over 75,000 engaged followers—including beauty enthusiasts and industry professionals—she provides brands with qualitative consumer insights, helping them navigate the complexities of packaging development with a deep understanding of consumer behavior and preferences. This direct connection has guided numerous brands to adopt innovative, value-aligned solutions that achieve both environmental and commercial success.

Her academic contributions include teaching university-level courses in cosmetic packaging development, where she equipped students with a rigorous framework to critically evaluate packaging, assess sustainability claims, and challenge greenwashing through critical thinking. Her approach fosters a new generation of professionals attuned to both technical and ethical considerations in packaging design. Her thought leadership extends beyond the classroom. She has been a featured speaker at prominent industry events, including Makeup New York, Supplier's Day, Beauty Accelerate, and the Florida Chapter of Cosmetic Chemists, and served as a packaging judge for Beauty Industry West's Emerge Awards. Additionally, her expertise has been featured in publications such as *Elle*, *Dazed*, *The Guardian*, *Dermascope*, *Glossy*, *PopSugar*, *Highsnobiety*, and *The Cut*.

She holds an MBA specializing in Management, a BA in Art History, and a Minor in Chicano Studies, along with certificates in Circular Economy and Sustainability Strategies, and Globalization. This diverse educational background has provided her with a unique foundation to appreciate the intersections of business, beauty, and sustainability, enabling her to approach cosmetic packaging challenges with both creativity and strategic insight. Motivated by a mission to demystify cosmetic packaging for both professionals and consumers, she blends academic rigor with actionable insights to advance sustainable and innovative practices in the beauty industry.

Valerie George is an award-winning cosmetic chemist and speaker, science communicator, educator, leader, and avid proponent of transparency in the beauty industry, she formulates for and lends expertise to hair care and hair color brands as CEO of Simply Formulas. Previously, she worked on the latest research in hair care and hair color formulations as Executive Vice President of R&D at John Paul Mitchell Systems, a leading salon brand. She is currently a national board member for the Society of Cosmetic Chemists and was the past chair for their Committee on Scientific Affairs. She additionally volunteers on the Governance Committee and Technical & Regulatory Committee for the Independent Beauty Association. She additionally teaches the Cosmetic Science Certificate Program at the University of

California Riverside. With more than a decade of beauty industry experience, she enjoys her time as a contributing writer at *Happi Magazine* and co-hosting the award-winning podcast, The Beauty Brains.

Paolo Giacomoni, PhD is as a consultant to the skin care industry and serves as the Head of R&D of L-Raphael. He was Chief Scientific Officer of Élan Rose International and served as VP of Skin Care World Wide R&D with Herbalife. As Executive Director for R&D with Estee Lauder, he contributed to the preparation of the best-selling products for Clinique and La Mer. He also served as scientific spokesperson for Clinique. In the frame of the European Society for Photo-biology, he was the Editor of two monographs on the effects of solar radiation on human skin. During his tenure at L'Oréal, he served as Head of the Department of Biology and then as scientific attaché to the Director of Applied Research. Results of his research were applied in products such as Primordial (Lancôme) and Plénitude (L'Oréal). During his work in academia, he was Maître de Conférences at the University of Paris and was Visiting Professor at the University of Milano. He is presently Editor-in-Chief of the *Journal of Cosmetic Science*. He received his PhD in Biochemistry from the University of Paris, Paris, France; his master's degree in Atomic Physics from the University of Milano, Milano, Italy; and has had Post-Doctoral Training at Deutsches Krebsforschungszentrum at Heidelberg, Germany, the University of Wisconsin, Madison, WI, and the University of California, San Diego, CA. He was one of the founders of the European Society for Photobiology as well as of the European Network for the Study of the Biology of Aging. He has authored 100+ publications and patents in Biochemistry, Molecular Biology, and Biology of Aging, focusing on DNA damage and the metabolic impairment induced by UV radiation, as well as on the positive effects of antioxidants.

Steve Herman currently serves as Scientific Advisor to Sino Lion USA. He was an Adjunct Professor in the Fairleigh Dickinson University Cosmetic Science Program from 1993 to 2018, teaching the Cosmetic Formulation Lab and Perfumery. Beginning in 2019 he teaches a distant learning Fragrance Science course for the University of Cincinnati. He has also conducted numerous webinars for the Center for Professional Advancement. He was a columnist for *GCI Magazine* for 17 years and has written a book, *Fragrance Applications: A Survival Guide*, and several book chapters. His Society of Cosmetic Chemists activities include service as Chairman of the NY Chapter in 1992 and 2013, election to Fellow status in 2002, and instructor in the Continuing Education Program.

Sanjay Jayaram is a prominent researcher in Pharmaceutics and is currently affiliated with the JSS College of Pharmacy at the JSS Academy of Higher Education and Research in Mysuru, India. With a strong academic background and a commitment to advancing pharmaceutical sciences, he has contributed to various research projects and published numerous articles in peer-reviewed journals. His research aims to bridge the gap between laboratory findings and clinical applications, striving to improve patient outcomes through enhanced pharmaceutical formulations. Through his dedication to education and research, he plays a vital role in shaping the future of pharmaceutical sciences in India and beyond.

Madhuchandra Kenchegowda is a research scholar in the Department of Pharmaceutics, JSS College of Pharmacy, JSS Academy of Higher Education and Research, Karnataka, India. He has contributed several book chapters and review articles. His work has been recognized for its significant impact on advancing the field of pharmaceutics and related areas. His research interests are diverse and interdisciplinary, focusing on pharmaceutics, pharmaceutical nanotechnology, and nanomedicine.

B. Mahalakshmi, MD, is Associate Professor of Microbiology at the Government Medical College, Kadapa, India, with over 15 years of teaching and clinical experience. She has published more than 10 research articles in the field of medical and environmental microbiology.

Maryam R. Mohamed is in her final year of the Bachelor of Pharmacy program at Ajman University, UAE. She is ranked among the best students in her class. She has been acknowledged for her outstanding academic achievements and has distinguished herself through her leadership skills, and participation in collaborative research projects.

Riyaz Ali M. Osmani, PhD, is an Assistant Professor in the Department of Pharmaceutics, JSS College of Pharmacy, JSS Academy of Higher Education and Research, Karnataka, India. He has authored more than 170 research publications (cumulative impact factor >400) and several book chapters and review articles. His research interests include pharmaceutics, pharmaceutical nanotechnology, nanomedicine, pre-formulation profiling, solid-state characterization, nanomaterials, microneedle technology, and polymer grafting.

Sonal Vijay Pande, MS, is a Doctoral Researcher specializing in Neuroscience with a focus on enhancing the oral bioavailability of polyphenols using metabolic adjuvants. She has extensive experience in developing diagnostic kits, molecular pharmacology, and regulatory toxicology. She has published more than nine research articles and two book chapters. She also has experience in preclinical studies, computational drug design, and pharmacokinetic profiling.

Chirag A. Patel, PhD, is an Assistant Professor in the Department of Pharmacology at L. M. College of Pharmacy. He has over 15 years of academic and research experience in the fields of pharmacology and toxicology. He has authored over 40 research papers, four books, and 10 book chapters. His research interests also include nutraceuticals, probiotics, neurodegenerative diseases, and antimicrobial resistance.

Ketan Ranch, PhD, is an Associate Professor in the Department of Pharmaceutics at L. M. College of Pharmacy. He has more than 19 years in teaching and research experience in formulation development. He has published more than 50 research/review articles in prestigious international journals and contributed to 9 book chapters and 1 book. He is a corporate trainer for Design Expert software and has conducted more than 25 training sessions across India, Bangladesh, and Singapore.

Doug Schoon is an internationally recognized research scientist, author, and educator with over 30 years of experience in the cosmetics, beauty, and personal care industries. He is a leading authority, renowned for his expertise in scientific research, regulatory compliance, and educational content in the cosmetics industry. He holds 17 patents for innovative cosmetic formulations and technologies.

He is the president of Schoon Scientific and has served as Co-chair of the Nail Manufacturers Council on Safety (NMC) under the Professional Beauty Association for 21 years. He was also CND's Vice President of Science and Technology for 20 years. He holds an MS in Chemistry from the University of California, Irvine.

He also serves as an expert witness in legal cases involving cosmetic products and services. He specializes in advanced testing of hair, nails, and cosmetic products, utilizing methodologies such as differential scanning calorimetry (DSC), scanning electron microscopy (SEM), high-resolution digital microscopy, tensile strength testing, gas chromatography-mass spectrometry (GC-MS), infrared spectroscopy (IR), high-performance liquid chromatography (HPLC), and thin-layer chromatography (TLC).

Keerthiraj Kalmantidoddi Shivaji is a research scholar in the Department of Pharmaceutics, JSS College of Pharmacy, JSS Academy of Higher Education and Research, Karnataka, India. He has made notable contributions to the development of innovative nanomaterials, with a focus on their application in advanced drug delivery systems. His research extends to the area of diabetic wound healing, where he is working on formulating nanotechnology-based interventions to enhance wound closure, reduce infection rates, and accelerate the repair process in diabetic patients.

A Global History of Makeup from Ancient Times to Today

Hillary Belzer

Abstract This chapter provides a general overview of makeup across multiple cultures from antiquity through the present day, defining makeup as pigment applied in a temporary fashion to the face and/or body for any number of purposes. As makeup's history is as broad and sweeping as other art forms, inevitably some areas and practices (theatrical and editorial makeup, various subcultures, nail painting, henna, and tooth blackening) are excluded. Also, the lack of accessible information on cosmetics in ancient civilizations outside of Greece, Rome, and Egypt, along with that of Indigenous peoples across the globe, means that makeup history for some regions or groups with rich cosmetic traditions is not as complete as for others.

Keywords Makeup history · Ancient makeup · Contemporary makeup · Makeup styles

This chapter provides an overview of makeup from prehistoric times through today. As makeup's history is as broad and sweeping as other art forms, inevitably some areas and practices (theatrical and editorial makeup, various subcultures, nail painting, henna, and tooth blackening) must be excluded. Also, the lack of accessible information on cosmetics in ancient civilizations outside of Greece, Rome, and Egypt, along with that of Indigenous peoples across the globe, means that makeup history for some regions or groups with rich cosmetic traditions is not as complete as for others.

H. Belzer (✉)
The Makeup Museum, Baltimore, MD, USA
e-mail: contact@makeupmuseum.org

K. Alexander, P. Romanowski (eds.), *Cosmetic Science*, AAPS Introductions in the Pharmaceutical Sciences 15, https://doi.org/10.1007/978-3-032-03596-7_1

Section 1: Prehistoric-ca. 600 CE

Makeup's story begins in prehistoric times. The oldest hard evidence for cosmetics usage is found in numerous artifacts from Egypt, the Middle East, and Europe. Shells dating to 50,000 years ago were found in southern Spain and their contents were analyzed in 2010.[1] The pigments—a yellow mineral derived from the jarosite family which was later used by Egyptians for both cosmetics and painting decorative objects, and a red powder flecked with ground pyrite to give a shiny, glitter-like effect—suggest that Neanderthals participated in the wearing of cosmetics.[2] In 2021, archaeologists studied the contents of 14 ceramic bottles dating to between 4350 and 4100 BCE that were excavated from a site in Slovenia formerly occupied by the Lasinja people.[3] Chemical analysis revealed that the bottles contained a mix of cosmetic ingredients: beeswax, plant oils, animal fat and the lead carbonate-based cerussite. As most of the containers had holes in their handles, they most likely were strung on rope and worn on the person for transport. Additionally, stone sticks were found nearby, which may have been used for application.[4]

Central and South America

While it is difficult to know exactly when Indigenous peoples of the Americas began using cosmetics, most likely their practices go back to antiquity. The Seri people of Northwest Mexico painted distinct patterns of lines and dots in red, blue, black, and white horizontally across the bridge of the nose onto the cheeks for a variety of purposes. Both men and women wore these designs for festivals, rituals, and to cure certain ailments. Women wore differing patterns to denote marital status as well. Interestingly, the blue pigment was prepared only by older girls and women; girls

---

[1] João Zilhão, Diego E. Angelucci, Ernestina Badal-García, et al., "Symbolic Use of Marine Shells and Mineral Pigments by Iberian Neanderthals," Proceedings of the National Academy of Sciences of the United States of America, volume 5, no. 3., January 11, 2010, 1023–1028. https://doi.org/10.1073/pnas.091408810

[2] Christopher Joyce, "Study: Neanderthals Wore Jewelry and Makeup," *NPR*, January 12, 2010. https://www.npr.org/2010/01/12/122466430/study-neanderthals-wore-jewelry-and-makeup

[3] Brine Kramberger, Christoph Bertheld, and Cynthianne Spiteri, "Fifth Millennium BC Miniature Ceramic Bottles from the South-Eastern Prealps and Central Balkans: A Multi-Disciplinary Approach to Study Their Content and Function." Journal of Archeological Science: Reports, Volume 38, August 2021.
https://doi.org/10.1016/j.jasrep.2021.102993

[4] David Bressan, "Stone-Age Cosmetics Based on Lead-Mineral Oldest of Its Kind," Forbes, July 15, 2021.
https://www.forbes.com/sites/davidbressan/2021/07/15/stone-age-cosmetics-based-on-lead-mineral-oldest-of-its-kind/

who had not yet reached menstruation as well as men and boys were forbidden from observing the process for fear that the pigment would not turn blue.[5]

Mayan murals and ceramics from the Pre-Classic era (ca. 1000 BCE) through their colonization by the Spanish depicted men and women wearing face and body paint. The body paint was usually monochromatic, painted in broad swaths of red or black on the shoulders, chest, and back. Face paint also usually consisted of one shade, although two or more colors appeared more frequently than in body paint. Sometimes the entire face was painted white as a base to make detailed patterns in red or black stand out. There did not appear to be any consistent system for face and body paint; the men and women in Mayan art—royalty, warriors, and attendants to the gods—were shown wearing a variety of designs that completely differed from each other, even within the same scene.[6] Thus, it is difficult to glean any particular meaning for cosmetics in terms of social status or religious purposes.

The Kayapó people of northern Brazil have a rich tradition of face and body painting. Their cosmetics primarily consist of red pigment made from the ground seeds of the achiote shrub, and a mixture of liquid from the genipa fruit and charcoal to produce black pigment. Painting is primarily performed by women, who learn the techniques at a young age by practicing on dolls carved from babassu nut.[7] All members of the culture and all ages—from infants to the elderly—are regularly adorned with face and body paint in intricate patterns both for aesthetic purposes and to indicate life stages or milestones, such as puberty or the birth of their first child.[8]

Middle East and South Asia

One of the earliest instances of lip color dates to somewhere between 2600 and 2450 BCE, discovered in the tomb of Queen Puabi in southern Iraq. She was buried with a variety of cosmetic containers, including shells which had remnants of various pigments. One of those pigments was comprised of red ochre and white lead, which were used as ingredients in lip color and blush.[9] In early 2024, scientists unearthed what is presumably a stone lipstick tube in Iran, dating to 1936–1687 BCE. While it is impossible to definitively identify the gender or social standing of the person it was buried with, the intricate incisions on the tube indicate

[5] Hillary Belzer, "Empire of Sand: The Seri (Comcaac) People," December 17, 2022. https://makeupmuseum.org/seri-comcaac-face-paint/

[6] Katherine Lukach and Jeffrey Dobereiner, "The Painted Body," in The Adorned Body: Mapping Ancient Mayan Dress. Austin: University of Texas Press, 2020, 32–38.

[7] "Kayapó Body Art," preview of exhibition at the Pitt Rivers Museum, Oxford, UK, August 17, 2017. https://www.meer.com/en/28769-kayapo-body-art

[8] Pinar Yolacan, "Interview: Kayapo Chief Tuire," Bomb Magazine. June 13, 2017. https://bombmagazine.org/articles/2017/06/13/kayap%C3%B3-chief-tuire/

[9] Penn museum.

a high degree of craftsmanship that may have been reserved for those on a high socioeconomic level.[10] Combined with the contents of Queen Puabi's tomb and the fact that cosmetic containers with red pigments are scarce relative to other cosmetic vessels and ingredients that have been found, lip color and blush may have just been for royalty and the upper classes within these eras and regions rather than a widely used cosmetic. More frequently found in this region were containers with traces of white minerals (cerussite, gypsum, calcite) dating anywhere between 3000 and 1000 BCE.[11]

Kohl, a type of eye makeup, was ubiquitous across social strata, genders and ages throughout the ancient Arab world and South Asia. Numerous vessels spanning from the first through fourth millenniums BCE with traces of different kohl ingredients have been unearthed in Jordan, Qatar, Yemen, Syria, India, and Turkey. The oldest known kohl pencil, discovered in Turkey, dates to approximately 6500 BCE.[12] While the word kohl is rooted in Arabic, it has many names and ingredients throughout various regions. In India, it is called kajal or *anjana* (collyrium) and traditionally made from soot gathered from a ghee lamp.[13] Anjana dates to the Vedic period (ca. 1500–500 BCE) and is mentioned several times in the *Rig-Veda*, a sacred Hindu text written during the early part of the era. (PubMed article) Surma is a powdered kohl that also originated in India and made from galena or antimony. In Persia, kohl was called surmeh or sormeh and derived primarily from soot from burning various types of nuts.[14] Two containers discovered in Western Iran dated to 4500 BCE were shown to have remnants of an eye makeup paste made from lapis lazuli.[15]

Kohl containers took on a variety of shapes and sizes and were usually made from glass, bronze, wood, or stone. Long, thin vessels resembling modern-day test tubes were common, as were shorter round pots. Some containers had holes drilled at the top so that a chain with an application stick could be attached.[16] Additionally, the design of the application tools and implements used to mix the substance varied by culture. The most common method involved grinding black powder (lead, lamp-

---

[10] Nasir Eskandari, Eugenio De Carlo, Federico Zorzi et al., "A Bronze Age lip-paint from Southeastern Iran," Scientific Reports, February 1, 2024. https://doi.org/10.1038/s41598-024-52490-w

[11] Massimo Vidale, Oliver Craig, François Desset et al., "A Chlorite Container Found on the Surface of Shahdad (Kerman, Iran) and its Cosmetic Content," Iran, vol. 50, January 2012, 29–54.
https://www.researchgate.net/publication/286213215_A_Chlorite_Container_Found_on_the_Surface_of_Shahdad_Kerman_Iran_and_its_Cosmetic_Content

[12] Oguz Buyukyildirim, "Archaeologists Discovered 8200-year-old Eyeliner in Türkiye's Yeşilova Höyük," Arkeonews. September 16, 2024.
https://arkeonews.net/archaeologists-discovered-8200-year-old-eyeliner-in-turkiyes-yesilova-hoyuk/

[13] Zahra Hankir, Eyeliner: A Cultural History. New York: Penguin Books, 2023, XXXVI.

[14] Hankir, Eyeliner, XXXVII.

[15] Fatema Soudavar Farmanfarmaian, "Haft Qalam Ārāyish: Cosmetics in the Iranian World." *Iranian Studies* 33, no. 3–4 (2022): 285–326. https://doi.org/10.1080/00210860008701984

[16] See Jolanda Bos, Paint It Black: A Biography of Kohl Containers. BLKVLD Publishers: Zandvoort, 2019.

black, etc.) mixing it with animal fat or grease using a type of mortar and pestle or spoon specifically reserved for cosmetics to make a paste. The paste was then stored in a container or dried into cakes or pencils. However, kohl was usually applied thickly with a separate stick along the top and bottom lash lines and water lines. While aesthetic uses were adopted later, kohl's earliest purpose was health-related: it had antibacterial, anti-inflammatory qualities and kept away disease-carrying insects. It also mitigated the harsh glare of the sun. Kohl is still widely used today in the Middle East, North Africa and India, as it is significant to various cultures' identity and heritage.

India

With the spread of Hinduism (ca. 2000 BCE) in India, forehead markings known as *tilaka* were used to differentiate between various sects and worn by all genders. Comprising various patterns, *tilaka* adorned the middle of the forehead, an area called the *ajna chakra*, one of the seven major spiritual energy centers in the body described in the Vedas (Hindu scripture written between 1500 and 500 BCE).[17] It was believed that placing a physical, visible representation of the "third eye" on the *ajna chakra* helped the wearer be more receptive to the spiritual realm.[18] The most common colors were red (made of vermilion or red lead), yellow (saffron or turmeric), white (cow dung, and for holy men, ashes of ritualistic fires[19]), and black. In later years, some forms of *tilaka* became a beauty aid for women. Sindoor, a red paste made from turmeric, lime, and occasionally mercury, was used to denote a woman's marital status. On their wedding day, the groom would apply sindoor to the center part of his bride's hair and she would apply it daily afterwards. Widows and single women were forbidden from wearing sindoor, as mercury was believed to increase libido and red was a symbol of fertility.[20] The bindi, a red dot made of sindoor, was worn by men and women for religious purposes, but in later times was primarily worn by married women as an additional symbol of their marital status.[21]

The *Kama Sutra* (ca. 200–300 CE) valued face and body painting so much that it was included as one of the 64 arts in the text. Women painted *viseaka*, decorative patterns in lac dye on their foreheads and cheeks using a special stick (*patrariguli*).

[17] Priyabala Shah, Tilaka: Hindu Marks on the Forehead. Ahmedabad: New Order Book Co., 1985.

[18] Syama Allard, "Five Things to Know about Tilaka," Hindu American Foundation. January 28, 2021. https://www.hinduamerican.org/blog/5-things-to-know-about-tilak

[19] https://web.prm.ox.ac.uk/bodyarts/index.php/temporary-body-arts/body-painting/191-hindu-facial-markings.html

[20] Bishnu Prasad Dahal, "Significance of *Sindoor* (Vermilion Powder) in Hindu Marriage Rituals," 77–78.

[21] https://www.britannica.com/topic/bindi

Sometimes designs were cut out from betel leaves.[22] The *Kama Sutra* also lists a recipe for a lash darkener of which lampblack is the main ingredient.[23] Other literature describes cosmetic practices as well. The *Jatakas* (ca. fourth century CE) discuss women's use of face powder, *anjana* and painting designs on the cheeks. For some occasions, a paste made from the red arsenic mineral realgar was smeared on the face. The *Brahmajala Sutta* (ca. sixth century CE) lists *anjana*, face powder, and rouge as well. Later texts describe similar routines.[24]

Egypt

Makeup was essential to the ancient Egyptians, so much so that they were buried with cosmetic vessels for use in the afterlife. Malachite and chrysocolla, minerals that produced a green hue, were the main ingredients for eye makeup in the predynastic era (5500–3100 BCE). While black pigment used for the eyes was found in some predynastic tombs, green was much more common as green copper ores were readily available.[25] During this time, green paint was mostly used to treat eye ailments; however, some researchers believe the ancient Egyptians linked green to new crops and fertility, or possibly associated green with the Eye of Horus and used the paint as a ritual or temple offering. It was not until the early dynastic period (ca. 3100 BCE) that black pigment took over as the most common eye makeup, most likely due to the fact that it has a lower melting point and was therefore easier to smelt.[26] From this time onwards, women and men of all social classes wore kohl for medicinal, aesthetic and ritual purposes, with the upper classes mostly using galena, a lead ore native to northern Egypt, while those further down the social ladder made do with soot.[27] Green was still occasionally used but most likely reserved for ritual use during the New Kingdom (ca. 1550–1070 BCE).[28] While no major kohl production center has been discovered, an analysis performed in 2023 of a late Bronze Age kohl box found in Turkey suggests that the Egyptians were such prolific users and makers of kohl that they traded it with other regions.[29] Additionally, a 2022 study

---

[22] Moti Chandra, "Cosmetics and Coiffure in Ancient India," Journal of the Indian Society of Oriental Art, vol. 8, 1940, 83.

[23] Maggie Angeloglou, A History of Makeup. London: MacMillan, 1970, 34.

[24] Chandra, "Cosmetics and Coiffure," 71; 91–92.

[25] Andrew Hardy and Gavyn Rollinson, "Green Eye Cosmetics in Antiquity," The Pharmaceutical Historian, vol. 39, no. 1, 2009, 2. https://www.researchgate.net/publication/26333151_Green_eye_cosmetics_of_antiquity

[26] Hankir, Eyeliner, 5.

[27] Hankir, Eyeliner, 6.

[28] Hardy and Rollinson, "Green Eye Cosmetics in Antiquity," 2.

[29] İsmail Tarhan and Fatma Şahin, "Characterization of a Late Bronze Age 'Kohl Box' from the Cilician Plain using FTIR and GC–MS," Journal of Archeological Science, vol. 50, no. 3, August 2023. https://doi.org/10.1016/j.jasrep.2023.104088

suggested that kohl contained more diverse ingredients than previously thought, revealing that the ancient Egyptians were quite skilled at cosmetic chemistry.[30] Less frequently worn than eye makeup was lip and cheek color, made from red ochre. Vessels containing red ochre or other ingredients used for lip and cheek cosmetics are scarce compared to the vast quantity of kohl containers. Additionally, there are only a handful of Egyptian artworks depicting the wearing of blush and lip tint.

In addition to the makeup itself, a key contribution of the ancient Egyptians to makeup history was the design of cosmetic containers and other tools. Contrasting with the simple shells found in previous eras, the Egyptians introduced a variety of elaborate styles in which to store and apply their cosmetics—at least, for royalty and upper classes. Ingenious swivel palettes and spoons shaped like animals to ornately carved or jeweled kohl containers demonstrate the creativity and attention to detail enjoyed by wealthy Egyptians. Complex cosmetic boxes held both makeup and application tools, along with hair styling/removal implements and fragrance. The Egyptians also utilized a wide variety of materials to construct these vessels, including alabaster, ivory, and blue faience.

East Asia

Multiple ancient Chinese literary texts convey the region's beauty standards. Pale white skin was favored for both men and women; in the latter, long dark eyebrows and rosy cheeks were also prized. Dating between the first and fifth centuries BCE, these texts discuss white face powder made from rice (*fen*) as well as a red-dyed version (*chengfen*) used for rouge. A red pigment derived from safflowers (*yanzhi*) was also used for rouge at least since the Han period (206 BCE-220 CE),[31] and there are several references to minimum derived from both mercury and lead.[32] Women used soot from burnt willow branches to darken and lengthen their eyebrows up until about 200 BCE, and then began shaving their brows and drawing them on during the Han dynasty.[33] Dai, a black pigment derived from graphite, was the primary source for eyebrow painting. During the early part of the Han period, long brows that drooped at the outer edges were popular, but by the second century CE, brows that formed a straight line became the preferred shape. The poet Zhuo Wenjun (175–121 BCE) was said to have painted long, thin eyebrows resembling "distant mountains," a style that became fashionable. Unlike the Middle East, North Africa,

[30] https://www.nature.com/articles/s41598-022-08669-0

[31] Cho Kyo, In Search of the Beautiful Woman: A Cultural History of Japanese and Chinese Beauty. Plymouth: Rowman & Littlefield, 2012, 24; 114–116.

[32] Edward H. Schafer, "The Early History of Lead Pigments in China," T'oung Pao, 1956, Vol. 44, 420–423.

[33] https://pubmed.ncbi.nlm.nih.gov/35262483/

and South Asia, lining the eyes did not appear to figure prominently into ancient East Asian makeup routines, especially in China where narrow eyes were valued.[34]

Hufen (lead powder) was noted as early as the time of King Zhou (ca. 1122 BCE) and was considered higher quality due to its longevity as compared to rice powder, making it the powder of choice for those who could afford it.[35] In 2022, scientists analyzed the lead white left in several bronze jars excavated from tombs in the nobility cemetery of Liangdaicun in Northeast China, which was established sometime during the Rui State (ca. 1100–476 BCE). The lead remnants dated to the eighth century BCE, several hundred years earlier than the synthetic lead used for cosmetics in ancient Greece, and they were created using a different process than that of the Greeks, indicating that the two ancient civilizations were producing and using lead-based makeup independently of each other. The bronze jars containing the lead white were found in the tombs of both male and female aristocracy, providing material evidence that both genders wore cosmetics to achieve white skin.[36] Historical accounts of the Han Dynasty also support the notion that court men wore face powder. In particular, scholar He Yan (ca. 195–249 CE) was known for his vanity and allegedly never seen without a powder puff.[37]

Generally speaking, makeup during the Zhou Dynasty was simple, consisting of white face powder, brow painting, and tinted lip grease. The Qin and Han periods witnessed more complex makeup rituals and imported ingredients due to increased trade. Red powder (*fufen*) and white powders were applied together as foundation, then rouge made from safflower or cinnabar (*dan*).[38] Eyebrow painting grew in popularity, most likely due to the story of Zhang Chang (d. 47 CE), a Han dynasty official who painted his wife's eyebrows each morning as a sign of love and affection.[39]

Two new makeup styles emerged during the Six Dynasties period (220–589 CE): slanting or oblique red (*xie hong*) and forehead yellow (*ehuang*). According to legend, a court maid named Xue Yelai was in the study of Wei Emperor Cao Pi (187–226 CE) as he was reading behind a crystal screen. Xue Yelai, new to the palace, did not see the glass and crashed into it, the shards forming deep gashes on each side of her temples. They healed into two crescents, which the Emperor likened to the morning sun just coming up over the horizon. The women of the court began imitating Xue Yelai's scars using red pigment, hoping to attract the emperor's attention.

---

[34] Cho Kyo, 18–19.

[35] Cho Kyo 24; 117.

[36] Bin Han, Bei Zhang, Jianrong Chong, et al., "Beauty and chemistry: the independent origins of synthetic lead white in east and west Eurasia," Humanities and Social Sciences Communications, 9, September 2022. https://doi.org/10.1057/s41599-022-01290-6

[37] Bret Hinsch, Passions of the Cut Sleeve: The Male Homosexual Tradition in China. United States: University of California Press, 1990, 65–66.

[38] https://www.globalizationpartners.com/2021/10/28/chinese-makeup-history/ - Shihi Tang.

[39] Li Yu.,The Fragrant Companions: A Play About Love Between Women. New York: Columbia University Press, 2022.

The rise of Buddhism across China, with its golden temple statues, influenced women to apply yellow paint or pre-cut yellow flakes with paste. It is not entirely certain what material was used for *ehuang*, but one source suggests a type of lead called massicot.[40] While it is unclear how common the practice was, the *Ballad of Muhan* (ca. fifth century CE) describes legendary Northern Dynasty warrior Hua Mulan as having adorned her forehead in this manner.[41] At the very least, the *Qimin Yaoshu*, an agricultural treatise completed between 533 and 544 CE that contained recipes and instructions for making face powder and rouge, suggests the widespread adoption of these basic cosmetics. Eyebrow painting continued into the sixth century CE, with green paint (*luo dai*) imported from Persia being the most popular with the court women of the Sui Dynasty (581–618).[42]

Korea and Japan

Like the Chinese, the people of ancient Korea valued pale skin free of any blemishes or pigmentation such as freckles. To achieve this, they primarily relied on skincare (cleansing water, facial scrubs, and moisturizers) made from natural ingredients such as rice and mung beans. However, makeup was also in use at least as early as the Three Kingdoms era (57 BCE-668 CE). Ground rice and millet were used as face powders and sometimes mixed with water or oil for better adhesion to the face.[43] Ground seeds of the mirabilis jalapa (nicknamed the "four o' clock flower" after the time of day it blooms) were also used for face powder and applied with a tool crafted from a silkworm cocoon[44]; however, given that the flower is native to South America, it likely was not introduced in Korea until the eighteenth century.[45] Several Goguryeo Dynasty murals in northern Korea tombs from the late fifth century depict noblewomen with blackened eyebrows and *yeonji* (rouge made of safflower).[46] The colorant was also used on the lips. However, the *Samguk Sagi*, a historical account of the Three Kingdoms era written in the twelfth century, notes

[40] Schafer, "Early Lead Pigments," 418–420.

[41] The Ballad of Mulan, Asia for Educators, Columbia University. The Flowering Plum and the Palace Lady: Interpretations of Chinese Poetry, by Han H. Frankel (New Haven: Yale University Press, 1976), 68–72. https://afe.easia.columbia.edu/ps/china/mulan.pdf

[42] Zhou Xun and Gao Chunming, 5000 Years of Chinese Costumes. San Francisco: China Books and Periodicals, 1987, 91.

[43] Chung Ah-young, "Tracing History of Cosmetics," *The Korea Times*, January 31, 2013. https://koreatimes.co.kr/www/news/culture/2013/01/135_129776.html

[44] Korea News Review, vol. 18, issue 52, 1989.

[45] Keith Pratt and Richard Rutt, Korea: A Historical and Cultural Dictionary. United Kingdom: Taylor & Francis, 2013, 130.

[46] Kyunghee Pyun and Minjee Kim, Dress History of Korea: Critical Perspectives on Primary Sources. London: Bloomsbury, 2023, 32.

that women in the Baekje kingdom did not wear rouge, only face powder.[47] The *Samguk Sagi* also describes young male warriors from the elite class of the Silla kingdom who were outfitted in elaborate clothing, jewelry, and makeup. These young men, known as *Hwarang* ("flower knights" or "flower boys") regularly wore face powder and rouge as part of their regalia.[48] It is not clear how the *Hwarang* started or the reason for their feminine attire. Some sources indicate that the *Hwarang* had shamanistic origins, i.e., as admirers of nature and music, cosmetics were a way of enhancing their physical appearance. Others suggest that two young women were selected to be "source flowers" whose beauty and musical talents the *Hwarang* would aspire to, and makeup was part of embodying the women's beauty.[49]

Based on the clay figurines with red pigment adorning their faces made during the Kofun period (300–710 CE), it seems the Japanese were using color cosmetics as early as the fourth century CE. However, this may have been more of a burial ritual than an everyday makeup practice.[50] Generally speaking, makeup in Japan was reserved for religious ceremonies and festivals. It was not until the seventh century when makeup became commonplace due to imports from Korea and China, and even then, worn primarily by the aristocracy.[51]

Greece and Rome

Ancient Greek women had little agency over their position in society. Their role was strictly limited to domestic activities, i.e., maintaining the household and raising children. As such, pale skin was valued as it signified she was fulfilling her expected duties and not outdoors. The ancient Greeks dictated that visible makeup was to be reserved for sex workers (*hetaerae*) to advertise their services.[52] Thus, it is probable that few Greek women wore makeup, but if they needed help achieving the ideal of fair skin and subtly rosy cheeks, they used a very light coating of ceruse (white lead powder) and a variety of ingredients for rouge, including red ochre, seaweed, alkanet and later, cinnabar. Large dark eyes and elongated brows that met above the nose were also highly valued, so eyes were defined with antimony, kohl, or a mixture of burnt cork and soot.[53]

---

[47] Han Mi-Ja, On the Scent of a Beautiful Life. London: Ebury Publishing, 2017.

[48] Stephen O. Murray, Homosexualities. Chicago: University of Chicago Press, 2000, 66.

[49] Roger Tennant. A History Of Korea. New York: Routledge, 2010, 35.

[50] https://web-japan.org/niponica/niponica23/en/feature/feature03.html

[51] Masako Isa and Eric Kramer, "Adopting the Caucasian 'Look': Reorganizing the Minority Face," in The Emerging Monoculture: Assimilation and the "Model Minority". Praeger: West Newport, 2003, 52.

[52] Terence McLaughlin, 45–46.

[53] Richard Corson, Fashions in Makeup from Ancient to Modern Times. London: Peter Owen Publishers, 2003, 56.

Women in ancient Rome had slightly more independence than their Greek predecessors. While beauty ideals remained the same as those of Greek culture, more women were able to participate in wearing makeup. Wealthy Roman women used enslaved or servant girls to assist with their daily beauty routine, including makeup application. They used the same ingredients as the Greeks, although some Roman women used chalk dust mixed with vinegar as an alternative to ceruse. Wine dregs and orchella weed (fucus) were also used for rouge.[54] Makeup was stored in containers called *pyxides*. Interestingly, there has been no material evidence of either the Greeks or Romans using lip color.[55]

The Roman era was also the first recorded mention of beauty patches or spots. Archaeological evidence of ancient beauty spots (*splenia*) has not been discovered yet, but there is literary evidence to suggest that they were worn by both men and women, usually to hide blemishes and scars and constructed from leather or cloth. There are also references in ancient Roman literature to formerly enslaved people wearing these spots to conceal branding marks they had gotten during their time as slaves.[56] In addition, beauty patches were worn not to conceal, but to highlight totally clear, pale skin—they made a very striking contrast, particularly if one was wearing whitening makeup. Another method of making skin appear whiter and give a translucent effect was to trace the veins with blue paint along one's temples; however, as this is only described in a poem by Prospertius, it is not clear whether Roman women actually participated in this practice.[57]

There is no concrete evidence to suggest that Greek men wore makeup. Some men wore makeup during the Roman empire, but it was generally frowned upon by their peers, who viewed "painted" men as being overly vain and therefore not suitable for working in particular fields such as law or politics.[58] Poets such as Juvenal mocked men who engaged in wearing makeup.[59]

Section 2: ca. 600–1900 CE

Western Europe

There is little material evidence of cosmetics after the dominance of Christianity began around 400 CE, as the tradition of burying the deceased with their belongings faded with its spread. Combined with the destruction and unrest caused by religious

[54] Kelly Olson, "Cosmetics in Roman Antiquity: Substance, Remedy, Poison," *Classical World*, vol. 102, no. 3, spring 2009, 296–297.

[55] Olson, "Cosmetics in Roman Antiquity," 299.

[56] Ortha L. Wilner, "Roman Beauty Culture," *The Classical Journal*, vol. 27, no. 1, October 1931, 32.

[57] A Dictionary of Greek and Roman Antiquities. Edited by William Smith. Oxford University, 1848, 551.

[58] Stewart, 47.

[59] Corson, Fashions in Makeup, 53–54.

wars, few cosmetic artifacts from the Middle Ages have survived. However, there is an abundance of literary and medical texts that point to cosmetic usage, along with recipes that were beginning to be recorded and passed down.[60] For the most part, beauty ideals of a fair, unblemished complexion and cheeks with a hint of color remained consistent from the Medieval period through the Renaissance. Chaucer's description of a lily-white complexion and rosy cheeks embodied the ideal face.[61] Eyes were not as emphasized as they had been previously, and in contrast to the darkened, elongated brows coveted by the Greeks and Romans, brows were plucked to achieve a high, perfectly smooth forehead.

Medical texts such as the *Trotula* (eleventh to twelfth century CE) and in particular the *Trotula Minor*, which provided recipes for makeup and body care, proved invaluable to our understanding of cosmetic ingredients used at the time.[62] Wheat flour or lily root were commonly used for face powder or paint; lead-based formulas were actually less common in the Middle Ages. Ground angelica leaves, brazilwood chips, and dried safflower were the primary coloring ingredients for blush and lip color, and occasionally crushed berries were used.[63] Some recipes for lip tint and blush called for red bryony, a plant that irritated the skin and caused redness.

The influence of Christianity cannot be overstated when it came to dictating what women could and could not wear on their faces. For the most part the Church viewed cosmetics as immoral and deceitful—to wear makeup was to indulge in the sin of vanity.[64] This was particularly true in England, where the Church had more of an impact on the average citizen; their French counterparts were bolder in their use of makeup.[65] However, clergy generally looked the other way if women wore cosmetics to disguise scars from disease or to attract a husband, or if she was married, to maintain his interest.[66] As long as the application was invisible (i.e., no one should detect that the woman is wearing makeup), women could get away with cosmetics. But heavy face paint was still considered an affront to God and largely associated with sex workers.[67]

As in earlier times, women of high standing were expected to care for their physical appearance and had access to better ingredients than the middle and lower classes. The court women of Edward IV most likely wore whitening powder and red lip salve, and a hint of kohl along the lashlines and brows.[68]

---

[60] Stewart, Painted Faces: A Colourful History of Cosmetics. Gloucestershire: Amberley Publishing, 2017, 63.

[61] Freja Gunn, The Artificial Face, 65.

[62] Stewart, Painted Faces, 68.

[63] Stewart, Painted Faces, 78–79.

[64] Corson, Fashions in Makeup, 65–67.

[65] Gunn, The Artificial Face, 67.

[66] Stewart, Painted Faces, 60.

[67] Sarah Jane Downing, Beauty and Cosmetics 1550–1950. Oxford: Shire Publications, 2012, 7.

[68] Sally Pointer, The Artifice of Beauty: A History and Practical Guide to Perfumes and Cosmetics. Gloucestershire: Sutton Publishing, 2005, 75.

As the Renaissance era progressed, increased trade and the debut of the printing press in 1440 meant that recipes could be more easily distributed. Indeed, makeup recipes in various "Books of Secrets" proliferated in Italy, where cosmetics were more acceptable than in other European countries.[69] As it had been for centuries, members of the royalty and aristocracy were the most influential in terms of makeup styles, and beauty standards themselves—pale, smooth, umblemished skin, rosy cheeks and lips—did not significantly change either. The rich used ceruse or alabaster for whitening the face and rouge made from red ochre or a lead base dyed with a host of other pigments such as vermilion, cochineal, or alkanet. Lip color was applied from a crude sort of "crayon" made from a dried paste of ground alabaster and various coloring agents. The entire face was then topped off with a layer of egg white to seal the makeup and give a shiny effect.[70]

While several portraits of Elizabeth I (1533–1603) depict her with a stark white visage, there is little evidence that the queen regularly wore ceruse, or at least not as thickly applied as has been assumed. One scholar suggests that the wearing of imported cosmetics shown in the Ditchley Portrait (1592) and the Armada Portrait (ca. 1588) is a reflection of Elizabeth's desire to convey English power and imperialism via whiteness.[71] Indeed, as colonization and use of enslaved people in the households of wealthy Europeans increased, so too did the value of white skin and its status as a marker of racial superiority. This is evidenced by the proliferation of recipes for skin-whitening creams and face powders throughout the continent.[72]

The average European man did not regularly use any type of makeup in the Medieval or Renaissance eras. There are a couple of anecdotes in which beggars used cosmetics in a more "special effect" way by disguising themselves as lepers.[73] Henry III of France was also said to have used whitening powder and rouge, while his attendants donned beauty patches.[74]

Despite an attempted ban on cosmetics by the Puritans, court ladies and upper-class women adopted the ritual of the toilette in seventeenth century England, in which visitors observed the lady of the house/palace be coiffed and made up by her servants.[75] Numerous anti-makeup treatises by male authors and clergymen did not deter women from wearing powder, rouge and eyebrow/lash darkener.[76] Due to the Stuarts' close relationship with the French court and increased trade within the Continent, cosmetics usage soared in seventeenth century England.[77] Venetian

---

[69] Stewart, Painted Faces, 90; 98–101.

[70] Downing, Beauty and Cosmetics, 13.

[71] Kimberly Poitevin, "Inventing Whiteness: Cosmetics, Race and Women in Early Modern England," Journal for Early Modern Cultural Studies, spring/summer 2011, vol. 11, no. 1, 62–63.

[72] Jill Burke, How to Be a Renaissance Woman: The Untold History of Beauty & Female Creativity. London: Pegasus Books, 2024, 84.

[73] Stewart, Painted Faces, 87.

[74] Angeloglou, A History of Makeup, 41.

[75] Downing, Beauty and Cosmetics, 19–21.

[76] Corson, Fashions in Makeup, 121–129.

[77] Gunn, The Artificial Face, 90.

ceruse and Spanish rouge (bits of wool or paper dyed and rubbed onto the cheeks) continued to be the most coveted articles in an urban, upper-class woman's makeup routine.[78] It was also around this time that beauty patches came into vogue. These faux beauty spots (known as *mouches* in France) were intended to make a striking contrast against white skin in addition to hiding smallpox and other scars, as they had been in ancient Rome. Beauty spots came in many shapes and sizes and sometimes were assigned certain meanings depending on their placement.[79] Several portraits of women in the Netherlands wearing large spots on their temples, coupled with medical texts from the time indicating that patches could help migraines, point to medicinal usage of patches.[80]

The beauty spot was easily the most villainized of any cosmetic at the time, with satirists implying they were hiding signs of venereal disease and therefore the mark of a prostitute. Others went so far as to state that the wearing of beauty spots was a practice borrowed from "barbarous nations".[81] Along with the fact that patches were used to emphasize whiteness, this statement reflects an aversion to them that was distinctly rooted in white supremacy.

While men were also mocked for wearing patches, some portraits from the seventeenth century suggest mouches had a more noble meaning for men. Paintings of several high-ranking British soldiers portray them with beauty spots ostensibly to hide battle scars; however, some scholars believe they were actually meant to highlight the sitters' valor and bravery, with the spots serving more as badges of honor and symbols of dedication to England.[82] As these are formal portraits, it is not clear whether the sitters wore them regularly or only for their portrait. And no matter the meaning of patches, the ornate gold and enameled boxes used to store them, the production of which proliferated in the 1700s, were certainly symbols of wealth and power.

France in the mid-late 1700s witnessed greatly increased production of cosmetics and fragrance. The origin of Paris as the most influential center for fashion and cosmetics in the Western world, a status it would maintain through the twentieth century, became firmly established. By the end of the 1700s, purchasing of makeup had become normalized for all social classes.[83] The ritual of the toilette flourished in the second half of the 1700s in both the royal court and non-aristocratic (but still well-to-do) households.[84] A heavily painted look was favored by the aristocracy, and

[78] Downing, Beauty and Cosmetics, 27.

[79] Downing, Beauty and Cosmetics, 22–23.

[80] Karen Hearn, "Revising the Visage: Patches and Beauty Spots in Seventeenth-Century British and Dutch Painted Portraits," Huntington Library Quarterly, Vol. 78, No. 4 (Winter 2015), 820–822. See also https://unframed.lacma.org/2018/02/01/beauty-eye-beholder-fashion-18th-century-mexico

[81] Digital Museum of Dress Accessories. https://dmda.york.ac.uk/the-patch-room/patch-02/

[82] Hearn, "Revising the Visage," 816–817.

[83] Morag Martin, Selling Beauty: Cosmetics, Commerce, and French Society, 1750–1830. Baltimore: Johns Hopkins Press, 2009, 30–31.

[84] Pointer, The Artifice of Beauty, 115.

both men and women wore white face powder, rouge, and patches. The French style of makeup was mocked by British commentators, whose opposition to the Parisians' excessive makeup was more a matter of politics than aesthetic preference. Despite their critiques, upper-class English women and royalty emulated French makeup styles, albeit in a slightly more subdued manner.[85] And although many physicians warned of the dangers of the lead-based ceruse, it continued to be worn with gusto by the upper class. Elizabeth and Maria Gunning, Irish sisters and socialites who became the darlings of London, were rumored to have serious health issues from their use of lead-based makeup. While darkening brows had been commonplace for centuries, there was little mention of eyelash treatments. One of the first instances of a beauty guide providing a recipe for a form of mascara appeared in 1772 in the *Toilet of Flora*.[86]

A rapid shift towards favoring a natural appearance across Western Europe took place after the French Revolution, as the "painted" look was associated with royal excess.[87] Obvious makeup quickly went out of style following the last decade of the 1700s and during the Napoleonic/Restoration periods, and beauty guides containing cosmetic recipes using home-grown (not imported), harmless ingredients proliferated. The trend of a natural look carried over to plant-based formulas free of lead and mercury, along with advice emphasizing the importance of making one's own cosmetics instead of purchasing from a seller.[88] Still, demand for makeup and other beauty products continued to increase among all social classes. And despite the pleas for lead-free makeup found in medical literature and other texts, ceruse continued to be worn through the mid-nineteenth century.[89]

The Victorian era carried on the preference for a natural look. Beauty guides, such as the one written by Irish dancer Lola Montez in 1858, mostly advised against the use of makeup; if one truly felt makeup was necessary, it should be absolutely undetectable.[90] Despite a defense of cosmetics by the poet Charles Baudelaire in 1863, visible makeup was still considered to be the domain of sex workers and actresses. "Respectable" women did wear makeup to attain the time-old standards of a pale, clear complexion, a hint of rosiness in the lips and cheeks and dark brows and lashes, but applied it subtly and secretly.[91] The 1800s saw the birth of the beauty salon, hundreds of which sprung up in major cities across Europe and the United States. (source) Both men and women could purchase premade cosmetics and toiletries from pharmacies such as Shiseido (1872), perfumeries including Guerlain

[85] Caroline Palmer, "Brazen Cheek: Face-Painters in Late Eighteenth-Century England," Oxford Art Journal, Vol. 31, No. 2 (2008), 205–206.

[86] Pointer, The Artifice of Beauty, 119.

[87] Corson, Fashions in Makeup, 291.

[88] Gunn, The Artificial Face, 129–131.

[89] McLaughlin, The Gilded Lily, 100.

[90] Lola Montez, The Arts of Beauty, or, Secrets of a Lady's Toilet. New York: Dick & Fitzgerald, 1858, 47–49.

[91] Rebecca Haffner, "Material Monarch: Queen Victoria and the Middle-Class Toilette," Master of Arts paper, 2018, 27–32.

(1828) and Rimmel (1834), or any of the theatrical makeup brands that had been established, such as Bourjois (1863) and Leichner (1873). Towards the end of the 1800s, ladies' magazines advertising cosmetics and department stores also contributed to makeup sales.[92]

The Victorian attitude towards obvious paint continued into the early twentieth century in Western Europe and the United States, but was quickly replaced by a widespread acceptance of makeup ushered in by technology and a new consumer culture.

East Asia

One of the most adventurous eras for makeup was China's Tang Dynasty (618–907 CE). Economic prosperity coupled with technological advances and increased exposure to Western cultures as well as China's own ethnic minorities contributed to growing availability and usage of makeup. Several major stylistic departures from previous eras included the increased use of red pigment, the adoption of *huadian* (a decorative element applied to the forehead), and rapid progressions through numerous trends. As in other cultures, royalty and the upper classes generally determined trends for the rest of the population. In China specifically, makeup changed to align with the emperor's preferences and what he perceived as beautiful.[93]

Tang women enjoyed the luxurious look of red makeup and used it in abundance. *Mianye* (also known as *zhuangye*) was a popular style consisting of a small red dot on each cheek next to the corners of the mouth, resembling dimples and similar to French *mouches*. According to an entry in the ninth century historical account *Youyang zazu*, *mianye* originated during the Han era, when Wu ruler Sun He accidentally injured his wife, Lady Deng. The medicine to treat her wound, which had not been properly mixed, left a red dot-shaped scar on her left cheek.[94] Other court women began dotting both cheeks as they thought the dimpled effect would please the emperor. Another use of red pigment came in the form of covering nearly the entire surface of both cheeks with red to imitate the flush one would get after drinking wine, a style known as *jiuyun* makeup. Less common but still notable styles developed quickly in the middle Tang years, including "tear" and "sob" (*shishi*) makeup. A poem recording the practices of the Kaiyuan and Tianbao eras (713–741 and 742–755 CE, respectively) describes women powdering their faces and crying, leaving spots of bare skin, a style that dated back to the late Han era and revitalized

92 Stewart, Painted Faces, 218.

93 Cao Chen, "How Cosmetics Were Created in Ancient China," China Daily, April 21, 2018. https://www.chinadaily.com.cn/a/201804/21/WS5ada295aa3105cdcf6519a30.html

94 Tattoo in Early China, 368.

during the mid-Tang period.[95] A later poem mentions a form of makeup that gave a mournful expression. Instead of tear traces, however, eyebrows were painted similarly to the character for the number 8 (an inverted V shape) and a black paste or oil was smeared on the lips; no powder or rouge was used.[96] Towards the end of the Tang Dynasty, a mural from the tomb of Emperor Zhao Yigong dated to 829 CE depicts several court women with eyebrows shaved off and red stripes painted above and below the eyes, a style known as *xie yun* ("blood halo") makeup, which mimics bloody scratches or wounds. Finally, Tibetan culture influenced the adoption of *zhemian* ("ruddy" or "ochre face"). While tradition says that China's royal court outlawed the practice as they felt it symbolized barbarism, there is little historic evidence to suggest that fashionable Chinese women stopped engaging in *zhemian*.[97]

According to a twelfth century poem, *huadian* is said to have originated with the story of Princess Shouyang of the Southern Dynasty (fifth century CE), who fell asleep under some plum blossom trees outside the palace. When she awoke, a plum blossom had settled on her forehead. Despite her attempts to brush it off, it remained stuck to her forehead for 3 days.[98] After the flower fell off, it left a mark which the other court ladies agreed made her even more lovely and charming, and soon they began decorating their foreheads similarly.

A multitude of *huadian* designs were introduced during the Tang Dynasty, ranging from simple dots to elaborate flower shapes, gold foil, kingfisher feathers, and even fish bones.[99] They were affixed through a glue made from fish maw (swim bladder) and washed off with hot water.[100] Examples of *huadian, jiuyun* makeup, *mianye*, as well as the "red slant" makeup that developed earlier, can be seen in various paintings and figurines from the Astana tombs in the Xinjiang region of northwest China.

Emphasis on eyebrows continued during the Tang Dynasty. Emperor Xuanzong (dates) ordered his court painters to depict no fewer than 10 different shapes to serve as a guide for the royal ladies, such as the "mandarin duck," the "hanging pearl," and the "five ridges".[101] During the entirety of the Tang era, roughly 16 shapes waxed and waned in popularity.[102]

---

[95] Kyo, In Search of the Beautiful Woman, 129.

[96] Kyo, In Search of the Beautiful Woman, 127.

[97] Li Yongxian, "Trade Between Tibet and the Tang Empire," Presentation for the University of British Columbia's Himilaya Program speaker series, March 31, 2017. https://himalaya.arts.ubc.ca/2017/03/31/trade-between-tibet-and-the-tang-empire/

[98] Shuen-Fu Lin, "*Ci* Poetry: Long Song Lyrics on Objects," in How to Read Chinese Poetry: A Guided Anthology. Edited by Zong-Qi Cai. New York: Columbia University Press, 2008, 294–295.

[99] Hua Mei, 32.

[100] Chunming Gao, 2010, 186.

[101] Cheng, Maria, Cheng and Wai Hung Tang, Essential Terms of Chinese Painting. Hong Kong: City University of Hong Kong Press, 2018, 231.

[102] Hua Mei, 33.

Makeup was applied more subtly in the Five Dynasties (907–960) and Song (960–1279) eras. Pale skin and red cheeks prevailed as beauty standards, but makeup used to achieve these was done with a lighter hand. The nomadic Khitan people who ruled the Liao Dynasty (907–1125) commonly wore a yellow paste made of snake gourd across their entire face. While this practice may have influenced the popular *ehuang* makeup style centuries before, most likely it was intended more as a means of protecting the skin in harsh winter conditions than a deliberate aesthetic choice.

Despite the overall simpler makeup, some styles for royalty remained fairly elaborate. Paintings of the female donors of the Cao family tomb (ca. 950–1000) show the women with *mianye* placed next to each side of the nose, rather than the corners of the mouth, and in the shapes of birds in place of plain dots. Official portraits of various Song empresses throughout the eleventh and twelfth centuries depict the rulers wearing pearls as *mianye* and *huadian*, a style that may have been reserved for Song royalty.[103] Additionally, a tomb belonging to a young female member of the Lu family excavated in 2010 in Lantian County in the Shaanxi province yielded a silver case containing red powder, later confirmed to be rouge,[104] along with a small porcelain container with white powder. Upon testing, the powder was discovered to be ground freshwater pearls. Although Tang empress Wu Zetian (625–705) was recorded as both applying and ingesting pearl powder for beauty purposes,[105] the discovery at Lantian provides the first material evidence of pearl powder being used as a cosmetic in China.[106] Finally, while it's unclear if it is pearl powder or lead/rice powder, several paintings from the Song era depict court ladies wearing white powder on the forehead, bridge of the nose and chin as an early form of highlighting. This style, *sanbai* ("three white") makeup, would continue well into the Ming Dynasty (1368–1644).

Two significant medical texts containing about 1000 cosmetic formulas were published during the Ming era, indicating the popularity of cosmetics throughout Chinese history.[107] A nineteenth century translation of one of the texts, the *Compendium Materia Medica* (ca. 1596) describes the four o'clock flower being

---

[103] Portraits of Empresses Cao, Zheng, Zhu, Wu, Xuanren, Chenshui all depict them wearing pearl *huadian*.

[104] Yu Fei, "Following a Grave Robber to His Noble Ancestor's Tomb," *Shanghai Daily*, August 17, 2010.

https://archive.shine.cn/feature/art-and-culture/Following-a-grave-robber-to-his-noble-ancestors-tomb/shdaily.shtml

[105] Souphiyeh Samizadeh, Non-Surgical Rejuvenation of Asian Faces. Switzerland: Springer International Publishing, 2022, 23.

[106] Z. R. Yu, X. D. Wang, B. M. Su, and Y. Zhang, "First Evidence of the Use of Freshwater Pearls as a Cosmetic in Ancient China: Analysis of White Makeup Powder From a Northern Song Dynasty Lv Tomb," Archeometry, volume 59, issue 4, August 2017.

https://onlinelibrary.wiley.com/doi/abs/10.1111/arcm.12268

[107] Nattaya Lourith and Karl W.K. Tsim, Phytocosmetics and Cosmetic Science. Boca Raton: CRC Press, 2021, 13.

used not for white face powder as in Korea, but for rouge.[108] Judging from visual art, makeup in the early part of the Qing era (1644–1912) appeared more subdued than in previous times.

Japan

Like the Chinese, the Japanese also removed their eyebrows and redrew them, a practice known as *hikimayu*. Described in several histories of Japan, including the Record of Ancient Matters (*Kojiki*, 712 CE) and Chronicles of Japan (*Nihon shoki*, 720 CE), *hikimayu* flourished during the Heian era. But unlike China's royal court, which cycled through dozens of eyebrow shapes placed on top of the original brows, the Japanese typically stuck to one style consisting of black smudges created by dipping the thumbs in soot (*haizumi*) and placed higher on the forehead, sometimes directly below the hairline. Additionally, during the Heian era, the practice was used as a right of passage to mark aristocratic girls' transition into adulthood rather than as a decorative element,[109] *Hikimayu* spread through the lower classes during the Edo period. Safflower arrived in Japan in the seventh century and was used for rouge and lip color as it had been in China. However, due to the difficulties of cultivation, it was extremely expensive and not widely used until the seventeenth century.[110] A white powder called *oshiroi* was used by the nobility during the Nara (710–794 CE) and Heian (794–1185 CE) periods. *Oshiroi* was traditionally made from rice flour and white clay[111]; however, lead-based powder became standard among the elite after a Buddhist monk presented Empress Jito with it in 692 CE. As in most other areas of the world, white skin was prized as a beauty standard according to the literature of the Heian era.[112] It was also during this time that the Japanese began establishing their own aesthetic and became less influenced by Chinese and Korean styles.[113] Additionally, from the Kamakura period starting in 1185 through the beginning of the 1600s, warriors and aristocratic men wore face powder, brow darkener, and rouge.[114]

---

[108] Frederick Porter Smith, Contributions Towards the Materia Medica & Natural History of China: For the Use of Medical Missionaries & Native Medical Students. Shanghai: American Presbyterian Mission Press, 1871, 188.

[109] https://papers.iafor.org/wp-content/uploads/papers/iicj2016/IICJ2016_26954.pdf

[110] Isa Kramer, "Adopting the Caucasian 'Look': Reorganizing the Minority Face," in The Emerging Monoculture: Assimilation and the "Model Minority," 52.

[111] Sandra Buckley, Encyclopedia of Contemporary Japanese Culture. Routledge: London, 2002, 90.

[112] Kyo, A Cultural History, 144–145.

[113] Stuart D.B. Picken, Historical Dictionary of Japanese Business. Lanham: Rowman & Littlefield, 2017, 99.

[114] Edna S. Levine and William Green, "The Cosmetic Mystique of Old Japan," Impressions, Winter/Spring 1980, no 4.

Makeup was widely used during the Muromachi period (ca. 1338–1573). The *Shichijuichiban shokunin uta-awase* (Poetry Contest Among Various Occupations in Seventy-One Rounds), published in the late fifteenth century, describes cosmetics merchants selling rouge and face powder.[115] Interestingly, it was listed as one of the few occupations led by women.[116] Cosmetics were also imported from Europe beginning in the late sixteenth century.

The modern Japanese cosmetic industry and aesthetics are rooted in the Edo period (1603–1868). The three main colors of red, white, and black that had been used for centuries were now officially listed in treatises and women's etiquette books, and more women could read them due to better printing technology and new temple schools which aided in literacy. In particular, the publication in 1813 of *Miyako Fuzoku Kewaiden* ("Cosmetic Manners and Customs in Edo") proved to be the foundation for modern cosmetic recipes and methods. Increased trading, industrial development and economic stability contributed to a firmly established cosmetic culture in Japan. As in the Western world, the number of cosmetics purveyors grew exponentially between the seventeenth and nineteenth centuries.[117]

Woodblock prints (*ukiyo-e*) from the early 1800s provide valuable visual records of makeup application among courtesans and geishas. While their lifestyles may not have been representative of the average woman, they were considered upper class and very influential in terms of fashion. Many are seen with white face powder from a brand called Bien Senjoko, possibly the first cosmetics company in Japan to advertise via woodblock print. The women are also shown applying the traditional lip color made from safflower known as *beni*.[118] While the plant was more easily cultivated than in the past, much of it was needed to produce adequate color, so quality *beni* remained expensive. In the first two decades of the 1800s, a trend for wealthy women emerged in which the lower lip was painted with multiple layers of *beni*, producing an iridescent greenish color. Given the quantity of scenes that depict meticulous details of the women's toilette, these ukiyo-e prints demonstrate that beauty routines had been elevated to ritual status and formed an important part of their everyday lives.[119]

The 1800s also saw a significant increase in calligraphy brush manufacturers that later specialized in cosmetics. The craft of *fude* ("brush") was brought to Japan from

[115] Japan Cosmetics Industry Association, "Cultural History of Cosmetics." https://www.jcia.org/en/info/history

[116] Haruko Wakita, Women in Medieval Japan: Motherhood, Household Management and Sexuality. Monash Asia Institute, 2006, 112–114.

[117] Chio Nishizawa, "Exploring Makeup Culture of Japanese Edo Period - Hiromi Yamamura, Makeup Culture Researcher," Work Mill, June 8, 2023.
https://workmill.jp/en/webzine/edo-makeup-culture-20230608/

[118] Pola Research Institute of Beauty and Culture Collection, Fashion and Makeup of Edo Beauties Seen in Ukiyo-e Prints. Tokyo:, 2018, 10–34.

[119] Young Won Park and Paul Hong, Cosmetics Marketing Strategy in the Era of the Digital Ecosystem: Revolutionizing Beauty in the New Market Frontier, Springer, 2024, 8.

China centuries earlier,[120] but it was during the 1800s that the small town of Kumano developed as the country's foremost brush production center.[121] It is not clear when companies began making brushes specifically for cosmetics, but given the widespread use of makeup and industrialization in the nineteenth century it is likely they started then.

Korea

Much of the makeup in Korea during the Goryeo Dynasty (918–1392) remained the same as in previous eras. Makeup for the wealthy was stored in celadon containers or wooden boxes with intricate mother-of-pearl inlay as these crafts grew and thrived.[122] Upper-class women tended to focus more on skincare to attain a desirable complexion, while lower-class working women wore heavier makeup. This was particularly true for entertainers known as *gisaeng* (or *kinyo). Gisaeng* were born into low-class status and trained to become professional entertainers. They wore layers of white face powder and drew thin, sharp eyebrows.[123] For everyday wear, upper-class women used a light shade of peach blush and only applied *yeonji* (red pigment made from safflower) on the lips.[124] This is corroborated by a Chinese envoy who visited Korea in 1123. In his account of his travels, he remarked that women did not wear rouge.[125] However, this would change in the Joseon era (1392–1910).

A portrait of the wife of Emperor Ha Yeon from the early part of the era indicates the application of red dots to cheeks and foreheads, a practice known as *gonji*. It is not clear when or why brides adopted this fashion, but it became a standard part of wedding day adornment during the Joseon era. Tradition holds that the red keeps away evil spirits, while one scholar believes the three dots represent three deities.[126]

Similar to Japanese ideals, red, black, and white were considered to be the main pillars of female beauty in Joseon times. However, makeup was still applied with a light hand due to the adoption of Confucianism, which called for restraint in self-adornment and a focus on inner beauty rather than physical appearance. "The three whites" (*sambaek*) consisted of the whiteness of the skin, teeth, and whites of the

---

[120] Sumiko Knudsen, Sumi-E: Japanese Ink Painting. Denmark: Books on Demand, 2018, 9.

[121] Art Research Center at Ritsumeikan University, "Kumano brush: Brushes Made in the 'Land of Brushes' Sustain Japan's Traditions," Google Arts and Culture.
https://artsandculture.google.com/story/kumano-brush-art-research-center-ritsumeikan-university/ZAVRpLJnc5nMKQ?hl=en

[122] Yi Song-mi, Fragrance, Elegance and Virtue, 132–136.

[123] Yi Song-mi, Fragrance, Elegance and Virtue, 131–132.

[124] https://asiasociety.org/korea/history-korean-beauty-part-2-goryeo-dynasty

[125] https://koreajoongangdaily.joins.com/2008/01/06/features/The-changing-face-of-Korean-makeup/2884756.html

[126] Yŏng-hak. Son, Handicrafts of the Korean People. South Korea: Dahal Media, 2004, 133–134.

eyes. "The three blacks" (*samheuk*) referred to jet black pupils, eyebrows, and hair. "The three reds" were for the rosiness of cheeks, lips, and fingernails.[127] *Gisaeng*, whose profession spread during the Joseon period, were now considered fashionable and many upper-class women emulated their style, albeit a toned-down version.[128] *Gisaeng* added some *yeonji* to their cheeks in addition to face powder and eyebrow ink, and used better quality ingredients than they had previously.[129]

White porcelain with intricately painted blue designs, sometimes imported from China or Japan, became the preferred cosmetic vessel in the eighteenth century among the aristocracy over the celadon of the Goryeo dynasty.[130] And while makeup was more subtle, royalty still engaged in wearing face powder and rouge, as evidenced by the containers excavated from the tomb of Princess Hwahyeop (1733–1752) in 2016.[131] As in Japan, cosmetics sellers were women (*maegunbu*) who went door-to-door.[132] Korea's cosmetic culture was further solidified with the publication of a women's encyclopedia called the *Gyuhap Chongseo* in 1809. Written by a female scholar, the book contained cosmetic recipes and types of makeup styles. Makeup storage also shifted towards wooden boxes that included mirrors, along with many drawers and compartments.[133]

Iran and India

As described previously, cosmetic practices during the Medieval and early Renaissance eras in Europe were fairly simple for most. In contrast, cultures further east embraced more involved routines. Iranian historical accounts describe a process of 7 cosmetic adornments called *haft galam jarayish*. The concept dates at least to the twelfth century when it was included in the first transcription of the romance

---

[127] Asia Society, "The History of Korean Beauty Part 3: Joseon Dynasty." https://asiasociety.org/korea/history-korean-beauty-part-3-joseon-dynasty

[128] https://koreatimes.co.kr/www/news/culture/2013/01/135_129776.html

[129] Asia Society, "The History of Korean Beauty Part 4: Gisaeng, Forgotten Artists." https://asiasociety.org/korea/history-korean-beauty-part-4-gisaeng-forgotten-artists

[130] Asia Society, "The Secret to Korean Women's Beauty."
https://asiasociety.org/korea/secret-korean-women%E2%80%99s-beauty

[131] Park Chan-kyong, "K-beauty history: How a princess's tomb shed light on eighteenth century cosmetic culture in Korea," South China Morning Post, October 20, 2019. https://www.scmp.com/week-asia/lifestyle-culture/article/3033621/how-18th-century-korean-princess-used-cosmetics-just-we

[132] Yim Seung-Hye, "A cosmetic line fit for a princess," Korea Joongang Daily, October 5, 2020.
https://koreajoongangdaily.joins.com/2020/10/05/culture/koreanHeritage/princess-hwahyeop-kbeauty-makeup/20201005175700564.html

[133] Yi Song-Mi, Fragrance, Elegance and Virtue, 144.

story *Samak-e Ayyar*; however, some of the items, such as kohl (*surma*) and white face makeup (*sefidab*), were used in early- and pre-Islamic times. In addition to *surma*, *haft galam jarayish* consists of the following: *nigar* (henna), *ghazah* (also known as *surkhab* or *surkhr*, literally "red water," denoting rouge), *wasma* or *vasma* (eyebrow darkener), *zarak* (gold dust sprinkled into the hair or gold paillettes adhered to the face), and *khal* (beauty spots). *Khal* was usually painted on with colloryium rather than in the form of an adhesive patch.[134] *Wasma* was made from walnut juice or indigo and sometimes applied to lengthen the brows so that they met above the nose.[135] The rouge was made from red hematite.[136] In addition to the uncertainty of the origins of *haft galam jarayish*, equally unclear is how often women partook or their class status. It appears probable that the full routine was limited to royalty and the upper class, while lower-class women did not use all seven items except for special occasions. However, as evidenced by numerous paintings during the Qarar period (1789–1925), we know that *haft galam jarayish* was practiced through the late 1800s.[137]

A similar list of items comprises *solah shringar*, an Indian list of 16 items of adornment. Originally, women would perform *solah shringar* for everyday wear, but over time it became only for brides to wear on their wedding day.[138] Of the 16 items, 6 are cosmetics and are much the same as *haft galam jarayish*, although they vary based on translation. Medieval texts list the bindi in place of *khal* and sindoor instead of *zarak*.[139]

Africa

Many cultures throughout Africa engage in painting the face and body. While most likely these practices can be traced to thousands of years ago, during the early modern period many tribes came together and formed the kingdoms and civilizations we know today.

---

[134] Niẓāmī Ganjavī. The Haft Paikar (The Seven Beauties): Containing the Life and Adventures of King Bahrām Gūr, and the Seven Stories Told Him by His Seven Queens. London: Arthur Probsthain, 1924, 106.

[135] Jamal Javānshīr Qarābāghī, Two Chronicles on the History of Karabagh: Mirza Jamal Javanshir's Tarikh-e Karabagh and Mirza Adigözal Beg's Karabagh-name. United States: Mazda, 2004, 149.

[136] Eldridge, Face Paint.

[137] See Mounia Chekhab Abudaya, Qajar Women: Images of Women in nineteenth-century Iran. Italy: Silvana Editoriale, 2017.

[138] Marie Anakee Miczak. Henna's Secret History: The History, Mystery & Folklore of Henna. Lincoln: Writers Club Press, 2001, 227.

[139] Nazim Husain and Mohd Khalid, "Aesthetic Significance of *Solah Shringar* (Sixteen Ornaments) in Unani Medicine," Journal of Complementary and Alternative Medical Research, 2021, 70–71.

In the sixteenth century, multiple nomadic tribes officially formed the Kuba kingdom and settled in what is now the Democratic Republic of Congo. The Kuba civilization uses tukula, a red pigment made from the camwood tree, on the face and body.[140] Tukula is also used to anoint the dead. The substance is stored in wooden boxes, usually in the shape of half-moons, onto which geometric patterns or human faces are carved.

The Wodaabe of Chad and Niger have existed for thousands of years, but accounts of their cosmetic practices were recorded only a few centuries ago. The Gerewol festival is an annual courtship ritual in which men apply makeup to compete in a beauty pageant of sorts. Faces are painted yellow or red and layered with intricate patterns of white or black on top. Kohl is applied and the lips painted black to emphasize the whiteness of the eyes and teeth, which are highly prized physical attributes in Wodaabe culture. The competing men form a long line and present themselves to the female judges, performing a dance in which they rise up and down on their toes, jump and stomp to show their physical stamina and height. They also make exaggerated facial expressions to show off their teeth and eyes.[141]

The Himba people are a formerly nomadic tribe that settled permanently in Namibia in the seventeenth century. Women cover their hair, faces, and bodies with *otjize*, a mixture of red ochre and butterfat, to both protect against the desert climate and as beautification. It is stored and carried in containers made from cattle horns called *ozonja*.[142] The Yoruba people of Nigeria adopted the tradition of kohl application to their eyes in the nineteenth century; called tiro, it was made from galena and used for health purposes by men and women.[143]

Section 3: 1900–2025

The last section of this chapter will cover makeup from the late 19th through the early twenty-first century, during which makeup became a mainstay as a common consumer good. The forces of industrialization, globalization, capitalism, and the

[140] Allen Wardwell, African Sculpture from The University Museum, University of Pennsylvania. Philadelphia: University of Pennsylvania Press, 1986, 117.

[141] Carol Beckwith and Angela Fisher, Painted Bodies: African Body Painting, Tattoos and Scarification. New York: Rizzoli, 2012, 290.

[142] Gregor Borg and Margaret Jacobsohn, "Ladies in Red – Mining and Use of Red Pigment by Himba Women in Northwestern Namibia," Conference paper, January 2013, 43–45.

[143] Dapo Adegoke, "Exploring *Tiro,* the Facial Ornamentation of the Yoruba in Western Nigeria," Journal of Humanities and Social Science, vol. 23, Issue 1, January 2018, 61.

strong influence of the Western world, particularly the United States, France, and the United Kingdom, blurred the geographic boundaries of specific ingredients and trends. As a result, beauty standards, makeup styles and attitudes towards makeup largely became homogenized across the globe. Therefore, this section will be loosely divided into decades rather than geographical areas and will focus on the United States.

The primary demographic remained largely the same throughout the twentieth century: middle to upper class white women between the ages of 18 and 40.

1900s and 1910s

The turn of the century until the end of the First World War saw a gradual acceptance of makeup. Visible cosmetics became permissible for the average woman, as long as they were applied solely for evening wear; the natural look was still preferred for daytime. "Beauty culturists" such as Helena Rubinstein, Elizabeth Arden, and Madam C.J. Walker opened salons and began selling their own brands of cosmetics, and dozens of small perfumers branched out into face powder scented with their proprietary fragrances. Door-to-door sales and mail orders proliferated across the United States, especially in rural areas and in regions with considerable Black populations.[144] In Korea, the nation's first mass-produced face powder was launched in 1915. Face powders both stateside and abroad were also being sold in a wider variety of shades. Shiseido was particularly innovative with the 1917 introduction of their "rainbow" powders tinted green and purple to both color correct one's complexion and also offset the effects of harsh electric lighting. The array of new items and their production on a larger scale were part of the formation of the mass market, which would take off after the end of the First World War.

1920s and 1930s

The cosmetics industry in the United States introduced thousands of products during the first two decades of the twentieth century to accommodate women's rapidly changing lives and a new consumer culture. Technological advances meant that makeup could be produced more easily and quickly on a large scale. They also meant increased magazine circulation and radio ads, which fueled a new consumerist society. Additionally, the appetite for makeup was fed by Hollywood, along with regular beauty columns that were now more widely distributed in newspapers and magazines. In short, changing social mores and technology led to a widespread

[144] Kathy Peiss, Hope in the Jar: The Making of America's Beauty Culture. New York: Metropolitan Books, 1998, 56–53.

acceptance of makeup.[145] White, middle-class women in the United States, where they had earned the right to vote, expressed a new sense of independence through cosmetics. Women became more visible in public and advertising directly linked cosmetics to the leisure activities that women were more fully participating in. Companies were designing products and packaging that would fit the idea of the new woman. The priority for cosmetics during this era was portability. Previously, makeup products were not designed to be carried. They were intended to sit on one's dressing table because the thought of using makeup on the go or in public was virtually unheard of. But as makeup gained in popularity, companies started introducing vanity cases. As the modern woman was out and about, she needed a sturdy case to take with her from daytime to nighttime activities that accommodated her makeup. By 1925, fashion magazines were touting the vanity case as a must-have accessory. Young fashionable women ("flappers") were the chief proponents of makeup usage and flaunted its application publicly.

The Black beauty industry expanded greatly during the interwar years. Black-founded companies and salons such as Overton's and Apex courted Black clientele, advertising their products and services as being able to suit the unique cosmetic needs of Black customers and promoting Black women as intrinsically beautiful. In doing so, they openly resisted the prevailing beauty ideals of white skin and straight hair.[146] The Black beauty industry also benefited Black society on a larger level, with founders becoming political leaders and distributing their wealth to the Black community.[147]

The interwar years were also the age of Art Deco, which had a tremendous impact on notions of modernity and by extension, the modern woman. The motifs commonly associated with Art Deco were incorporated into packaging after the 1925 Paris Exposition. As the United States entered the 1930s, compact manufacturers and makeup companies scrambled to update their packaging or add new designs to their existing lineup, and incorporate the word "modern" or variations of it into their ads and products. Makeup became critical to the notion of the "modern girl," a concept that spread beyond the West.[148] Additionally, advertising now presented makeup as a necessity for women as the ultimate signifier of their femininity. As such, men were stigmatized for wearing makeup. Seen as sexual deviants, men who dared carry cosmetics on their person could be arrested if they were found.[149] This general societal attitude towards men in makeup (outside of the theater and film) continued well into the 1960s.

---

[145] Emma Fay, "The Victims of Beauty: How Women Paid the Price of the Industrial Revolution, Science, and Hollywood," 143.

[146] Susannah Walker, Style and Status: Selling Beauty to African-American Women, 1920–1975. Lexington: University Press of Kentucky, 2007, 11–45.

[147] Cheryl Woodruff-Brooks, Golden Beauty Boss: The Story of Madame Sara Spencer-Washington and the Apex Empire. Mechanicsburg: Sunbury Press, 2020, 43–45.

[148] Jessica Guerra.

[149] Matt Houlbrook, "The Man with the Powder Puff in Interwar London," The Historical Journal, Vol. 50, no. 1, March 2007, 147–148.

In terms of the products themselves, long-lasting, waterproof makeup and a variety of shades were at the forefront of cosmetic chemistry. Makeup according to type took the guesswork out of shades. During the interwar years, "color harmonies" based on complexion, hair and eye color were developed, along with colors for all types of lighting. Along these lines, as travel increased, particularly to warm, sunny destinations, along with the rising popularity of tans, companies also created shades either to obtain a fake tan or that would suit tanned skin.

The 1940s saw much of the same products and advertising tactics as the previous two decades: portable and long-wear makeup, travel sets, and color coordination. Packaging segued from slim, geometric Art Deco lines to curvier contours, more elaborate designs. At the beginning of the decade, as the United States entered the war, there was new pressure on women to maintain their beauty care. As the Art Deco obsession with modernity waned, a stronger emphasis on femininity emerged. In the United Kingdom, which had a higher proportion of uniformed women than the United States at the start of the war, Elizabeth Arden made beauty kits specifically for military women and publications advised on which makeup coordinated with one's uniform.[150] As the war progressed, more American women entered roles traditionally held by men. As such, women were expected not to neglect their beauty care and wear makeup to reinforce their femininity.[151] Rather than forgoing lipstick and powder altogether, women were encouraged to refill their existing metal lipstick tubes and compacts or use paper and plastic versions that companies had launched in the face of metal shortages. The use of makeup to look feminine while doing men's work and remaining fresh-faced was presented as a morale booster critical to the war effort, with some products and ads directly tying makeup to victory and American tradition.[152]

Throughout the 1940s and 1950s, Black beauty salons continued to open across the United States despite segregation and discriminatory financial practices such as prohibiting Black entrepreneurs to secure small business loans. The salons of this era, particularly the Rose Meta House of Beauty and Beauty Salon and Carmen Murphy's House of Beauty provided "head-to-toe" services for a luxurious experience.[153] Black beauty culture was further solidified through the establishment of lifestyle magazines such as *Ebony*, *Tan*, and *Jet*.

[150] Britannia publication.

[151] Melissa A. McEuen, Making War, Making Women: Femininity and Duty on the American Home Front, 1941–1945. Athens: University of Georgia Press, 2010, 6.

[152] Peiss, Hope in a Jar, 245.

[153] Walker, Style and Status, 115–119.

1950s and 1960s

Postwar prosperity, the rise of the suburbs and accompanying shopping centers meant a fully solidified and ever-growing consumer culture.[154] A significant change was market segmentation. Targeting certain consumer categories began in the 1940s but flourished in the 1950s. Segmentation in cosmetics marketing further delineated specific roles and identities for women. In particular, the teen market grew exponentially as companies realized they had more purchasing power than ever before.[155] Fashion in print magazines was elevated by innovative photographers such as Richard Avedon and Irving Penn, who captured the emerging role of the celebrity model. In turn, a new symbiotic relationship between fashion and makeup was born.[156] Packaging-wise, compacts grew in size from the tiny, wafer-thin silhouettes of the Art Deco era, and novelties such as miniature furniture replicas were popular. The atomic age encouraged a new modern style as well. Revlon's Futurama refillable lipstick promised that a new tube could be inserted in 3 s, unlike older refillable lipsticks, while the name implied it was the lipstick of the future. Hollywood continued to be influential, especially with Max Factor's continuation of ad campaigns featuring leading starlets, while television offered a new medium for advertising.

While the previous decade's red lips and feline-esque eyeliner comprised the dominant makeup style at the start of the 1960s, by 1965 more emphasis was placed on eyes rather than lips. Most models appearing in magazines did their own makeup in previous years, but in the early and middle parts of the decade the editorial makeup artist emerged. Usually, this was an artist who worked for a particular brand, lending their talent in creating various styles and dispensing beauty advice. Pablo Manzoni, working for Elizabeth Arden, and William Lowe for Charles of the Ritz became the makeup trendsetters of the decade due to their avant-garde approach towards eye makeup. Manzoni famously used feathers, rhinestones, lace and other unconventional materials to create fantastical looks while Lowe drew on Op-Art principles, painting elaborate trompe l'oeil designs on models' eyes. Towards the end of the decade, the role of the editorial makeup artist expanded to creative director, the position responsible for developing new products and visuals for the accompanying advertising. One example is Serge Lutens, who devised the iconic imagery for Dior's makeup line from 1968 through 1980.

In general, lips were frosted/pale, false eyelashes and exaggerated liner gave a wide-eyed look made popular by Twiggy and swinging London.[157] Hippie culture, which embraced body positivity, brought about a brief body painting fad among

---

[154] See Lizbeth Cohen, A Consumer's Republic: The Politics of Mass Consumption in Postwar America. New York: Vintage Books, 2003.

[155] "Makeup for Teen-Agers Still Needs to Be Given a Big Fashion Push," *Women's Wear Daily*, vol. 81, issue 107, December 1, 1950, 10.

[156] Kate de Castelbajac, The Face of the Century: 100 Years of Makeup and Style. London: Thames & Hudson, 1995, 113–117.

[157] Corson, Fashions in Makeup, 565–567.

young women, as well as sheer coverage in base products marketed as "naked" or "nude." Packaging had largely turned to plastic and single-use items rather than refillable metal containers. In the late 1960s and early 1970s, Biba and Mary Quant led the way in youth-oriented fashion and released nontraditional colors such as brown lipstick and marketed them as genderless. Towards the end of the decade, the Black is Beautiful movement spurred Black entrepreneurs to launch lines catering to Black women.[158] Second-wave feminists spoke out against makeup and beauty culture, declaring both. Influenced by the emerging consumer rights movement, the 1960s saw a new wave of reporting on animal testing and toxic ingredients in cosmetics, such as the book "Cosmetics: Trick or Treat?" published by investigative journalist Toni Stabile in 1966.

1970s and 1980s

In the early 1970s, the trend of the "natural" face achieved using sheer products continued. This interest in natural in turn fueled the demand for plant- or fruit-derived ingredients rather than synthetic chemicals.[159] Male musicians such as David Bowie experimented with bold makeup. All genders in the emerging goth and punk subcultures sometimes participated in the wearing of heavy eyeliner and black lip color.[160] However, the average American man still did not engage in makeup. Moreover, nontraditional colors were not in demand among the general population and were sold primarily through specialty stores or drugstores.

Fashion Fair Cosmetics was launched by *Ebony* founder Eunice Johnson in 1973 as a prestige line for women of color. White-owned mainstream brands also began pursuing Black customers, wanting to get a piece of their high purchasing power. Companies started integrating Black models in their ads and hired Black executives to create collections intended for Black customers, such as Revlon's Polished Ambers line, which debuted in 1975.

By the end of the decade, disco-influenced makeup had taken over. The nightlife stars of New York City's famous Studio 54 dictated the trends of hyperglossy lips and strong contouring. The late 1970s witnessed the rise of celebrity makeup artists such as Joey Mills and Way Bandy, along with the development of seasonal color stories consistent with the fashion displayed in runway shows.

Contouring and heavy blush stayed strong throughout the decade. In an era where "greed is good," maximalist, loud colors were favored. Seasonal colors stories became more firmly entrenched, and a spate of fashion designers—Ralph Lauren, Diane von Furstenburg, and Calvin Klein, among others—launched

---

[158] Gabriel, Joyce. "Beauty Is Indescribable Thing, Says Executive in National Cosmetics Firm." *Kitsap Sun*, Oct. 23, 1970, 5.

[159] Corson, Fashions in Makeup, 590–591.

[160] Corson, Fashions in Makeup, 598–599.

cosmetic lines. Canadian company MAC (Makeup Art Cosmetics), whose tagline was "all ages, all races, all sexes" (now "all genders") was founded in 1984, one of the first brands that directly advertised inclusivity.[161] Male musicians continued wearing outlandish makeup, but again, the average American man did not partake.[162] The role of celebrity/editorial makeup artist continued to gain prestige, along with the film makeup artist: in 1981 makeup artistry finally earned its own Oscars category.[163]

1990s

The minimal makeup of the first half of the decade served as a response to the bright shades and heavy textures of the 80s. Neutrals and brown tones were particularly popular, possibly due to the influence of the newly established coffee culture. Mainstream giants—among them Revlon, Maybelline, the Body Shop, and Clinique—engaged in an expensive battle to secure the "multicultural" market, producing special lines for non-white customers or extending their shade ranges to better suit deep skintones.[164] They also competed with smaller Black-founded brands such as Iman.

The 1990s saw the emergence of the makeup artist turned entrepreneur. Makeup professionals who created the looks for runway shows began launching their own brands. In an effort to capture the youth market, companies appropriated various subcultures such as Riot Grrrl and hip-hop. The idea of makeup as a tool of self-expression and empowerment rather than beautification was also beginning to be commodified through brands such as Stila, whose name comes from the Italian verb *stilare* (to pen), as founder Jeanine Lobell believed one's makeup should be as unique as their signature. Third-wave feminists took the reins from the second wave in terms of critiquing normative beauty standards and advertising, only to have their efforts sold back to them in the guise of feminism and "girl power".[165] Indeed, a 1997 study indicated that most professional women felt that makeup was a requirement in the workplace, let alone optional or empowering.[166] The second half of the decade saw a somewhat abrupt shift towards more playful makeup, such as frosted eye shadow and body glitter. Both packaging and color palettes became dominated

[161] Michael DeVellis, "Milestone: 40 Years of MAC" On Makeup Magazine, October 21, 2024. https://www.onmakeupmagazine.com/milestone-40-years-of-mac/

[162] Corson, Fashions in Makeup, 601.

[163] https://www.oscars.org/collection-highlights/rick-baker

[164] Ginger Lundy, "Makeup Blends with Skin Color," *Daily World*, April 20, 1992.

[165] Elizabeth A. Kreydatus, "Marketing to the 'liberated' woman: Feminism, social change, and beauty culture, 1960—2000," College of William and Mary, 2005, 279–280.

[166] Kirsten Dellinger and Christine L. Williams, "Makeup at Work: Negotiating Appearance Rules in the Workplace," Gender and Society, vol 11, no. 2, April 1997, 151.

by silver tones as a response to the looming new millennium. Online retail disrupted the traditional brick-and-mortar stores and mail-order shopping.[167]

2000s

As a new millennium dawned, futuristic packaging and trends continued. However, some brands countered this outlook by creating vintage-inspired designs.[168] Collaborations with artists and the popularity of Japanese brands in the United States also increased in the 2000s. As a possible side effect of the "metrosexual" trend, further attempts at men's makeup were made with the launch of Jean Paul Gaultier's Le Male line (2003) and Guerlain's Terracotta Bronzer, but failed to gain traction in the United States. Makeup, it seemed, was still largely the domain of women and a handful of male musicians. Skin lightening creams and whitening powders had largely been phased out in the United States, but large multinational corporations (Unilever, L'Oreal, and Shiseido) continued promoting these products across Asia and the Middle East.[169]

The 2010s were much a continuation of the previous decade, but there were several significant shifts. Social media engendered the rise of "influencers," many of whom were makeup hobbyists and enthusiasts (not professional artists), who were either paid by companies to promote their products or provided free products for review. Social media also made it easier for small indie brands to gain a foothold in the business. The industry's longtime tactic of appropriating certain cultures in advertising and product names—in particular South American, East Asian and ancient Egyptian—was beginning to be more widely recognized by consumers.

The broader cultural interest in health and wellness led to a growing demand for sustainable, natural formulas free of synthetic chemicals and potential irritants and allergens such as talc, sulfates, and parabens, which the industry dubbed "green" beauty.[170] Companies also began marketing "cruelty-free" beauty, which commonly refers to products not currently tested on animals. However, like green beauty, the phrase has no legal definition and is not regulated by the Food and Drug Administration.

The popularity of Korean beauty in the Western world exploded in the 2010s, fueling the demand for high-quality yet affordable products and new formulations

[167] Vaerie Seckler, "E-Commerce Called 'Future of Fashion': 'E-Tail Or Die,' Says Retail Analyst." *Women's Wear Daily,* vol. 178, no. 12, July 19, 1999, 2–4.

[168] Hillary Belzer, "Friday Fun: Too-Faced Quickie Chronicles Artwork, Solved (part 1)," The Makeup Museum, November 2014.

[169] Evelyn Nakano Glenn, "Yearning For Lightness: Transnational Circuits in the Marketing and Consumption of Skin Lighteners," Gender and Society, vol. 22, no.3, June 2008, 296.

[170] Rachel Brown, "Au Naturale: Green-Based Beauty Brands are Booming as Consumers Seek Products Befitting their Increasingly Health-Conscious Lifestyles." *Women's Wear Daily,* September 2, 2015, 54–59.

such as cushion compacts and BB (beauty balm) creams.[171] K-beauty was also responsible for the renewed interest in skincare, a product category that ended up surpassing color cosmetics in sales by 2020. The desire for glowing skin, a byproduct of K-beauty, resulted in a series of complexion-based trends that relied more heavily on skincare than base products. "No-makeup makeup" became intertwined with "athleisure" makeup, both of which promoted a look that was meant to appear natural, effortless, and suitable for casual activities.[172] The other side of the coin, however, was a fully made-up "beat" face intended for Instagram and other social media platforms. Commodification of larger cultural movements continued, with some companies selling rainbow-themed products and donating a portion of the proceeds in an attempt to appear supportive of the LGTBQ+ community. Some also pivoted to marketing their products as ones that would contribute to general health and wellness. The 2017 release of musician Rihanna's brand Fenty, which boasted a whopping 40 shades of foundation, spurred many other companies to expand their shade ranges. The "Fenty effect," as it became known, strengthened consumer expectations for brands at all price points to offer an inclusive range of colors.[173]

The protests surrounding a long-running streak of racial violence in June 2020 ignited a demand for cultural change in the United States. As part of the broader call for change, many companies pledged to hire Black employees at the senior level, support emerging Black entrepreneurs and expand their shade ranges to be inclusive of deep skintones.[174] More generally, consumers have come to expect inclusive makeup lines and advertising across all categories—race, gender, age, disability, and culture. Progress seems to have been made in recent years with brands offering more shades and new genderless brands being launched.[175] A greater proportion of American men are more accepting of makeup. However, a 2022 analysis of cosmetic advertising in the United States shows that much work is to be done in terms of inclusive and diverse brands and campaigns.[176] Companies have also had a mixed response to other recent consumer expectations of sustainability and beneficial ingredients. Not only must the product have wasteless, refillable, or recyclable packaging and be free of potentially harmful ingredients, but it also needs to offer

[171] Elise Hu, Flawless: Lessons in Looks and Culture from the K-Beauty Capital. New York: Dutton, 2023, 58–59.

[172] Rina Rafael, February 6, 2017. https://www.fastcompany.com/3067727/athleisure-makeup-has-the-cosmetics-industry-breaking-a-sweat

[173] Carson Miller, "The Fenty Effect: A Case Study on the Fusion of Celebrity, Luxury, And Inclusivity." BA thesis, Clark Honors College, 2022, 44–47.

[174] Liz Flora, "No longer silent: Beauty industry joins wave of corporate support for Black Lives Matter," Glossy, June 2, 2020.

https://www.glossy.co/beauty/no-longer-silent-beauty-industry-joins-wave-of-corporate-support-for-black-lives-matter/

[175] Leah Chernikoff, "The Rise of Gender-Inclusive Beauty," Business of Fashion, February 14, 2020.

https://www.businessoffashion.com/articles/beauty/the-rise-of-non-binary-beauty/

[176] Lisa Alli, Diversity and Inclusion in Beauty and Cosmetic Advertising and its Impact on Corporate Reputation, May 2022, 23–27.

health benefits. Many brands now offer refillable packaging and formulas made with ingredients proven to be beneficial.

The boom in intellectual property agreements, which allow brands to use a certain character, artist or product, shows no signs of abating.[177] Everything from Frosted Flakes cereal to Betty Boop to Van Gogh's *Starry Night* has worked with beauty brands on their own licensed makeup collection. Stylistically, the early part of the pandemic encouraged emphasis on eye makeup as masks covered up the lower half of the face, but also brought a more relaxed or even experimental attitude toward makeup, one that at times promoted deliberately "ugly" application and colors that defied traditional beauty norms.[178] Additionally, while makeup artists are still influential, seasonal color stories have largely disappeared. However, social media platforms, particularly Tik-Tok, have filled the void left by traditional trendsetting mechanisms.

---

[177] Noor Lobad, "Why Beauty Loves Collabs," *Yahoo! Life*, July 16, 2024. https://uk.style.yahoo.com/why-beauty-loves-collabs-050000192.html

[178] Katy Kelleher, "Ugly Makeup: The Trend Highlighting What's Beyond Conventional Beauty," *The Guardian*, May 8, 2020. https://www.theguardian.com/fashion/2020/may/08/ugly-makeup-trend

Physiology of Skin, Hair, and Nails as Differentiated by Sex and Age

Chirag A. Patel, Ketan Ranch, Sonal Vijay Pande, B. Mahalakshmi, Tasnem H. Abdelfattah Allam, Maryam R. Mohamed, and Sai H. S. Boddu

Abstract The physiology of skin, hair, and nails is influenced by sex and age, reflecting distinct biological processes and health implications. Sex differences result in varying skin thickness, sebum production, and collagen density, which contribute to divergent aging patterns and susceptibility to dermatological conditions. Hair physiology also varies by sex, with hormonal influences like androgens in males contributing to patterns of hair growth and loss that differ from those observed in females. Aging affects hair follicle activity, leading to thinning and reduced pigment production, which presents differently across sexes. Nail physiology reveals further differentiation, as variations in the growth rate, strength, and texture between sexes can be attributed to hormonal fluctuations and genetic predispositions. Age-related changes in nails, including slower growth and increased brit-

C. A. Patel (✉) · S. V. Pande
Department of Pharmacology, L. M. College of Pharmacy, Ahmedabad, India
e-mail: chiragapatel@lmcp.ac.in; sonal.pande@lmcp.ac.in

K. Ranch
Department of Pharmaceutics and Pharmaceutical Technology, L. M. College of Pharmacy, Ahmedabad, India
e-mail: ketan.ranch@lmcp.ac.in

B. Mahalakshmi
Department of Microbiology, Government Medical College & Hospital, Kadapa, Andhra Pradesh, India

T. H. Abdelfattah Allam · M. R. Mohamed
Department of Pharmaceutical Sciences, College of Pharmacy and Health Sciences, Ajman University, Ajman, United Arab Emirates
e-mail: 202110187@ajmanuni.ac.ae; 202020011@ajmanuni.ac.ae

S. H. S. Boddu
Department of Pharmaceutical Sciences, College of Pharmacy and Health Sciences, Ajman University, Ajman, United Arab Emirates

Center of Medical and Bio-Allied Health Sciences Research, Ajman University, Ajman, United Arab Emirates
e-mail: s.boddu@ajman.ac.ae

K. Alexander, P. Romanowski (eds.), *Cosmetic Science*, AAPS Introductions in the Pharmaceutical Sciences 15, https://doi.org/10.1007/978-3-032-03596-7_2

tleness, are universal but manifest differently across genders. This chapter highlights the physiological variations occurring in the skin, hair, and nails differentiated based on sex and age, while offering insights into their implications for clinical practice and personal care. By understanding these differences, strategies for maintaining dermatological health across the lifespan can be effectively tailored to individual needs.

Keywords Skin · Hair · Nails · Physiology · Cosmetics · Age · Sex

Introduction

The skin, hair, perspiration and oil glands, nails, and sensory receptors make up the integumentary system. This system is essential for controlling body temperature, defending the body from outside threats, and giving sensory input about the surroundings [1]. The skin, also known as the integument or cutaneous membrane, covers the body's exterior and is the largest organ in terms of weight and surface area, making up around 16% of the body's total weight. The skin performs a variety of essential activities. These include vitamin D production, thermoregulation, blood storage, defense, feelings, excretion, and absorption [2]. Hormone variations between males and females significantly affect skin structure and function. Androgens in males stimulate sebum production, leading to thicker skin and potential acne issues. Estrogen in females influences collagen synthesis, resulting in thinner skin, while fluctuations during pregnancy and menopause can affect skin sensitivity and pigmentation. Furthermore, hormonal changes throughout life also play a role. Puberty brings hormonal surges affecting sebum production and acne development [3]. Declining hormone levels in older adults contribute to skin thinning, dryness, and decreased wound healing. Studying skin physiology based on the sex and age is essential for valuable insights into its diverse functions, paramount to understanding the pathophysiology of skin diseases. This allows personalized care, early detection of health issues, development of targeted interventions, and promotes overall well-being [4].

Hair is a defining characteristic of mammals, boasting many functions including protecting against external elements, releasing pheromones, apocrine sweat, and sebum. These substances affect social and sexual dynamics, help with thermoregulation, and store important stem cells. Hair has two primary parts, derived from the epidermis: the hair shaft and the follicle. A thorough understanding of the intricate aspects of hair physiology based on sex and age is valuable for understanding the pathophysiology of hair diseases. Also, it holds the potential to revolutionize approaches to hair care and treatment interventions and enhance overall hair health and appearance [5]. Nails make up of densely packed, tough, keratinized cells that create a transparent, sturdy layer covering the upper surfaces of the fingers and toes. A nail includes a free edge, nail root, and nail body. Nails prevent the hypertrophy of the tissue along the distal wall, a condition that can happen after losing the toenail on the big toe, and apply counterpressure to the fingertip pulp, which is necessary for tactile sensitivity in the fingers [6]. While there are a few significant sex-based differences in nail physiology compared to skin and hair, studying nail physiology stays valuable for promoting healthy nails and overall well-being throughout life [7].

Skin, hair, and nails are central to the cosmetic industry. The rationale behind skin, hair, and nail care products is rooted in maintaining and enhancing the health and appearance of these vital parts of the body. These products serve various purposes, ranging from basic hygiene and protection to addressing specific cosmetic or medical concerns. Whether for practical, medical, or aesthetic reasons, these products play a significant role in personal care and well-being. Understanding the unique characteristics of different demographics aids in the development of effective and safe products. For example, hair texture varies widely among different ethnic groups, and a product designed for one type of hair may not suit another. Similarly, skincare formulations need to consider factors like skin sensitivity, pigmentation, and environmental exposure, which can vary across demographics [8]. This chapter serves as a comprehensive compilation of skin, hair, and nail physiology, unraveling the intricate interplay between sex and age in shaping their characteristics and behaviors.

Skin Anatomy and Physiology

General Structure of the Skin

The skin and its supporting tissues are included in the integumentary system. The skin is made up of two main layers. Superficial epidermis, which is made up of specialized cells called keratinocytes that synthesize keratin, and the epithelial tissue. The dermis layer, found beneath the epidermis, is mostly made up of collagen, a structural protein that forms the skin's skeleton. The subcutaneous layer, not seen as a skin, is beneath the dermis (Fig. 2.1). This layer, also referred to as the hypodermis, is made up of adipose and areolar tissues.

Epidermis: Keratinized stratified squamous epithelium is a strong, layered tissue that makes up the epidermis, the outermost layer of our skin. Four main cell types—keratinocytes, melanocytes, Langerhans cells, and Merkel cells—are found in this layer, and each is essential to the other [9]. Keratinocytes are approximately 90% of the epidermal cells. As they migrate from the birthplace, the basal layer, towards the skin surface, they undergo a transformation called keratinization. They generate the fibrous protein known as keratin, which protects the skin from chemicals, heat, and bacteria. In addition, they secrete a lipid-rich layer that acts as a water repellent and guards against outside chemicals and dehydration [10]. Melanocytes are pigment-producing cells developed from the embryo and form 8% of epidermal cells. They have long protrusions to transfer melanin granules to keratinocytes. In addition to giving skin its color, this yellow-brown pigment shields the underlying DNA from

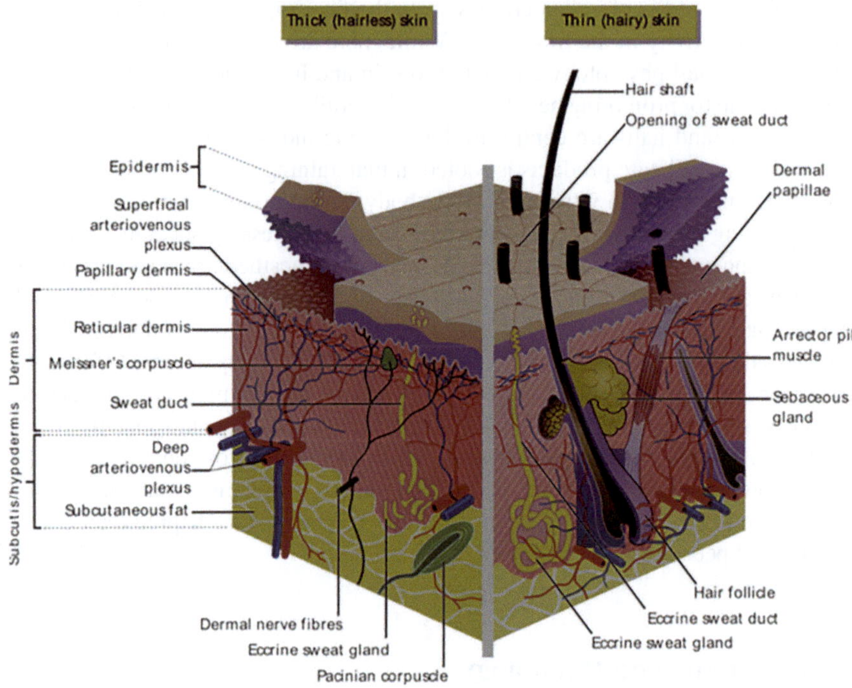

Fig. 2.1 Skin anatomy. (Reproduced with permission from: https://en.wikipedia.org/wiki/File:Skin_layers.png#filelinks)

ultraviolet (UV) rays [11]. Langerhans cells make up a tiny part of the epidermis. They originate from bone marrow and are involved in immunological responses. They help other immune cells recognize and destroy invading microbes but are themselves vulnerable to UV damage. Merkel cells produce Merkel discs when they attach to sensory nerve fibers in the epidermis lowest layer. These specialized structures detect touch sensations.

At various stages of development, the epidermis is further separated into discrete layers of keratinocytes. The epidermis is composed of four layers in most areas: the stratum corneum (cornified or horny cell layer), stratum granulosum (granular cell layer), stratum spinosum (squamous cell layer), and stratum basale (basal cell layer). An extra layer is present in high-friction areas like the palms and soles: lucidum stratum: a thin, translucent layer found in skin that is thicker [8]. Stratum basale consists of a single row of cuboidal or columnar keratinocytes, and it is the deepest layer of the epidermis. Some of the cells in this layer function as stem cells, dividing to produce new keratinocytes continuously. Tonofilaments, a protein that eventually transforms into keratin in the deeper layers of the epidermis, cling to desmosomes and connect stratum basale cells to one another and to the adjacent stratum spinosum cells. This layer is also called stratum germinativum as it produces new cells. Stratum spinosum consists of 8–10 closely spaced layers of polygonal keratinocytes located between stratum basale and stratum granulosum layers. The organelles found in stratum basale cells are also present in these keratinocytes. Every spiny projection seen in a piece of tissue that has been processed shows the location where tonofilament bundles are inserted into a desmosome to firmly hold the cells together. The skin is made both strong and flexible by this structure. Stratum granulosum consists of 3–5 layers of flattened keratinocytes going through apoptosis. This layer is situated about halfway into the epidermis. Tonofilaments become more noticeable in this layer as the cell's nucleus and other organelles start to degenerate. Darkly stained granules having keratohyalin, a protein that promotes the conversion of tonofilaments into keratin, are a distinctive feature of cells in the stratum granulosum. Furthermore, the keratinocytes include membrane-enclosed lamellar granules that secrete a lipid-rich fluid. This secretion acts as a water-repellent sealant to fill the gaps between the stratum lucidum, stratum granulosum, and stratum corneum cells, preventing water loss and foreign material penetration. Stratum corneum consists of flattened, dead keratinocytes usually making up of 25–30 levels. Deeper layer cells constantly replace the cells in the top layer as they shed. The main material inside these cells is keratin. This layer functions as an efficient water-repellent barrier because of the lipids from lamellar granules between the cells. It has several layers of dead cells that act as a barrier against microbial invasion and damage to deeper layers. Calluses are an abnormal thickening of the stratum corneum caused by prolonged friction contact to the skin [12]. Stratum lucidum is only found in the thick skin of places like the soles, palms, and fingertips. It includes 3–5 layers of flattened, clear, dead keratinocytes having abundant keratin and thickened plasma membranes.

Dermis: The second and deeper layer of skin, called the dermis, is made up of strong connective tissue that is high in collagen and elastic fibers. With its exceptional tensile strength, this complex web of fibers can withstand pressures due to pull or stretch. The dermis can also expand and contract. The dermis houses a sparse population of cells, primarily fibroblasts, along with some macrophages and a few adipocytes near its boundary with the subcutaneous layer. Blood arteries, nerves, glands, and hair follicles are epithelial invaginations of the epidermis that are embedded within the dermal layer. Important structural and functional interactions are established by the dermis and surrounding layers, which are critical to the survival of the epidermis [13]. The papillary region is made up of areolar connective tissue, which has thin collagen and fine elastic fibers, and makes up around one-fifth of the overall layer thickness. The papillary region's surface area is greatly increased by dermal papillae, which are tiny, finger-like structures that protrude into the epidermis' underside. While some dermal papillae have capillary loops, others have free nerve terminals that are responsive to a range of sensations including tactile receptors called Meissner corpuscles or corpuscles of touch [14].

The reticular region is attached to the subcutaneous layer and is made up of dense, irregular connective tissue that has bundles of collagen, coarse elastic fibers, and fibroblasts. In the reticular region, collagen fibers weave together in a pattern like a net. The spaces between these fibers are occupied by adipose cells, hair follicles, nerves, sebaceous (oil) glands, and sudoriferous (sweat) glands (Fig. 2.1). The reticular region's collagen and elastic fibers work together to give the skin its strength, extensibility, and elasticity—the ability to stretch and then return to its original shape. In cases of obesity, around joints, and during pregnancy, skin extensibility is most noticeable. Exertion can cause tiny tears in the dermis, which can then result in the creation of striae, or stretch marks, which appear as red or silvery-white streaks on the skin surface [13].

Functions of the Skin

Numerous processes are carried out by the skin, such as blood storage, warmth regulation, protection, feeling reception, excretion, and absorption, and the production of vitamin D [14].

Thermoregulation: The two main ways that the skin aids in thermoregulation are through controlling blood flow in the dermis and sweating on the skin's surface. Sweat from eccrine sweat glands increases in reaction to elevated temperatures outside or heat from physical exertion, which helps perspiration evaporate from the skin's surface to reduce body temperature. Furthermore, the dermis's blood vessels widen to allow more blood to pass through it, which improves the body's ability to expel heat. Conversely, in response to lower environmental temperatures, sweat production decreases to conserve heat, and blood vessels in the dermis constrict, reducing blood flow through the skin and minimizing heat loss from the body [15].

Blood Reservoir: The dermis is a blood reservoir with a vast network of blood vessels that carry 8–10% of an adult's total blood flow at rest.

Protection: Tightly packed keratinocytes prevent microbial invasion, while keratin shields underlying tissues from heat, chemicals, abrasion, and pathogens. Lamellar granules produce lipids that stop water from evaporating from the skin surface, protecting against dehydration and preventing water from entering the skin when taking a shower or a swim. Sebum from sebaceous glands has antibacterial compounds and keeps skin and hair from drying out. Sweat's acidic pH prevents microorganisms from growing, and melanin protects the skin from UV radiation. Dermal macrophages phagocytize bacteria and viruses, while epidermal Langerhans cells notify the immune system of potentially dangerous microorganisms [10].

Cutaneous Sensations: Thermal and tactile sensations, such as warmth and coolness, pressure, vibration, and tickling, are examples of cutaneous sensations. Pain is typically a sign of imminent or existing tissue damage, and the skin's many nerve endings and receptors mediate those sensations.

Excretion and Absorption: Although the skin plays a little part in these processes, it does help the body rid itself of toxins and allow things to enter body cells. Even though the stratum corneum is waterproof, about 400 mL of water evaporates through it every day in addition to perspiration loss. Moreover, urea, ammonia, carbon dioxide, and salts are expelled through sweat. Water-soluble compounds are absorbed through the skin little, while lipid-soluble items, including some medications, vitamins that are fat-soluble, and gases like carbon dioxide and oxygen, can pass through the skin. On the other hand, hazardous substances such as organic solvents and heavy metal salts can potentially enter the body through the skin and present health concerns. Topical steroids, being lipid-soluble, can easily penetrate the skin effects by inhibiting histamine production [16].

Vitamin D Synthesis: Sunlight exposure activates a precursor molecule in the skin, which in tusrn leads to the production of vitamin D. The activated molecule is further modified by the liver and kidney enzymes, which produces calcitriol, the strongest form of vitamin D. As a hormone, calcitriol helps the gastrointestinal tract absorb calcium from meals and transfer it to the bloodstream. Only little UV light exposure is necessary for adequate vitamin D synthesis, usually 10–15 minutes at least twice a week [17].

Skin Type Classification

The skin classification can be categorized based on various factors including the hydration level of the stratum corneum, the nature of the hydrolipidic film, response to sunlight, pigmentation, skin tone, sensitivity, and signs of aging.

Hydration Level of the Stratum Corneum: A stratum corneum with approximately 10% water content is considered optimal hydration, which in turn leads to skin being tough, flexible, glowing, soft, and velvety. However, when the water level drops, the skin looks uneven, becomes inflexible, and might show signs of flaking and dehydration lines.

Nature of the Hydrolipidic Film: The hydrolipidic film consists of a combination of sweat and sebum. Its composition, both qualitative and quantitative, differs from person to person, thereby affecting the balance between the aqueous and lipid phases, which in turn determines the type of skin [18]. Sebum production varies with age, gender, and topographical variations of the skin. For instance, oily skin has a more lipophilic hydrolipidic film because of greater sebum secretion. Various factors such as skin tightness after washing, pore size and number, daily oiliness, and makeup maintenance are all used to characterize skin as oily or dry.

Response to Sunlight: The reaction of the skin to sunlight is influenced by factors such as sensitivity to solar exposure, the capacity to tan, and the frequency of solar erythema occurrences. These attributes are essential in identifying the skin phototype.

Pigmentation, Skin Tone, Sensitivity, and Signs of Aging on the Skin: Skin color results from the interplay of melanin and hemoglobin. Melanin imparts a yellowish-brown color, while hemoglobin contributes to red color. The genetic makeup of an individual primarily dictates skin color, which is influenced by the distribution of melanin within the epidermis. Additionally, environmental elements, such as exposure to sunlight, along with hormonal influences, can cause variations in melanin levels on the skin's surface. Skin sensitivity is characterized by the emergence of uncomfortable sensations, including stinging, burning, pain, itching, and tingling, which may or may not be accompanied by redness. These sensations occur in response to stimuli that would typically not provoke such reactions in healthy skin [19]. An aging skin is characterized by wrinkles, uneven pigmentation and texture, and lack of elasticity.

The Fitzpatrick Scale: Developed in 1975 by Dr. Thomas Fitzpatrick, the Fitzpatrick Scale has established itself as the authoritative standard for skin typing. Initially, the scale was designed to assess the skin's capacity to withstand ultraviolet (UV) light and physicians utilized this scale to evaluate a patient's likelihood of experiencing sunburn or tanning, which correlates with the quantity of melanin produced in the skin. This information enabled healthcare providers to determine the appropriate level of UV light therapy required for treating skin conditions such as psoriasis and eczema, while minimizing the risk of phototoxicity or irritation to the patient. The scale comprises a series of inquiries regarding natural skin, hair, and eye color, in addition to factors such as ethnicity and the skin's reaction to unprotected sun exposure. Individuals with a lower Fitzpatrick score, indicative of lighter skin tones, are more susceptible to sunburn and should limit their UV light exposure. Conversely, those with a higher Fitzpatrick score possess more melanin and are more prone to tanning. While individuals who tan easily can endure greater UV exposure, they may also experience heightened sensitivities due to sun exposure.

While the Fitzpatrick Scale is widely recognized, several other skin typing systems exist to assess skin reactivity. The Lancer Ancestry Scale, created by Dr. Lancer in 1998, enhances the Fitzpatrick Scale by incorporating a more detailed analysis of a client's ancestry. The Glogau Photoaging Scale evaluates the extent of skin aging due to sun exposure specifically in Caucasian skin. This scale serves as a predictor for potential treatment outcomes, operating under the premise that a higher level of skin damage correlates with a lower likelihood of improvement. This scale has not gained widespread acceptance due to its limited scope. Later in 2008, Dr. Wendy Roberts designed the Roberts Skin Type Classification system based on the Fitzpatrick Scale to forecast the skin's reaction to injury and inflammation. This classification considers four key elements: the Fitzpatrick Scale category, the Glogau Scale score, the propensity for hyperpigmentation, and the likelihood of scarring [20].

Fitzpatrick Phototype Scale: This is a 6-point classification system that assesses skin types based on their response to sun exposure, ranging from type I to type VI (FZ1:white skin, always burns, never tans; FZ2: white skin, always burns, minimal tan; FZ3: white skin, burns minimally, tans moderately and gradually; FZ4: light brown skin, burns minimally, tans well; FZ5: brown skin, rarely burns, tans deeply; FZ6: dark brown/black skin, never burns, tans deeply).

Glogau Scale: This 4-point system categorizes the degree of photoaging in the skin by examining the presence of wrinkles, classified from type I to type IV (G1: no wrinkles, early photoaging; G2: wrinkles in motion, early to moderate photoaging; G3: wrinkles at rest, advanced photoaging; G4: only wrinkles, severe photoaging).

Roberts Hyperpigmentation Scale: A 7-point scale that evaluates post-inflammatory pigmentation and the likelihood of developing pigmentation issues, with classifications as follows: type 0 indicates hypopigmentation; type I represents minimal and transient hyperpigmentation; type II denotes minimal and permanent hyperpigmentation; type III signifies moderate and transient hyperpigmentation; type IV indicates moderate and permanent hyperpigmentation; type V reflects severe and transient hyperpigmentation; and type VI represents severe and permanent hyperpigmentation.

Roberts Scarring Scale: This 6-point scale assesses scarring patterns by analyzing the skin's response to injury and inflammation, with classifications including type 0 for atrophy; type I for no scar; type II for macule; type III for plaque; type IV for keloid; and type V for keloidal nodule [20].

This classification system is applicable to individuals of all skin tones and proves beneficial in clinical settings, as it can anticipate the skin's reaction to injury and inflammation following aesthetic interventions or as a result of complications. This predictive capability aids in formulating treatment strategies. The Roberts Skin Type Classification System facilitates the selection and planning of both medical and cosmetic procedures, as well as the appropriate treatment products. Additionally, it assists in the management of wound care and skin recovery, while also evaluating the short-term and long-term outcomes of these treatments and procedures [20, 21].

Sex-Based Differences

Noticeable differences have been observed in male and female skin types. These differences are highlighted in Table 2.1. For instance, females have thicker subcutaneous tissues than males, while males have thinner skin. Female skin thins more quickly than male skin does with age, and postmenopausal women in particular see a decrease in skin thickness, which is an indicator of estrogen involvement in skin maintenance. Skin thickness can be altered by sexual steroids, ovariectomy (linked to skin thinning), and estrogen therapy (thickens the skin). Skin pH is hypothesized to affect the flora of organisms living on stratum corneum layer, which serves as the skin's barrier function. Males have a greater variety of aerobic biotypes and flora than females, even in the absence of any obvious quality differences [22]. According to Man et al., men aged 13 to 70 had more sebum on their foreheads than women of the same age [23]. Pillai et al. measured the collagen, dermal thickness, and collagen density of the skin on the forearms of healthy subjects. This study reported that

Table 2.1 Changes in the skin structure based on gender and age

Skin layers/ physiological parameters	Categories		Differences between categories
Epidermis	Age	Newborns	Significantly thinner compared to adults, making it more susceptible to injury [63]
		Infants	Begins to thicken compared to newborns, but it is still thinner than adult skin [64] Skin barrier continues to develop, becoming less permeable to water and irritants [64]
		Adults	Has optimal thickness and structural integrity [65] Skin barrier is fully functional, providing excellent protection against water loss, irritants, and pathogens [65]
		Elderly	Thin due to age-related changes in cell turnover and collagen production [66] Skin barrier weakens, leading to increased dryness, sensitivity, and a higher risk of skin infections [66]
Dermis	Age	Newborns	Significantly thinner than that of adults, contributing to their delicate and sensitive skin [67]
		Infants	Continues to grow and mature, but it remains thinner than that of adults [64]
		Adults	Fully developed with well-organized collagen and elastin fibers, which provide skin with strength and elasticity. However, aging starts to affect the dermis around the third decade of life, with gradual loss of collagen and elastin contributing to wrinkles and reduced skin elasticity [68]
		Elderly	Marked decline in collagen and elastin, thinning of the dermal layer, and reduced skin hydration Increased wrinkling, sagging, and reduced elasticity. The rate of collagen degradation increases with age, and there is also a reduction in the density of dermal fibroblasts [69]
Stratum corneum	Age	Newborns	More hydrated than that of adults
		Infants	Thick and more cohesive, though it remains more permeable than in adults [70]
		Adults	Fully developed, with well-organized lipid layers that provide an effective barrier function [71, 72]
		Elderly	Structure and function of the SC are compromised in older people with elevated skin pH due to changes in serine protease activity, as well as inhibition of lamellar body secretion [64]
	Sex	Male	Thicker compared to females likely due to higher androgen levels (testosterone) in males, which promote epidermal proliferation and an increase in skin thickness [73, 74]
		Female	Thinner stratum corneum along with less dense collagen, can contribute to females having more sensitive skin and higher water loss [75]

(continued)

Table 2.1 (continued)

Skin layers/ physiological parameters	Categories		Differences between categories
Sebum secretion	Age	Newborns	Strong androgenic stimulation of sebum secretion prior to birth Sebum levels are high during the first week of life and then fall Skin surface is dry and rough at birth compared to older children
		Infants	Lower total lipid content compared to adults, which correlates with low sebum levels
		Elderly	Sweat and sebum production is reduced, which further reduces the skin's ability to act as a buffer [74]
	Sex	Male	Sebum production peaks during adolescence and remains higher in men than in women throughout adulthood With age, sebum production decreases but tends to remain higher in older men than in women of the same age [75]
		Female	Produce less sebum than males, and production is more variable due to hormonal fluctuations during menstrual cycles, pregnancy, and menopause Sebum production in females significantly declines after menopause, leading to drier skin compared to men, especially in the later stages of life
Skin pH	Age	Newborns	Higher at birth, which can affect enzyme activity and barrier maturation [76]
		Infants	Decreases over time, promoting a better barrier function and lower susceptibility to irritation [77]
		Adults	Adult skin pH may fluctuate due to factors such as diet, skincare products, diseases (e.g., eczema), and environmental stressors [78] Acidic environment helps maintain the skin's protective barrier property, preventing bacterial overgrowth, and retaining moisture [77]
		Elderly	Aging causes structural and functional changes in the skin, including reduced sebaceous gland activity and slower skin cell turnover, both of which can alter skin pH [79]
Subcutaneous Tissues	Sex	Male	Thinner skin, skin thickness started to decline linearly from age 20 Gradual decline in subcutaneous fat, leading to increased skin wrinkling and loss of skin elasticity, though this occurs slower than in women [68]
		Female	Thicker subcutaneous tissues but it thins more quickly than male skin does with age and in postmenopausal women in particular a decrease in skin thickness is seen, which is an indicator of estrogen involvement in skin maintenance Skin thickness remained constant in women until around age 50

(continued)

Table 2.1 (continued)

Skin layers/ physiological parameters	Categories		Differences between categories
Skin collagen	Sex	Male	Decline in skin collagen is slower Men start losing collagen in their 30s, with a more gradual reduction over time [80]
		Female	Skin collagen levels decline with age Women can lose up to 30% of their skin collagen in the first 5 years after menopause, contributing to more rapid aging in female skin compared to male skin [81]
Skin elasticity	Sex	Male	Less elasticity
		Female	More elasticity, but rebounds from stretching more quickly

skin collagen declined with age and was lower in female participants across the board. In addition, this study demonstrated that skin thickness in men started to decline linearly from age 20, while remained constant in women until around age 50 [24]. Epidermal thickness of males and females did not significantly differ, except for forehead skin which is significantly thinner in old females than in males [25]. Researchers discovered that while the difference in skin elasticity between male and female respondents was not statistically significant (male: 0.27; female: 0.273), it was greater in female subjects. In addition, the researchers found that although women's skin is more elastic than men's, it rebounds from stretching more quickly [26].

Age-Related Changes

Skin maturation starts during embryogenesis with intercellular and intracellular signals between different tissue layers. With increasing gestational age, the barrier develops, and the epidermis fully matures by 34 weeks of age. Collagen fibrils are smaller in preterm neonates than in term newborns or adults, the paracellular dermis beneath the dermoepidermal junction is edematous, and the anchoring structures have fewer and wider spaces between them [27]. Due to a strong androgenic stimulation of sebum secretion prior to birth, sebum levels are high during the first week of life and then fall. Compared to adult skin, infant skin has lower total lipid content, which correlates with lower sebum levels at 6 months of age [28]. Compared to older children, the skin surface is drier and rougher at birth. Smoother skin is associated with more hydrated skin in the first 30 days of life. Over the next 3 months, the stratum corneum becomes more hydrated than that of adult tissue. The stratum corneum of newborns aged 3 to 12 months is notably more hydrated than that of adults. Infant skin has a higher pH than adult skin, normally classified as between 5 and 5.5.

Elderly Skin: Prior research has indicated a slight increase in pH with age. A clinical trial demonstrated the beneficial effects of topically applied oil-in-water emulsions with varying pH values of 3.5 or 4.0: After applying either emulsion to a pH of less than 5.0 for more than 7 hours, the pH of individuals over 80 years old significantly decreased [29]. The structure and function of the SC are compromised in older people with elevated skin pH due to changes in serine protease activity, as well as inhibition of lamellar body secretion. This may result in more desquamation, which manifests as rough skin. Furthermore, sweat and sebum production is reduced in the elderly, which further reduces the skin's to act as a buffer. Our skin's ability to function as a buffer is reduced, which makes us less resistant to changes in both endogenous and exogenous pH [29]. Consequently, aged skin is more susceptible to allergens and contact irritants, which helps to explain why the prevalence of contact dermatitis in the elderly is on the rise. Ageing skin thus shows increased sensitivity to mechanical stress as a result of atrophic changes as well as an elevated pH level. In addition, older people's skin has a rough, dry appearance that can occasionally cause itching, which is likely caused by an elevated pH [30].

The face, neck, upper chest, and extensor regions of the hands and forearms exhibit a decrease in epidermal thickness, especially in females. On average, thickness declines by 6.4% every 10 years, and this is accompanied by a decline in the quantity of epidermal cells. As skin ages, keratinocytes change shape, getting fatter and shorter, while corneocytes enlarge due to a reduction in epidermal turnover. There's a reduction in the skin's natural emulsion of fat and water, and the stratum corneum's water content is also reduced. The aged skin exhibits a reduction in global lipid content of up to 65%. Age-related changes in the amino acid composition of the skin may result in less cutaneous natural moisturizing factor and less ability for the skin to bind water [31].

Hair Anatomy and Physiology: A Journey of Growth and Transformation

Hair covers the body's surfaces, including the scalp, eyebrows, axillae (armpits), and external genitalia, with distribution largely influenced by genetics and hormones. Hair has limited protective barrier properties and aids in decreasing heat loss. Eyebrows, eyelashes, and hair in nostrils and the external ear canal provide protection against foreign particles. Touch receptors associated with hair follicles, known as hair root plexuses, detect light touch [32].

Anatomy of Hair

Extracellular proteins hold the dead, keratinized epidermal cells that make up each hair together. The medulla, cortex, and cuticle are the three concentric layers that make up the shaft, which protrudes above the skin's surface, and the root, which penetrates the dermis and occasionally the subcutaneous layer (Fig. 2.2). The cuticle is the outermost layer of the hair follicle made up of flat, highly keratinized cells, which are stacked like shingles. The cortex makes up most of the shaft with elongated cells and produces the medulla, which is unevenly formed and often absent in thin hair. The hair follicle has a thick dermal root sheath and internal/external root sheaths surrounds the root. The hair papilla and the hair matrix, a germinal layer in charge of hair development and replacement, are found in the bulb at the base of each hair follicle.

Associated Structures: Hairs are connected to the sebaceous gland and the muscles of the arrector pili (Fig. 2.2). Under stress, the arrector pili, which runs from the superficial dermis to the dermal root sheath, contracts, forcing the hair shafts to stand upright and giving rise to "goose bumps." Hair root plexuses are the dendrites of neurons that surround each hair follicle. They are touch-sensitive and produce nerve impulses when the hair shaft moves [33].

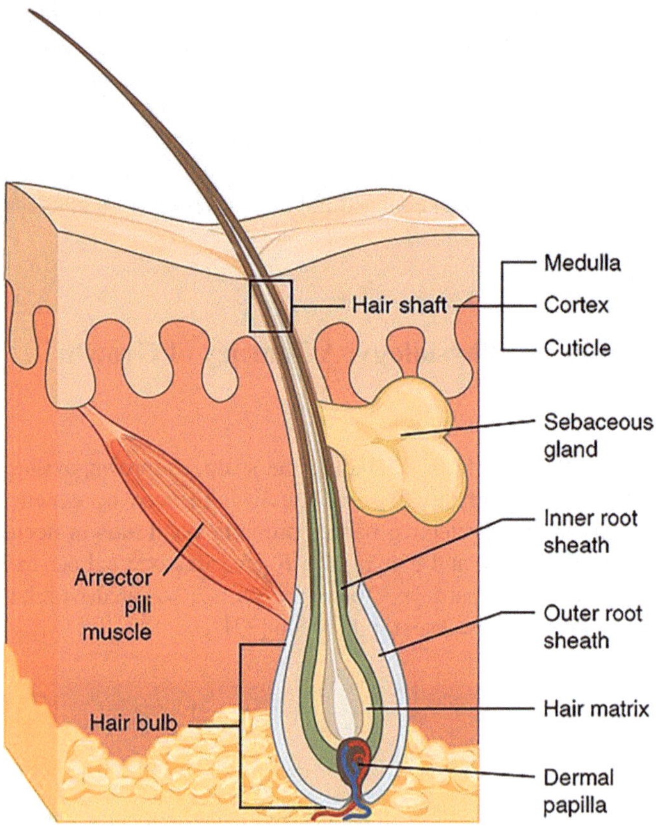

Fig. 2.2 Hair structure. (Reproduced with permission from: https://commons.wikimedia.org/wiki/File:506_Hair.jpg)

Hair Growth: Every hair follicle goes through three phases in its growth cycle: anagen (growth), catagen (regression), and telogen (resting). Hair grows because of cell division in the hair matrix during the anagen stage, where new cells push the older ones upward. Cells keratinize and finally perish as they ascend. During the catagen stage, hair development stops followed by follicle atrophy. The follicle enters telogen, a resting phase, following catagen. Later, a fresh growth cycle starts, causing the old hair root to shed and a new one to sprout. Scalp hair normally spends 2 to 6 years in the anagen phase, 2 to 3 weeks in the catagen phase, and roughly 3 months in the telogen phase. Approximately 85% of scalp hairs are in the growth stage at any given time. Though visible hair is dead, portions of its root within the scalp remain alive until pushed out by new hair growth. On an adult scalp, 70 to 100 hairs typically fall out each day. The growth rate and replacement cycle can be altered by several conditions, including disease, radiation therapy, chemotherapy, age, heredity, gender, and extreme emotional stress. Diets designed to lose weight quickly, especially ones that are low in protein or calories, can worsen hair loss. Furthermore, shedding rates could rise for a few months following childbirth. Alopecia, which is the loss of some or all of one's hair, can be caused by aging, endocrine abnormalities, skin diseases, chemotherapy, or hereditary susceptibility [33].

Anagen: The secondary epithelial germ in the telogen hair follicle between the dermal papilla and club hair begins to undergo mitosis at the beginning of the anagen phase. The active growth phase known as anagen is when the hair follicle grows, takes on its original structure, and generates hair fibers. Most scalp hairs—between 85 and 90 percent—are in the anagen phase [34]. There are six sections in the anagen stage. Hair stem cells divide from anagen I to V, covering the dermal papilla, penetrating deep into the skin, and initiating the growth of the hair shaft and inner root sheath (IRS), in that order. As a result, melanocytes in the hair matrix start producing pigment and the structure of the hair shaft begins to take shape. The new hair shaft appears from the skin during anagen VI, after the hair bulb next to the dermal papilla has finished forming. This stage in hair follicles can last for 6–8 years [35]. During the anagen phase pigmentation and hair shaft synthesis take place. The final hair structure's curvature is determined by the degree of axial symmetry in the hair bulb. The duration of the anagen phase or the actively growing period of the follicle often correlates with the fiber's length. Bone morphogenetic proteins (BMPs), sonic hedgehog, different WNT proteins and receptors, insulin-like growth factor-1 (IGF-1), fibroblast growth factor-7, hepatocyte growth factor (HGF), and vascular endothelial growth factor are important regulatory proteins in the anagen phase [36].

Catagen: After anagen, matrix cells stop mitosis, and the follicle moves into the carefully regulated involuntary phase known as catagen. In humans, catagen usually lasts for 2 weeks, depending on the location and kind of follicle. The proximal part of the hair shaft keratinizes during catagen [37]. Catagen is divided into eight phases. The termination of melanogenesis in the hair bulb is the first sign of catagen. Apoptosis and cyclic differentiation are seen in cellular populations found in the mesenchyme, follicular epithelium, neuroectodermal cells, perifollicular vascular, and neural systems. However, because suppressor bcl-2 is expressed in the dermal papilla, apoptosis does not occur there [35]. Catagen is an example of a bulbar involution stage. The vitreous membrane thickens as the perifollicular sheath compresses. Eventually, the dermal papilla gets closer to the bulge as the lower part of the hair follicle shrinks to an epithelial strand. The epithelial strand gradually lengthens and eventually reaches slightly below the pilar muscle insertion. The epithelial strands start to involute and gradually shorten after the presumed club hair has been keratinized. This is followed by the condensation and upward movement of the papilla, which finally settles behind the bulge. After a while, the column narrows to a nipple, which is where the secondary hair germ appears beneath the club. The cortical and cuticle cells that make up the club hair itself are the only ones that have pigmentation. During the catagen phase, dermal papilla migration toward the bulge area fails in part due to the existence of a hairless gene mutation. FGF5 is a critical inducer of catagen, as shown by the extended anagen phase observed in FGF5-deficient animals. Besides FGF5, other factors such as TGF-β1, IL-1b, neurotrophins NT-3 and NT-4, BMP2/4, and TNF-α have also been implicated in inducing catagen [38].

Telogen: This stage occurs between the end of follicular regression and the start of the anagen phase that follows. Ten to 15 percent of all hair is in the telogen stage, which lasts for 2 to 3 months on average. The hair shaft changes into a club hair during telogen and eventually sheds. The follicle stays in this stage until the hair germ shows increased proliferative and transcriptional activity in late telogen, resulting in the commencement of anagen. The hair germ reacts to anagen-initiating signals from the dermal papilla [39]. One of the hair cycle's main goals is the telogen, which is regulated by substances including thyroid hormones, prolactin, adrenocorticotropic hormone (ACTH), retinoids, and androgens. Estrogen receptor expression has been shown to be restricted to telogen papilla fibroblasts, despite the lack of distinct molecular markers linked to the telogen follicle. Fibroblast growth factor-5 (FGF-5) and basonuclin are also expressed by telogen follicle germ cells. As a growth factor, bone morphogenic protein-4 (BMP-4) is essential for inhibiting follicular development and differentiation during the telogen stage [33]. The macroenvironment that envelops the hair follicle has a role in controlling changes in the cycle. Follicles in the subcutaneous fat are kept in a "refractory" telogen state with the help of BMPs. The follicle can advance to a "competent" telogen phase, which is marked by a hair germ that is receptive to anagen-initiating signals and capable of entering a new anagen phase, when BMPs stop their inhibitory function [34].

Exogen: The process of hair loss receives less attention, but from the patient's point of view, it is undoubtedly the most vital part of hair development. Human telogen hairs often survive multiple follicular cycles, showing that the anagen and exogen stages are separate processes. It is believed that the shedding phase occurs actively and is unrelated to telogen and anagen [35].

Sex-Based Differences

Men's hair tends to grow rapidly, is generally shorter, and shows a lesser extent and a more forward pattern of baldness compared to women's hair, which tends to thin out and follow a circular pattern along the front. Male hair was found to be entirely black, while human female hair is brown at the distal end and black at the proximal end [40]. In females, the proximal end is present but absent in males. The cuticle is absent in males and present in females. According to a study by Lan et al., normal women typically have advanced linear hair growth rate (LHGR) than usual men [41]. Hair width and linear hair growth rate (LHGR) have a positive correlation ($P < 0.006$), and slower growth rates are linked to global thinning. Together, male and female patients with patterned hair loss showed lower rates of hair growth when compared to controls with hair of the same thickness. Growth rates continued to decline in males with pattern hair loss as the clinical severity increased. However, statistical analysis of linear hair growth rate in severely affected females was not possible due to sample size limitations. The face is one area of the body where hormones alter hair growth. A lot of androgen reactions happen gradually, and some follicles take years to fully express their response. During puberty, the weight of the beard grows significantly, and it keeps growing until the mid-30s [42].

Hair loss is the most common cause of male pattern baldness, also known as male androgenetic alopecia. Hair loss is a gradual condition. Baldness is caused by the scalp being gradually drained of hair as terminal hairs gradually turn into vellus hairs in a very predictable way. While nearly all young bald men (98.1%) and older bald men (92.3%) carried the androgen receptor gene restriction fragment length polymorphism, only 77% of non-bald men did. The development of male androgenetic alopecia appears to require this polymorphism, even if the presence of it in non-bald men indicates that it is necessary but not sufficient to generate the phenotypic [37].

Many investigations into the possible functions of prolactin have been conducted recently, and it is now proved that prolactin influences angiogenesis, lactation, reproduction, osmoregulation, and hair growth in addition to these other processes. Female pattern hair loss in women of reproductive age is caused by excess androgen. However, a lot of women who have hyperandrogenism also show signs of scalp hair loss and complain about it, suggesting that androgens play a part in hair loss. Individuals who have an excess of androgen typically experience female pattern hair loss in their early adult years. The etiology of female pattern hair loss in postmenopausal women is more complex, also involve an inadequate amount of estrogen. Additionally, genetic factors play a part in the female pattern hair loss [37]. Female pattern hair loss and facial hirsutism are common throughout menopause. Hair density and diameter decrease noticeably during the perimenopausal period or the menopausal transition. It is believed that modifications in the properties of hair growth stem from the inhibition of ovarian estrogen synthesis and complex interactions with growth factors, cytokines, and other hormones [35].

Age-Related Changes

Infant Hair: The hair that grows on babies inside the mother's womb after 7 to 8 months of pregnancy is called neonate hair. Around 3 to 4 months into the pregnancy, the first hair, or lanugo, starts to grow throughout the entire body of the developing embryo inside the uterus and falls out 7 or 8 months later [41, 43]. The lanugo is the first hair follicle to form and lacks melanin granules and a medulla. The hair growing near the newborn's scalp has a diameter of about 34 μm and a slightly rough surface. Additionally, the diameter of the space between the scale and the surface was 10–12 μm, which was smaller than the space between the scale and the hair's middle or terminal end [44, 45].

Elderly Hair: Hair pigmentation maintenance requires melanocytes, which are supported by stem cells in the bulge area of the hair follicle. The demise of melanocytes and melanocyte stem cells is associated with the loss of pigmentation in human hair that happens as people age. Studies on pMel17 and the microphthalmia-associated transcription factor revealed a decrease in the number of unpigmented melanocytes in the bulge region of the hair follicle. Growing older is also linked to a decrease in the overall number of hair follicles on the scalp and body and an increase in the proportion of telogen hair follicles [45, 46]. Research has shown that women's hair diameter increases from the middle of the following decade to the end of the second decade. Later, the hair diameter starts to progressively shrink. As a result, in a typical adult, thinning does not become noticeable until later in life. A decline in pigmentation is another feature of ageing hair. One widely accepted theory states that by the time they are 50 years old, half of the population will have grey hair. Melanocyte damage can result from oxidation, such as that caused by ultraviolet (UV) radiation, which can create free radicals. If the damage to these cells reaches a certain point, they will be compelled to undergo apoptosis, which will reduce the number of melanocytes in the hair follicle, corresponding to a decrease in pigment formation. In addition to hair thinning and greying, there are still a plethora of distinctions between healthy and ageing hair that might not be as obvious as the characteristics listed above. For example, as we age, damage to the cuticle layer's scale-like pattern with irregular, jagged edges causes the hair fiber to become rougher [47, 48]. The age- and sex-related differences in the growth of hair are highlighted in Table 2.2.

Nail Anatomy and Physiology: A Story of Strength and Protection

Nail Anatomy

The components of the fingernail include nail matrix, nail plate, nail bed, nail sinus, nail fold, nail root, nail cuticle (eponychium), and nail lunula (Fig. 2.3). The nail matrix is an active tissue that generates cells which harden as they move outward from the nail root to the nail plate.

Table 2.2 Changes in hair follicles based on gender and age

Hair feature	Category		Differences between categories
Growth speed and length	Sex	Male	Hair grows quickly, is shorter in length, and exhibits a lower overall portion and forward flow pattern of baldness than women's hair [48]
		Female	Overall thinning and whorl pattern of frontal hair [44]
Hair color		Male	Entirely black [44]
		Female	Brown at the distal end and black at the proximal end [44]
Proximal ends		Male	Absent
		Female	Present
Cuticle		Male	Absent
		Female	Present
Hair loss		Male	Hair loss begins at the temples and hairline, gradually receding A bald spot often forms on the crown of the head [82]
		Female	Hair thins overall, often starting with a widening of the part or a thinning crown. The hairline usually remains intact [83, 84] Hair density and diameter decrease noticeably during the perimenopausal period or the menopausal transition
Hair changes	Age	Newborns	Hair that grows out of the uterus after 7 to 8 months of pregnancy is called neonate hair Three to 4 months into the pregnancy, the first hair, or lanugo, starts to grow throughout the entire body of the developing embryo inside the uterus and falls out 7 or 8 months later Hair growing near the newborn's scalp had a diameter of about 34 μm and a slightly rough surface [45]
		Elderly	Loss of pigmentation in human hair that happens as people age Decrease in the overall number of hair follicles on the scalp and body and an increase in the proportion of telogen hair follicles Damage to the cuticle layer's scale-like pattern with irregular, jagged edges causes the hair fiber to become rougher with age [46]

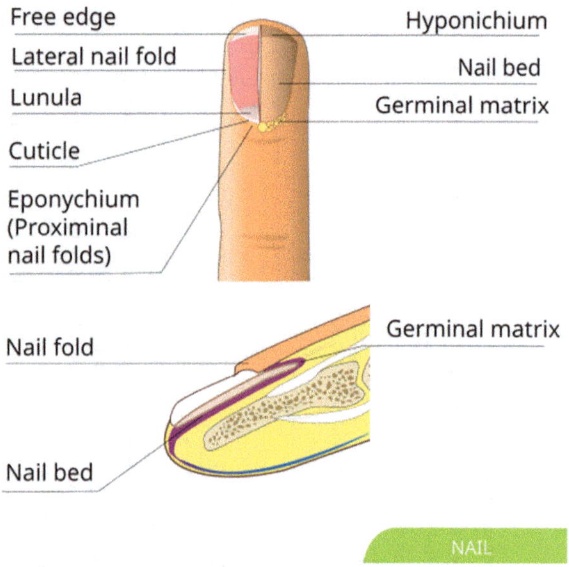

Fig. 2.3 Nail anatomy showing internal detail. (Reproduced with permission from: https://commons.wikimedia.org/wiki/File:Nail_svg_hariadhi.svg)

Nail Fold: A nail fold secures the lateral and proximal margins of the nail plate and holds the nail in place. It is located next to the distal interphalangeal joint at its proximal margin. Biological and surgical processes affecting these two structures are relevant to the 1.2 mm location of this joint's extensor tendon insertion [6, 49].

Nail Plate: The matrix overlying the nail bed is the nail plate, which provides a layered keratinized structure. The nail bed is typically described as having three layers (trilaminar). However, its physical properties suggest it behaves more like a structure with only two layers (bilaminar). This suggests that an additional layer, which contributes to the trilaminar structure, might be a superficial attachment to the nail bed [49]. The nail plate has a curved, longitudinal, and transverse axes. The free edge is a strong attachment point and a useful tool due to its ability to become lodged in nail folds at both the proximal and lateral margins. The lateral edges of the nail and matrix go halfway around the great toe's terminal phalanx. This provides the foot with the necessary strength. The top of the nail plate has a variety of ridges that are mostly smooth but become clearer as people age [50].

Lateral Nail Folds: The cushioned cutaneous lateral margins of the nail are provided by the lateral nail folds. They usually show up more in the toes than the fingers, given their role in the nail's strong adherence to the nail bed.

Proximal Nail Fold: When the nail origin forms a pocket on the dorsum during week 14 of embryogenesis, a lip of skin known as the proximal nail fold is formed. It adheres to the nail plate's dorsal aspect and hides the nail matrix, which is known as the "lunula" in clinical terminology. It works in tandem with the nail plate to create a barrier to guard the matrix [51].

Cuticle (Eponychium): A layer of epidermis called the cuticle extends after the proximal nail fold to the dorsal feature of the nail plate. Cuticle loss is often an early sign of chronic paronychia and can be caused by repeated manipulation or manicures, inflammation, or infection [51].

Nail Matrix and Lunula (Nail Root): The matrix (15–25%) and the nail bed (75–85%) are the two groups of tissues that lie beneath the nail. On an average, the proximal element extends half the distance between the distal interphalangeal joint's (DIPJ) central crease and the nail fold. The pale "half-moon" known as the lunula is typically visible on the thumb, but less visible in other ulnar digits [52].

Distal Groove: A cutaneous ridge that separates the finger pulp from subungual structures [6].

Hyponychium: It is the skin under the free edge of the nail. The distal groove forms the distal boundary of the cutaneous margin that underlies the free nail.

Nail Bed: The nail bed reaches the hyponychium from the distal edge of the lunula. A complementary set of longitudinal epidermal ridges extends to the lunula beneath the nail plate upon avulsion of the nail, leading to the description of the nail being led up the nail bed as if on rails. The dermis beneath the nails is thin, lacking in follicular or sebaceous appendages, firm collagenous adherence to the underlying periosteum, and little fat [52].

Growth Rate and Influencing Factors

Bean et al. examined his left thumbnail and noticed a slight but steady decline in the rate of nail growth with age. This study was conducted over a period of 35 years. Interestingly, the nail growth rate was 2.85 mm/month at the end of the study (at age 67) compared to 3.69 mm/month at the beginning of the study (at age 32) [53]. Furthermore, Hamilton et al. discovered that only in the 15–19 age range did male nails grow more quickly than female nails. The results were not statistically significant in the other age groups [54].

Environment: There is an ongoing debate regarding the impact of seasonal variation on nail growth rates. In comparison to the temperate region, the mean daily growth of nails was found to be slower in the Arctic environment.

Trauma: Numerous studies have shown that trauma, such as nail biting, accelerates the growth of nails. Furthermore, there was no discernible impact of mild stress, like that encountered during final exam periods, on the nail growth rates.

Disease: Evidence suggests that acute severe illnesses are linked to a slowing of nail growth, even though there is no change in nail growth with acute minor illnesses. High-flow vascular conditions like arteriovenous shunts and hyperthyroidism are linked to faster nail growth [55].

Medication: Two studies have shown that itraconazole accelerates the growth of nails. In one trial, individuals with onychomycosis were treated with a 400 mg daily pulse regimen for 1 week per month for a total of 3 to 4 months. The authors calculated that the patients receiving itraconazole had an average growth rate of 2.4–2.7 mm per month, while the unaffected group had an average growth rate of 1.1 mm per month [56, 57].

Sex-Based Differences in Structure and Growth

According to a study by Manuela et al., female nails have more sulfur ($p < 0.001$) and less nitrogen ($p = 0.001$) than male nails, while the carbon content of both are the same (a low N/S ratio) [58]. A decrease in nitrogen content ($p = 0.05$, females) and an increase in carbon content ($p < 0.01$, both sexes) with age lead to an increased C/N ratio ($p < 0.001$). A study by Paulina et al. reported the structural differences in the fingernail keratin between women and men. In comparison to female fingernail proteins, male fingernail proteins had significantly more β-sheet and random coil structural regions and fewer α-helix structural regions. Furthermore, the same study reported that fingernail α-keratin from males had more disulfide bonds than from females [59]. The Cys content of male nails was 0.6 mole% lower than that of the female nails, or approximately 6 Cys residues per 1000 amino acid residues. Due to hormonal differences, specifically testosterone, which promotes thick nail growth, nails of men may be slightly stronger and thicker than nails of women. The average thickness of a man's fingernail and toenail is 1.65 mm, while a woman's nail plate is typically 0.5 mm. Nail plates in men are often thicker than women. In 2000, Rodushkin et al. examined elemental characteristics in an attempt to differentiate between genders [60]. They demonstrated that males have higher levels of Na, K, and Mn and lower levels of bismuth and silicon than females. Delvin conducted a comparison of the major mineral concentration in males and females. This study demonstrated that males have larger levels of nitrogen (N) and potassium (K), females had higher quantities of zinc (Zn), iron (Fe), chromium (Cr), and manganese (Mn) [61].

Age-Related Changes

Newborns: Newborns have thin, soft nails, and children's nails grow at a rate that is comparable to that of young adults, with the fastest rates of nail growth (1.5 mm per day) occurring between the ages of 10 and 14. During the first two decades of life, the nail plate thickness and width rapidly increase. It is also well known that nails have embryos [52]. During the neonatal stage, the nails are extremely thin, and minor trauma can easily separate the distal edge, as in the case of thumb sucking. With ageing, this modification most likely disappears. Newborns' fingernails typically have a flat curvature or an oval shape [55]. In 50% of babies, there is no lunula. Nails on toes can be round or triangular. The overlapping is caused by a force applied by the thin, triangular nail plate, which pushes the lateral folds down. As a result, 73% of newborns have proximal and lateral nail-fold pseudohypertrophy, which resembles a pseudo-ingrown nail.

Infant Nail: The formation of nail unit begins around 9 weeks. The first sign is an invasion of the primitive epidermis, which forms the nail field (the area where the nail will grow in the future) by creating a groove on the dorsum of each fingertip. Epithelium cells divide and proliferate into two halves. The superficial portion creates the proximal nail fold, and the deeper section develops the matrix. Week 11 is when the fingers begin to develop. The nail plate is visible at 13 weeks, located distal to the proximal nail fold. The plate glides across the nail bed until 22 weeks, at which point it becomes formed. Up until 22 weeks, when the plate is constituted, it moves over the nail bed.

Elderly Nail: Older individuals often experience various nail changes, including both age-related modifications and conditions prevalent in this population. These changes can be attributed to several factors such as inadequate circulation in the distal extremities, poor biomechanics, infections, neoplasms, and systemic or skin disorders that manifest as nail disease. Such factors can affect not only the nail plate but also other components of the nail unit, including the matrix, nail bed, hyponychium, or nail folds, leading to secondary abnormalities in the nail plate [62]. Common nail conditions in the elderly include brittle nail syndrome, characterized by transverse and lamellar splitting of the free edge and distal portion of the nail plate; onychauxis, marked by discoloration, subungual hyperkeratosis, and loss of translucency; and onychoclavus, which presents as a tender, dark area beneath the distal nail plate of the great toe. Additionally, paronychia involves tender erythematous swelling of the nail fold, while onychocryptosis results in inflammation of the lateral nail fold, granulation tissue, and secondary infection. Subungual hematomas, which cause painful discoloration of the nail that transitions from red to blue, and onychomycosis, characterized by nail thickening and discoloration, are also notable conditions affecting elderly individuals [62]. The age- and sex-related differences in the growth of nails are highlighted in Table 2.3.

Conclusion

The physiology of skin, hair, and nails varies significantly by sex and age, impacting their structure, function, and responses to external factors. Recognizing these differences is essential for personalized dermatological care and effective prevention and treatment strategies. Ongoing research into the molecular and hormonal mechanisms driving these variations will improve our ability to address age-related and gender-specific conditions. By exploring how hormonal, genetic, and

Table 2.3 Changes in nails based on gender, age, and other factors

Category		Differences between categories	
Sex	Male	Significantly more β-sheet and random coil structural regions, fewer α-helix structural regions, and the fingernails α-keratin has more disulfide bonds Cys content of the male nails was 0.6 mole% lower than that of the female nails, or approximately 6 Cys residues per 1000 amino acid residues Due to hormonal differences, specifically testosterone, men's nails are slightly stronger and thicker The average thickness of a man's fingernail and toenail is 1.65 mm Males have higher levels of Na, K, and Mn and lower levels of bismuth and silicon than females Males have larger levels of nitrogen (N) and potassium (K) [85, 86]	
	Female	Females' nails have more sulfur and less nitrogen than males' nails, resulting in a lower nitrogen/sulfur ratio, when the carbon content is the same for both sexes Woman's nail plate is typically 0.5 mm Higher quantities of zinc (Zn), iron (Fe), chromium (Cr), and manganese (Mn)	
Age	Newborns	Nails formation	*Week 9:* Formation of nails begins. Primitive epidermis forms the nail field with a groove on each fingertip. Epithelium cells divide into 2 halves, superficial portion forms the proximal nail fold, deeper section forms the matrix [52, 55] *Week 11:* Formation of fingers begins *Week 13:* Nail plate is visible, located distal to the proximal nail fold *Weeks 13–22:* Plate glides across the nail bed until 22 weeks, when it becomes fully formed it moves over the nail bed [52]
		Nails characteristics	Nails are thin and soft Prone to separation at the distal edge due to minor trauma (e.g., thumb sucking) Fingernails have flat or oval curvature; 50% of newborns lack a lunula Toenails may be round or triangular 73% of newborns have proximal and lateral nail fold pseudohypertrophy, resembling pseudo-ingrown nails due to thin, triangular nail plates pushing down the folds [60]
	Children	Nail growth rate is similar to young adults Fastest growth (1.5 mm per day) occurs between ages 10 and 14 Rapid increase in nail plate thickness and width during the first two decades of life, since nails have embryos Thin nails in newborns become thicker as children age, and susceptibility to trauma decreases over time [59]	

	Elderly	Age-related changes	Factors influencing nail changes: poor circulation, biomechanics, infections, neoplasms, systemic or skin disorders Nail changes can affect the nail plate, matrix, nail bed, hyponychium, or nail folds, leading to secondary abnormalities
		Common nail conditions in older adults	*Brittle nail syndrome:* Characterized by transverse and lamellar splitting of the free edge and distal nail plate *Onychauxis:* Discoloration, subungual hyperkeratosis, and loss of nail translucency *Onychoclavus:* Tender, dark area beneath the distal nail plate of the great toe *Paronychia:* Tender, erythematous swelling of the nail fold *Onychocryptosis (ingrown nail):* Inflammation of the lateral nail fold with granulation tissue and possible secondary infection *Subungual hematomas:* Painful nail discoloration transitioning from red to blue *Onychomycosis*: Nail thickening and discoloration due to fungal infection [62]
Other influencing factors	Physiology	A slight but consistent decrease in nail growth rate was observed with increasing age [87] Male nails grew faster than female nails only during the 15–19 age range [54]	
	Environment	The daily nail growth is slower in Arctic environment compared to temperate region	
	Trauma	Trauma, like nail biting, accelerates nail growth, while mild stress during examination, for instance, has no significant effect [50]	
	Disease	Acute severe illnesses slow nail growth [57] High flow vascular conditions like arteriovenous shunts and hyperthyroidism increase nail growth [55]	
	Medication	Itraconazole accelerates nail growth [56, 57]	

environmental factors influence skin elasticity, hair growth, and nail strength, and integrating advances in genomics and systems biology, future research promises to develop more targeted and effective solutions for maintaining skin, hair, and nail health throughout life.

Glossary

Alopecia A condition resulting in the loss of hair, which can be caused by genetics, aging, or hormonal imbalances

Anagen phase The active growth phase of the hair cycle

Androgens Hormones like testosterone that influence hair growth and skin characteristics, leading to differences between male and female skin and hair physiology

Apocrine glands Sweat glands associated with hair follicles in areas such as the armpits and groin, often involved in scent and pheromone release

Arrector pili Small muscles attached to hair follicles that cause hair to stand upright when contracted, resulting in 'goosebumps'

Brittle nail syndrome A condition seen in the elderly, characterized by transverse and lamellar splitting of the nail plate

Catagen phase The short transitional phase in the hair growth cycle when the hair follicle shrinks and detaches from the dermal papilla

Collagen A structural protein in the skin that provides tensile strength and elasticity. It declines with age, leading to wrinkles and sagging

Collagen fibrils Small fibers that make up collagen, providing structure and support to the skin

Cutaneous membrane Another term for the skin, consisting of the epidermis and dermis

Cuticle (eponychium) A layer of epidermis extending from the proximal nail fold to the dorsal aspect of the nail plate, offering protective functions

Dermis The layer beneath the epidermis, composed of collagen and elastic fibers, housing hair follicles, blood vessels, and glands. It provides strength and flexibility to the skin

Distal groove A cutaneous ridge separating the finger pulp from subungual structures

Disulfide bonds Bonds between sulfur atoms in keratin proteins, contributing to the strength of male nails compared to female nails

Embryogenesis The developmental stage where the nail unit begins to form around week 9, leading to the formation of the proximal nail fold and matrix

Epidermis The outermost layer of the skin, consisting primarily of keratinocytes. It functions as a barrier and is responsible for skin tone

Hair follicle A structure in the dermis from which hair grows. It undergoes cycles of growth (anagen), regression (catagen), and rest (telogen)

Hair root plexus Nerve endings around hair follicles that detect hair movement and provide sensory input

Hyperkeratosis Excessive thickening of the outer layer of the skin or nail plate, often seen in nail conditions like onychomycosis

Hypodermis Also known as the subcutaneous layer, this is found beneath the dermis and is composed of adipose tissue that insulates the body and provides cushioning

Hyponychium The skin under the free edge of the nail, forming a distal boundary for the nail

Integumentary system The system comprising the skin, hair, nails, glands, and sensory receptors, which serves functions like temperature regulation, protection, and sensation

Keratin A fibrous protein forming the main structural component of the nail plate, contributing to its strength

Keratinization The process where keratinocytes, as they move toward the skin's surface, fill with keratin, die, and form a protective layer

Lamellar granules Organelles within keratinocytes that release lipids, contributing to the water barrier function of the skin

Langerhans cells Immune cells in the epidermis that help detect and fight off pathogens

Lateral nail folds Cushioned cutaneous margins of the nail, providing strong adherence to the nail bed, particularly in the toes

Lunula The pale "half-moon" visible at the base of the nail, representing the visible portion of the nail matrix

Melanocytes Pigment-producing cells in the epidermis responsible for skin color. These cells transfer melanin to keratinocytes, providing protection against UV radiation

Merkel cells Specialized cells found in the epidermis that are involved in touch sensation

Nail avulsion The process of nail removal, exposing the underlying nail bed, which has a set of longitudinal ridges

Nail bed The structure beneath the nail plate, described as having two or three layers, extending from the lunula to the hyponychium

Nail fold Secures the lateral and proximal margins of the nail plate, holding the nail in place

Nail matrix The tissue underneath the base of the nail that generates cells for nail growth

Nail plate The visible hard part of the nail composed of keratin, produced by the nail matrix

Onychauxis Thickening of the nail, often accompanied by discoloration, hyperkeratosis, and loss of nail translucency in elderly individuals

Onychoclavus A tender, dark area under the distal nail plate, typically seen in the great toe, caused by pressure

Onychocryptosis Ingrown nail, where the lateral nail fold becomes inflamed and infected, often due to nail growth into the surrounding skin

Onychomycosis Fungal infection causing nail thickening, discoloration, and separation of the nail from its bed

Paronychia Inflammation and infection of the nail fold, leading to tenderness and swelling, commonly caused by bacteria or fungi

Proximal nail fold A lip of skin formed during embryogenesis, covering the dorsal aspect of the nail plate and protecting the nail matrix

Sebaceous glands Glands associated with hair follicles that secrete sebum. They are more active in males due to androgens, leading to differences in skin oiliness

Sebum An oily secretion produced by sebaceous glands, which helps to lubricate and protect the skin

Stratum corneum The outermost layer of the epiderfmis, composed of dead keratinocytes, which provides a barrier against environmental damage

Striae Also known as stretch marks, these are streaks that form on the skin due to rapid stretching and tearing of the dermis

Subungual hematoma Blood collection under the nail caused by trauma, resulting in painful nail discoloration

Telogen phase The resting phase in the hair growth cycle where the hair stops growing and eventually sheds

Thermoregulation The process by which the skin helps regulate body temperature through sweating and blood flow adjustments

Trilaminar/bilaminar Refers to the structure of the nail bed, which appears to have three layers but may function more like a two-layered structure

References

1. Bielitzki, J. T. (1998). *Integumentary system. Nonhuman primates in biomedical research* (pp. 363–375). Elsevier.
2. Mauldin, E. A., & Peters-Kennedy, J. (2016). Integumentary system. In *Jubb, Kennedy & Palmer's pathology of domestic animals: Volume 1* (p. 509). Elsevier.
3. Diegel, K. L., Danilenko, D. M., & Wojcinski, Z. W. (2018). The integumentary system. In *Fundamentals of Toxicologic Pathology* (pp. 791–822). Elsevier.
4. Walker, P. G., & Niehaus, A. J. (2022). Integumentary system. In *Medicine and surgery of camelids* (pp. 232–302). Wiley.
5. Elliott, D. G. (2000). Integumentary system. In *The laboratory fish* (pp. 271–306). Elsevier.
6. de Berker, D. (2013). Nail anatomy. *Clinics in Dermatology, 31*, 509–515.
7. Johnson, M., Comaish, J., & Shuster, S. (1991). Nail is produced by the normal nail bed: A controversy resolved. *The British Journal of Dermatology, 125*, 27–29.
8. Kolarsick, P. A., Kolarsick, M. A., & Goodwin, C. (2011). Anatomy and physiology of the skin. *Journal of the Dermatology Nurses' Association, 3*, 203–213.
9. Benson, H. A. (2012). Skin structure, function, and permeation. In *Topical and transdermal drug delivery: Principles and practice* (pp. 1–22). Elsevier.
10. Jalolidinovna, I. Z. (2023). Morphology and histology of skin. *Texas Journal of Medical Science, 16*, 52–56.
11. McLafferty, E., Hendry, C., & Farley, A. (2013). The integumentary system: Anatomy, physiology and function of skin. *Nursing Standard, 2012*(27), 35.

12. Lai-Cheong, J. E., & McGrath, J. A. (2013). Structure and function of skin, hair and nails. *Medicine (Baltimore), 41*, 317–320.
13. Samadi, A., Yazdanparast, T., Shamsipour, M., Hassanzadeh, H., Hashemi Orimi, M., Firooz, R., et al. (2022). Stratum corneum hydration in healthy adult humans according to the skin area, age and sex: A systematic review and meta-analysis. *Journal of the European Academy of Dermatology and Venereology, 36*, 1713–1721.
14. Menon, G. K. (2014). *Skin basics; structure and function. Lipids skin health* (pp. 9–23). Springer.
15. Monteiro-Riviere, N.A. (2020). Comparative anatomy, physiology, and biochemistry of mammalian skin. In: *Dermal and ocular toxicology* (pp. 3–71).
16. Liu, Y., Li, X., Yang, H., Zhang, P., Wang, P., Sun, Y., et al. (2023). Skin-interfaced superhydrophobic insensible sweat sensors for evaluating body thermoregulation and skin barrier functions. *ACS Nano, 17*, 5588–5599.
17. Dupuis, M. L., Pagano, M. T., Pierdominici, M., & Ortona, E. (2021). The role of vitamin D in autoimmune diseases: Could sex make the difference? *Biology of Sex Differences, 12*, 1–12.
18. Oliveira, R., Ferreira, J., Azevedo, L. F., & Almeida, I. F. (2023). An overview of methods to characterize skin type: Focus on visual rating scales and self-report instruments. *Cosmetics, 10*, 14.
19. Lev-Tov, H., & Maibach, H. I. (2012). The sensitive skin syndrome. *Indian Journal of Dermatology, 57*, 419–423.
20. Roberts, W. E. (2009). Skin type classification systems old and new. *Dermatologic Clinics, 27*, 529–533.
21. Gupta, V., & Sharma, V. K. (2019). Skin typing: Fitzpatrick grading and others. *Clinics in Dermatology, 37*, 430–436.
22. McKnight, G., Shah, J., & Hargest, R. (2022). Physiology of the skin. *Surgery (Oxford), 40*, 8–12.
23. Lefèvre-Utile, A., Braun, C., Haftek, M., & Aubin, F. (2021). Five functional aspects of the epidermal barrier. *International Journal of Molecular Sciences, 22*, 11676.
24. Pillai, S., Manco, M., Oresajo, C., & Baalbaki, N. (2022). Epidermal barrier. In *Cosmetic dermatology: Products and procedures* (pp. 1–15). Wiley.
25. Sivesind, T. E., & Maibach, H. I. (2022). Biophysical characteristics of skin: relation to skin of color, age, gender, and site. In *Handbook of cosmetic science and technology* (pp. 1–17). CRC Press.
26. Bellomo, R., Hegarty, C., Story, D., Ho, L., & Bailey, M. (2012). Association between a chloride-liberal vs chloride-restrictive intravenous fluid administration strategy and kidney injury in critically ill adults. *Journal of the American Medical Association, 308*, 1566–1572.
27. Moore, E. R., Bergman, N., Anderson, G. C., & Medley, N. (2016). Early skin-to-skin contact for mothers and their healthy newborn infants. *Cochrane Database of Systematic Reviews, 2016*, CD003519.
28. Oranges, T., Dini, V., & Romanelli, M. (2015). Skin physiology of the neonate and infant: Clinical implications. *Advances in Wound Care, 4*, 587–595.
29. Schreml, S., Kemper, M., & Abels, C. (2014). Skin pH in the elderly and appropriate skin care. *The European Medical Journal Dermatology, 2*, 86–94.
30. Reszke, R., Pełka, D., Walasek, A., Machaj, Z., & Reich, A. (2015). Skin disorders in elderly subjects. *International Journal of Dermatology, 54*, e332–e338.
31. Ramos-e-Silva, M., & da Silva Carneiro, S. C. (2007). Elderly skin and its rejuvenation: Products and procedures for the aging skin. *Journal of Cosmetic Dermatology, 6*, 40–50.
32. Buffoli, B., Rinaldi, F., Labanca, M., Sorbellini, E., Trink, A., Guanziroli, E., et al. (2014). The human hair: From anatomy to physiology. *International Journal of Dermatology, 53*, 331–341.
33. Cooper, G. A. A. (2015). Anatomy and physiology of hair, and principles for its collection. In *Hair analysis in clinical and forensic toxicology* (pp. 1–22). Elsevier.
34. Wolff, H., Fischer, T. W., & Blume-Peytavi, U. (2016). The diagnosis and treatment of hair and scalp diseases. *Deutsches Arzteblatt International, 113*, 377.

35. Hirt, P. A., & Paus, R. (2019). Healthy hair (Anatomy, Biology, Morphogenesis, Cycling, and Function). In *Alopecia* (pp. 1–22). Elsevier. https://doi.org/10.1016/B978-0-323-54825-0.00001-6
36. Peereboom-Wynia, J. D. R. (1982). *Hair root characteristics of the human scalp hair in health and disease*. Erasmus University Rotterdam, Rotterdam. http://hdl.handle.net/1765/31387
37. Aly, D. (2014). Hair: A review of prominent scalp hair changes throughout a females' life. *Journal of Pigmentary Disorders, 1*, 2.
38. Randall, V. A., & Botchkareva, N. V. (2009). The biology of hair growth. In *Cosmetics applications of laser & light-based systems* (pp. 3–35). William Andrew Publishing.
39. Buckley, S. T., Bækdal, T. A., Vegge, A., Maarbjerg, S. J., Pyke, C., Ahnfelt-Rønne, J., et al. (2018). Transcellular stomach absorption of a derivatized glucagon-like peptide-1 receptor agonist. *Science Translational Medicine, 10*, eaar7047.
40. Rantala, M. J., Pölkki, M., & Rantala, L. M. (2010). Preference for human male body hair changes across the menstrual cycle and menopause. *Behavioral Ecology, 21*, 419–423.
41. Olsen, E. A. (2001). Female pattern hair loss. *Journal of the American Academy of Dermatology, 45*, S70–S80.
42. Westgate, G. E., Botchkareva, N. V., & Tobin, D. J. (2013). The biology of hair diversity. *International Journal of Cosmetic Science, 35*, 329–336.
43. Ostrea, E. M., Jr., Bielawski, D. M., Posecion, N. C., Jr., Corrion, M., Villanueva-Uy, E., Jin, Y., et al. (2008). A comparison of infant hair, cord blood and meconium analysis to detect fetal exposure to environmental pesticides. *Environmental Research, 106*, 277–283.
44. Furdon, S. A., & Clark, D. A. (2003). Scalp hair characteristics in the newborn infant. *Advances in Neonatal Care, 3*, 286–296.
45. Rexbye, H., Petersen, I., Iachina, M., Mortensen, J., McGue, M., Vaupel, J. W., et al. (2005). Hair loss among elderly men: Etiology and impact on perceived age. *The Journals of Gerontology. Series A, Biological Sciences and Medical Sciences, 60*, 1077–1082.
46. Villani, A., Ferrillo, M., Fabbrocini, G., Ocampo-Garza, S. S., Scalvenzi, M., & Ruggiero, A. (2022). Hair aging and hair disorders in elderly patients. *International Journal of Trichology, 14*, 191.
47. Arentze, T., Timmermans, H., Jorritsma, P., Olde Kalter, M.-J., & Schoemakers, A. (2008). More gray hair—But for whom? Scenario-based simulations of elderly activity travel patterns in 2020. *Transportation, 35*, 613–627.
48. Hordinsky, M., Sawaya, M., & Roberts, J. L. (2002). Hair loss and hirsutism in the elderly. *Clinics in Geriatric Medicine, 18*, 121–133.
49. Cecchini, A., Montella, A., Ena, P., Meloni, G. B., & Mazzarello, V. (2009). Ultrasound anatomy of normal nails unit with 18 mhz linear transducer. *Italian Journal of Anatomy and Embryology, 114*, 137–144.
50. Farren, L., Shayler, S., & Ennos, A. (2004). The fracture properties and mechanical design of human fingernails. *The Journal of Experimental Biology, 207*, 735–741.
51. Stern, D. K., Creasey, A. A., Quijije, J., & Lebwohl, M. G. (2011). UV-A and UV-B penetration of normal human cadaveric fingernail plate. *Archives of Dermatology, 147*, 439–441.
52. De Berker, D., André, J., & Baran, R. (2007). Nail biology and nail science. *International Journal of Cosmetic Science, 29*, 241–275.
53. Bean, W. B. (1980). Nail growth: Thirty-five years of observation. *Archives of Internal Medicine, 140*, 73–76.
54. Hamilton, S. W., & Baird, K. S. (2009). The treatment of established non-union of the proximal humerus using the Polarus locking intramedullary nail. *International Journal of Shoulder Surgery, 3*, 53.
55. Hillman, R. W. (1955). Fingernail growth in the human subject: Rates and variations in 300 individuals. *Human Biology, 27*, 274.
56. Yaemsiri, S., Hou, N., Slining, M., & He, K. (2010). Growth rate of human fingernails and toenails in healthy American young adults. *Journal of the European Academy of Dermatology and Venereology, 24*, 420–423.

57. Yu, H., Kwon, H., Oh, D., & Kim, J. (2004). Is slow nail growth a risk factor for onychomycosis? *Clinical and Experimental Dermatology, 29*, 415–418.
58. Dittmar, M., Dindorf, W., & Banerjee, A. (2008). Organic elemental composition in fingernail plates varies between sexes and changes with increasing age in healthy humans. *Gerontology, 54*, 100–105.
59. Brzózka, P., & Kolodziejski, W. (2017). Sex-related chemical differences in keratin from fingernail plates: A solid-state carbon-13 NMR study. *RSC Advances, 7*, 28213–28223.
60. Rodushkin, I., & Axelsson, M. D. (2000). Application of double focusing sector field ICP-MS for multielemental characterization of human hair and nails. Part II. A study of the inhabitants of northern Sweden. *Science of The Total Environment, 262*, 21–36.
61. Delvin, E., & Levy, E. (2020). Trace elements: Functions and assessment of status through laboratory testing. In *Contemporary practice in clinical chemistry* (pp. 851–864). Elsevier.
62. Rao, S., Banerjee, S., Ghosh, S. K., Gangopadhyay, D. N., Jana, S., & Mridha, K. (2010). Nail changes and nail disorders in the elderly. *Indian Journal of Dermatology, 55*, 301.
63. IMF, d.-S., GLN, V., & ZSN, R. (2019). Skin thickness dimensions in histological section measurement during late-fetal and neonatal developmental period: A systematic review. *Skin Research and Technology, 25*, 793–800.
64. Rahma, A., & Lane, M. E. (2022). Skin barrier function in infants: Update and outlook. *Pharmaceutics, 14*, 433.
65. Kanwar, A. J. (2018). Skin barrier function. *The Indian Journal of Medical Research, 147*, 117–118.
66. *The effects of aging on skin: Dry skin, loose skin, and more* [Internet]. [Cited 2024 Sept 7]. Available from: https://www.webmd.com/beauty/cosmetic-procedures-aging-skin
67. Khalid, K. A., Nawi, A. F. M., Zulkifli, N., Barkat, M. A., & Hadi, H. (2022). Aging and wound healing of the skin: A review of clinical and pathophysiological hallmarks. *Life, 12*, 2142.
68. Rittie, L., & Fisher, G. J. (2015). Natural and sun-induced aging of human skin. *Cold Spring Harbor Perspectives in Medicine, 5*, a015370–a015370.
69. Huang, Y., Zhao, H., Shi, X., Liu, J., Lin, J.-M., Ma, Q., et al. (2023). GRB2 serves as a viable target against skin fibrosis in systemic sclerosis by regulating endothelial cell apoptosis. *Journal of Dermatological Science, 111*, 109–119.
70. *Dissecting the formation, structure and barrier function of the stratum corneum*. International Immunology, Oxford Academic [Internet]. [Cited 2024 Sept 7]. Available from: https://academic.oup.com/intimm/article/27/6/269/678859?login=false
71. van Smeden, J., Janssens, M., Gooris, G. S., & Bouwstra, J. A. (2014). The important role of stratum corneum lipids for the cutaneous barrier function. *Biochimica et Biophysica Acta-Molecular and Cell Biology of Lipids, 1841*, 295–313.
72. Rawlings, A. V. (2014). The stratum corneum and aging. In M. A. Farage, K. W. Miller, & H. I. Maibach (Eds.), *Textb aging skin* (p. 1–25) [Internet]. Springer [Cited 2024 Sept 7]. . Available from: https://doi.org/10.1007/978-3-642-27814-3_7-2. .
73. Zouboulis, C. C., & Degitz, K. (2004). Androgen action on human skin -- From basic research to clinical significance. *Experimental Dermatology, 13*(Suppl 4), 5–10.
74. Rahrovan, S., Fanian, F., Mehryan, P., Humbert, P., & Firooz, A. (2018). Male versus female skin: What dermatologists and cosmeticians should know. *International Journal of Women's Dermatology, 4*, 122–130.
75. Madison, K. C. (2003). Barrier function of the skin: "la raison d'être" of the epidermis. *The Journal of Investigative Dermatology, 121*, 231–241.
76. Schmid-Wendtner, M.-H., & Korting, H. C. (2006). The pH of the skin surface and its impact on the barrier function. *Skin Pharmacology and Physiology, 19*, 296–302.
77. Ali, S., & Yosipovitch, G. (2013). Skin pH: From basic Science to basic skin care. *Acta Dermato-Venereologica, 93*, 261–267.
78. Lambers, H., Piessens, S., Bloem, A., Pronk, H., & Finkel, P. (2006). Natural skin surface pH is on average below 5, which is beneficial for its resident flora. *International Journal of Cosmetic Science, 28*, 359–370.

79. Hachem, J.-P., Crumrine, D., Fluhr, J., Brown, B. E., Feingold, K. R., & Elias, P. M. (2003). pH directly regulates epidermal permeability barrier homeostasis, and stratum Corneum integrity/cohesion. *The Journal of Investigative Dermatology, 121*, 345–353.
80. Shuster, S., Black, M. M., & McVITIE, E. (1975). The influence of age and sex on skin thickness, skin collagen and density. *The British Journal of Dermatology, 93*, 639–643.
81. Chowdhury, A., Nosoudi, N., Karamched, S., Parasaram, V., & Vyavahare, N. (2021). Polyphenol treatments increase elastin and collagen deposition by human dermal fibroblasts; implications to improve skin health. *Journal of Dermatological Science, 102*, 94–100.
82. Ho, C. H., Sood, T., & Zito, P. M. (2024). *Androgenetic alopecia.* StatPearls [Internet]. Treasure Island: StatPearls Publishing [cited 2024 Sept 8]. Available from: http://www.ncbi.nlm.nih.gov/books/NBK430924/
83. *Female pattern baldness Information.* Mount Sinai - New York [Internet]. Mt. Sinai Health Syst. [Cited 2024 Sept 8]. Available from: https://www.mountsinai.org/health-library/diseases-conditions/female-pattern-baldness
84. *Female pattern baldness: Causes, treatment, and prevention* [Internet]. 2018 [Cited 2024 Sept 8]. Available from: https://www.medicalnewstoday.com/articles/321527
85. Brzózka, P., & Kolodziejski, W. (2017). Sex-related chemical differences in keratin from fingernail plates: A solid-state carbon-13 NMR study. *RSC Advances* [Internet]. [Cited 2024 Sept 8], *7*, 28213–28223. Available from: https://xlink.rsc.org/?DOI=C7RA03487C
86. Dittmar, M., Dindorf, W., & Banerjee, A. (2008). Organic elemental composition in fingernail plates varies between sexes and changes with increasing age in healthy humans. *Gerontology* [Internet]. [cited 2024 Sept 8], *54*, 100–105. Available from: https://karger.com/GER/article/doi/10.1159/000128269
87. Bean, W. B. (1980). Nail growth: Thirty-five years of observation. *Archives of Internal Medicine* [Internet]. [Cited 2024 Sept 8], *140*, 73. Available from: http://archinte.jamanetwork.com/article.aspx?doi=10.1001/archinte.1980.00330130075019

Chirag A. Patel, Ph.D. is an Assistant Professor in the Department of Pharmacology at L. M. College of Pharmacy. He has over 15 years' academic and research experience in the fields of pharmacology and toxicology. Dr. Patel has authored over 40 research papers, 4 books, and 10 book chapters. His research interests also include nutraceuticals, probiotics, neurodegenerative diseases, and antimicrobial resistance

Ketan Ranch, Ph.D. is an Associate Professor in the Department of Pharmaceutics at L. M. College of Pharmacy. He has more than 19 years in teaching and research experience in formulation

Development. He has published more than 50 research/review articles in prestigious international journals and contributed to 9 book chapters and one book. Dr. Ranch is a corporate trainer for Design Expert software and has conducted more than 25 training sessions across India, Bangladesh, and Singapore

Sonal Vijay Pande, M.S., Doctoral Researcher specializing in Neuroscience with a focus on enhancing the oral bioavailability of polyphenols using metabolic adjuvants. She has extensive experience in developing diagnostic kits, molecular pharmacology, and regulatory toxicology. Ms. Pande has published more than 9 research articles and 2 book chapters. She also has experience in preclinical studies, computational drug design, and pharmacokinetic profiling

Dr. Mahalakshmi B, MD is an Associate Professor of Microbiology at the Government Medical College, Kadapa, India, with over 15 years of teaching and clinical experience. She has published more than 10 research articles in the field of medical and environmental microbiology

Ms. Tasnem H. Abdelfattah Allam is in the last year of the Bachelor of Pharmacy program at Ajman University, UAE. She is recognized for consistently excelling academically and being among the top students in her class. She has received recognition for exceptional academic accomplishments and has set herself apart through her leadership abilities, guidance of younger students, and involvement in joint research initiatives.

Ms. Maryam R. Mohamed is in her final year of the Bachelor of Pharmacy program at Ajman University, UAE. She is ranked among the best students in her class. She has been acknowledged for her outstanding academic achievements and has distinguished herself through her leadership skills, and participation in collaborative research projects.

Sai H. S. Boddu, Ph.D. is a Professor of Pharmaceutics in the College of Pharmacy and Health Sciences, Ajman University (AU), UAE. Boddu has over 20 years of pharmaceutical experience in both academic and industrial settings. Dr. Boddu has developed several pharmaceutical formulations and novel nanocarrier systems for Rx and OTC drugs. He holds 2 patents and published more than 120 research and review articles in various peer-reviewed journals. Boddu is a co-Editor of the book titled *Excipient Applications in Formulation Design and Drug Delivery*, published by Springer. Dr. Boddu serves as an Associate Editor of Cancer Reports and Scientific Advisor on the *Journal of Pharmaceutical Sciences*. He is an active member of the American Association of Pharmaceutical Scientists (AAPS)

[illegible]

[illegible]

[illegible]

[illegible]

Skincare for All Ages: The Science, Advances and Future of Skincare from Infancy to Adulthood

Andrea Echeverry

Abstract This chapter explores the evolution of skincare, tracing its transformation from apothecary formulations to industrial-scale products influenced by petrochemical and pharmaceutical advancements where consumer demand for safety, sustainability, and efficacy is driving new developments. We discuss the impact of environmental stressors such as ultraviolet radiation, urban pollution, and climate variability on the skin's barrier function and aging, addressing the physiological needs of skin from infancy through adulthood. Data from randomized controlled trials and regulatory programs are used to support discussions on efficacy, tolerability, and formulation optimization highlighting regional differences in ingredient approval and safety data requirements particularly around sunscreens and over-the-counter actives. Biomimetic ingredients are emphasized for their multifunctional role in supporting hydration, skin barrier function, inflammation reduction, and photoprotection. Special attention is given to the rise of biocosmetics with a focus on the therapeutic potential of probiotic, prebiotic, postbiotic, synbiotic, and bacteriophage-based skincare. These biocosmetics offer promising strategies to modulate skin health in infants, children, pregnant women, the elderly, and in people with conditions like acne or eczema. By blending ancient ethnobotanical practices with modern biotechnological insights, we propose that the skin microbiome may be both a tool and a target for innovation in evidence-based skincare. In parallel, microbiome sequencing platforms and artificial intelligence technologies are revolutionizing ingredient and formulation development, enabling highly personalized product design and predictive modeling of efficacy and tolerance. These advances represent a paradigm shift from traditional antimicrobial agents toward biologically derived, targeted modulators of skin homeostasis reflecting the interdisciplinary nature of modern skincare research.

A. Echeverry (✉)
Cosmetic Tattooist (Scalp & Scar Micropigmentation Artist), Marathon, FL, USA
e-mail: andrea.echeverry@trichologyassociation.org

K. Alexander, P. Romanowski (eds.), *Cosmetic Science*, AAPS Introductions in the Pharmaceutical Sciences 15, https://doi.org/10.1007/978-3-032-03596-7_3

Keywords Acne · Atopic Dermatitis · Bacteriophage · *Cuticubacterium acnes* · Eczema · *Nitrosomonas eutropha* · Pediatric skincare · Postbiotic · Prebiotic · Probiotic · Sensitive Skin Syndrome · Skin microbiome · Synbiotic · *Staphylococcus aureus*

Abbreviations

ACD	Allergic contact dermatitis
AAD	The American Academy of Dermatology Association
AHAs	α-hydroxy acids
AI	Artificial intelligence
AMPs	Antimicrobial peptides
AOB	Ammonia oxidizing bacteria
BHA	β-hydroxy acid
CIR	Cosmetic Ingredient Review
DALY	Disability-adjusted life-years
DDAC	Didecyldimethylammonium salts
EGF	Epidermal growth factor
EPS	Exopolysaccharides
FDA	Food and Drug Administration
FM	Fragrance mix
FOS	Fructooligosaccharides
GHK-Cu	Glycyl-L-Histidyl-L-Lysine Copper
GLA	Gluconolactone
GRASE	Generally recognized as safe and effective
HA	Hyaluronic acid
INCI	International Nomenclature Cosmetic Ingredient
ICD	Irritant contact dermatitis
IUPAC	International Union of Pure and Applied Chemistry
ITS	Internal transcribed spacer
NEA	National Eczema Association
OTC	Over-the-counter
PA	Protection grade of UVA
PBS	Phosphate-buffered saline
PIH	Post-inflammatory hyperpigmentation
PHA	Polyhydroxy acid
QACs	Quaternary ammonium compounds
RA	Retinoic acid
RCT	Randomized controlled trial
ROS	Reactive oxygen species
SA	Salicylic acid

SCFA	Short-chain fatty acids
SPF	Sun protection factor
Spp	Species
SOD1	Superoxide dismutase 1
TEWL	Transepidermal water loss
TiO_2	Titanium dioxide
UPF	Ultraviolet protection factor
UV	Ultraviolet
UVA	Ultraviolet A
UVB	Ultraviolet B
VCRP	Voluntary Cosmetic Registration Program
XOS	Xylo oligosaccharides
ZnO	Zinc oxide

The evolution of skincare products intertwines with the history of human migration across the planet, cultural exchanges, and our evolving relationship with the physical world. Elders, healers, and shamans formulated ointments, pomades, salves, scrubs, and tinctures to soothe insect bites, scrapes, cuts, lesions, rashes, and even control infections, caused by chronic exposure to the natural elements. Other formulations evolved simply to cleanse or enhance our physical appearance. For millennia, humans have been experimenting with natural materials like aloe vera (aloe vera juice and gel), fruits (coconut), flower petals (jasmine, rose, calendula), seeds (hemp, grape, sesame, olive, jojoba), grains (oatmeal, rice), beans (coffee), legumes (soy), bee-produced materials (honey, beeswax, propolis), ferments (milk, yogurt, vinegar, rice), and even the elements (clay, mud, crushed rocks and crystals) [1]. As civilizations have come and gone, many of the natural resources that have persisted over time still make up the foundation of cosmetic products available today [2–4]. Humans have also gravitated towards sources of water with claimed healing properties like seawater, and natural cold and hot springs [5]. We still see this natural predilection in those living away from these environments with the commercially available sprayable mineral waters, packaged muds, and scrubbing clays [6, 7]. This evolution reflects the journey from essential practices for hygiene and protection against the elements to health-promoting beauty rituals.

Skincare treatments in the last few centuries were mostly under the control of apothecaries and pharmacies. In the nineteenth century, early pioneers like Pears and Yardley in the United Kingdom laid the groundwork for modern personal care with the first mass-market soap bars [8]. The solid bar soap, a simple mixture of fats and lye, was not exactly revolutionary at that time, since solid bar soaps had been in existence for thousands of years [9], but the mass-marketing process that came from that product is certainly reflective of humanity's need for creating some universal standards of hygiene and personal grooming [8]. Heavily fragranced bar soaps were favored since perfumes were usually out of financial reach for most people, so the bar soap became the utilitarian cleanser and deodorizer of the contemporary human living in urbanized conditions [9]. Despite the availability for centuries of materials from natural sources, the development and refinement of new ingredients from petrochemicals defined the skincare industry of the nineteenth century. However, it would not be until the twentieth century that hand soaps, body cleansers, and shampoos would become accessible in large volumes and advertised for the whole family. That expansion would not have occurred without the petrochemical industry which developed technologies and methods still in use today. Understanding this evolution provides insight into the current landscape of skin cosmetic science, where tradition meets innovation and classical skincare products and ingredients have become revolutionary again.

One of the most notable trends in modern skincare is the increasing emphasis on sustainability and environmental impact. As consumers become more conscious of their choices, brands are responding with waterless formulations, biodegradable materials, and recyclable packaging. This shift towards eco-friendly practices is accompanied by a growing demand for products that cater to sensitive skin to protect it against intrinsic and extrinsic stressors. If this book had been written just a decade ago, this chapter would tell quite a different story, reflecting how rapidly cosmetic science has evolved in response to society's pursuit of hygiene, beauty, and self-expression. As this chapter explores, advancements in microbiome-based technologies are dramatically shifting the skincare industry. Where skincare products are no longer just for cosmetic purposes used to mask our humanity, but are powerful tools driven by a deeper understanding of skin biology and technological innovation.

The Role of the Environment in Skincare Product Development and Formulation

The impact of environmental factors on skincare product development is significant as formulations have been evolving to meet the modern demands of human migration across the world, with the accompanying exposures to different altitudes, humidity fluctuations, temperature extremes, and urban pollution [10]. One of the most important environmental factors that contributes to reduced skin immune function, skin damage, and signs of aging is sunlight exposure [11]. Moisturizing formulations like sunscreens try to minimize the detrimental effects of ultraviolet (UV) radiation on skin including sunburn, inflammation, reactive oxygen species production, and DNA damage associated with potential skin cancer development. Between the 1930s and the 1970s, ingredients that had photoprotective functions like benzyl salicylate and aminobenzoic acid (PABA) came to market [12]. In the 1970s, concerns about the destruction of the ozone layer by chlorofluorocarbons (CFCs) and potential increased exposure to UV rays from the sun, led to the discovery and manufacturing of additional chemicals to reduce photodamage and provide photoprotection [13]. In recent years, exposure to blue light (450–495 nm) emitted from solar radiation, computers, and handheld devices has been gaining interest as another source of photodamage [14]. Sunscreens, pivotal in protecting the skin from UV damage, come in compositions like lotions, gels, oils, solid sticks, non-CFC aerosols, continuous sprays, and powders, reflecting innovation and adaptability of skincare science.

Sunscreens can be categorized as organic (chemical), inorganic (physical), and mixed containing both organic and inorganic agents. The classification of a chemical vs physical agent is not accurate since all sunscreens either absorb or disperse UVB/UVA radiation based on specific chemical properties of the sunscreen ingredients. The FDA does not allow the term sunblock as a marketing tool since there is no chemical or composition that can "block" or reflect UV-radiation from the sun. These ingredients can target UVA (315–400 nm), UVB (280–315 nm) or both wavelengths. Those that target both wavelengths are known as broad-range or broad-spectrum sunscreens. In vitro models demonstrated that UVB rays induce damage in the epidermis leading to sunburn, while UVA rays penetrate deeper into the dermal layer placing UVA rays as the potential driving force behind skin cancer development [13]. For this reason, broad spectrum sunscreens are ideal for all age groups [15]. Sunscreen regulations in the USA differ dramatically compared to regulations in other countries and continents. Sunscreens in the USA carry the label of Sun Protection Factor (SPF) of 5–100 depending on their ability to protect only against sunburn caused by UVB. The SPF label does not reflect protection against UVA rays. In Asia, especially Japan, sunscreens use a different rating system to represent how much UVA protection a product offers labeled as protection grade of UVA (PA). In this rating system, PA+ blocks 40–50% of UVA rays, PA++ blocks 60–70% of UVA rays, PA+++, blocks up to 90% of UVA rays, PA++++, blocks up to over 95% of UVA rays. In the United States, sunscreens often do not include the PA rating although the product may offer UVA protection.

All active ingredients used to formulate sunscreens are FDA-regulated and therefore any products with these ingredients must be rigorously tested and sold as over-the-counter (OTC) medications, which makes the development of new ingredients and their incorporation into new formulations a long and expensive process for manufacturers. Therefore, the United States has a limited availability of UVA sunscreen ingredients. Organic sunscreens include avobenzone (butyl methoxydibenzoylmethane), homosalate (homomethyl salicylate), meradimate (menthyl anthranilate), octinoxate (octyl methoxycinnamate), octisalate (octyl salicylate), octocrylene, oxybenzone (benzophenone-3), padimate O (octyldimethyl-PABA), and sulisobenzone (benzophenone-4). In Europe, ecamsule (terephthalylidene dicamphor sulfonic acid) [16] and drometrizole trisiloxane [17] have been used as broad-spectrum sunscreen ingredients since they were patented in 1982 and 1989, respectively. Inorganic ingredients include zinc oxide (ZnO) and titanium dioxide (TiO_2). Mixed compositions will combine organic and inorganic ingredients, but they currently make up a small proportion of available sunscreens. Nano ZnO and nano TiO_2 are inorganic ingredients with particulate sizes between 50 and 100 nm, compared to non-nano ZnO and non-nano TiO_2 which are 200–300 nm in size.

In 2019, systemic absorption of sunscreen actives avobenzone (2–3%), oxybenzone (4–6%), octocrylene (2.35–10%), and ecamsule (0–2%) using four different formulations (2 sprays, 1 lotion, and 1 cream) was measured in humans [18]. This randomized controlled trial (RCT) revealed that the plasma concentration of avobenzone far exceeded the threshold allowed by the FDA [18]. The plasma levels of oxybenzone and octocrylene were also above the safety threshold. More importantly, this study brought to light the gaps in human data regarding systemic exposure to sunscreens [18]. In 2021, chemical screenings of 293 sealed and 368 consumer-provided sunscreens reported a high incidence of benzene (≥2.0 ppm), a known carcinogen [19], especially in spray and after-sun care formulations [20]. This led to the voluntary removal or recall of sunscreens by some manufacturers in the USA [20] The limitations of the study include the fact that benzene levels were not measured in humans, not every sunscreen in the market was screened for benzene, and products were provided by consumers unsealed and/or past their expiration date [20]. This reignited the ongoing controversy about the potential toxicity of organic sunscreen ingredients since most products with benzene levels above 2.0 ppm were in this category [20]. This study also brought to light that different batches of the same branded product had either no detectable benzene or had benzene in the upper limit [20]. This may point to potential issues in the manufacturing chain of ingredients underscoring the importance of batch-to-batch quality control testing.

Shortcomings with the SPF rating system and the need for systemic exposure safety data for broad-spectrum sunscreen led the FDA to issue an ordinance for sunscreen manufacturers in the United States [21]. The FDA now requires additional safety data to establish if sunscreens listed in the Stayed 1999 Final Monograph as Generally Recognized as Safe (GRAS) may need to be re-categorized [21]. An OTC monograph or rule sets the guidelines and conditions under which specific OTC drugs can be marketed without requiring a new drug application if these drugs are Generally Recognized as Safe and Effective (GRASE) and are not misbranded. This rule classifies sunscreen active ingredients as category I (GRASE and not misbranded), category II (not GRASE or to be misbranded), or category III (additional data needed). First, the FDA order would disallow manufacturers to label products SPF of 30 or below as they may not provide sufficient sunburn protection, which would only allow manufacturers to label products SFP 60+ (and up to SPF 80+) [21]. Under this proposal, cinoxate, dioxybenzone, ensulizole, homosalate, meradimate, octinoxate, octisalate, octocrylene, padimate O, sulisobenzone, oxybenzone, and avobenzone would be classified as Category III ingredients, requiring manufacturers to submit new safety data on sunscreen formulations that include these ingredients [21]. Second, dosage formats like oils, lotions, creams, gels, butters, pastes, ointments, sticks, and sprays would receive category I status. However, for spray sunscreens to be GRASE, it would be required to test health risks associated with involuntary inhalation which may restrict aerosolized particle size and flammability [21]. Powdered sunscreens and make-up products containing mainly TiO_2 and ZnO and usually rated SPF 10–15, would be classified in Category III and subject to restrictions based on particle size just like spray sunscreens [21]. Third, products advertised as sunscreens in the form of wipes, towelettes, body washes, and shampoos would be reclassified as new drugs because these products were not in the market and data were not available on them prior to 1972 [21]. Finally, current sunscreen formulations with up to 25% ZnO and/or TiO_2 rated SPF 60+ would continue to be category I (GRASE). Part of this proposal was originally established in 1999 but never implemented and the new ruling has not taken effect as of 2024. Implementation of this ruling could initially reduce the number of available sunscreens in the US market, making some well-known brands unable to differentiate their sunscreens from other brands, since essentially all sunscreens in the USA may end up having similar formulations to be considered GRASE and available under the OTC monogram [22].

Concerns over skin irritation and sensitivity to aminobenzoic acid (PABA), one of the first sunscreen ingredients in the market, led to the PABA-free labeling in skincare products and sunscreens. PABA is no longer on the list of GRASE ingredients, so it has been rebranded as a Category II ingredient (Not GRASE or to be misbranded). As of September 2024, the following sunscreen ingredients have been added to Category II (not GRASE or to be misbranded): diethanolamine methoxycinnamate, digalloyl trioleate, ethyl 4-[bis(hydroxypropyl)] aminobenzoate, glyceryl aminobenzoate, lawsone with dihydroxyacetone, and red petrolatum.

Sunscreen Efficacy: Beyond Ingredients and Formulations

Melanomas of the skin are one of the top six diagnosed cancers in the United States with prevalence rates that increased from 15 per 100,000 people (1999) to 23 per 100,000 people (2021) [23], so the need for comprehensive sun protection strategies is more urgent than ever. This includes applying sunscreen correctly but also reapplying regularly [24] and using formulations suited to specific environmental conditions [25]. Such efforts are essential in reducing skin cancer risk and fostering long-term protective habits among consumers. The efficacy of topical sunscreens relies on the application of the product as an even layer since any gaps in coverage minimize their protective effect. Factors that affect the efficacy of sunscreens are poor understanding by consumers about UV-related skin damage, infrequent or uneven application of sunscreens, using products past their expiration date, and the effects of sweat, water, and sand contact with sunscreens [26]. This is why topical sunscreens in lotion form are favored over sprayable or powder forms as it is easier to accomplish even layering of the ingredients on the skin with a lotion. Topical sunscreen labeled water-resistant may offer protection for either 40 or 80 minutes in the presence of sweat and water. The water-resistant property of sunscreens is due to emulsifiers like sorbitan sesquioleate and other sorbitol derivatives (sorbitan monolaurate, sorbitan monopalmitate, sorbitan monostearate, sorbitan tristearate, sorbitan monooleate, and sorbitan trioleate) [27]. The FDA does not allow waterproof or sweatproof claims in consumer advertising since it is misleading, and lack of reapplication reduces the effectiveness of sunscreens [26]. One of the reasons behind the current melanoma rates, despite the wide availability of sunscreens, may be the expectation by consumers that sunscreen alone is sufficient to protect against UVA rays [26].

With children being considered a special population when it relates to skin safety and absorption of chemicals [28], topical sunscreen marketed for children will more likely contain ZnO at 20% than any other sunscreen ingredient. These formulations may also combine ceramides and prebiotics for a more biomimetic composition. The FDA has issued a recommendation that infants or children under 6 months old should not receive topical sunscreens due to greater risk for rashes or sensitization in this population [29]. Instead, the recommendation is for caretakers of children under 6 months old to limit direct sun exposure during peak UV index times and instead rely on physical barriers like shaded areas and clothing made with fabrics rated as having ultraviolet protection factor (UPF) of at least 50 [22]. Recently, a randomized controlled trial (RCT) in 10–12-year-old Iranian school children using an online-based application (Sunshine and Skin Health App) for 8 days educating them on the long-term effects of UV-induced photoaging and skin cancer were found to be more likely to reapply sunscreen only after hand and face washing compared to children who did not receive this education [30]. Daily application of sunscreen creams did not improve between groups even with the information provided through the app, highlighting potential limitations on skincare product availability and affordability since children must rely on a caretaker to procure these products and young people may need longer interventions than 8 days to sediment the idea of daily topical product application [30]. Studies like this underscore the critical roles that regulations, innovative technology, consumer education, and consistent messaging have in promoting sunscreen use, emphasizing the importance of compliance for effective sun protection [24].

Revolutionizing Sun Care: Advanced Ingredients for Enhanced Protection and Repair

The search for UV-resistant agents highlights novel approaches to sun protection, mirroring the complexity of environmental challenges and consumer needs. Beyond the current suncare ingredients available, the inclusion of new-generation technologies and ingredients like nanostructured lipid encapsulated UV filters [31], UV protectants derived from the algae *Anacystis nidulans* [32], and botanical antioxidant extracts like *Rubus idaeus* (raspberry) seed oil (RSO) reflect ongoing innovation [33]. By incorporating chemical and physical UV-blocking agents alongside molecules targeting photorepair such as the flavoprotein photolyase from *Anacystis nidulans*, it is possible to minimize immediate UV radiation effects while providing DNA repairing mechanisms using visible lights spectra for ingredient activation [34]. Sunscreens with *Anacystis nidulans* photolyase have been available in the EU since 2008 demonstrating improvements in mild actinic keratosis [35], the most common skin lesions that precede the diagnosis of skin cancer. Efforts to develop sunscreens with RSO, ZnO, and niacinamide (vitamin B3), have shown that there is a dose-dependent improvement in SPF value with these ingredients, although further improvements to the encapsulation and ingredient concentration especially ZnO may yield optimal formulations with SPF values closer to regulation values [36]. The SPF of 10% RSO is 20, a formulation with 10% RSO and 5% niacinamide has SPF 23.6, a formulation with 10% RSO and 10% ZnO has SPF 24.0, and a formulation with 10% RSO, 10% ZnO, and 5% niacinamide has SPF 25.4 [36]. These advances reflect a nuanced approach to the balance between efficacy, safety, and aesthetics.

Production of reactive oxygen species is part of the skin's response to solar radiation, so it continues to be a target for suncare products. Superoxide dismutase (SOD1) is an enzyme that functions as an antioxidant in human cells, but it is also produced by microbes, cyanobacteria, chromista, marine plants, terrestrial plants, and fruits [37]. This copper- and zinc-dependent enzyme serves as a natural scavenger of superoxide (O_2^-) radicals, neutralizing their oxidative effects [38]. The skin commensal *Staphylococcus epidermidis* produces in vitro the highest levels of SOD compared to other native skin bacteria including *S. aureus*, *C. acnes*, *Micrococcus*, *Streptococcus*, and *Bacillus* and culturing *S. epidermidis* in the presence of manganese/zinc ions amplifies this effect [39]. Consequently, SOD is now commercially available for daily use on face and body in a sprayable format. SOD has been complemented with copper carbonate hydroxide mineral extracted from malachite rock, which serves as a pH balancer and provides stability to the formulation. Future research may focus on optimizing SOD1 production from native skin bacteria like *S. epidermidis*, potentially enhancing its antioxidant capacity through the incorporation of minerals such as copper, manganese and zinc.

Extremolytes: Skincare Ingredients from Microbes Surviving Extreme Environments

Extremolytes are a rising category of ingredients with applications in the food, pharmaceutical, and skincare industries. Microbes thriving in the most extreme environments on earth-like deserts, salt lakes and seawater produce extremolytes to survive under these conditions [40]. Halophiles are microbes that thrive in high salinity environments like salt lakes and seawater which has made them an ideal source of extremolytes [41]. Ectoine, originally discovered in 1985 in an Egyptian desert salt lake, prevented UVA photoaging in in vitro assays [42]. In humans, 2% ectoin topical cream promotes skin hydration [43] while countering the effects of temperature or humidity extremes [44]. Microbes also excrete exopolysaccharides (EPS) into their surrounding environment to help them form biofilms and shield them against extreme environments [45]. EPS from the marine bacterium *Lactiplantibacillus plantarum* EI6 shows in vitro the ability to increase wound healing [46]. Recently, the marine bacterium *Halomonas malpeensis* YU-PRIM-29 ^{T}

was shown in rats to induce faster burn wound closure by exopolysaccharide (EPS-H29) containing uronic acid [47]. Uronic acid is a nonfermentable sugar that contributes to the formation of glycosaminoglycans, which are essential components of connective and skin tissue. Uronic acids are also involved in the synthesis of polysaccharides and serve as precursors to produce ascorbic acid, essential in the skin's biological function. Liquid skincare applications in the market already include EPS from *Pseudoalteromonas* in formulations designed to help combat environmental challenges [40]. This targeted evolution in skincare underscores the industry's commitment to responding to the complex interplay of environmental stressors affecting skin health in people of all ages.

The Rise of Biomimetic and Biocosmetic Formulations: From Hero Ingredients to Hero Formulations

Cosmeceuticals

The skincare cosmetic industry has relied for decades on marketing formulations around the presence of a "hero" ingredient. A hero ingredient is a standout, star ingredient that plays a key role in the effectiveness of a product and highlighted in marketing strategies because of proven dermatological benefits, efficacy, safety, and consumer recognition. Formulations with these ingredients are referred to as "cosmeceuticals," a term coined in the 1960s, which implies that a product has both cosmetic and pharmaceutical properties [48]. For the FDA, cosmeceuticals or dermocosmetics do not exist as a class of products since a formulation can only be a cosmetic or a pharmaceutical, not both [49]. Sunscreens, acne medications, and treatments for seborrheic dermatitis, dandruff, and psoriasis are among the nonprescription drug categories covered by OTC rules [50]. It is not surprising that skincare brands have taken advantage of the OTC moniker to make products available that otherwise would not be without medical prescription. For this reason, medical professionals have become involved in the cosmeceutical market by creating and endorsing products as dermatologist certified. Skincare formulations with these regulated pharmaceutical ingredients are available OTC, most times at lower concentrations than the prescription drugs, while still delivering acceptable cosmetic outcomes such as improving skin texture and tone and preventing acne breakouts [51].. Dermatologists and cosmetologists use pharmaceutical ingredients at higher concentrations (>20%) than those found in cosmeceuticals (2–10%) because they can be unsafe without medical or professional supervision [52, 53]. This is most prevalent in products containing ingredients to counter the effects of environmental stressors like UV-radiation and side effects from skin conditions like acne vulgaris which result in scars and post-inflammatory hyperpigmentation (PIH) [51].

Retinoids

One of the most recognizable classes of active ingredients are vitamin A derivatives, such as retinoic acid (RA) (retinol), retinaldehyde, and retinyl palmitate [54]. When retinoids activate nuclear retinoic receptors (RARα, RARβ, and RARγ), they trigger gene transcription, influencing the growth and differentiation of skin cells thereby promoting cell turnover and collagen production, supporting anti-aging claims of reducing fine lines, wrinkles, and acne control [54]. In 1969, topical vitamin A (retinoid acid or tretinoin) was compared to sulfur-resorcinol, and benzoyl peroxide as an acne vulgaris treatment [55]. RA was postulated to decrease the rate of keratinization, one of the first steps in the induction of comedones, and development of acne lesions by reducing the production of horny cells in the follicular canal [56]. In 1971, the FDA approved RA for the treatment of acne and as of 2024 it is still available by prescription only trademarked as Retin-A™. Side effects such as local irritation, low tolerability, and suboptimal patient adherence are common with all topical retinoids [54, 57]. Therefore, retinoids are usually employed as adjuvant acne treatments when combined with oral retinoids and antibiotics, and/or topical antibiotics with benzoyl peroxide [54, 58, 59]. In 1997, the FDA approved adapalene as a treatment for acne [54]. It is a third-generation synthetic retinoid with a more selective retinoid receptor activity (RAR-β, RAR-γ), but similar effectiveness to tretinoin with reduced side effects [59]. Adapalene appears to exert anti-inflammatory effects by downregulating the expression of toll-like receptor 2 and IL-10 in epidermal keratinocytes which may allow more efficient immune clearance of microbes present in acne lesions [60]. In the United States, adapalene has been available OTC since 2016 in 0.1% gel for patients over 12 years old [59]. Adapalene 0.1% lotion, cream, and 0.3% gel are available by prescription only [54]. The mildest retinoid available in the market is retinyl palmitate usually found in skincare products with antiaging or anti-wrinkle claims. These moisturizers will often include retinyl palmitate, vitamin C, hyaluronic acid, and palmitic acid usually derived from coconut oil or palm oil routinely identified as "pro-retinol," implying that the formulation may induce retinoid-like effects. There are at least 16 essential free fatty acids that have been identified in human skin, with palmitic acid being the most abundant saturated fatty acid in human sebum, followed by linoleic acid, and α-linolenic acid [61]. Linoleic acid is an essential omega-6 polyunsaturated fatty acid in skin that humans must obtain from foods. Arachidonic acid, a proinflammatory fatty acid, accumulates in skin only if linoleic acid deficiency occurs [61].

Skincare formulations containing active ingredients often face stability issues such as oxidation and degradation leading to poor performance [62]. By combining traditional ingredients with delivery systems and encapsulation technologies such as hydrogels, microemulsions, liposomes, polymeric microparticles, solid lipid nanoparticles, formulations help prevent oxidation and ingredient degradation while improving targeted delivery of ingredients [62–64]. For instance, vitamin C (ascorbic acid, sodium ascorbate, or ascorbyl palmitate) is highly susceptible to degradation when exposed to air, aqueous solutions, metal ions, and high pH which reduces its effectiveness and shelf life [65]. Therefore, stabilizers such as ferulic acid, glutathione, and sodium metabisulfite (an anti-chelating agent) produce higher viscosity formulations while lowering oxidation rates, improving the stability of vitamin C [66–69]. Encapsulating vitamin C in phosphatidylcholine liposomes has been shown to improve both its stability and skin penetration, offering enhanced antioxidant protection, and reducing photodamage [70].

Although retinoids are popular in the OTC market, they do come with side effects such as irritation, pain, burning, itching, and increased photosensitivity, especially without sunscreens use [62, 71]. With the introduction of adapalene to the OTC market and our current knowledge that inactive ingredients in cosmeceutical formulations contain compounds that ameliorate some of these side effects [72], scientists have tested formulations with botanical extracts with antioxidant and antimicrobial activity against adapalene in humans [73, 74]. The combination of 0.03% retinol, 0.7% rose extract, and 0.05% hexamidine diisethionate was as effective as 0.1% adapalene gel in reducing severity of mild-to-moderate in acne vulgaris patients, but with improved tolerability at 2 weeks compared to 4 weeks for 0.1% adapalene gel [73]. Hexamidine diisethionate is a diamidine salt used since the 1950s for its broad-spectrum antimicrobial efficacy that has been gaining interest in skin research [75]. One of the botanicals that has been with humans for thousands of years is the rose, so rose petal oil and plant extracts are commonplace in the cosmeceutical market, despite its high cost of production [76]. For this reason, *Rosa canina L.* (rosehip) oil is available in formulations between 0.01% and 1%. The *Rosa damascena* flower contains flavonoids, terpenes, anthocyanin, and glycosides [2, 76], while *Rosa canina L.* (rosehip) extract and oil contain antioxidants such as vitamin C, quercetin, and ellagic acid [77]. Altogether, this supports the idea that one strategic aim for cosmetic scientists should be to generate formulations to improve the tolerance to drugs that expected to become available OTC in upcoming years.

Hydroxy Acids

The most popular OTC ingredients are hydroxy acids such as α-hydroxyacids (AHA), β-hydroxy acid (BHA), and polyhydroxy acids (PHA) [78, 79]. Although the mechanisms of action between these compounds differ, AHAs in general reduce hyperkeratinization, improve hydration, collagen and elastin synthesis, promote cell turnover leading to exfoliation, thereby reducing wrinkles, fine lines, and hyperpigmentation [53, 80–82]. The formulation of choice for these ingredients is as a serum for topical application, but gels and ointments, whole-body cleansers, and sprayable forms are available. AHAs include glycolic acid, L-lactic acid, tartaric acid, mandelic acid, citric acid, and malic acid [83]. Glycolic acid (hydroxyacetic acid), and lactic acid are in OTC products at concentrations between 2% and 10% [53]. Malic and citric acid may act as pH adjusters, chelating agents, or fragrances [84]. Malic acid is a fragrance at low concentrations (0.008–0.4%) but used as a skin conditioning or humectant in leave-on skincare formulations up to 2.1%, while products diluted for baths can be as high as 50% [85]. Citric acid functions as an AHA or a BHA depending on its chemical configuration [84]. The inactive ingredients in these formulations may also have cosmetic benefits by helping relief the irritative potential of exfoliating AHAs and SA [72, 86].

Salicylic acid (SA) has been used since the nineteenth century for medicinal purposes as an antiseptic and for the treatment of infectious diseases and skin disorders including eczema of the face and scalp [87, 88]. Currently, manufacturers advertise salicylic acid as a BHA [89], chemically it is not, but the BHA label may help SA be more recognizable to consumers [52]. Although, the mechanism of action of SA is still not fully understood in humans [52], SA is thought to function as a desmolytic agent by disrupting desmosomal proteins, including desmogleins, which are crucial for the cohesion of epidermal cells [52, 90]. This disruption leads to exfoliation by weakening cellular tight junctions rather than breaking keratin filaments [91]. SA also exerts antimicrobial activity which helps explain its effectiveness in treating acne lesions [92]. Synthetic salicylic acid is the primary compound in OTC formulations with concentrations between 0.5% and 2%. Salicine is the naturally occurring salicylate from S*alix alba* (willow bark) found in formulations as an extract between 0.5% and 5%. Current OTC formulations with SA will use cellulose derivatives, plant polysaccharides, as a matrix for the sustained release of SA [93].

PHAs are carboxylic acids such as lactobionic acid and gluconolactone (GLA) [3,4,5-trihydroxy-6-(hydroxymethyl)oxan-2-oneor] [79]. GLA has gained popularity because it is less likely to cause irritation than AHAs, offering chelating, antioxidant and moisturizing benefits, and protecting elastin fibers from UV-induced damage, making them a suitable option for sensitive skin types, including those with rosacea or atopic skin [53]. A study evaluated the effects of 10% and 30% GLA chemical peels on skin parameters such as pH, transepidermal water loss (TEWL), and sebum levels using a split-face model on 16 female subjects [94]. Results showed a significant reduction in pH across all four facial areas treated and decreased TEWL around the eyes, on the left forehead, and right cheek. Sebum levels on the cheeks reduced at both concentrations [94]. According to data from the 2022 FDA's Voluntary Cosmetic Registration Program (VCRP), GLA was in a total of 312 formulations, with 195 being leave-on and 117 rinse-off products [95]. GLA was used at concentrations up to 15% in leave-on skincare formulations. GLA was also reported to be used in eye lotions, eye makeup removers, and other eye makeup preparations (up to 0.075%), and in three baby products, though specific concentrations were not provided for all products [95]. Additionally, mucous membrane exposure is possible with feminine wipes using GLA concentrations up to 0.56% [95]. A double-blind clinical trial on acne patients assessed skin tolerance to a 14% GLA aqueous lotion, compared to the placebo aqueous lotion and 5% benzoyl peroxide [71]. Skin tolerance outcomes included burning, stinging, erythema, scaling, pruritus, and dryness. There were no significant differences found between the treatment groups for erythema, pruritus, burning, or stinging. About 24% of patients using GLA reported side effects, with slightly higher occurrences of erythema, burning, stinging, pruritus, and scaling compared to the placebo [71].

Controlling Acne

Hair follicles and associated sebaceous glands are everywhere on the human body except palms, bottom of the feet, and lips. Another treatment for acne is benzoyl peroxide, which became available OTC in 2011 in topical washes, creams, gels, and ointments between 2.5% and 10% concentrations, showing the greatest antimicrobial activity at the highest concentration [96]. Due to its exfoliative and skin drying effects, formulations at all concentrations of benzoyl peroxide usually include other ingredients to help control skin irritation and dehydration such as bisabolol, *Matricaria recutita* (chamomile) extract, *Camellia sinensis* (Green tea) extract, and *Calendura officinalis* (Marigold) [86]. Benzoyl peroxide has recently come under scrutiny due to the presence of benzene in OTC formulations [97]. Temperature and UV exposure appear to be factors in this process, with low temperature (2 °C) preventing benzene detection, while high temperature (50 °C) even in encapsulated formulations leading to detectable benzene levels even in the surrounding air [97]. The impact from this report may affect OTC status and formulation availability for benzoyl peroxide in the USA, although the FDA has not issued any notices yet.

Brands offering products designed for acne sufferers have also integrated mini hydrocolloid patches to dress individual acne lesions [98]. Hydrocolloid dressings are composed of two layers: an inner colloid layer and an outer water-impermeable layer. The outer layer provides protection and prevents the spread of microbes, but it can also create hypoxic conditions that may promote the growth of *Cuticubacterium acnes* [98]. Alternative bilayer patches made from gelatin/chitosan containing Cortex *Phellodendron amurense*/*Centella asiatica* extracts were shown to be more effective compared to patches with hydrocolloid, and they were as effective as patches with chlorhexidine diacetate at neutralizing *C. acnes* in vitro [99].

Since the hair follicle is one of the major sites for lipophilic bacteria like *C. acnes* [100], oils could function as occlusives with potential comedogenic activity. Comedones are hair follicle lesions that do not present with inflammation but cosmetically they give the skin the uneven complexion that exfoliating cosmeceuticals attempt to address [101]. The face and trunk (back and chest) are areas where they occur. They are plugs described as either closed (whitehead) or open (black head), made up of sebum secretions and keratinized dead skin cells (corneocytes). Cosmetic habits like not removing make-up can promote their appearance [101]. A nanoemulsion of *Melaleuca alternifolia* (tea tree) oil, and a mixture of essential oils (tea tree, lavender, eucalyptus, and lavender) at 2% (w/w) demonstrated similar reduction of comedones in male subjects compared to *Melaleuca alternifolia* (tea tree) oil alone [102]. The chemical composition of the four oils was: *Melaleuca alternifolia* (tea tree) oil (41% terpin-4-ol; 21% γ-Terpinene; 9% α-Terpinene; 1% 1,8-cineole), *Lavandula angustifolia* (lavender) oil (34% linalool; 38% linalyl acetate), *Eucalyptus globulus* (eucalyptus) oil (83% 1,8-cineole; 9% d-limonene), and *Citrus reticulata* (tangerine) oil (86% d-limonene; 7% γ-Terpinene). Therefore, natural oils and essential oil blends show promise in reducing comedones, though their effectiveness appears to be enhanced by formulation improvements.

Skincare products containing silicone (cyclomethicone, dimethicone), polymers of siloxane (cyclopentasiloxane), and glycerin are oil-free and expected to be inert [86]. Oil-free formulations may be advantageous for people prone to hair follicle disorders like acne since the resident microbes are not able to utilize polymers of siloxane for their growth [86]. To circumvent the limitations and ecological implications of using synthetic ingredients while at the same time nurturing a balanced skin microbiota, lipid-base serums use squalane, linoleic acid, phytosterols, mono/diglycerides/triglycerides of capric and caprylic acids, lecithin, and phosphatidyl serine [61, 63, 86, 103, 104]. Synthetic human sebum formulations have also been developed containing 12.4% squalene, 17% fatty acids, 25% wax monoester from *Simmondsia Chinensis* (jojoba) oil, and 44.7% triglyceride [105]. Amphiphilic lipids like tocopherols (dioleyl tocopheryl methylsilanol, potassium ascorbyl tocopheryl phosphate, tocophersolan, tocopheryl Acetate, tocopheryl Linoleate, tocopheryl linoleate/oleate, tocopheryl nicotinate, tocopheryl succinate) function as antioxidants or skin conditioning agents in skincare products extending the shelf life of moisturizers, serums, cleansers, and diluted bath solutions [63, 106]. Analysis of data from The FDA's Voluntary Cosmetic Registration Program in 2014 provided to the Cosmetic Ingredient Review (CIR) board revealed that the use of tocopherol in cosmetic products increased from 1072 (1998) to 6635 uses (2014) including 5366 leave-on and 23 baby products, and the use of tocopheryl acetate increased from 1322 (1998) to 9677 uses (2014) including 7590 leave-on and 56 baby products [106]. In contrast, the reported use of tocopheryl linoleate decreased from 279 (1998) to 102 (2014) [106]. It is reasonable to propose that these preservatives by virtue of their antioxidant function may prevent peroxidation of skin lipids supporting the skin's barrier, in turn, this would help support and maintain the resident microbiota in a balanced state, thereby reducing inflammation, and inducing skin stability and rejuvenation capacity as has been proposed for lycopene microemulsions [107].

Not Dead nor Alive: Bacteriophages in Skincare

Acne vulgaris is a complex hair follicle disorder characterized by overgrowth of *C. acnes*, the most prevalent skin bacterial strain. Strains of *C. acnes* associated with acne lesions carry a linear plasmid particularly phylotype IA1 [108]. These strains produce virulence factors, such as light-sensitive porphyrins and hyaluronate lyases and lipases, which are associated with inflammatory responses in the skin [109]. Colonization of the sebaceous gland deep within the hair follicle [100], an environment rich in lipids and where oxygen levels are low, makes treating this biofilm with standard treatments like antimicrobials and benzoyl peroxide challenging. Despite all this, *C. acnes* are not exclusively pathogenic as the maintenance of skin and hair follicle health relies on a balanced diversity of *C. acnes* strains and other skin microbes [108]. Notably, *C. acnes* can regulate the presence of *Staphylococcus aureus* [110] and *Staphylococcus epidermidis* [111]. At the same time, *S.*

epidermidis can modulate *S. aureus* [112] and *C. acnes* levels [113]. These microbial populations engage in a very intricate dance where healthy vs disease state rests on their balanced levels, rather than the complete elimination of one specific population. The skin surface and pilosebaceous unit are also home to viruses and related bacteriophages [114, 100]. Bacteriophages are not dead nor alive in a strict sense since they rely on the presence of specific strains of bacteria for their replication and without them, they do not exist on their own like P100D and P101A which are strict *C. acnes* phages [100, 114]. This intricate relationship has led to the realization that bacteriophages could be a powerful tool in the effort to control pathogenic *C. acnes* strains, while helping rebalance the levels of non-pathogenic *C. acnes* strains, other bacteria or fungi on the skin. Consequently, *C. acnes* bacteriophage PHIT-101 is now available in the USA in a cosmetic formulation for treating all forms of acne [115]. This is a patented product that requires refrigeration for stability [116]. The company has formulated an accompanying gel cleanser with 2% SA and *Melaleuca alternifolia* (Tea Tree) Oil and a moisturizer with squalane and Niacinamide to help maintain the integrity of the bacteriophage and support resident nonpathogenic *C. acnes* and other microbial populations [117]. Therefore, bacteriophages by virtue of not being chemical antimicrobial agents, but rather biological ones, may prove to be an effective product for women to use during pregnancy (another stage in life where acne appears) which could help control pathogenic *C. acnes* strains, helping to restore the mother's skin microbiome balance and be potentially passed on to the next generation of newborns. Genetic sequencing of *C. acnes* bacteriophages has revealed that siblings share the same bacteriophage type suggesting that close living proximity may enable this process [114]. Altogether, microbial-derived skincare solutions may serve pregnant women, young children and teenagers the most, whose treatment options have been limited until now [57].

Controlling Hyperpigmentation

Melanogenesis is a complex process regulated by various cellular factors [118], with the copper-dependent tyrosinase serving as the key enzyme responsible for catalyzing the rate-limiting step in melanin production [119]. This enzyme has become a critical target for managing pathological hyperpigmentation disorders such as PIH, and cosmetic flaws like freckles and age spots. Consequently, there has been significant research leading to the identification of inhibitors that specifically act on tyrosinase to control excess melanin production and address pigmentation issues [1, 120–123]. One of the first skin lighteners identified was azaleic acid, which is naturally produced by the human skin yeast commensal, *Malassezia furfur*, but it can be derived from multiple sources for skincare preparations [124]. Azaleic acid is thought to dampen ROS production by neutrophils leading to reduced oxidative damage and hyperpigmentation at the site of inflammation [125]. It has many applications including improvement of erythema and hyperpigmentation of acne lesions when used as a 15% azaleic acid gel [126]. Hydroquinone was available

OTC for facial and body moisturizers at 2–3% concentrations until 2020, when the FDA restricted the OTC sale of hydroquinone formulations due to concerns about its side effects. It is available by prescription only at 4% [1]. Therefore, US consumers are advised to verify ingredient list of imported skincare products that may still contain hydroquinone. *Glycyrrhiza Glabra* or *Glycyrrhiza inflata* (Licorice) root extract contains active compounds which help to limit melanin production and reduce the appearance of dark spots and hyperpigmentation, such as glabrene, isoliquiritigenin, licuraside, isoliquiritin, and licochalcone A [127]. A double-blind, within-person, study was performed in mild-to-moderate acne vulgaris patients to evaluate the effect of a gel containing multiple actives including licochalcone A (7% glycolic acid, 1% salicylic acid, 2% gluconolactone, 0.05% licochalcone A, 0.1% adapalene) compared to 0.1% adapalene gel [74]. Subjects applied 0.25 g of each gel to different sides of the face at night for 28 days. Erythema, scaling, and lesion scores were measured, and digital photographs were taken with Visia® camera once a week. Most patients reported very mild erythema and scaling. However, only the group receiving the multiple active gel had significantly greater burning sensation at all points compared to the adapalene gel only. Both groups had similar improvements of active lesions and hydration with reduction of comedones and TEWL. Unfortunately, this study had limitations including the lack of a multiple-active gel control without adapalene, and the treatment length as the subjects treated with the multiple active gel started to show lower active lesion scores at 28 days compared to adapalene alone, but the study was not designed to go beyond a month [74]. Properly controlled randomized human trials with adequate numbers of participants are imperative when comparing formulations with multiple OTC active ingredients as this is the only way to guarantee the full recognition of risks and benefits of these ingredients.

Toners in liquid or gel format may be used daily after cleansing to remove any remaining traces of cleanser or impurities, and prepare the skin for subsequent skincare products, but they are also used in combination with exfoliants (AHAs or BHAs) to help minimize hyperpigmentation [86]. Hydrating, soothing and astringent benefits can be achieved by including humectants, botanical extracts, and antioxidants such as *Aloe barbadensis* (Aloe vera) leaf Juice, *Rosa damascena* (Rose) water, *Centaurea Cyanus* (cornflower) flower water, and *Hamamelis virginiana* (witch hazel) extract. Sodium trehalose sulfate (trehalose), a naturally occurring disaccharide composed of two glucose molecules, has emerged as a promising ingredient in toners acting as a potent humectant by increasing water content in the stratum corneum and reducing TEWL [128].

Biomimetic Ingredients in Skincare

Most inactive ingredients in cosmeceutical skincare products are biomimetic. Biomimetic skin formulations represent an expanding category in skincare. Biomimetic ingredients are those that have a similar chemical structure and functionality to chemicals naturally produced by human skin. At the core of biomimetic formulation lie ingredients and technologies that mirror the skin's natural lipid composition, hydration mechanisms, and protective functions [61]. These formulations provide optimal compatibility with the skin, potentially reducing the likelihood of adverse reactions and irritation which may be considered a priority in the formulation of skincare products for infants, children, and people with sensitive skin.

The most prominent biomimetic ingredients include cholesterol and derivatives, ceramides, glycosaminoglycans like hyaluronic acid, squalane, sodium 5-oxo-2-pyrrolidinecarboxylate (sodium PCA), urea, glycerin, ubiquinone (coenzyme Q10), peptides including Glycyl-L-Histidyl-Lysine (GHK-Cu) [epidermal growth factors, phospholipids, sphingolipids, and vitamins like niacinamide or nicotinamide (vitamin B3) and panthenol (vitamin B5) [129–133]. Since these ingredients are biocompatible, they provide similar benefits to the human-produced equivalents such as reinforcing and supporting skin barrier integrity, promoting hydration and elasticity, aiding wound healing, and reducing inflammation [134, 135]. Hyaluronic acid (HA) and derivatives like sodium hyaluronate or hydrolyzed sodium hyaluronate, sodium butyroyl hyaluronate, and sodium hyaluronate crosspolymer are humectants with a wide range of molecular weights that attract and retain moisture, helping to hydrate and plump the skin making it one of the most common ingredients in moisturizers, eye creams, serums, toners, micellar water, and foaming facial cleansers [136]. A double blind RCT in women who applied a topical gel containing sodium hyaluronate crosspolymer and mixture of 5 HA with different molecular weights (1 kDa hyaluronic acid, 5 kDa hydrolyzed sodium hyaluronate, 50 kDa hydrolyzed hyaluronic acid, 200 kDa and 2000 kDa sodium hyaluronate) demonstrated improved volumizing, and wrinkle depth reduction resulting in face recontouring and lifting of the cheeks within 7 days with continued improvements weekly for 1 month [136]. The volumizing effect on lips was statistically significant just after 3 hrs of application compared to placebo [136]. Another study demonstrated that 0.1% HA of either 50 kDa or 130 kDa is efficacious in reducing wrinkles around the eye [137]. Additionally, oral consumption of 100–200 mg of HA first improves skin dryness by lowering TEWL, followed by improvements in skin tone, and lastly improving epidermal thickness after 12 weeks [138, 139], which demonstrates the true versatility of this compound as an essential topical and oral cosmetic ingredient.

Ceramides are one of the lipids that constitute sphingomyelin, a key component of the lipid bilayer in the cell membranes of keratinocytes found in the epidermis [140]. Due to their lipophilic nature, ceramides provide a better consumer experience by easily integrating onto the stratum corneum without penetrating deeper layers which makes it an ideal ingredient in sunscreens, and body moisturizers for people of all ages [141]. Another ingredient that is commonly paired with ceramides is niacinamide. When applied topically, nicotinamide elevates ceramide and free fatty acid levels in the stratum corneum while reducing TWEL in dry skin [142]. Niacinamide enhances the synthesis of glucosylceramide and sphingomyelin by increasing the activity of the rate-limiting enzyme in sphingolipid synthesis, serine palmitoyltransferase which simultaneously promotes ceramide synthesis [142].

These properties have made niacinamide a standard ingredient in skincare cosmetics. Squalene is one of the primary emollients made by the sebaceous gland in human skin [61, 132]. Squalane has become a central ingredient in products with anti-aging claims for face and body including facial oils, serums, sunscreens, deodorants, hair care products, and oil-based makeup removers [143]. It is also marketed as supporting a balanced skin microbiome including products for children and people afflicted with acne or psoriasis, potentially making squalane one of the most versatile biomimetic ingredients in the market [86, 143]. Analysis of data from the FDA's Voluntary Cosmetic Registration Program (VCRP) in 2018–2019 provided to the CIR revealed that cosmetic products containing squalane and squalene increased from 595 products in 2001 to 2785 in 2019 for squalane, and from 29 products (2001) to 527 (2019) for squalene [143]. The representation for squalane in baby products also increased from 2 (2001), to 11 products (2019), and squalene which was not reported to be in use in 2001, was reported in 2 baby products in 2019 [143],. Squalane concentrations in leave-on formulations was 0.0001–97%, and for rinse-off formulations was 0.0001–34.9%. Squalene leave-on concentrations were lower at between 0.0045% and 0.7% [143]. The livers of deep-sea sharks were the original source of squalane, though manufacturers can now source this ingredient from olive oil, corn, and yeast fermentation of sugarcane, thereby reducing pressure on marine ecosystems and environmental concerns [144]. Olive-derived squalane may not be suitable for people with oral sensitivity to olive oil, making alternative sourcing through biotechnology and fermentation favored approaches to manufacturing this ingredient [144].

Exosomes, the small vesicles naturally generated by stem cells have been gaining notoriety in the global skincare market due to their fast-acting activity and potency through the action of the growth factors, peptides, and microRNA they can deliver directly to cells [131]. Skincare formulations with exosomes target inflammatory and skin rejuvenation pathways with remarkable results [145]. Due to their specialized mechanism of action and biocompatibility, exosomes and other cell-free nanoparticles are expected to continue gaining momentum in the skincare space in the next decade. However, current regulations in the USA prevent the use of exosomes for all therapeutic and cosmetic purposes, which will delay their integration in the US market [145]. Overall, biomimetic skin formulation represents one critical paradigm shift in skincare by prioritizing harmony with the skin's natural biology. As the science of biomimicry continues to evolve, these ingredients and formulations hold immense potential for the future of skincare. Nanoencapsulation and delivery complex systems will be covered more in depth in another chapter.

Management of Dry Skin, Itch, and Eczema

Eczema, also known as atopic dermatitis, is a chronic skin condition characterized by recurrent episodes of inflamed patches of dry, itchy skin that can appear anywhere on the body [146]. Genetic factors and immune system deregulation are associated with the condition [147], requiring ongoing management with moisturizers and medications and by avoiding triggers which vary from person to person [148, 149]. Flare-ups may be triggered by UV radiation, environmental allergens, chemical and cosmetic irritants, stress and pollution [148, 147, 149]. Diagnosis of eczema is more common in infants and children (<4 years old) with a decrease in prevalence during adolescence and young adulthood (10–24 year olds) followed by a slight prevalence increase in people over 50 years old [150] with females being slightly more affected than males [151]. The most common reported side effects include poor developmental growth in children, lack of sleep, depression, and anxiety [146], making eczema a significant physiological and psychological burden to children, the elderly, and caretakers. The Global Burden of Disease study conducted between 1990 and 2017, reported that the global prevalence of eczema measured by disability-adjusted life-years (DALYs) remained unchanged from 121 in 1990 to 123 in 2017 with Sweden, the UK, and Iceland ranking on top [150]. In 2013, The International Study of Asthma and Allergies in Childhood (ISAAC), surveyed 1.2 million schoolchildren across the world on the prevalence of allergic diseases [151]. The average rate of eczema worldwide in 6–7 years old was 7.9% (3–17%) with the highest rates in Latin America and Oceania (17%); while in 13–14 years old, the average was 7.3% (3.2–13.7%) with the highest rates in Africa (13.7%). This analysis may be limited by the lack of socioeconomic status of each surveyed child and the confounding factor that scabies, a skin infestation caused by the mite *Sarcoptes scabiei*, can be mistaken for eczema due to the chronic itch that it causes which could account for the increased prevalence of self-reported eczema in areas where scabies is endemic such as Asia and South America [152]. Specific socioeconomic and geographical conditions, climate, and environmental pollution factors may also contribute to the differences reported in eczema between geographical centers. Overall, eczema rates continue to increase in middle and low-income countries because of urban growth and shifts in lifestyle that come with improving socioeconomic conditions [153].

Grooming Guidelines for the Atopic Skin

International guidelines for infants and children with sensitive or eczema-prone include bathing for 5–20 minutes with warm water and gentle cleanser as needed, followed by immediate application of emollients to support skin hydration and prevent TEWL [154]. The American Academy of Dermatology Association (AAD) recommends eczema patients avoid soaps and use fragrance-free, low to neutral pH and hypoallergenic cleansers [155]. Unfortunately, most products for infants under 1 year old labelled as hypoallergenic and dermatologist-recommended may have ingredients considered contact irritants or allergens [156]. The only consistent recommendation between the USA and the rest of the world is moisturizing as often as needed to alleviate dryness to help prevent flare ups [154, 155, 157]. Since 2014, the

AAD in the USA has not updated its guidelines about bathing frequency or intervention such as addition of oils or emollients to bathing water due to poor efficacy of these products in eczema patients [155, 158]. Bath oils are bathing water additives that may deposit lipids forming a film on the skin surface after bathing. While bath oils and moisturizers can improve skin hydration and reduce TEWL in healthy individuals with chronic dry skin [159], the effectiveness of bath additives formulations containing 63% light liquid paraffin, 85% *Glycine soya* (soy) oil, gels with liquid paraffin or isopropyl myristate, or creams with disinfectants such as chlorocresol in soft white paraffin or liquid paraffin appears to be limited in preventing or treating eczema in infants and children [158, 160, 161]. In addition, caretakers may perceive bath oils as an alternative to moisturizing with emollients following a bath which may work against the benefits of daily moisturizing.

For the first 2 months of life, healthy infants at low risk for eczema can be bathed daily just using water and limiting cleansers to twice a week followed by moisturizing [162, 154]. However, in 6-month to 11 year old children with moderate to severe eczema, bathing twice a week for 10 minutes followed by moisturizing was less effective in reducing eczema severity than bathing twice a day for 15–20 minutes followed by moisturizing [157]. In the Japanese WASH RCT in 2–8 year-old children with controlled eczema, children chose between two well-known Japanese brand cleansers with different pH levels to use on one side of the body, while the other side only received water. The children were to use their hands to lather gently and avoid scrubbing with any tools. Here, cleansing with either water or cleansers during the fall/winter season [163], or summer [164] did not induce flare ups or worsening recurrence of eczema. In a 2023 US prospective study, researchers set out to explore the impact of showering and moisturizing habits on the severity of eczema in adults [165]. The results revealed that showering or bathing more than once a day led to worse eczema outcomes, while showering/bathing less frequently than daily had no effect. Consistent or inconsistent moisturizing after bathing was associated with reduced eczema severity [165]. Shower duration also had no impact on eczema outcomes. This study indicates that adults with eczema may not need to be advised on shower/bathing duration but instead the focus should be on keeping the frequency to once a day or less followed by moisturizing [165]. Since all these studies were carried out in urban centers, it is likely that soaking in water removes environmental contaminants transferred to the skin by handling of food, door handles, clothing, household items, and previously applied cosmetic products [166, 167]. The beneficial effect of bathing does not appear to be related to changes in S. *aureus* levels in eczema-prone [157] or healthy infants [162]. The other possibility is that the skin in general favors aquatic environments like the placental atmosphere humans spend months in before being born [168]. Optimistically, high-quality RCTs will continue identifying specific grooming behaviors and formulations with more precision in vulnerable groups (neonates, infants, children, elderly) and help guide worldwide grooming guidelines to reduce the burden of human skin diseases.

Skincare Challenges for Sensitive and Eczema-Prone Skin

Sensitive skin syndrome (SSS) is increasingly recognized as a global phenomenon, with over 50% of people reporting sensitive skin [169]. Dermatological symptoms associated with SSS often overlap with chronic metabolic conditions such as obesity, diabetes, and autoimmune disorders [170]. Despite pharmacological developments in immunologically targeted biologics for management of eczema [171, 172], at least 10% of the population still faces challenges when selecting skincare products [173]. Therefore, the skincare market has seen a significant growth in products designed for sensitive-skin or eczema-prone skin with organizations like the National Eczema Association (NEA) providing a Seal of Acceptance™ to products that do not include ingredients known or proposed to be skin contact sensitizers in moisturizers, cleansers, hair products, disposable wipes, deodorants, lip care, sunscreens and OTC-drugs [174]. The most current list of skincare products with the NEA's Seal of Acceptance™ includes 219 moisturizers in multiple compositions (balms, lotions, gels, ointments, sticks, serums, and oils) and 79 cleansers, with 33 moisturizers (15%), and 13 cleansers (16%) as skincare products advertised suitable for children with eczema. The most well-known global skincare brands, and the AAD adhere to the NEA's product recommendations platform [148].

The NEA's Ecz-clusion list includes the two most popular fragrance mixes used in cosmetic products: fragrance mix 1 (FM1) which contains α-amyl cinnamaldehyde, cinnamyl alcohol, cinnamaldehyde, geraniol, eugenol, hydroxy citronellal, isoeugenol, and oak moss absolute; and fragrance mix 2 (FM2) which contains Lyral® (hydroxyisohexyl 3-cyclohexene carboxaldehyde), citral, citronellol, farnesol, coumarin and α-hexyl-cinnamic aldehyde [175]. Other ingredients include fragrance/emollients (*Citrus aurantifolia* (lime) peel powder, *Citrus aurantium dulcis* (orange) peel oil, Citrus extract, *Citrus limon* (lemon) peel powder), preservatives (formaldehyde, methylchloroisothiazolinone, methyl dibromo glutaronitrile, methylisothiazolinone, triclosan), *Myroxylon balsamum* (Balsam of Tolu), *Myroxylon pereirae* resin (Balsam of Peru), propolis and shellac [174]. OTC drugs containing the antibiotics (bacitracin, neomycin sulfate, polymyxin B sulfate), local anesthetics (benzocaine, lidocaine), and all organic sunscreen ingredients (except Zn, TiO_2, and iron) also belong to this list [173, 174]. Based on this list, preservatives and fragrances may be the most problematic for atopic skin [176].

Efforts to reduce the impact of preservatives on both human health and the environment have led to an increased demand for "clean beauty"products with emphasis on the use of "natural" preservatives with effective antimicrobial properties and lower risks of skin sensitization [177]. Environmental contamination is another concern with the preservatives triclocarban, chloroxylenol, methylisothiazolinone, and benzalkonium chloride because of the impact these chemicals have on water quality and aquatic life since all cosmetic ingredients eventually end up in wastewater treatment plants and once sewage is processed and released into environments, they may contaminate surface water, muds, and soils [178]. Triclosan [5-chloro-2-(2,4 dichlorophenoxy) phenol], a notable preservative in the NEA's Ecz-clusion list was introduced to the global market in 1969 as a potent broad-spectrum antibacterial and antiviral [179]. In 2016, the FDA issued a rule to remove triclosan and triclocarban from OTC nonprescription antiseptic products including liquid, foam and gel hand soaps, bar soaps, and hand/body washes due to concerns about liver carcinogenicity, metabolic disruption, skin sensitization effects, and the ability to induce antimicrobial resistance even at low concentrations [179]. That rule came into effect in the USA and the EU in 2017 [180].

Natural or synthetic fragrances mask potentially unpleasant smells in cosmetic formulations, making it more appealing to consumers. Although the percentage of people who have a positive skin reaction to FM1 or FM2 may be low (3–7%) [175], the advent of fragrance-free marketing highlights the pervasiveness in consumer's perception that products with fragrance, parfum or perfume may be harmful, while those that include botanical extracts may be beneficial [181]. Products labeled "unscented" may not be fragrance-free either. A major caveat when discussing fragrances is that one single ingredient can contain several individual chemicals [182]. For example, crude *Myroxylon pereirae* resin (Balsam of Peru) or MP is a complex mixture of hundreds of compounds that was removed as a fragrance in the 1970s [182]. However, MP shares a dozen aromatic compounds with propolis, an ingredient produced by bees in the construction and maintenance of their hives, which has gained popularity in recent years in the "natural" skincare space [183]. These shared compounds include benzoic acid, benzyl alcohol, benzyl benzoate, benzyl caffeate, benzyl cinnamate, benzyl ferulate, benzyl isoferulate, caffeic acid, cinnamic acid, cinnamyl alcohol, coniferyl benzoate, farnesol, nerolidol, and vanillin [182]. Cosmetic brands advertising fragrance-free formulations have introduced essential oils and botanical extracts as fragrance alternatives to FM1 or FM2. This strategy aims to introduce multifunctional compounds that can serve as emollients, fragrances, and preservatives [181]. The most encountered multifunctional oils include *Melaleuca alternifolia* var. *alternifolia* (tea tree) oil, *Rosmarinus officinalis* (rosemary) oil, *Eucalyptus globules Labill.* (eucalyptus) oil, *Thymus vulgaris L.* (thyme) oil, *Lavandula angustifolia* (lavender) oil, *Mentha piperita L.* (peppermint) oil, and *Cinnamomum zeylanicum J. Presl* (cinnamon) oil [184].

Food products can contain the same preservatives and fragrances found in skincare items, so repeated exposure through mucous membranes may eventually trigger a skin sensitization response [182]. Assumptions about the chemical complexity of skincare ingredients understate the importance of careful biochemical analysis of ingredient alternatives that may share chemicals with food additives. Some of these botanical extracts include *Fragaria Vesca* (Strawberry) Fruit Extract, *Pyrus Malus* (Apple) Fruit Extract, *Actinidia Chinensis* (Kiwi) Fruit Extract, *Citrus Aurantifolia* (Lime) Peel Water, and Peel Powder from *Citrus Jabara*, *Citrus Junos*, and *Citrus Unshiu* [185–187]. The CIR expert panel determined in 2021 that 47 *Citrus*-derived peel oils were safe for leave-on and rinse-off products when the final product concentration of the phototoxic 5-methoxypsoralen (5-MOP) does not exceed 0.0015% (15 ppm) and if the entire formulation is non-sensitizing or irritating [187]. 48-hr human skin sensitivity testing using *Citrus Aurantium Amara* (Bitter Orange) peel extract (1.55% in formulation from 20% water solution) and *Citrus Reticulata* (Tangerine) peel extract (3.06% in formulation) induced low frequency of reactions (3.3–6.7%) which completely subsided 24 hrs after patch removal [187]. It is noteworthy that these ingredients were tested under occlusion, suggesting that they may not be suitable for formulations with occlusive ingredients such as lanolin, petrolatum and/or mineral oil. This indicates that layering multiple products at once may increase the absorption of ingredients that are nonirritating on their own but can become more reactive when combined with occlusive formulations. It is conceivable that sensitivity, irritation, and erythema to some cosmetic products could be in fact caused by the occlusion of ingredient residues by other cosmetic products or even clothing like gloves, socks or pillowcases akin to how clinical skin patch tests are conducted [188, 189].

The resurgence in skincare of essential oils and botanical extracts underscores the importance of both single-ingredient and complete formulation safety and assessment of sensitivity or irritation to different routes of exposure in humans of all ages [176, 186]. The bimodal prevalence of eczema (childhood and late adulthood) may be an indicator of specific requirements needed to take into consideration such as changes in skin barrier composition and loss of hydration retention of late adulthood [134, 173]. Another special consideration regarding ingredient safety and exposure assessments in newborn, infants, and children is the larger skin surface area relative to body mass and potential increased dermal permeability from birth up to 2 years of age, compared to adults [28]. Liver enzyme systems mature quickly in newborns, with the main exception being enzymes involved in glucuronidation reactions [190]. Enzymes responsible for other types of reactions, such as sulfotransferases, can significantly increase by around 6 months of age and may even surpass the biotransformation rates seen in adults during childhood [191]. These differences in liver detoxification places infants and children in a unique subpopulation when conducting safety assessments of cosmetic ingredients like fragrances and botanical extracts.

Diaper Rash Ointments: Adapting to New Demands

Diaper rash ointments are OTC drugs for infants and adult diaper users with ZnO as the active ingredient with concentrations between 10% and 40% [192]. The most classic diaper rash formulation contains 40% ZnO along with petrolatum, cod liver oil, lanolin, talc, glycerin, sorbitan sesquioleate, beeswax, tocopheryl acetate, and fragrance. Cod liver oil is also the main ingredient in A and D ointment or cream since it is a natural source of Vitamin A and D [192]. Talc is a recent example of a classical ingredient that has been phased out in the skincare space. Originally introduced in 1894 as the main ingredient in the classical baby talcum powder, talc was replaced globally with cornstarch in 2023 [193, 194]. In the last decade, diethyl phthalate (DEP) has also been phased out from baby and adult shampoo and other cosmetic products [195]. Removal of these ingredients signals a shift in the skincare industry that aims to integrate the complexity of skin biochemistry, immunology, and microbiology that did not exist when these ingredients were first introduced into the skincare market [156, 196–198]. Lanolin, another inactive ingredient, is a complex wax derived from the sebaceous secretion of sheep's wool, which humans have used for multiple purposes for thousands of years reflecting the coevolution of humans and farm animals [199]. Lanolin is another controversial cosmetic ingredient with The American Contact Dermatitis Society naming lanolin, the 2023 Allergen of the Year [200]. Adherence to veganism has led to a rise in vegan-friendly skincare products where animal-sourced ingredients such as lanolin, tallow, and natural beeswax are forbidden. Tallow, the rendered fat or suet of cattle and bison, has recently regained momentum in the natural skincare space since it is rich in the fat-soluble vitamins A, D, K, E, as well as oleic acid, palmitic acid, stearic acid, and linoleic acid [201]. Vegetable alternatives to lanolin, petrolatum, mineral oil, tallow or beeswax, include bis-diglyceryl polyacyladipate-2, and bis-diglyceryl polyacyladipate-1 [202], and ingredients derived from olive oil and *Butyrospermum parkii* (shea) butter [203, 204]. *Butyrospermum parkii* (shea) butter or oil comes from the nut of the shea tree which grows naturally in Ghana, West Africa. It unfortunately has environmental and societal impacts that are worth considering [205]. The extraction process utilizes woodfire which may contribute to local deforestation and generates considerable wastewater, without this process being financially profitable to the processors, majority of them being women [205]. It is considerations like this that highlight some of the challenges the cosmetic industry faces when trying to identify new ingredients that may be considered eco-friendly. Regardless, the cosmetic industry continues to address these issues with innovative research to cater to growing consumer demands.

Lanolin and its derivatives are emollients or emulsifiers widely used in thousands of cosmetic formulations including hair, make-up, perfumes and even moisturizers to soothe nipples on lactating mothers [206, 207]. Eight additional skincare ingredients are derived from lanolin: acetylated lanolin, acetylated lanolin alcohol, hydrogenated lanolin, hydroxylated lanolin, lanolin acid, lanolin alcohol, lanolin oil, and lanolin wax [206]. Analysis of data from the FDA's VCRP in 2023 provided to the CIR clearly shows that lanolin-derived ingredients have been losing favoritism as emollients despite lanolin itself being a weak sensitizer [206]. Between 2002 and 2023, the frequency of total cosmetics for adults and children containing lanolin and its derivates showed a 76% decrease in prevalence from a high of 2571 products in 2002 to 623 products in 2023 [206]. Skincare products containing lanolin and derivatives had a 65% reduction in prevalence from 1015 products in 2002 to 351 products in 2023. Lanolin was reported in 2023 in only one in the baby lotion/oil/powder/cream category compared to three products in 2002. Acetylated lanolin, acetylated lanolin alcohol, lanolin alcohol, lanolin oil are not in use anymore in skincare products for infants [206]. Acetylated lanolin alcohol and lanolin wax are the only two lanolin derivatives to have increased in skincare moisturizers with acetylated lanolin alcohol having a five-fold increase, from 33 products (2002) to 167 products (2023) [206]. Acetylated lanolin alcohol was present in four face and neck moisturizers in 2002 and 2023, suggesting that these formulations have not changed and may favor this lanolin derivative. Therefore, it appears that acetylated lanolin alcohol has replaced acetylated lanolin as a moisturizer ingredient because out of 111 products in 2002 with acetylated lanolin, only one was reported in 2023 with acetylated lanolin [206].

Lanolin skin sensitivity tests have traditionally used 30% lanolin alcohol, but Amerchol® L-101 is now preferred. This is because Amerchol® L-101, when used under occlusive conditions, can identify about 25% more positive reactions [208]. 100% Amerchol® L-101 is standardized to contain 100% active lanolin fatty alcohols, while 50% Amerchol® L-101 is 50% lanolin alcohol and 50% diluent, usually petrolatum [208]. Retrospective analysis of data collected in the USA and Canada between 2001 and 2018 by the North American Contact Dermatitis Group reported positive lanolin skin reactions to 30% lanolin alcohol and Amerchol® L-101 50% in petrolatum in 4.5% of children and 3.2% of adults [209]. People with a history of eczema or hay fever were more likely to have a lanolin allergy, and to be non-Hispanic white females younger than 40 years old [209]. In Europe, contact allergy to lanolin has been reported to be 0.4% [199]. Since lanolin is an animal-derived ingredient, pesticide residue and heavy metal contamination with copper, chromium, and nickel especially in hydrogenated lanolin may have been the sources of contact dermatitis, rather than the lanolin extract itself [206, 210]. With hydrogenated lanolin having been reported in only four skin care products in 2022 compared to 62 skin care products in 2002, none of them being for babies, the potential for nickel sensitivity being the source of skin sensitizing may continue to decrease overtime [206]. Given the high quality of final lanolin ingredients following the refinement process, pesticides levels have been reported to be below current safety levels in nipple care products which also minimizes oral exposure in lactating infants [211, 206]. On the other hand, beeswax and *Citrus*-derived oils can exceed safety levels of pesticides which also require further refinement and purification [212]. While lanolin sensitivity remains a concern, advancements in ingredient refinement processes have significantly reduced contamination risks, making lanolin-based products safer, especially for vulnerable populations like infants.

Preservatives in Skincare: A Formulators Challenge and an Opportunity

Undoubtedly, the cosmetic industry would not exist without preservatives. Preservatives can be synthetic or natural, acting as antimicrobial agents in cosmetics, personal care products, and pharmaceuticals to inhibit the growth of bacteria and fungi, extending the shelf life of formulations. Preservatives are subject to regulations and recommended usage limits by health authorities worldwide which may differ between regions in the world [213]. With the advent of microbiome-friendly marketing, formulating skincare products to simultaneously preserve a product from microbial contamination and minimizing a negative impact to the skin microbiome is clearly one of the major challenges faced by cosmetic companies and formulators [4].

The use of preservatives is especially important in aqueous-based skincare since microbes have the potential to contaminate a skincare product at any point during the manufacturing process and potentially cause human disease. Since the presence of water allows the growth of aerobic bacterial and fungal populations [214], this challenge has been met by formulators with alternatives like packaging actives in skincare capsules, adding expiration dates to the package labeling, requiring refrigeration of the product, or reducing water content [215]. This may be one reason soap bombs or effervescent bath products have gained popularity since they came to market in 1998 [216]. However, in the 1980s soap bars became less desirable because once water encounters it, it promotes the growth of bacteria since microbes can grow under limited nutritional conditions, and in the presence of surfactants [41, 217, 218]. Despite the negligible risk of bacterial contamination to consumers re-using soap bars [219], this could still pose a significant risk to people with a compromised skin barrier in a world where antibiotic-resistant microbes continue to rise [220]. In the mid-2000s, a substantial shift in the industry led to the paraben-free market due to potential negative impacts on hormonal health from formulations with parabens and formaldehyde donors [178, 221, 222]. Since then, alternative ingredients like phenoxyethanol (2-phenoxyethanol), methylisothiazolinone, methylchloroisothiazolinone, benzyl alcohol, sodium benzoate and ethylhexylglycerin have gained popularity [223, 215]. Dehydroacetic acid and lactic acid are pH adjusters which may also serve as preservatives [122]. Analysis of a partial list of skin care products in the USA showed that from 2007 to 2010, Phenoxyethanol had the greatest fold increase compared to butylparaben, ethylparaben, butylparaben, isobutylparaben, methylisothiazolinone (MIT), chloromethylisothiazolinone (CMIT/MIT), DMDM Hydantoin and imidazolidinyl urea [224]. Methylparaben and propylparaben were still the top two most popular preservatives with formulations using phenoxyethanol almost equaling those of propylparaben [224]. Phenoxyethanol is a broad-spectrum antimicrobial used in cosmetic products at 1% concentrations including those for children [215, 225, 226].

In the USA, analysis of 5144 cosmetic products recalled between 2004 and 2016 due to adverse events, revealed that products marketed for newborns and infants (creams, lotions, shampoos, oils, and powders) comprised the largest recall product category, followed by unclassified, personal cleanliness, hair care, hair coloring, and skincare products [227]. Another analysis of cosmetic product recalls between 2002 and 2016 also found that baby cosmetic products accounted for 24% of all recalls, with microbial contamination accounting for 76% (237/313) of recalled products [228]. The microbial contamination of 500 open skincare containers partially used by consumers, also identified contamination by antibiotic-resistant aerobic bacteria [229]. This suggests that despite the presence of preservatives, skincare products with preservatives that have a limited record may pose a significant risk for manufacturers and consumers as the timeline from testing to production and eventual product recall could be years.

Sulfate-Free and Surfactant Alternatives in Skincare

It is now well established that dry and atopic skin have defective barrier permeability and water-retention functions as measured by higher TEWL, a sign of a compromised skin barrier [230]. Surfactants help solubilize cosmetic ingredients in formulations and in contact with water or other ingredients, form micelles, which attract and remove dirt, oil, makeup, and impurities from the skin. Sodium lauryl sulfate (SLS) is a potent surfactant used for decades in facial and body cleansers for people of all ages [189]. Human studies on the effects of direct skin exposure to topical SLS application showed weak reactions to 0.12% SLS [231]; and levels between 0.5% and 1% SLS serve as models of contact dermatitis by a surfactant [230]. SLS has been shown to affect thickness and increase TEWL in normal skin [232], and to increase permeability of this ingredient into the dermis even in areas not involved in atopic lesions [233]. Addition of surfactants like benzalkonium chloride can mitigate the irritative properties of SLS [234], although it does not completely abolish it [189]. Sodium laureth sulfate (SLES) is like SLS but considered to be gentler and less irritative [189].

Efforts towards developing sulfate-free (SLS-free and SLES-free) cleansing systems and vegetable-derived surfactants have yielded a variety of cosmetic ingredients defined as syndets [9]. Syndets demonstrate hydrating and moisturizing properties without the extreme lathering effect that made classical soap bars, and SLS-based products such potent cleansers [9, 235] Coconut or palm oil-derived ingredients have been favored because of low cost of raw materials [189]. Decyl glucoside and alkyl glucosides are nonionic surfactants derived from coconut or palm oil and glucose found in adult and baby facial and body cleansers and moisturizers [236]. Analysis of data from The FDA's VCRP survey data in 2011 revealed that the most prevalent alkyl glucosides ranked by highest number of products, listed in parenthesis, were: decyl glucoside (492), cetearyl glucoside (477), lauryl glucoside (399), and coco-glucoside (350) [236]. The least prevalent were arachidyl glucoside (75), caprylyl/capryl glucoside (58), C12–20 alkyl glucoside (58), ethyl glucoside (24), and miristyl glucoside (5) [236]. The most frequently reported surfactants in baby products were Decyl glucoside (25) and coco-glucoside (11) [236]. Decyl glucoside was in 421 rinse-off products including 96 in hair coloring/noncoloring products [236]. Cetearyl glucoside was in 445 leave-on products [236]. Products diluted for bath use were most likely to have lauryl glucoside and coco-glucoside [236].

Betaine and its derivates, cocoamidopropyl betaine (CAPB), behenyl betaine, cetyl betaine, babassuamidopropyl betaine and coco-betaine, are surfactants that may also preserve formulations by serving as pH adjusters [237, 238]. Additional ingredients include disodium cocoamphodiacetate, sodium methyl cocoyl taurate, sodium cocoyl isethionate, sodium cocoyl glutamate, and disodium cocoyl glutamate. Dissolvable shower gel tablets with 7.5% sodium coco-sulfate and 5% glycerin were found to be hydrating and non-irritating [239]. Other alternatives to the classical SLS and CAPB mixture are alkyl olefin sulfonate (AOS), alkyl polyglucoside (APG) and lauryl hydroxysultaine (Sultaine) [240]. AOS and Sultaine at a 1:1 ratio show optimal foaming function, and the ternary surfactant system showed optimal foaming at 4.5% AOS: 3% APG: 7.5% Sultaine, making these ingredient combinations ideal sulfate-free surfactant systems [240]. Silk fibroin, derived from Mulberry-fed silkworm fibers, has emerged as a gentle surfactant. When combined with capryl glucoside, silk fibroin forms an effective co-surfactant system [241]. This combination has better foamability, surface tension, and cleansing properties compared to combined surfactants SLES and CAPB.

Syndets in Modern Cleansing, Grooming and Baby Care Products

The need for emulsions without allergenic and irritative surfactants is still vital because irritant contact dermatitis (ICD) and allergic contact dermatitis (ACD) attributed to surfactants continue to be reported [189, 196]. ICD has been mostly associated with SLS and benzalkonium chloride [189]. However, decyl glucoside, cocamidopropyl betaine, oleamidopropyl dimethylamine, 3-dimethylaminopropylamine, amidoamine, and cocamide diethanolamine have been associated with ACD, a more enduring type of immune reaction due to repeated allergen exposure [189]. With skincare manufacturers favoring decyl glucoside and cocoamidopropyl betaine as surfactants even in baby products, consumers and formulators should keep in mind the potential for ACD with these ingredients [236, 237].

Facial, body, and baby wet wipes (towelettes) soaked in syndets along with emollients and preservatives may leave behind remnants of these ingredients on the skin [242]. People who use micellar water may also experience skin issues, since micellar water is usually not washed off from the face. Wiping has the potential to weaken the stratum corneum due to mechanical friction on the skin potentially leading to deeper penetration of other ingredients present in the wipe or products applied after wiping [243, 242]. The concentration of surfactants in wet wipes ranges from 0.10% to 0.45% (bis-PEG/PPG-16/16 PEG/PPG-16/16 dimethicone), to under 0.5% (coco-glucoside, decyl glucoside, lauryl glucoside, coco-betaine, polysorbate 20), to under 0.80% PEG-40 hydrogenated castor oil, and 1–2% Glyceryl stearate [236, 244, 245]. Facial wipes surveyed in the Unites States hold an average of 17 ingredients including many natural products that could be potential skin irritants and allergens [246, 176]. The preservatives, methylisothiazolinone (MIT) and methylchloroisothiazolinone (MCI), were in 2.2%, and 1.1% of adult facial wipes surveyed [246]. However, ACD from wet wipes containing MIT was reported in the United States for the first time in 2014 in different body parts of six children [247], and ACD has been reported in meta-analysis of baby wet wipes [242]. Irritation on the palms of caretakers using baby wet wipes occurs in a grip-like pattern (thumb, index, middle finger) indicating that baby wipes were the source of ACD in these cases; and suspected cases of hand eczema may share a similar pattern [248].

Therefore, the most proactive steps to take at the first signs of irritation is to stop wet wipe use, and to inspect other grooming products with comparable ingredients, since skin allergy testing without clues may be a demanding and impractical process [248, 249]. Altogether, this is an appealing area for future research which may advance ingredient innovation for children and children wiping products.

New Generation Syndets

Sustainable sources of surfactants, emulsifiers, and emollients with multifunctional properties are still essential for pushing the existing boundaries in the skincare industry. Biosurfactants and bioemulsifiers from plants and microbes may help bridge this need for truly natural solutions [250–253]. The ideal emulsions for adults would help regulate sebum production in the face and scalp, while reducing dryness in areas with low hair follicle density like elbows, knees, forearms, and legs, the sites most affected in the atopic skin [250, 254]. Those for children should work in harmony with the microbial and immunological changes that occur during infancy [255]. For example, facial cleansers and make up removers using isolated sugar derivatives and *Persea gratissima* (avocado) esters are safe alternatives to classical surfactant-based make up removers acting as both cleansing agents and toners [256]. Formulations with fatty alcohol ethylhexanoic acid esters, isoamyl alcohol lauric acid ester, catalyzed avocado butyl alcohol ester, and gallic acid ester may help break down complex formulations with pigments, silicones and UV filters [256]. Gallic acid ester serves as both a cleansing agent and antioxidant, aiding in the removal of impurities while providing oxidative protection [257]. Glycol ethers or 1-dodecylazacycloheptan-2-1 (laurocapram) may be added to enhance the absorption of subsequent facial products enhancing the effect of topical serums.

Plants produce a variety of organic compounds such as saponins and phytobiotics, which are not essential for their survival but offer evolutionary advantages, like defense against microorganisms and predators [258, 259]. Saponins derive their name from the *Quillaja Saponaria* (soapbark) tree, and they consist of a glycone (either a triterpene or steroid) linked to a sugar chain [253]. Alternative sources of saponins with potential skincare applications [251] have been identified in the *Agave* plant [260], and *Camellia* (tea) seed cakes [261]. Saponins are a clear example of a traditional skin and hair care ingredient that has regained popularity. Skin microbiome and skin function analysis with a saponin-containing body wash formulation demonstrated that it may be superior to formulations with current syndets [250]. In the Japanese WASH RCT [164], one of the cleansers had dipotassium glycyrrhizinate which is derived from Glycyrrhizin, a saponin from *Glycyrrhiza Glabra* (Licorice) root [127]. More efficient saponin identification methods may expand the availability of these ingredients by reducing the amount of raw material needed for exploration [262]. In a Brazilian RCT [162], one of the moisturizer ingredients was naringenin, a grapefruit phytobiotic, which has been available in

German skincare products for decades but has recently gained popularity in the USA [263]. This highlights that not only individual ingredients, but formulations with good safety profiles already exist around the world.

Probiotics: Beyond Cosmetic Skin Therapeutics

Despite all the cosmetic products available to temporarily improve the appearance of wrinkles, dryness, oiliness, and skin sensitivity, pharmaceutical or cosmetic chemicals may not be sufficient to reduce the burden of suffering in people struggling with chronic skincare conditions, especially children. The skin and hair follicles host a complex and interdependent ecosystem of arachnids, bacteria, fungi, archaea, viruses, and bacteriophages forming a dynamic biofilm-like structure referred to collectively as the skin microbiome [264, 265]. Since there is an association between imbalances in skin microbiota populations with eczema flare-ups in children [266, 267], human studies have looked at the role and impact of probiotics on the relief of skin symptoms associated with eczema [268, 269]. Probiotics are either live or dead microorganisms that can be used orally or topically with the aim of supporting or restoring balance to the skin [4, 270, 271]. In 2008, a meta-analysis from five RCT found no statistically significant reduction in eczema symptoms in children treated orally with probiotics compared to placebo [268]. The probiotics used included bacterial species of the genus *Lactobacilli* or *Bifidobacteria* and fungal *Saccharomyces*, suggesting that either the route (oral) or timing of exposure, or the specific probiotics or probiotic combinations used, may have limited clinical application for skin relief in eczema. In 2014, a comprehensive meta-analysis started to elucidate the conditions under which probiotics could influence eczema relief. When probiotic blends were given orally to pregnant women, infants at birth and up to 3 months of age, they provided risk reduction in disease development [269]. The use of single strain probiotics of either *Lactobacilli* or *Bifidobacteria* did not appear to be effective in children over the age of 2, suggesting that both timing of exposure (prenatal and early infancy) and bacterial strain composition complexity (multi-strain) may provide the greatest risk reduction and skin symptoms benefits [269].

Advancements in the understanding of skin physiology and the essential role that the skin microbiome has on overall skin and host function [272] have led us to the realization that attempting to control microbes with antibiotics, stripping away at the skin with detergents and complicated grooming routines, may partially explain the current rates of skin disease [8, 220, 273]. Skin microbiome analysis from indigenous populations that have never encountered pharmaceutical and skincare products, demonstrated that their skin microbiota is very diverse and completely different to those of US subjects [274]. The main reflection from these studies is that the urbanized human skin may have lost the genetic diversity provided by these microbial populations which may explain why skin conditions that afflict people in modernized societies, do not occur in indigenous populations [275]. This could be considered a microbial extinction event [220, 276, 277]. Evidence for this can be appreciated when children with atopic dermatitis who applied topically the skin commensal *Roseomonas mucosa* had dramatic improvements in skin barrier and reduction in pathogenic skin microbiota populations including *S. aureus* leading to reduction in disease severity, itch and improvement in the quality of life for the children and caretakers [254].

Another example of this microbial extinction phenomenon can be appreciated with the bacteria *Nitrosomonas* [278]. Commonly found in soil and aquatic environments [279, 280], *Nitrosomonas* are ammonia-oxidizing bacteria (AOB), essential in the nitrogen cycle, specializing in the conversion of ammonia present in sweat into bioactive metabolites such as nitric oxide and nitrite [281]. Nitric oxide is a known regulator of inflammation and vasodilation, while nitrite has antibacterial properties [281]. The establishment of modern societies that have moved humans away from natural AOB's sources like farm soil [279], and natural bodies of water may be partly responsible for the decline of *Nitrosomonas* spp. from the modern human's skin. AOB strains, including *N. eutropha* B244 and D23, have been patented [282, 283]. AOB modulates immune responses by downregulating the inflammatory cytokines IL-4, IL-5, IL-13, and IL-31 linked to atopic dermatitis and pruritus [284]. Clinical trials using live *N. eutropha* B244 in sprayable form buffered in disodium phosphate and magnesium chloride [41, 285]. Phase 2 trials for moderate-to-severe acne and rosacea have been completed. In a double-blind placebo RCT for moderate-to-severe pruritus in mild-to-moderate atopic dermatitis, the AOB showed statistically significant reductions in itch and clinical presentations of atopic dermatitis by 4 weeks, allowing it to enter phase 3 clinical trials [285]. This probiotic shows high efficacy at increasing doses with similar adverse reactions to control (PBS buffer) [285]. It also has cosmetic benefits such as reduction in facial wrinkle depth and severity with improvements in skin pigmentation [286]. Based on its safety and efficacy, the related *N. eutropha* D23 strain became commercially available in the United States in 2016, under the name AO+ Mist and Mother Dirt AO+ Restorative Mist. As a live probiotic suspended in liquid, it can be sprayed anywhere in the body, but it is especially recommended for the scalp, face, armpits, feet, and urogenital area [287]. It has a refrigerated shelf-life of 6 months, while remaining viable at room temperature for 30 days [287]. It became temporarily unavailable in the United States skincare market in 2022, but it came back in Q4 of 2024 under the brand name D23 Ammonia Oxidizing Bacteria Topical Spray [288]. A licensing agreement to make the B244 strain available in Japan will make this product more widely available in Asia. All these studies combined may be compelling indicators of the need for targeted biocosmetics and practices to help improve and not worsen skin conditions in pregnant women, their newborns and infants [289]. Developers of skincare formulations may need to consider the interactive nature of the mother-newborn unit when formulating new products or improving on existing products for pregnant women or newborns as this is the most critical window of time for proper skin tissue and immune tolerance development in newborns.

Biocosmetics in Skincare: Ferments, Lysates, and Filtrates

The recent recognition of the skin's microbiome role in supporting overall skin health has led the cosmetic industry and the public to embrace biocosmetic formulations [4271]. Manufacturers already rely on microbial fermentation to isolate skincare ingredients for moisturizing, conditioning, and thickening purposes [45, 290, 291]. Xanthan gum, biosaccharide gum-1, sclerotium gum, beta-glucan, and gellan gum are among the most common microbial-derived ingredients [291]. The bacteria *Xanthomas campestris* and *Pseudomonas elodea* produce the exopolysaccharides xanthan gum and gellan gum, respectively, which are used as thickeners in skin care and food products [45, 291]. In 2012, xanthan gum was reported to the FDA's VCRP to be in almost all cosmetic categories with 3470 cosmetic products including 29 in the baby category [291]. The fungus *Sclerotium rolfsii* produces sclerotium gum, a polysaccharide used as an emulsion stabilizer that also acts as a fast-absorbing silicone-like skin conditioning agent [291]. Unlike these refined microbial ingredients, microbial ferments are the complete mixture from the fermentation process which is then physically shattered by sonication or heat-inactivated to produce ferment extracts or lysates [4]. The final product has non-viable probiotic fragments along with complex byproducts from fermentation such as peptides [271]. Ferment extracts or lysates may provide benefits as live probiotics do, but with greater shelf-life stability [271]. *Lactobacillus, Lactococcus, Lactiplantibacillus, Bacillus* ferment, and *Bifidobacterium longum* extract (*Bifida* ferment lysate) appear to be the most common microbial ferments in skincare preparations [4]. This has been inspired by clinical trials in humans using similar probiotics for their ability to enhance the skin's natural barrier function, soothe irritation, and reduce skin sensitivity [270, 271, 292]. For instance, Greek yogurt and *Lactobacillus* ferment lysate appear in skincare products for their conditioning properties along with the ability of lactic acid to function as a pH adjuster with a multitude of benefits once applied on the skin [4, 122, 293]. As of 2023, Yeast Extract was the most prevalent fungal-derived ingredient reported in 398 cosmetic products, including 280 face, neck, and body moisturizers, but none for babies [294].

Ferment filtrates are the aqueous fractions from fermentation where high molecular weight compounds are filtered out, so they do not have live or dead bacterial/fungal cells from the fermentation process [295, 294]. For this reason, ferment filtrates may not be considered probiotics or postbiotics since these ferment filtrates could be chemically characterized as done with any other cosmetic ingredient [4]. Filtrates from fungal *Saccharomyces* and *Galactomyces* ferments are used as emollients and humectants for skin and hair products [294]. As of 2023, *Saccharomyces* lysate extract and *Galactomyces* ferment filtrate were present in 88, and 77 cosmetics, respectively, but none for babies [294]. *Galactomyces* ferment filtrate has been shown to increase epithelial tight junctions improving the skin barrier, while reducing inflammatory signaling in the skin [296]. *Leuconostoc*/Radish Root Ferment Filtrate, the filtered fraction from Daikon radishes (*Raphanus sativus*) fermented by the bacterium *Leuconostoc kimchi*, received patented status in 2010 as a preservative for leave on products such as mascara due to broad-spectrum antimicrobial function against *Enterobacter*, *Pseudomonas*, *S. aureus*, yeast, and mold [297]. As of 2022, *Leuconostoc*/Radish Root Ferment Filtrate was reported to be in 254 cosmetic products, but only 1 baby product [295]. Marketed under the name Leucidal®, the filtrate is claimed to have sodium salicylate and methyl salicylate, proposed to be a potential acne therapy due to in vitro inhibition of *C. acnes* [298]. Unfortunately, chemical and carbon dating analysis of commercially available *Leuconostoc*/Radish Root Ferment Filtrates found that the antimicrobial activity in these fermentation filtrates came from didecyldimethylammonium salts (DDAC) and SA rather than from chemicals generated by microbial fermentation [299]. DDAC, a common quaternary ammonium disinfectant used in the processing of plant materials, poses significant risks to humans [300]. It is possible that the contamination of *Leuconostoc*/Radish Root Ferment Filtrates with DDAC comes from processing of raw materials used the fermentation process which would be expected to accumulate, rather than be used during fermentation. This report brings to light the potential issue of adulteration of microbial-derived ingredients. Since these ingredients still have no standardized regulations, manufacturers are not yet required to catalogue the complete composition of chemical compounds after each production batch of microbial ferments and filtrates, so it would not be surprising if this gap in standardization could lead to adulteration of these materials [4]. This could have detrimental effects on this new branch of the skincare industry by creating distrust in biocosmetics and eroding customer brand loyalty.

Synbiotics: Emerging Skincare Solutions

Postbiomic technology focuses on optimizing the production of microbial-derived ingredients for cosmetic purposes. Synbiotics, a mixture of live probiotics, prebiotics and postbiomic ingredients and technologies, are proposed to have synergistic effects on the skin as they have been proposed for the gut [301, 302]. Through the effect of short chain fatty acids (SCFAs), peptides, polysaccharides, enzymes, and organic acids, a skincare product could modulate skin health through anti-inflammatory, antioxidant and antimicrobial properties of its microbial-derived ingredients [47, 110, 267, 303, 304, 111]. Prebiotics, sugars/fibers such as fructooligosaccharides (FOS), alpha-glucan oligosaccharide, and xylooligosaccharides (XOS) support the proliferation of microbes [112, 305]. The FOS Inulin derived from *Cichorium intybus* (chicory) root serves as an antioxidant and skin conditioning agent in facial cleansers, body wash gels, and moisturizers [302, 305]. FOS and XOS have been studied extensively in the context of intestinal microbiome health for their role in inducing short-chain fatty acid (SCFA) production but have just recently been studied in vitro on skin microbes [112]. FOS promotes superior production of the SCFA's acetic acid and isovaleric acid from *S. epidermidis* compared to inulin [112]. This leads to growth of *S. epidermidis*, while simultaneously suppressing the proliferation and biofilm formation of *S. aureus* [112]. *C. acnes* can also be induced in vitro to produce SCFA's in the presence of glycerol [110] which also regulates the biofilm potential of *S. epidermidis* [111]. The identification of effective SCFA's in vitro is expected to give rise to new skincare products to help regulate the skin's microbial ecosystem in vivo. Recently, *Bacillus* spores from *B. subtilis*, *B. licheniformis*, *B. megaterium*, and *B. amyloliquefaciens* mixed with 1% inulin, and *Polygonum aviculare* Plant Extract in a skincare formulation were effective at modulating skin bacterial levels, reducing facial oiliness and acne lesions in humans [302]. Therefore, the biocosmetic skincare field is evolving towards minimalist formulations and sustainable production methods, exemplified by Postbiomic™ fermentation and BioIntact™ cold emulsion where moisturizers and serums could be created by microbes from as few as three ingredients [306, 303]. By removing heat from the production process, the integrity and potency of the emulsion may be preserved, potentially providing improved skin benefits while reducing the carbon footprint during manufacturing. The key may lie in the naturally-occurring emulsifiers in *Bacillus*, and *Lactobacillus/Lactiplantibacillus* spp., which may lead the production of microbiologically optimized, self-stabilizing, or self-preserving skincare formulations [303]. As consumers continue to demand cleaner, more sustainable, and biologically active skincare solutions, we can predict a surge in products that not only support the skin microbiome but continue to harness its full therapeutic potential [4]. Overall, microbiome-based technologies are transforming skincare by shifting the focus from ingredients that may harm the skin's natural barrier to strategies that protect it at a microbial level, promoting a balanced ecosystem that strengthens the skin's defense against pathogens and environmental stressors.

Skin Microbiome Fingerprinting for Skincare Product Selection

The future of microbiome skincare looks promising, with innovations aiming to personalize skincare products to help manage skin aging and skin conditions like acne, rosacea, and eczema with unprecedented precision [307]. One of the emerging directions in skincare science is the development of skin microbiome testing platforms to identify an individual's unique bacterial ecosystem using artificial intelligence (AI) driven diagnostics, 16S ribosomal RNA sequencing, Internal Transcribed Spacer, and whole genome metagenomic shotgun sequencing (WGS) [308, 307]. Multiple companies now offer direct-to-consumer at-home microbiome testing kits, to compare a person's microbial fingerprint to the baseline microbial fingerprint of people without skin conditions to design tailored skincare regimens [309–311]. Cosmetic, pharmaceutical, and textile companies can also evaluate ingredients, formulations, and products in either human clinical trials or standardized in vitro microbial assays to measure favorable or detrimental effects on specific skin bacterial and fungal spp. from multiple body sites, utilizing the collective scientific knowledge on microbiota populations [309, 312, 313]. One of these companies, established in Lichtenhein with subsidiaries in Germany, Norway, United States,

Australia, New Zealand, South Korea, Japan, and Thailand, has trademarked Microbiome-friendly® in 40 countries [314]. Following testing, the company provides a Microbiome-friendly® certification to products that meet the microbiological standards established for rinse-off, leave-on or textile products [315, 316]. Once certified, these ingredients and products are added to an online database of single ingredients and formulations, providing a directory for individuals and formulators seeking these alternatives [317]. With this certification, companies may substantiate marketing claims including that the product is gentle on the microbiome or that it preserves the natural balance and integrity of the skin's microbiome, both on the surface and in deeper layers [315]. These certification standards aim to create a framework that can be improved and expanded. Standardized approaches are redefining consumer experience by prioritizing personalized skincare ingredients and routines. Chemical analysis using mass-spectrometry in adults showed that individual ingredients from a popular whole-body baby cleanser, a facial moisturizer, and a deodorant, could be detected in other body parts of the participants for weeks after interruption of the products [307]. The ability to measure persistence and longevity of skincare ingredients on the skin could become an integral tool in the search for multifunctional skincare products. Integration of chemical analysis, microarrays, metabolomics, and microbial fingerprinting could help find skincare solutions suitable for any age and skin type [302, 318]. The integration of AI and machine learning in the skincare space could revolutionize our understanding of the complex interactions within the skin's microbial ecosystem, leading to more effective, science-backed formulations guiding product and ingredient selection with the most accuracy.

Final Remarks

The impact of the environment on skincare product development extends well beyond traditional UV protection. The growing demand for products catering to sensitive skin reflects the increased awareness that external stressors, especially urban pollution, water quality, and environmental extremes have on the proper functioning of the skin barrier [319, 149]. The emergence of biomimetic ingredients aimed at replicating the skin's natural structure and function has redefined the market, offering formulations that provide compatibility, and efficacy to address the complex interplay of modern life stressors on skin health [131]. The rise of microbiome-centric skincare reflects an increased appreciation of the skin's intricate relationship with its microbial inhabitants and the key role the microbiome plays in human health. The realization that microbes are a critical functional part of the skin that can be used to our advantage will continue to shape the skincare industry in the next decades [320]. As scientists continue discovering resources from microbes living in extreme and marine environments, the skincare industry will provide even more sophisticated protection strategies, ensuring consumers can face the environmental challenges of today and the future [40, 46, 47]. With the current resurgence of solid skincare products like soap bombs, dissolvable shower gels, beauty face masks, and hair cleansing bars, there is a clear shift towards more experiential formulations signaling a renaissance of methods and principles from the past. The recognition that cosmetic ingredients used in topical applications also have benefits when consumed orally highlights even more the need for comprehensive whole-body approaches to skin care [40, 139, 321]. Skincare protocols using both topical and oral formulations may be more synergistic, targeting dryness, oiliness, wrinkles, hyperpigmentation while also helping manage skin disorders such as eczema, psoriasis, acne, and skin cancer from the inside out [322]. By embracing the implementation of twenty-first century technologies such as AI-driven diagnostics and microbiome analyses catering to specific skin concerns and unique skin microbial profiles, the skincare industry is primed to redefine beauty standards, focusing on long-term health and resilience rather than superficial solutions. The challenge and opportunity lie in continuing to push the boundaries, creating the next generation of skincare that is truly in harmony with our skin's natural ecosystem.

Glossary

Acne vulgaris Chronic hair follicle disorder characterized by the presence of blackheads, whiteheads, inflamed papules, pustules, painful nodules, or cysts. Usually appears on the face, chest, and shoulders during puberty. Scarring of the lesions is a common side effect.

α-Hydroxy Acid A group of natural acids found in fruits that are often used in skincare products to promote cell turnover and improve skin texture. Common α-Hydroxy Acids include glycolic acid, lactic acid, and citric acid.

α-Linoleic Acid A liquid omega-6 polyunsaturated fatty acid derived from linoleic acid. It is considered an essential fatty acid, because the human body cannot synthesize it and must obtain it through diet. Found in plant-based oils like flaxseed oil, canola oil, and soybean oil.

Allergic contact dermatitis An immune response to a small molecule or hapten after skin or oral contact in a sensitized individual. Also known as type 4 or delayed-type hypersensitivity. Increased frequency of exposure leads to a faster immune response to the sensitizing agent. Examples of sensitizing agents include nickel and formaldehyde.

Antimicrobial peptides Short peptides produced by skin cells and/or microbial cells to control the proliferation of other microbes.

Antioxidants Compounds that help protect tissues from oxidative damage caused by free radicals.

Apothecary Old-fashioned term for a professional who prepared and dispensed medicine. The role of the apothecary was gradually replaced by pharmacists and drugstores as modern medicine advanced.

Artificial intelligence is a general term that describes machines and technologies that can learn and perform tasks that were previously done by humans. Artificial intelligence is a field of computer science that aims to simulate human-like intelligence.

Arachidonic acid a polyunsaturated fatty acid found in high concentrations in animal fats. It is considered essential in human nutrition. It may have proinflammatory functions since it is a precursor of prostaglandins which mediate inflammation and pain. It is also present in some plant oils, such as sunflower oil and soybean oil.

Atopic dermatitis A chronic skin condition characterized by dry, itchy, and inflamed skin. Also known as eczema.

Atopic skin Skin that is prone to irritation or allergy leading to the development of dermatitis. It is often associated with other conditions such as asthma, and hay fever.

Bacillus A genus of Gram-positive rod-shaped bacteria that are usually aerobic but may survive in low-oxygen environments like soil. Characterized by the presence of endospores (spores), which may be released into the surrounding environment during times of stress like desiccation or low nutrient availability.

Bacteriophage A virus that infects and destroys bacteria, explored in skincare for its potential to target acne-causing *Cutibacterium acnes*.

β-Hydroxy Acid A type of hydroxy acid that is oil-soluble and penetrates deep into the pores to exfoliate the skin.

Bifidobacterium (Bifida) A genus of Gram-positive, anaerobic, non-motile, usually rod-shaped bacteria that are part of the normal human intestinal and vaginal flora. They produce lactic and acetic acid as byproducts of carbohydrate fermentation.

BioIntact™ cold emulsion A technology that allows the creation of emulsions without heat, preserving ingredient potency.

Biocosmetics Skincare formulations that use microbial-derived ingredients, including probiotics, prebiotics, and postbiotics.

Biofilm a community of microorganisms that can form on living or nonliving surfaces by adhering to surfaces. These microorganisms are enclosed in a protective coating made of extracellular polymeric substances, primarily polysaccharides. Biofilms can protect microorganisms from harmful conditions, such as antibiotics, and thermal stress, and may pose a medical problem, especially if they contain pathogenic forms of microorganisms.

Biome-friendly (see microbiome-friendly) describes an ingredient or formulation that does not harm the skin microbiota and may promote the growth of existing beneficial skin bacteria and fungi. It may also suppress the proliferation of pathogenic microorganisms.

Biomimetic Ingredients that mimic the skin's natural lipid composition and hydration mechanisms, such as ceramides and phospholipids.

Broad-spectrum sunscreens Sunscreens that protect against both ultraviolet A and ultraviolet B rays.

Ceramides Are lipid molecules found in the stratum corneum, the outermost layer of the epidermis, where they occupy the spaces between corneocytes. They play a crucial role in forming a lipid barrier that helps retain moisture, preventing water loss and protecting the skin from environmental damage. Acting as the "mortar" between the "bricks" of skin cells, ceramides work alongside cholesterol and free fatty acids to create a structured lipid matrix. Structurally, ceramides consist of a sphingosine base linked to a fatty acid chain.

Chelating Prevents metal ions from reacting with other substances in a cosmetic formulation. Chelating agents are chemical compounds that react with metal ions to form a stable, water-soluble complex. Anti-chelating agents do not react with metal ions but may serve as antioxidants in a formulation.

Cholesterol A waxy fat found in all animal tissues, especially in the brain, spinal cord, and adipose tissue. It's produced primarily in the liver but may be derived from dietary sources. It is a component of cell membranes and serves as the building block for vitamin D and steroid hormones like estrogen. It supports the skin's barrier function, enhancing moisture retention and resilience.

Chromista Biological kingdom not related to plants, composed of both unicellular and multicellular eukaryotic organisms that share common characteristics in their photosynthetic structures, known as plastids. This group encompasses all protists whose plastids contain chlorophyll c, including various types of algae, diatoms, kelp, mildews, oomycetes (parasitic fungi), and certain protozoans.

Cosmetic ingredient review A program that assesses the safety of ingredients in cosmetics and personal care products. It is a nonprofit program that operates independently of the cosmetics industry. It is supported by the Personal Care Products Council, the US Food and Drug Administration, and the Consumer Federation of America. The Cosmetic Ingredient Review Expert Panel is made up of independent experts in dermatology, toxicology, and chemical sciences.

Coenzyme Q10 (see ubiquinone) A fat-soluble molecule that helps generate energy in mitochondria. Naturally present in all human cells, acting as an antioxidant by reducing oxidative stress intracellularly. It is also known as CoQ10, Q10, vitamin Q10, or ubiquinone.

Commensal An organism that obtains food or other benefits from another without damaging or benefiting it. Ecological relationship implicated in biofilm formation, fermentation, and nitrification.

Comodone (follicular comodone) Non-inflammatory lesions of hair follicle, sebaceous glands and skin pores classified as closed or open comodones. Closed comedones or whiteheads are small, white, or flesh-colored bumps formed when sebum, dead skin cells, and protein block the follicle opening. The blockage remains beneath the skin surface. Open comedones are enlarged pores filled with excess oil, dead skin cells, and protein. Exposure to air causes the fatty acids to oxidize, giving them a dark appearance which is why they are called blackheads.

Contact dermatitis A reaction by contact with a substance that causes an allergic or irritant response. See Allergic Contact Dermatitis and irritant contact dermatitis.

Corneocyte Mature differentiated keratinocytes. As keratinocytes move towards the skin surface, they turn into corneocytes or flat, dead cells which are eventually shed.

Cosmeceuticals Skincare products that may have both cosmetic and pharmaceutical properties. These products often contain active ingredients that provide therapeutic benefits.

Cutibacterium acnes is an important species of Gram-positive rod-shaped anaerobic bacteria found in human skin and hair follicles. It is considered a human commensal, but it is associated with acne since it inhabits hair follicles. Prior to 2016, it appears in the scientific literature by the name *Propionibacterium acnes*.

Desmogleins are cell adhesion molecules that help maintain tissue integrity and cell communication. Desmogleins are components of desmosomes, which are protein structures that connect adjacent cells. They help form stable junctions between cells.

Disability-adjusted life-years quantifies the overall burden of a disease by representing the loss of 1 year of full health. It is calculated as the sum of two components: Years of Life Lost due to premature death and Years Lived with Disability, which account for the time spent living with a disease or health condition within a population.

Eczema A chronic skin condition characterized by dry, itchy, and inflamed skin. Also known as atopic dermatitis.

Emollient Substances that help prevent moisture loss by creating a barrier on the skin's surface. Found in creams, lotions, ointments, salves, and balms and include ingredients such as oils, lipids, and fatty acids. May help manage dry, rough, scaly, and itchy skin conditions.

Emulsifier Ingredients that allow oil and water to blend into a homogeneous mixture that does not separate. They are used in cosmetic formulations to enhance functionality and consumer perception of the formulation.

Encapsulation technology A method used in skincare to protect active ingredients from degradation and improve delivery into the skin.

Epidermal growth factor Polypeptide hormone that plays a role in wound healing, and tissue regeneration. Protein that helps maintain skin cell growth and repair, helping to improve skin texture and reduce signs of aging.

Erythema Redness of the skin caused by increased blood flow to the capillaries.

Exfoliation describes the process of removing dead skin cells from the surface of the skin, through physical methods like scrubbing with a washcloth or chemical methods like using exfoliating acids.

Exopolysaccharides Polysaccharides excreted by microbes into their surrounding environment to help them form protective biofilms and shield them against extreme environments.

Exosomes Small vesicles that are released from stem cells that can carry proteins, metabolites, nucleic acids, and other cellular waste between cells. They are also known as extracellular vesicles.

Extremolytes A category of ingredients produced by microbes thriving in extreme environments like deserts, salt lakes, and seawater. These compounds help the microbes survive under harsh conditions and are used in skincare for their protective properties.

Fatty acids Essential building blocks of the skin's lipid barrier, which help maintain hydration and support the growth of beneficial bacteria.

Ferment filtrate The liquid fraction obtained from microbial fermentation, rich in beneficial microbial metabolites that does not contain fragments of microbial cells.

Fermentation The enzyme-catalyzed anaerobic breakdown of an energy-rich compound such as a carbohydrate by the action of microorganisms that occur naturally and is commonly used in the production of food, alcoholic, and pharmaceuticals. Fermentation byproducts can be modulated by changing the fermentation conditions, the substrate or by controlling the rate of enzymatic activity.

Fibroin A water-insoluble fibrous protein which makes up over 60% of the content of natural silk, with the remainder being the water-soluble protein sericin.

Filaggrin A protein essential for skin barrier function, aiding in moisture retention. Mutations in the Filaggrin gene have been associated with a higher risk of developing atopic dermatitis.

Galactomyces a genus of yeast-like fungus that is used in the production of fermented foods and drinks like sake. *Galactomyces* Ferment Filtrate is a widely used ingredient in skincare cosmetics trademarked under the name Pitera.

Glycosaminoglycans Long unbranched chains of polysaccharides (carbohydrates) that are essential extracellular components of connective and skin tissue involved in maintaining collagen and elastin fibers. Hyaluronic acid is a glycosaminoglycan that attracts and retains moisture, improving skin hydration and elasticity. They are also known as mucopolysaccharides.

Glucuronidation A detoxification process that attaches glucuronic acid to a chemical, making it water-soluble and easier to excrete out of the body. It is a key part of metabolizing drugs, pollutants, hormones, fatty acid derivatives, retinoids, and other substances in the body.

Halophile Microbes that thrive in high salinity environments like salt lakes and seawater. Skin bacteria such as *Staphylococcus* species are halophiles.

Hapten Molecules smaller than 500 Daltons in size that can promote an immune response. Most common examples are found in vaccines, but they are also implicated in Allergic Contact Dermatitis to skincare products.

Hay fever An acute seasonal allergic reaction to pollens that presents itself with sneezing, itching, watering of the eyes, and nasal congestion and discharge.

Humectants Hydrating ingredients that attract moisture to the skin, helping to keep it hydrated and plump.

Hydrocolloid patch Wound dressings that form a gel matrix that can deliver therapeutic agents composed of an inner colloid layer and an outer water-impermeable layer, used to treat acne lesions and skin abrasions.

Hyperpigmentation Excess pigmentation or melanin accumulation in a tissue like the skin.

IL-10 (Interleukin-10) A cytokine produced by white blood cells and other immune cells that helps regulate the body's immune response to microbes. IL-10 has anti-inflammatory properties and helps prevent tissue damage. It reduces inflammation by blocking the production of pro-inflammatory cytokines thereby controlling allergies and asthma.

Incidence refers to the number of new occurrences of a disease, injury, symptom, or death within a specific time frame. It is typically expressed as a rate, such as the number of new cases per 1000 individuals in a population. Incidence measures the probability that an individual within a population will develop a condition over a defined period. Accurately measuring incidence can be challenging due to changes in the at-risk population caused by births, deaths, or migration, as well as difficulties in distinguishing new cases. Incidence rates are valuable for comparing disease risk across different populations.

Inorganic sunscreens Sunscreens that contain inorganic minerals like zinc oxide and titanium dioxide.

Internal transcribed spacer A specific DNA region within ribosomal RNA genes exhibiting high variability between different fungal species used to identify and differentiate different species of fungi, making it a valuable tool for fungal microbiome analysis.

In vitro Tests or assay done outside a living body and in an artificial environment like a test tube.

In vivo Test or assay done in a living being such as a plant, rodent, or human.

Irritant contact dermatitis The most common type of contact dermatitis that manifests itself with itching, redness, swelling, and blistering. It is a non-allergic skin reaction that occurs when a chemical damages the skin's outer layer due to repeated exposure to mild irritants. Aerosolized particles in solvents and sprays can also induce it.

Keratin A protein produced by keratinocytes that forms the main structural constituent of hair, feathers, hoofs, claws, and horns. In skin keratin, the amino acids glycine, serine, glutamic acid, arginine, and threonine help keratin interact with water and maintain proper hydration levels in the skin. In human hair keratin, cysteine is the most important amino acid for keratin formation, as its sulfur-containing side chain forms disulfide bonds with other cysteine residues, creating strong cross-links that give keratin its rigidity and strength. The amino acids valine, leucine, and isoleucine contribute to the protein's structural stability and help form the hydrophobic core of keratins. The amino acids proline, methionine, and lysine are also found in human skin and hair keratins.

Keratinization Process that occurs as cells move from one epithelial layer to the next, where they accumulate more and more keratin. May give a rough character to epithelial tissue in many parts of the body, making skin less likely to tear or scratch. Acne may be a manifestation of keratinization disorder.

Keratinocyte A skin cell formed in the basal epidermal layer above the dermis that produces keratin. It divides into more keratinocytes and ultimately migrates into the outer protective layers of the skin where it undergoes final differentiation into a corneocyte. Melanin produced by melanocytes is eventually contained within keratinocytes. The amount of melanin in keratinocytes determines skin tone.

Lactiplantibacillus A genus of rod-shaped, Gram-positive, and non-spore-forming lactic acid bacteria that was previously part of the *Lactobacillus* genus. In April 2020, the name was changed when the *Lactobacillus* genus was split into 25 different genera.

Lactobacillus A genus of rod-shaped, Gram-positive, and non-spore-forming bacteria lactic acid bacteria that includes over 200 species. It is the largest group of lactic acid bacteria. Singular is *Lactobacilli.*

Linoleic acid A liquid omega-6 polyunsaturated fatty acid in skin.

Lipophilic Microbes that thrive by breaking down fatty acids. Also, it refers to molecules that have an affinity for areas with high fatty acid content.

Liposomes Microscopic artificial vesicles composed of one or more concentric phospholipid bilayers used to deliver microscopic substances (such as active ingredients or DNA) enhancing their stability and delivery into the skin.

Lysate Preparation obtained after breaking up microbial cells using chemical reagents and enzymes, mechanical disruption, or osmotic changes. The preparation can be further separated to isolate soluble compounds to generate cell-free ingredients.

Melanocytes Mature cells that produce melanin, providing protection against oxidative damage by ultraviolet radiation. They are found in the skin, hair follicles, mucosa, iris, cochlea, and brain.

Microbiome-friendly (See Biome-friendly) describes a product or ingredient that does not harm the skin microbiota and may promote the growth of existing beneficial skin bacteria and fungi.

Microbiome-Friendly® Certification A certification granted to skincare products or ingredients that support and maintain the skin's microbiome balance.

Microbiome fingerprinting The process of analyzing an individual's skin microbiome to tailor skincare ingredient and recommendations.

MicroRNA A cellular RNA fragment that prevents the production of a particular protein by binding to and neutralizing the messenger RNA that would have produced the protein.

Nano titanium dioxide Ultrafine inorganic sunscreen ingredient made up of titanium dioxide particulate sizes between 50 and 100 nm.

Nano zinc oxide Ultrafine inorganic sunscreen ingredient made up of zinc oxide particulate sizes between 50 and 100 nm.

Niacinamide A powerful antioxidant that helps improve skin barrier function, reduce inflammation, and regulate oil production. Also known as nicotinamide or vitamin B3.

Nitrosomonas A genus of spherical to rod-shaped bacteria found in soil and water studied for its role in the nitrogen cycle. They obtain energy for growth by oxidizing ammonia to nitrites.

Non-nano titanium dioxide Inorganic sunscreen ingredient with titanium dioxide particulate sizes between 200 and 300 nm.

Non-nano zinc oxide Inorganic sunscreen ingredient with zinc oxide particulate sizes between 200 and 300 nm.

Occlusive Ingredients that form a protective barrier on the skin's surface to prevent moisture loss.

Occlusion The process of forming a protective barrier on the skin's surface to prevent moisture loss using an ingredient or material such as a wound dressing.

Organic sunscreens Sunscreens that contain carbon-containing chemicals for ultraviolet protection.

Palmitic Acid A waxy crystalline saturated fatty acid occurring free or in the form of esters (as glycerides) in most fat and fatty oils and in many essential oils and waxes. The body can synthesize palmitic acid and does not require it from the diet to function properly. It is naturally found in human sebum, essential for skin barrier function. Derived from coconut oil or palm oil for skin care purposes. Also called hexadecanoic acid.

Panthenol A humectant that improves skin hydration and soothes irritation. It is commonly known as Vitamin B5.

Pathogen Microorganism with potential to cause disease.

Pathogenic Capable of causing disease.

Peptides Short chains of amino acids that signal the skin to produce collagen and elastin, helping to firm and rejuvenate the skin.

Phosphate-buffered saline It is an isotonic solution commonly used in biomedical research because its osmolarity and ion concentration match those of human cells. It maintains a stable pH of around 7.4 making it non-toxic to most cells, allowing for various applications in laboratory and clinical settings.

Photodamage Damage to skin or DNA caused by exposure to ultraviolet radiation.

Photolyase A flavoprotein derived from the algae *Anacystis nidulans*, used in sunscreens to minimize immediate UV radiation effects while providing DNA repairing mechanisms. An enzyme activated by light that helps repair DNA damage caused by ultraviolet radiation.

Photoprotect Refers to the action of protecting against damage caused by light, particularly ultraviolet (UV) radiation. Photoprotection is used in the context of biological and physical processes that help organisms shield themselves from harmful effects of excessive light. Sunscreens, protective clothing and sunglasses are photoprotective measures for humans.

Phospholipids Lipids that help stabilize the skin's barrier and maintain its integrity, promoting smoothness and softness.

Pilosebaceous unit Unit of the skin made up of a hair follicle, hair shaft, a sebaceous gland, and an erector (arrector) pili muscle.

Plant Extracts Botanical ingredients that have antimicrobial, anti-inflammatory, and antioxidant properties. Examples include aloe vera, chamomile, and calendula.

Plant Stem Cells Progenitor cells derived from botanical sources that produce growth factors that help protect and regenerate the skin.

Plasmid Extrachromosomal piece of DNA especially of bacteria that replicate autonomously. They can be circular or linear.

Postbiotic Metabolic byproducts produced by microbes that have beneficial effects on the skin. They do not contain live microorganisms.

Post-inflammatory hyperpigmentation Skin condition where areas of darkened skin develop after an injury or inflammation caused by the skin producing excess melanin. It is more noticeable in people with darker skin tones and can be triggered by acne, eczema, burns, or other skin trauma.

Prebiotic Compounds that promote the growth or activity of beneficial microorganisms such as bacteria and fungi. In skincare, prebiotics are used to support the skin's microbiome. Examples include inulin, fructooligosaccharides, and Xylo oligosaccharides.

Prevalence Represents the proportion of a population affected by a condition at a specific point in time. It may include both new and existing cases. Prevalence is calculated by multiplying the incidence of a condition or disease by amount of time.

Probiotic Beneficial live microorganisms, such as *lactobacillus* and *bifidobacterium* species that can be found in fermented foods like yogurt and kefir, and in dietary supplements.

Prospective study A study design where researchers collect data such as symptoms, observations, and questionnaire results at regular time intervals to identify or measure the prevalence of specific outcomes as time passes. It may also gather other data that might relate to the development or worsening of a medical condition.

Protection grade of UVA (PA) A rating system used in Japan to represent how much Ultraviolet A ray protection a sunscreen product offers.

Pruritus Severe itching of the skin that may reflect a skin condition or another ailment. Tingling or irritation that makes you want to scratch a body part.

Randomized controlled trial A study design in which participants are randomly assigned to different groups to compare the effects of various treatments. It can be single-blind meaning the participant does not know if they're receiving treatment or a placebo, or double-blind where both investigators and participants are unaware of the treatment conditions. It is considered the highest level of scientific evidence since it is expected to minimize experimental bias and the placebo effect.

Reactive oxygen species Chemically reactive molecules containing oxygen. In the context of skincare, reactive oxygen species are produced as part of the skin's response to solar radiation and are a target for suncare and skincare products.

Retinoids Vitamin A derivatives used in skincare for their anti-aging and acne treatment properties. Common retinoids include retinol, retinaldehyde, retinyl palmitate, and retinoic acid.

Retrospective analysis A research design that analyzes past data to identify associations between prior exposures or events and a particular outcome. Unlike prospective studies, which track participants over time, retrospective studies examine existing records, such as medical histories, registries, or survey responses, without actively collecting new data at the time of the study. Often structured as a case-control study, where individuals with a condition (cases) are compared to those without it (controls) to assess prior exposures.

Rosacea Chronic inflammatory skin disorder typically involving the nose, cheeks, forehead, and chin that is characterized especially by redness, flushing, telangiectasia (spider vessels), inflamed lesions, and burning sensation, and sometimes by eye dryness and irritation and thickening of tissues especially of the nose.

Saccharomyces Genus of unicellular yeasts that are distinguished by their sparse or absent mycelium and by their facility in reproducing asexually by budding. Brewer's yeast is the most studied.

Salve A semisolid healing ointment for application to wounds or sores generally made by infusing beeswax, cocoa or shea butter, or oils (coconut, olive, sesame, sunflower, sweet almond oil) with herbal oils and essential oil.

Sebaceous gland Small saclike exocrine glands of the skin located in the dermis, typically open into an associated hair follicle. It contains cells which produce and secrete sebum called sebocytes.

Seborrheic dermatitis A skin condition that presents itself with yellow to red, scaly, itchy patches affecting areas with many large sebaceous glands like the face, scalp, beard, and chest. The condition is thought to be driven by the overgrowth of multiple species of the skin fungus *Malassezia*.

Sebum Fatty lubricant matter secreted by sebaceous glands of the skin. Specialized cells called sebocytes within the sebaceous gland accumulate lipids and then rupture, releasing their contents. Sebum plays a crucial role in maintaining the health and appearance of the skin and hair by providing lubrication, protection, and anti-inflammatory properties.

Sonication Act of breaking-up cells by exposure to high-frequency sound waves.

Sorbitan An emulsifier used in diaper-rash ointments and water-resistant sunscreens derived from sorbitol.

Sphingolipids Lipids that possess a chemical backbone of sphingosine or derivatives that help regulate cell growth and differentiation, intracellular cell signaling, and cell death. In the skin, ceramides are the most well-studied sphingolipids.

Spore A dormant body produced by plants, fungi, and some microorganisms like *Bacillus* bacteria that are resistant to environmental challenges like drying or heat. Also has reproductive functions as it is capable of development into a new individual either directly or after fusion with another spore.

Staphylococcus Gram-positive round-shaped bacteria that inhabit mucosal and skin tissues. *Staphylococcus epidermidis*, *Staphylococcus aureus*, and *Staphylococcus hominis* are the most prevalent in human skin. *S. aureus* can cause skin and hair follicle infections and is often linked to atopic dermatitis.

Stratum corneum Outermost top layer of the epidermis made up of corneocytes (dead keratinocytes) and skin microbes.

Sun Protection Factor A measure of how well a sunscreen protects against Ultraviolet B rays. Sun Protection Factor is calculated by dividing the time it takes for skin to redden with sunscreen by the time it takes to redden without sunscreen.

Sunscreens Products that protect the skin from UV radiation. They can be categorized as organic, inorganic, or mixed.

Sulfotranferases Enzymes that transfer a sulfo group to an acceptor molecule. Sulfotransferases are involved in the metabolism of drugs, toxins, steroid hormones and other substances.

Superoxide dismutase Antioxidant enzyme that neutralizes superoxide radicals to protect cells from oxidative stress. Produced by microbes, cyanobacteria, chromista, marine plants, terrestrial plants, and fruits.

Surfactants Used as wetting ingredients in cosmetics. They function by lowering the water's surface tension. Surfactants are classified in four categories: anionic, nonionic, cationic, and amphoteric. Cationic surfactants have a positive charge, anionic surfactants have a negative charge, and nonionic surfactants have no charge. Amphoteric surfactants have both a negative and positive charge.

Synbiotic A formulation that combines live probiotics and prebiotics that exert a beneficial effect on gut or skin health.

Syndet A term for detergents that are used in cleansing products like soaps and shampoos. It is a combination of the words synthetic and detergent. Surfactants derived from plant origin are common in syndets providing gentler cleansing options compared to classical soap bars.

Synergistic Having the capacity to interact such that the total effect is greater than the sum of the individual effects.

16S ribosomal RNA sequencing A biotechnology method used for identifying bacteria and fungi in a sample by sequencing the highly conserved 16S ribosomal RNA gene. The gene has variable regions that can help differentiate and classify bacteria and fungi into specific species and strains.

Talc Naturally occurring soft mineral composed of magnesium, silicon, oxygen, and hydrogen used in cosmetics for its absorptive properties. It is also known as talcum powder.

Titanium dioxide A mineral UV filter commonly used in sunscreens for broad-spectrum UV protection.

Toll-like receptor 2 A membrane-bound receptor that helps the innate immune system recognize and respond to foreign substances. It is a member of a group of pattern recognition receptors found on immune and skin cells that recognize lipoproteins from Gram-positive bacteria and fungi.

Transepidermal water loss The process of moisture loss from the skin, used as a measure of skin barrier function.

Ubiquinone A fat-soluble molecule that helps generate energy in the mitochondria. Naturally present in all human cells, acting as an antioxidant by reducing oxidative stress intracellularly. It is also known as Coenzyme Q10, CoQ10, Q10, vitamin Q10.

Uronic acid A nonfermentable sugar that contributes to the formation of glycosaminoglycans and is involved in the synthesis of polysaccharides.

Ultraviolet light Light particle with wavelengths between 10 and 400 nm that falls on the electromagnetic spectrum between visible light and X-rays. Human eyes cannot detect ultraviolet light because of its high frequency.

Ultraviolet A rays Type of ultraviolet radiation from the sun with wavelengths between 315 and 400 nm. It is also called blacklight or long-wave UV. It is the closest UV radiation to visible light. Most of the ultraviolet radiation that makes it through the earth's atmosphere is ultraviolet A. It penetrates deeper into the skin than ultraviolet B rays.

Ultraviolet B Rays Type of ultraviolet radiation from the sun with wavelengths between 280 and 315 nm. Ultraviolet B rays cause sunburn because they carry more energy than ultraviolet A rays.

Ultraviolet protection factor Measures the ratio of ultraviolet radiation that reaches the skin with and without a fabric. Dark or bright colors absorb ultraviolet rays better than lighter shades. Ultraviolet protection factor ratings are similar to Sun Protection Factor ratings for sunscreen.

Virulence factor Molecules produced by microbial pathogens that contribute to their pathogenicity or ability to promote colonization or infection of eukaryotic cells or tissues.

Whole genome metagenomic shotgun sequencing Method that analyzes the DNA of an entire community of microorganisms. It is a sequencing technique that provides a more authentic view of the target microbial community by allowing detection of low abundance members within microbial communities. It can be used to study complex systems with microbial communities that contain thousands of organisms.

Zinc oxide An inorganic ingredient that provides broad-spectrum UV protection and anti-inflammatory functions.

References

1. Liu, J. K. (2022). Natural products in cosmetics. *Natural Products and Bioprospecting, 12*(1), 40. https://doi.org/10.1007/s13659-022-00363-y
2. Wang, H. (2024). Beneficial medicinal effects and material applications of rose. *Heliyon., 10*(1), e23530. https://doi.org/10.1016/j.heliyon.2023.e23530
3. Karagounis, T. K., Gittler, J. K., Rotemberg, V., & Morel, K. D. (2019). Use of "natural" oils for moisturization: Review of olive, coconut, and sunflower seed oil. *Pediatric Dermatology, 36*(1), 9–15. https://doi.org/10.1111/pde.13621
4. Puebla-Barragan, S., & Reid, G. (2021). Probiotics in cosmetic and personal care products: Trends and challenges. *Molecules, 26*(5), 1249. https://doi.org/10.3390/molecules26051249
5. Huang, A., Seité, S., & Adar, T. (2018). The use of balneotherapy in dermatology. *Clinics in Dermatology, 36*(3), 363–368. https://doi.org/10.1016/j.clindermatol.2018.03.010
6. Woolery-Lloyd, H., Andriessen, A., Day, D., et al. (2023). Review of the microbiome in skin aging and the effect of a topical prebiotic containing thermal spring water. *Journal of Cosmetic Dermatology, 22*(1), 96–102. https://doi.org/10.1111/jocd.15464
7. Antonelli, M., & Donelli, D. (2018). Mud therapy and skin microbiome: A review. *International Journal of Biometeorology, 62*(11), 2037–2044. https://doi.org/10.1007/s00484-018-1599-y
8. Hamblin J. Clean: The new science of skin. ; 2020.
9. Mijaljica, D., Spada, F., & Harrison, I. P. (2022). Skin cleansing without or with compromise: Soaps and syndets. *Molecules, 27*(6), 2010. https://doi.org/10.3390/molecules27062010
10. Drissi, M., Carr, E., & Housewright, C. (2022). Sunscreen: A brief walk through history. *Baylor University Medical Center Proceedings, 35*(1), 121–123. https://doi.org/10.1080/08998280.2021.1966602
11. Krien, P. M., & Moyal, D. (1994). Sunscreens with broad-spectrum absorption decrease the *trans* to *cis* photoisomerization of urocanic acid in human *stratum corneum* after multiple UV light exposures. *Photochemistry and Photobiology, 60*(3), 280–287. https://doi.org/10.1111/j.1751-1097.1994.tb05105.x

12. Urbach, F. (2001). The historical aspects of sunscreens. *Journal of Photochemistry and Photobiology. B, 64*(2–3), 99–104. https://doi.org/10.1016/S1011-1344(01)00202-0
13. Bernerd, F., & Asselineau, D. (2008). An organotypic model of skin to study photodamage and photoprotection in vitro. *Journal of the American Academy of Dermatology, 58*(5), S155–S159. https://doi.org/10.1016/j.jaad.2007.08.050
14. Kumari, J., Das, K., Babaei, M., Rokni, G. R., & Goldust, M. (2023). The impact of blue light and digital screens on the skin. *Journal of Cosmetic Dermatology, 22*(4), 1185–1190. https://doi.org/10.1111/jocd.15576
15. Moyal, D. D., & Fourtanier, A. M. (2008). Broad-spectrum sunscreens provide better protection from solar ultraviolet–simulated radiation and natural sunlight–induced immunosuppression in human beings. *Journal of the American Academy of Dermatology, 58*(5), S149–S154. https://doi.org/10.1016/j.jaad.2007.04.035
16. Séite, S., Moyal, D., Richard, S., et al. (1998). Mexoryl® SX: A broad absorption UVA filter protects human skin from the effects of repeated suberythemal doses of UVA. *Journal of Photochemistry and Photobiology. B, 44*(1), 69–76. https://doi.org/10.1016/S1011-1344(98)00122-5
17. Moyal, D. (2004). Prevention of ultraviolet-induced skin pigmentation. *Photodermatology, Photoimmunology & Photomedicine, 20*(5), 243–247. https://doi.org/10.1111/j.1600-0781.2004.00111.x
18. Matta, M. K., Zusterzeel, R., Pilli, N. R., et al. (2019). Effect of sunscreen application under maximal use conditions on plasma concentration of sunscreen active ingredients. *Journal of the American Medical Association, 321*(21), 2082. https://doi.org/10.1001/jama.2019.5586
19. Baan, R., Grosse, Y., Straif, K., et al. (2009). A review of human carcinogens-part F: Chemical agents and related occupations. *The Lancet Oncology, 10*(12), 1143–1144. https://doi.org/10.1016/S1470-2045(09)70358-4
20. Hudspeth, A., Zenzola, N., Kucera, K., Wu, Q., & Light, D. (2022). Independent sun care product screening for benzene contamination. *Environmental Health Perspectives, 130*(3). https://doi.org/10.1289/EHP10386
21. U.S. Food and Drug Administration. (2019). *Sunscreen drug products for over-the-counter human use. 84*(38). Accessed September 9, 2024. https://www.govinfo.gov/content/pkg/FR-2019-02-26/pdf/2019-03019.pdf
22. Abdel Azim, S., Bainvoll, L., Vecerek, N., DeLeo, V. A., & Adler, B. L. (2024). Sunscreens part 2: Regulation and safety. *Journal of the American Academy of Dermatology*. https://doi.org/10.1016/j.jaad.2024.02.066
23. U.S. Cancer Statistics Working Group. (2024, June). *U.S. Cancer Statistics Data Visualizations Tool*. US Department of Health and Human Services, Centers for Disease Control and Prevention and National Cancer Institute.
24. Heerfordt, I. M., Philipsen, P. A., & Wulf, H. C. (2020). A handful of sunscreen for whole-body application. In *Sunlight, vitamin D and skin cancer* (pp. 381–385). Springer. https://doi.org/10.1007/978-3-030-46227-7_19
25. Bennett, S. L., & Khachemoune, A. (2022). Dispelling myths about sunscreen. *The Journal of Dermatological Treatment, 33*(2), 666–670. https://doi.org/10.1080/09546634.2020.1789047
26. Diffey, B. (2001). Sunscreen isn't enough. *Journal of Photochemistry and Photobiology. B, 64*(2–3), 105–108. https://doi.org/10.1016/S1011-1344(01)00195-6
27. Rebecca, L. S., & Yamarik, T. A. (2002). Cosmetic ingredient review panel. Final report on the safety assessment of sorbitan caprylate, sorbitan cocoate, sorbitan diisostearate, sorbitan dioleate, sorbitan distearate, sorbitan isostearate, sorbitan olivate, sorbitan sesquiisostearate, sorbitan sesquistearate, and sorbitan triisostearate. *International Journal of Toxicology, 21*(1_suppl), 93–112. https://doi.org/10.1080/10915810290096414
28. CIR Expert Panel. (2014). *Dermal penetration, absorption, and other considerations for babies and infants in safety assessments*. Accessed September 23, 2024. https://www.cir--safety.org/sites/default/files/Infskn092014rep-%20final.pdf

29. Should you put sunscreen on infants? Not usually. (2024, May 9). *Consumer Updates*. Accessed September 23, 2024. https://www.fda.gov/consumers/consumer-updates/should-you-put-sunscreen-infants-not-usually
30. Okati-Aliabad, H., Hosseini, E. S., Sharifabad, M. A. M., Mohammadi, M., Ardakani, M. E., & Talebrouhi, A. H. (2024). Efficacy of a facial-aging web app on sun protection behaviors among primary school students in Iran: A randomized controlled trial. *BMC Public Health, 24*(1), 737. https://doi.org/10.1186/s12889-024-18241-2
31. de Araújo, M. M., Schneid, A. C., Oliveira, M. S., et al. (2024). NLC-based sunscreen formulations with optimized proportion of encapsulated and free filters exhibit enhanced UVA and UVB photoprotection. *Pharmaceutics., 16*(3), 427. https://doi.org/10.3390/pharmaceutics16030427
32. Sancar, A. (2003). Structure and function of DNA photolyase and cryptochrome blue-light photoreceptors. *Chemical Reviews, 103*(6), 2203–2238. https://doi.org/10.1021/cr0204348
33. Ispiryan, A., Bobinaite, R., Urbonaviciene, D., et al. (2023). Physico-chemical properties, fatty acids profile, and economic properties of raspberry (*Rubus idaeus L.*) seed oil, extracted in various ways. *Plants., 12*(14), 2706. https://doi.org/10.3390/plants12142706
34. Puig, S., Granger, C., Garre, A., Trullàs, C., Sanmartin, O., & Argenziano, G. (2019). Review of clinical evidence over 10 years on prevention and treatment of a film-forming medical device containing photolyase in the management of field cancerization in actinic keratosis. *Dermatology and Therapy, 9*(2), 259–270. https://doi.org/10.1007/s13555-019-0294-1
35. Puig, S., Puig-Butillé, J., Díaz, M., Trullas, C., & Malvehy, J. (2014). Field cancerisation improvement with topical application of a film-forming medical device containing photolyase and UV filters in patients with actinic keratosis, a pilot study. *Journal of Clinical & Experimental Dermatology Research, 05*(03). https://doi.org/10.4172/2155-9554.1000220
36. Ghosh, D., & Dhandha, M. M. (2024). Investigation of sun protection factor (SPF) of raspberry seed oil, Niacinamide and zinc oxide in combination for their possible use in a sunscreen formulation. *SKIN The Journal of Cutaneous Medicine., 8*(5), 1807–1814. https://doi.org/10.25251/skin.8.5.4
37. Rosa, A. C., Corsi, D., Cavi, N., Bruni, N., & Dosio, F. (2021). Superoxide dismutase administration: A review of proposed human uses. *Molecules, 26*(7), 1844. https://doi.org/10.3390/molecules26071844
38. Zheng, M., Liu, Y., Zhang, G., Yang, Z., Xu, W., & Chen, Q. (2023). The applications and mechanisms of superoxide dismutase in medicine, food, and cosmetics. *Antioxidants., 12*(9), 1675. https://doi.org/10.3390/antiox12091675
39. Kaneko, A., Kondo, S., Akiyama, J., & Okabe, K. (1997). A new cosmetic SOD delivery system using skin surface resident *Staphylococcus epidermidis* by lotions containing Mn/Zn ions. *Drug Delivery System., 12*(5), 339–345. https://doi.org/10.2745/dds.12.339
40. Alves, A., Sousa, E., Kijjoa, A., & Pinto, M. (2020). Marine-derived compounds with potential use as cosmeceuticals and nutricosmetics. *Molecules, 25*(11), 2536. https://doi.org/10.3390/molecules25112536
41. Brown, A. D. (1976). Microbial water stress. *Bacteriological Reviews, 40*(4), 803–846. https://doi.org/10.1128/br.40.4.803-846.1976
42. Buenger, J., & Driller, H. (2004). Ectoin: An effective natural substance to prevent UVA-induced premature photoaging. *Skin Pharmacology and Physiology, 17*(5), 232–237. https://doi.org/10.1159/000080216
43. Heinrich, U., Garbe, B., & Tronnier, H. (2007). *In vivo* assessment of ectoin: A randomized, vehicle-controlled clinical trial. *Skin Pharmacology and Physiology, 20*(4), 211–218. https://doi.org/10.1159/000103204
44. Marini, A., Reinelt, K., Krutmann, J., & Bilstein, A. (2014). Ectoine-containing cream in the treatment of mild to moderate atopic dermatitis: A randomised, comparator-controlled, intra-individual double-blind, multi-center trial. *Skin Pharmacology and Physiology, 27*(2), 57–65. https://doi.org/10.1159/000351381

45. Ahuja, V., Bhatt, A. K., Banu, J. R., et al. (2023). Microbial exopolysaccharide composites in biomedicine and healthcare: Trends and advances. *Polymers (Basel).*, *15*(7), 1801. https://doi.org/10.3390/polym15071801
46. Zaghloul, E. H., & Ibrahim, M. I. A. (2022). Production and characterization of exopolysaccharide from newly isolated marine probiotic *Lactiplantibacillus plantarum* EI6 with in vitro wound healing activity. *Frontiers in Microbiology, 13*, 13. https://doi.org/10.3389/fmicb.2022.903363
47. Nagaraj, A., Subramaniyan, Y., Surya, S., & Rekha, P. D. (2024). Burn wound healing abilities of a uronic acid containing exopolysaccharide produced by the marine bacterium *Halomonas malpeensis* YU-PRIM-29 T. *Applied Biochemistry and Biotechnology*. https://doi.org/10.1007/s12010-024-04966-8
48. Reed, R. E. (1963). The definition of "cosmeceutical". *Journal of the Society of Cosmetic Chemists, 54*(I11), 103–106.
49. Draelos, Z. D. (2009). Cosmeceuticals: Undefined, unclassified, and unregulated. *Clinics in Dermatology, 27*(5), 431–434. https://doi.org/10.1016/j.clindermatol.2009.05.005
50. U.S. Food and Drug Administration. (2024, October 18). *Format and content requirements for over-the-counter (OTC) drug product labeling*. Code Fed Regul. 201.66. Accessed October 21, 2024. https://www.ecfr.gov/current/title-21/section-201.66
51. Crous, C., Pretorius, J., & Petzer, A. (2024). Overview of popular cosmeceuticals in dermatology. *Skin Health and Disease, 4*(2), e340. https://doi.org/10.1002/ski2.340
52. Arif, T. (2015). Salicylic acid as a peeling agent: A comprehensive review. *Clinical, Cosmetic and Investigational Dermatology, 26*, 455–461. https://doi.org/10.2147/CCID.S84765
53. Kornhauser, A., Coelho, S. G., & Hearing, V. J. (2010). Applications of hydroxy acids: Classification, mechanisms, and photoactivity. *Clinical, Cosmetic and Investigational Dermatology, 3*, 135–142. https://doi.org/10.2147/CCID.S9042
54. Baldwin, H., Webster, G., Stein Gold, L., Callender, V., Cook-Bolden, F. E., & Guenin, E. (2021). 50 years of topical Retinoids for acne: Evolution of treatment. *American Journal of Clinical Dermatology, 22*(3), 315–327. https://doi.org/10.1007/s40257-021-00594-8
55. Kligman, A. M., Fulton, J. E., & Plewig, G. (1969). Topical vitamin A acid in acne vulgaris. *Archives of Dermatology, 99*(4), 469–476. https://doi.org/10.1001/archderm.1969.01610220097017
56. Heilman, E. R., & Lavker, R. M. (1983). Abnormal follicular keratinization. *Dermatologic Clinics, 1*(3), 353–359. https://doi.org/10.1016/S0733-8635(18)31021-0
57. Tobiasz, A., Nowicka, D., & Szepietowski, J. C. (2022). Acne vulgaris—Novel treatment options and factors affecting therapy adherence: A narrative review. *Journal of Clinical Medicine, 11*(24), 7535. https://doi.org/10.3390/jcm11247535
58. Kligman, A. M. (1997). The treatment of acne with topical retinoids: One man's opinions. *Journal of the American Academy of Dermatology, 36*(6), S92–S95. https://doi.org/10.1016/S0190-9622(97)70049-5
59. Rusu, A., Tanase, C., Pascu, G. A., & Todoran, N. (2020). Recent advances regarding the therapeutic potential of adapalene. *Pharmaceuticals., 13*(9), 217. https://doi.org/10.3390/ph13090217
60. Tenaud, I., Khammari, A., & Dreno, B. (2007). *In vitro* modulation of TLR-2, CD1d and IL-10 by adapalene on normal human skin and acne inflammatory lesions. *Experimental Dermatology, 16*(6), 500–506. https://doi.org/10.1111/j.1600-0625.2007.00552.x
61. Moore, E. M., Wagner, C., & Komarnytsky, S. (2020). The enigma of bioactivity and toxicity of botanical oils for skin care. *Frontiers in Pharmacology, 11*, 11. https://doi.org/10.3389/fphar.2020.00785
62. Zhong, J., Zhao, N., Song, Q., Du, Z., & Shu, P. (2024). Topical retinoids: Novel derivatives, nano lipid-based carriers, and combinations to improve chemical instability and skin irritation. *Journal of Cosmetic Dermatology, 23*(10), 3102–3115. https://doi.org/10.1111/jocd.16415

63. Kumar, A., Kushwaha, V., & Sharma, P. K. (2014). Pharmaceutical microemulsion: Formulation, characterization and drug deliveries across skin. *International Journal of Drug Development and Research., 6*(1), 1–21. Accessed September 25, 2024. https://www.researchgate.net/publication/285026890_Pharmaceutical_Microemulsion_Formulation_Characterization_and_Drug_deliveries_across_skin
64. Kouassi, M. C., Grisel, M., & Gore, E. (2022). Multifunctional active ingredient-based delivery systems for skincare formulations: A review. *Colloids and Surfaces. B, Biointerfaces, 217*, 112676. https://doi.org/10.1016/j.colsurfb.2022.112676
65. Caritá, A. C., Fonseca-Santos, B., Shultz, J. D., Michniak-Kohn, B., Chorilli, M., & Leonardi, G. R. (2020). Vitamin C: One compound, several uses. Advances for delivery, efficiency and stability. *Nanomedicine, 24*, 102117. https://doi.org/10.1016/j.nano.2019.102117
66. Sheraz, M., Ahmad, I., Vaid, F. H. M., Ahmed, S., Shaikh, R., & Iqbal, K. (2011). Formulation and stability of ascorbic acid in topical preparations. *Systematic Reviews in Pharmacy, 2*(2), 86. https://doi.org/10.4103/0975-8453.86296
67. Lin, F. H., Lin, J. Y., Gupta, R. D., et al. (2005). Ferulic acid stabilizes a solution of vitamins C and E and doubles its photoprotection of skin. *The Journal of Investigative Dermatology, 125*(4), 826–832. https://doi.org/10.1111/j.0022-202X.2005.23768.x
68. Tournas, J. A., Lin, F. H., Burch, J. A., et al. (2006). Ubiquinone, idebenone, and kinetin provide ineffective photoprotection to skin when compared to a topical antioxidant combination of vitamins C and E with ferulic acid. *The Journal of Investigative Dermatology, 126*(5), 1185–1187. https://doi.org/10.1038/sj.jid.5700232
69. Maia, A. M., Baby, A. R., Pinto, C. A. S. O., et al. (2006). Influence of sodium metabisulfite and glutathione on the stability of vitamin C in O/W emulsion and extemporaneous aqueous gel. *International Journal of Pharmaceutics, 322*(1–2), 130–135. https://doi.org/10.1016/j.ijpharm.2006.05.038
70. Milara, J., Serrano, G., Almudever, P., et al. (2015). Phosphatidylcholine liposomes as carriers to improve topical ascorbic acid treatment of skin disorders. *Clinical, Cosmetic and Investigational Dermatology*, 591. https://doi.org/10.2147/CCID.S90781
71. Hunt, M. J., & Barnetson, R. S. (1992). A comparative study of gluconolactone versus benzoyl peroxide in the treatment of acne. *The Australasian Journal of Dermatology, 33*(3), 131–134. https://doi.org/10.1111/j.1440-0960.1992.tb00100.x
72. Lodén, M. (2003). Role of topical emollients and moisturizers in the treatment of dry skin barrier disorders. *American Journal of Clinical Dermatology, 4*(11), 771–788. https://doi.org/10.2165/00128071-200304110-00005
73. Lee, H. E., Ko, J. Y., Kim, Y. H., et al. (2011). A double-blind randomized controlled comparison of APDDR-0901, a novel cosmeceutical formulation, and 0.1% adapalene gel in the treatment of mild-to-moderate acne vulgaris. *European Journal of Dermatology, 21*(6), 959–965. https://doi.org/10.1684/ejd.2011.1546
74. Kantikosum, K., Chongpison, Y., Chottawornsak, N., & Asawanonda, P. (2019). The efficacy of glycolic acid, salicylic acid, gluconolactone, and licochalcone A combined with 0.1% adapalene vs adapalene monotherapy in mild-to-moderate acne vulgaris: A double-blinded within-person comparative study. *Clinical, Cosmetic and Investigational Dermatology, 12*, 151–161. https://doi.org/10.2147/CCID.S193730
75. Parisi, N., Matts, P. J., Lever, R., Hadgraft, J., & Lane, M. E. (2017). Hexamidine salts—Applications in skin health and personal care products. *International Journal of Cosmetic Science, 39*(4), 361–365. https://doi.org/10.1111/ics.12392
76. Boskabady, M. H., Shafei, M. N., Saberi, Z., & Amini, S. (2011). Pharmacological effects of *Rosa damascena*. *Iranian Journal of Basic Medical Sciences, 14*(4), 295–307. Accessed September 21, 2024. https://www.ncbi.nlm.nih.gov/pmc/articles/PMC3586833/
77. Tumbas, V. T., Čanadanović-Brunet, J. M., Četojević-Simin, D. D., Ćetković, G. S., Đilas, S. M., & Gille, L. (2012). Effect of rosehip (*Rosa canina* L.) phytochemicals on stable free radicals and human cancer cells. *Journal of the Science of Food and Agriculture, 92*(6), 1273–1281. https://doi.org/10.1002/jsfa.4695

78. Green, B. A., Yu, R. J., & Van Scott, E. J. (2009). Clinical and cosmeceutical uses of hydroxyacids. *Clinics in Dermatology, 27*(5), 495–501. https://doi.org/10.1016/j.clindermatol.2009.06.023
79. Grimes, P. E., Green, B. A., Wildnauer, R. H., & Edison, B. L. (2004). The use of polyhydroxy acids (PHAs) in photoaged skin. *Cutis, 73*(2 Suppl), 3–13. Accessed September 20, 2024. https://pubmed.ncbi.nlm.nih.gov/15002656/
80. Rawlings, A. V., Davies, A., Carlomusto, M., et al. (1996). Effect of lactic acid isomers on keratinocyte ceramide synthesis, stratum corneum lipid levels and stratum corneum barrier function. *Archives of Dermatological Research, 288*(7), 383–390. https://doi.org/10.1007/BF02507107
81. Berardesca, E., Distante, F., Vignoli, G. P., Oresajo, C., & Green, B. (1997). Alpha hydroxyacids modulate stratum corneum barrier function. *The British Journal of Dermatology, 137*(6), 934–938. Accessed September 23, 2024. https://pubmed.ncbi.nlm.nih.gov/9470910/
82. Okano, Y., Abe, Y., Masaki, H., Santhanam, U., Ichihashi, M., & Funasaka, Y. (2003). Biological effects of glycolic acid on dermal matrix metabolism mediated by dermal fibroblasts and epidermal keratinocytes. *Experimental Dermatology, 12*(s2), 57–63. https://doi.org/10.1034/j.1600-0625.12.s2.9.x
83. Tang, S. C., & Yang, J. H. (2018). Dual effects of alpha-hydroxy acids on the skin. *Molecules, 23*(4), 863. https://doi.org/10.3390/molecules23040863
84. Fiume, M. M., Heldreth, B. A., Bergfeld, W. F., et al. (2014). Safety assessment of citric acid, inorganic citrate salts, and alkyl citrate esters as used in cosmetics. *International Journal of Toxicology, 33*(2_suppl), 16S–46S. https://doi.org/10.1177/1091581814526891
85. Burnett, C. L., Bergfeld, W. F., Belsito, D. V., et al. (2022). Amended safety assessment of malic acid and sodium malate as used in cosmetics. *International Journal of Toxicology, 41*(3_suppl), 69–76. https://doi.org/10.1177/10915818221117535
86. Chularojanamontri, L., Tuchinda, P., Kulthanan, K., & Pongparit, K. (2014). Moisturizers for acne: What are their constituents? *The Journal of Clinical and Aesthetic Dermatology, 7*(5), 36–44. Accessed September 18, 2024. https://www.ncbi.nlm.nih.gov/pmc/articles/PMC4025519/
87. Squibb, E. R. (1875). Note on salicylic acid. *Buffalo Medical and Surgical Journal, 14*(9), 337–344. Accessed September 21, 2024. https://www.ncbi.nlm.nih.gov/pmc/articles/PMC9398296/?page=8
88. White, J. W. (1882). Salicylic acid: Its history and uses. *Glasgow Medical Journal, 17*, 253–258. Accessed September 20, 2024. https://www.ncbi.nlm.nih.gov/pmc/articles/PMC5900674/?page=2
89. Kligman, D., & Kligman, A. M. (1998). Salicylic acid peels for the treatment of Photoaging. *Dermatologic Surgery, 24*(3), 325–328. https://doi.org/10.1111/j.1524-4725.1998.tb04162.x
90. Abdel-Motaleb, A. A., Abu-Dief, E. E., & Hussein, M. R. (2017). Dermal morphological changes following salicylic acid peeling and microdermabrasion. *Journal of Cosmetic Dermatology, 16*(4). https://doi.org/10.1111/jocd.12315
91. Lin, F. A. N., & Nakatsui, T. (1998). Salicylic acid revisited. *International Journal of Dermatology, 37*(5), 335–342. https://doi.org/10.1046/j.1365-4362.1998.00452.x
92. Sykes, E. M. E., White, D., McLaughlin, S., & Kumar, A. (2024). Salicylic acids and pathogenic bacteria: New perspectives on an old compound. *Canadian Journal of Microbiology, 70*(1), 1–14. https://doi.org/10.1139/cjm-2023-0123
93. Raeisi Estabragh, M. A., Sajadi Bami, M., Dehghannoudeh, G., Noudeh, Y. D., & Moghimipour, E. (2023). Cellulose derivatives and natural gums as gelling agents for preparation of Emulgel-based dosage forms: A brief review. *International Journal of Biological Macromolecules, 241*, 124538. https://doi.org/10.1016/j.ijbiomac.2023.124538
94. Jarząbek-Perz, S., Dziedzic, M., Rotsztejn, H., & Kołodziejczak, A. (2023). Evaluation of the effects of 10% and 30% gluconolactone chemical Peel on sebum, pH, and TEWL. *Journal of Cosmetic Dermatology, 22*(12), 3305–3312. https://doi.org/10.1111/jocd.15864

95. Cherian P. (2022). *Safety assessment of glycolactones as used in cosmetics*. Accessed September 23, 2024. https://cir-reports.cir-safety.org/view-attachment/?id=75069b71-f8e2-422b-a5fb-c6c1b6dabbf0
96. Blaskovich, M. A. T., Elliott, A. G., Kavanagh, A. M., Ramu, S., & Cooper, M. A. (2019). *In vitro* antimicrobial activity of acne drugs against skin-associated bacteria. *Scientific Reports, 9*(1), 14658. https://doi.org/10.1038/s41598-019-50746-4
97. Kucera, K., Zenzola, N., Hudspeth, A., et al. (2024). Evaluation of benzene presence and formation in benzoyl peroxide drug products. *Journal of Investigative Dermatology*. https://doi.org/10.1016/j.jid.2024.09.009
98. Qothrunnadaa, T., & Hasanah, A. N. (2021). Patches for acne treatment: An update on the formulation and stability test. *International Journal of Applied Pharmaceutics*, 21–26. https://doi.org/10.22159/ijap.2021.v13s4.43812
99. Kuo, C. W., Chiu, Y. F., Wu, M. H., et al. (2021). Gelatin/chitosan bilayer patches loaded with *cortex Phellodendron amurense/Centella asiatica* extracts for anti-acne application. *Polymers (Basel)., 13*(4), 579. https://doi.org/10.3390/polym13040579
100. Hall, J. B., Cong, Z., Imamura-Kawasawa, Y., et al. (2018). Isolation and identification of the follicular microbiome: Implications for acne research. *The Journal of Investigative Dermatology, 138*(9), 2033–2040. https://doi.org/10.1016/j.jid.2018.02.038
101. Wong, C. M., Guo, C., Scheufele, C. J., et al. (2024). Presentations of cutaneous disease in various skin pigmentations: Acne vulgaris - Comedonal acne. *HCA Healthcare Journal of Medicine., 5*(1). https://doi.org/10.36518/2689-0216.1634
102. Infante, V. H. P., Darvin, M. E., & Maia Campos, P. M. B. G. (2023). Characterization and efficacy of essential oil-based cosmetic formulations for acne-prone skin. *Cosmetics, 10*(6), 158. https://doi.org/10.3390/cosmetics10060158
103. Imokawa, G., Akasaki, S., Hattori, M., & Yoshizuka, N. (1986). Selective recovery of deranged water-holding properties by stratum Corneum lipids. *The Journal of Investigative Dermatology, 87*(6), 758–761. https://doi.org/10.1111/1523-1747.ep12456950
104. Raut, S., Azheruddin, M., Kumar, R., Singh, S., Giram, P. S., & Datta, D. (2024). Lecithin Organogel: A promising carrier for the treatment of skin diseases. *ACS Omega, 9*(9), 9865–9885. https://doi.org/10.1021/acsomega.3c05563
105. Wertz, P. W. (2009). Human synthetic sebum formulation and stability under conditions of use and storage. *International Journal of Cosmetic Science, 31*(1), 21–25. https://doi.org/10.1111/j.1468-2494.2008.00468.x
106. Fiume, M. M., Bergfeld, W. F., Belsito, D. V., et al. (2018). Safety assessment of tocopherols and tocotrienols as used in cosmetics. *International Journal of Toxicology, 37*(2_suppl), 61S–94S. https://doi.org/10.1177/1091581818794455
107. Lopes, L. B., VanDeWall, H., Li, H. T., et al. (2010). Topical delivery of lycopene using microemulsions: Enhanced skin penetration and tissue antioxidant activity. *Journal of Pharmaceutical Sciences, 99*(3), 1346–1357. https://doi.org/10.1002/jps.21929
108. O'Neill, A. M., Cavagnero, K. J., Seidman, J. S., et al. (2024). Genetic and functional analyses of *Cutibacterium acnes* isolates reveal the Association of a Linear Plasmid with skin inflammation. *The Journal of Investigative Dermatology, 144*(1), 116–124.e4. https://doi.org/10.1016/j.jid.2023.05.029
109. Yu, Y., Champer, J., Agak, G. W., Kao, S., Modlin, R. L., & Kim, J. (2016). Different *Propionibacterium acnes* Phylotypes induce distinct immune responses and express unique surface and secreted proteomes. *The Journal of Investigative Dermatology, 136*(11), 2221–2228. https://doi.org/10.1016/j.jid.2016.06.615
110. Shu, M., Wang, Y., Yu, J., et al. (2013). Fermentation of *Propionibacterium acnes*, a commensal bacterium in the human skin microbiome, as skin probiotics against methicillin-resistant *Staphylococcus aureus*. *PLoS One, 8*(2), e55380. https://doi.org/10.1371/journal.pone.0055380
111. Nakamura, K., O'Neill, A. M., Williams, M. R., et al. (2020). Short chain fatty acids produced by *Cutibacterium acnes* inhibit biofilm formation by *Staphylococcus epidermidis*. *Scientific Reports, 10*(1), 21237. https://doi.org/10.1038/s41598-020-77790-9

112. Shao, L., Li, T., Yang, S., et al. (2024). The prebiotic effects of fructooligosaccharides enhance the growth characteristics of *Staphylococcus epidermidis* and enhance the inhibition of *Staphylococcus aureus* biofilm formation. *International Journal of Cosmetic Science*. https://doi.org/10.1111/ics.13020
113. Lai, Y., Cogen, A. L., Radek, K. A., et al. (2010). Activation of TLR2 by a small molecule produced by *Staphylococcus epidermidis* increases antimicrobial defense against bacterial skin infections. *The Journal of Investigative Dermatology, 130*(9), 2211–2221. https://doi.org/10.1038/jid.2010.123
114. Liu, J., Yan, R., Zhong, Q., et al. (2015). The diversity and host interactions of *Propionibacterium acnes* bacteriophages on human skin. *The ISME Journal, 9*(9), 2078–2093. https://doi.org/10.1038/ismej.2015.47
115. *Phylabiotics. Clinical study*. (2024). Accessed October 1, 2024. https://phylabiotics.com/pages/clinical-study
116. Varma, Y., & Van Prooyen, N. (2024, February 20). *Compositions comprising Propionibacterium acnes bacteriophages for treating acne*. Accessed October 1, 2024. https://patentimages.storage.googleapis.com/1c/06/1c/53028413783e9d/US11903984.pdf
117. *Phylabiotics. 3 Step acne fighting phage system*. (2024). Accessed October 1, 2024. https://phylabiotics.com/products/3-step-acne-fighting-phage-system
118. Wang, Y., Viennet, C., Robin, S., Berthon, J. Y., He, L., & Humbert, P. (2017). Precise role of dermal fibroblasts on melanocyte pigmentation. *Journal of Dermatological Science, 88*(2), 159–166. https://doi.org/10.1016/j.jdermsci.2017.06.018
119. Iozumi, K., Hoganson, G. E., Pennella, R., Everett, M. A., & Fuller, B. B. (1993). Role of tyrosinase as the determinant of pigmentation in cultured human melanocytes. *The Journal of Investigative Dermatology, 100*(6), 806–811. https://doi.org/10.1111/1523-1747.ep12476630
120. Chang, T. S. (2012). Natural melanogenesis inhibitors acting through the down-regulation of tyrosinase activity. *Materials., 5*(9), 1661–1685. https://doi.org/10.3390/ma5091661
121. Zaid, A. N., & Al, R. R. (2019). Depigmentation and anti-aging treatment by natural molecules. *Current Pharmaceutical Design, 25*(20), 2292–2312. https://doi.org/10.2174/1381612825666190703153730
122. Huang, H. C., Lee, I. J., Huang, C., & Chang, T. M. (2020). Lactic acid bacteria and lactic acid for skin health and Melanogenesis inhibition. *Current Pharmaceutical Biotechnology, 21*(7), 566–577. https://doi.org/10.2174/1389201021666200109104701
123. Tung, X. Y., Yip, J. Q., & Gew, L. T. (2023). Searching for natural plants with Antimelanogenesis and Antityrosinase properties for cosmeceutical or Nutricosmetics applications: A systematic review. *ACS Omega, 8*(37), 33115–33201. https://doi.org/10.1021/acsomega.3c02994
124. Sood, A., Upadhyay, R., & Maurya, S. K. (2022). Sustainable production of azelaic acid from vegetable oils over a heterogeneous catalyst. *Industrial Crops and Products, 186*, 115139. https://doi.org/10.1016/j.indcrop.2022.115139
125. Akamatsu, H., Komura, J., Asada, Y., Miyachi, Y., & Niwa, Y. (1991). Inhibitory effect of azelaic acid on neutrophil functions: A possible cause for its efficacy in treating pathogenetically unrelated diseases. *Archives of Dermatological Research, 283*(3), 162–166. https://doi.org/10.1007/BF00372056
126. Shucheng, H., Zhou, X., Du, D., Li, J., Yu, C., & Jiang, X. (2024). Effects of 15% azelaic acid gel in the Management of Post-Inflammatory Erythema and Post-Inflammatory Hyperpigmentation in acne vulgaris. *Dermatology and Therapy, 14*(5), 1293–1314. https://doi.org/10.1007/s13555-024-01176-2
127. Cerulli, A., Masullo, M., Montoro, P., & Piacente, S. (2022). Licorice (*Glycyrrhiza glabra*, *G. uralensis*, and *G. inflata*) and their constituents as active cosmeceutical ingredients. *Cosmetics, 9*(1), 7. https://doi.org/10.3390/cosmetics9010007
128. Maeda, K., Zhou, Z., Guo, M., Zhang, J., Chen, L., & Yang, F. (2024). Functional properties and skin care effects of sodium Trehalose sulfate. *Skin Research and Technology, 30*(4). https://doi.org/10.1111/srt.13666

129. Bogdan Allemann, I., & Baumann, L. (2008). Antioxidants used in skin care formulations. *Skin Therapy Letter, 13*(7), 5–9. Accessed September 25, 2024. https://www.skintherapyletter.com/aging-skin/antioxidants/
130. Husein el Hadmed, H., & Castillo, R. F. (2016). Cosmeceuticals: Peptides, proteins, and growth factors. *Journal of Cosmetic Dermatology, 15*(4), 514–519. https://doi.org/10.1111/jocd.12229
131. Yi, K., Winayanuwattikun, W., Kim, S., et al. (2024). Skin boosters: Definitions and varied classifications. *Skin Research and Technology, 30*(3). https://doi.org/10.1111/srt.13627
132. Mijaljica, D., Townley, J. P., Spada, F., & Harrison, I. P. (2024). The heterogeneity and complexity of skin surface lipids in human skin health and disease. *Progress in Lipid Research, 93*, 101264. https://doi.org/10.1016/j.plipres.2023.101264
133. Dou, Y., Lee, A., Zhu, L., Morton, J., & Ladiges, W. (2020). The potential of GHK as an anti-aging peptide. *Aging Pathobiology and Therapeutics, 2*(1), 58–61. https://doi.org/10.31491/APT.2020.03.014
134. Tzellos, T. G., Klagas, I., Vahtsevanos, K., et al. (2009). Extrinsic ageing in the human skin is associated with alterations in the expression of hyaluronic acid and its metabolizing enzymes. *Experimental Dermatology, 18*(12), 1028–1035. https://doi.org/10.1111/j.1600-0625.2009.00889.x
135. Shin, J. W., Kwon, S. H., Choi, J. Y., et al. (2019). Molecular mechanisms of dermal aging and antiaging approaches. *International Journal of Molecular Sciences, 20*(9), 2126. https://doi.org/10.3390/ijms20092126
136. Nobile, V., Buonocore, D., Michelotti, A., & Marzatico, F. (2014). Anti-aging and filling efficacy of six types hyaluronic acid based dermo-cosmetic treatment: Double blind, randomized clinical trial of efficacy and safety. *Journal of Cosmetic Dermatology, 13*(4), 277–287. https://doi.org/10.1111/jocd.12120
137. Pavicic, T., Gauglitz, G. G., Lersch, P., et al. (2011). Efficacy of cream-based novel formulations of hyaluronic acid of different molecular weights in anti-wrinkle treatment. *Journal of Drugs in Dermatology, 10*(9), 990–1000. Accessed September 24, 2024. https://pubmed.ncbi.nlm.nih.gov/22052267/
138. Kawada, C., Yoshida, T., Yoshida, H., et al. (2014). Ingested hyaluronan moisturizes dry skin. *Nutrition Journal, 13*(1), 70. https://doi.org/10.1186/1475-2891-13-70
139. Gao, Y., Wang, R., Zhang, L., et al. (2023). Oral Ddministration of hyaluronic acid to improve skin conditions via a randomized double-blind clinical test. *Skin Research and Technology, 29*(11). https://doi.org/10.1111/srt.13531
140. Imokawa, G., Abe, A., Jin, K., Higaki, Y., Kawashima, M., & Hidano, A. (1991). Decreased level of ceramides in stratum corneum of atopic dermatitis: An etiologic factor in atopic dry skin? *The Journal of Investigative Dermatology, 96*(4), 523–526. https://doi.org/10.1111/1523-1747.ep12470233
141. Burnett, C. L., Boyer, I. J., Bergfeld, W. F., et al. (2020). Safety assessment of ceramides as used in cosmetics. *International Journal of Toxicology, 39*(3_suppl), 5S–25S. https://doi.org/10.1177/1091581820958692
142. Tanno, O., Ota, Y., Kitamura, N., Katsube, T., & Inoue, S. (2000). Nicotinamide increases biosynthesis of ceramides as well as other stratum corneum lipids to improve the epidermal permeability barrier. *The British Journal of Dermatology, 143*(3), 524–531. https://doi.org/10.1111/j.1365-2133.2000.03705.x
143. Fiume, M., Bergfeld, W. F., Belsito, D. V., et al. (2023). Squalane and squalene. *International Journal of Toxicology, 42*(3_suppl), 107S–109S. https://doi.org/10.1177/10915818231204276
144. Kim, S. K., & Karadeniz, F. (2012). Biological importance and applications of squalene and squalane. *Advances in Food and Nutrition., 65*, 223–233. https://doi.org/10.1016/B978-0-12-416003-3.00014-7
145. Yousefian, F., Espinoza, L., Yadlapati, S., Lorenc, Z. P., & Gold, M. (2024). A comprehensive review of the medical and cosmetic applications of exosomes in dermatology. *Journal of Cosmetic Dermatology, 23*(4), 1224–1228. https://doi.org/10.1111/jocd.16149

146. Tokura, Y., Yunoki, M., Kondo, S., & Otsuka, M. (2024). What is "eczema"? *The Journal of Dermatology*. https://doi.org/10.1111/1346-8138.17439
147. Thyssen, J. P., McFadden, J. P., & Kimber, I. (2014). The multiple factors affecting the association between atopic dermatitis and contact sensitization. *Allergy, 69*(1), 28–36. https://doi.org/10.1111/all.12358
148. Sidbury, R., Tom, W. L., Bergman, J. N., et al. (2014). Guidelines of care for the management of atopic dermatitis. *Journal of the American Academy of Dermatology, 71*(6), 1218–1233. https://doi.org/10.1016/j.jaad.2014.08.038
149. Zeldin, J., Ratley, G., Shobnam, N., & Myles, I. A. (2024). The clinical, mechanistic, and social impacts of air pollution on atopic dermatitis. *The Journal of Allergy and Clinical Immunology, 154*, 14. https://doi.org/10.1016/j.jaci.2024.07.027
150. Laughter, M. R., Maymone, M. B. C., Mashayekhi, S., et al. (2021). The global burden of atopic dermatitis: Lessons from the global burden of disease study 1990–2017. *The British Journal of Dermatology, 184*(2), 304–309. https://doi.org/10.1111/bjd.19580
151. Mallol, J., Crane, J., von Mutius, E., Odhiambo, J., Keil, U., & Stewart, A. (2013). The international study of asthma and allergies in childhood (ISAAC) phase three: A global synthesis. *Allergol Immunopathol (Madr), 41*(2), 73–85. https://doi.org/10.1016/j.aller.2012.03.001
152. Karimkhani, C., Colombara, D. V., Drucker, A. M., et al. (2017). The global burden of scabies: A cross-sectional analysis from the Global Burden of Disease Study 2015. *The Lancet Infectious Diseases, 17*(12), 1247–1254. https://doi.org/10.1016/S1473-3099(17)30483-8
153. Arents, B. W. M., van Zuuren, E. J., Hughes, O., Fedorowicz, Z., & Flohr, C. (2023). The future is now: The global atopic dermatitis atlas (GADA). *The British Journal of Dermatology, 189*(6), 761–763. https://doi.org/10.1093/bjd/ljad286
154. Pagliaro, M., Pecoraro, L., Stefani, C., Pieropan, S., Piacentini, G., & Pietrobelli, A. (2024). Bathing in atopic dermatitis in pediatric age: Why, how and when. *Pediatric Reports, 16*(1), 57–68. https://doi.org/10.3390/pediatric16010006
155. Eichenfield, L. F., Tom, W. L., Berger, T. G., et al. (2014). Guidelines of care for the management of atopic dermatitis. *Journal of the American Academy of Dermatology, 71*(1), 116–132. https://doi.org/10.1016/j.jaad.2014.03.023
156. Dumycz, K., Kunkiel, K., & Feleszko, W. (2019). Cosmetics for neonates and infants: Haptens in products' composition. *Clinical and Translational Allergy, 9*(1), 15. https://doi.org/10.1186/s13601-019-0257-8
157. Damron, J. M., & Shaker, M. (2020). Frequent versus infrequent bathing in pediatric atopic dermatitis: A randomized clinical trial. *Pediatrics, 146*(Suppl 4), S340–S340. https://doi.org/10.1542/peds.2020-023861FF
158. Santer, M., Ridd, M. J., Francis, N. A., et al. (2018). Emollient bath additives for the treatment of childhood eczema (BATHE): Multicentre pragmatic parallel group randomised controlled trial of clinical and cost effectiveness. *BMJ*, k1332. https://doi.org/10.1136/bmj.k1332
159. Kottner, J., Kanti, V., Dobos, G., et al. (2017). The effectiveness of using a bath oil to reduce signs of dry skin: A randomized controlled pragmatic study. *International Journal of Nursing Studies, 65*, 17–24. https://doi.org/10.1016/j.ijnurstu.2016.10.010
160. Skjerven, H. O., Rehbinder, E. M., Vettukattil, R., et al. (2020). Skin emollient and early complementary feeding to prevent infant atopic dermatitis (PreventADALL): A factorial, multicentre, cluster-randomised trial. *The Lancet., 395*(10228), 951–961. https://doi.org/10.1016/S0140-6736(19)32983-6
161. Bradshaw, L. E., Wyatt, L. A., Brown, S. J., et al. (2024). Emollient application from birth to prevent eczema in high-risk children: The BEEP RCT. *Health Technol Assess (Rockv), 28*, 1–116. https://doi.org/10.3310/RHDN9613
162. Garcia Bartels, N., Scheufele, R., Prosch, F., et al. (2010). Effect of standardized skin care regimens on neonatal skin barrier function in different body areas. *Pediatric Dermatology, 27*(1), 1–8. https://doi.org/10.1111/j.1525-1470.2009.01068.x

163. Inuzuka, Y., Natsume, O., Matsunaga, M., et al. (2020). Washing with water alone versus soap in maintaining remission of eczema. *Pediatrics International, 62*(6), 663–668. https://doi.org/10.1111/ped.14216
164. Katoh, Y., Natsume, O., Yasuoka, R., et al. (2024). Skin care by washing with water is not inferior to washing with a cleanser in children with atopic dermatitis in remission in summer: WASH study. *Allergology International, 73*(3), 428–435. https://doi.org/10.1016/j.alit.2024.01.007
165. Rakita, U., Kaundinya, T., & Silverberg, J. I. (2023). Associations between shower and moisturizing practices with atopic dermatitis severity: A prospective longitudinal cohort study. *Dermatitis, 34*(5), 425–431. https://doi.org/10.1089/derm.2022.29020.jis
166. Petras, D., Nothias, L. F., Quinn, R. A., et al. (2016). Mass spectrometry-based visualization of molecules associated with human habitats. *Analytical Chemistry, 88*(22), 10775–10784. https://doi.org/10.1021/acs.analchem.6b03456
167. Akdis, C. A. (2022). The epithelial barrier hypothesis proposes a comprehensive understanding of the origins of allergic and other chronic noncommunicable diseases. *The Journal of Allergy and Clinical Immunology, 149*(1), 41–44. https://doi.org/10.1016/j.jaci.2021.11.010
168. Shamsnajafabadi, H., & Soheili, Z. S. (2022). Amniotic fluid characteristics and its application in stem cell therapy: A review. *International Journal of Reproductive BioMedicine (IJRM)*. https://doi.org/10.18502/ijrm.v20i8.11752
169. Do, L. H. D., Azizi, N., & Maibach, H. (2020). Sensitive skin syndrome: An update. *American Journal of Clinical Dermatology, 21*(3), 401–409. https://doi.org/10.1007/s40257-019-00499-7
170. Chen, L., & Zheng, J. (2021). Does sensitive skin represent a skin condition or manifestations of other disorders? *Journal of Cosmetic Dermatology, 20*(7), 2058–2061. https://doi.org/10.1111/jocd.13829
171. Bieber, T. (2022). Atopic dermatitis: An expanding therapeutic pipeline for a complex disease. *Nature Reviews. Drug Discovery, 21*(1), 21–40. https://doi.org/10.1038/s41573-021-00266-6
172. Drucker, A. M., Lam, M., Prieto-Merino, D., et al. (2024). Systemic immunomodulatory treatments for atopic dermatitis. *JAMA Dermatology, 160*(9), 936. https://doi.org/10.1001/jamadermatol.2024.2192
173. Silverberg, J. I., & Hanifin, J. M. (2013). Adult eczema prevalence and associations with asthma and other health and demographic factors: A US population–based study. *The Journal of Allergy and Clinical Immunology, 132*(5), 1132–1138. https://doi.org/10.1016/j.jaci.2013.08.031
174. National Eczema Association. (2024). *The Ecz-clusion list*. Accessed September 1, 2024. https://nationaleczema.org/eczema-products/the-ecz-clusion-list/
175. Mowitz, M., Svedman, C., Zimerson, E., Isaksson, M., Pontén, A., & Bruze, M. (2017). Simultaneous patch testing with fragrance mix I, fragrance mix II and their ingredients in southern Sweden between 2009 and 2015. *Contact Dermatitis, 77*(5), 280–287. https://doi.org/10.1111/cod.12834
176. Ortiz, K. J., & Yiannias, J. A. (2004). Contact dermatitis to cosmetics, fragrances, and botanicals. *Dermatologic Therapy, 17*(3), 264–271. https://doi.org/10.1111/j.1396-0296.2004.04027.x
177. Chen, X., Sullivan, D. A., Sullivan, A. G., Kam, W. R., & Liu, Y. (2018). Toxicity of cosmetic preservatives on human ocular surface and adnexal cells. *Experimental Eye Research, 170*, 188–197. https://doi.org/10.1016/j.exer.2018.02.020
178. Nowak-Lange, M., Niedziałkowska, K., & Lisowska, K. (2022). Cosmetic preservatives: Hazardous micropollutants in need of greater attention? *International Journal of Molecular Sciences, 23*(22), 14495. https://doi.org/10.3390/ijms232214495
179. Luo, N., Chen, J., Chen, X., et al. (2024). Toxicity evolution of triclosan during environmental transformation and human metabolism: Misgivings in the post-pandemic era. *Environment International, 190*, 108927. https://doi.org/10.1016/j.envint.2024.108927

180. U.S. Food and Drug Administration. (2024). Skip the antibacterial soap; use plain soap and water. *Consumer Updates*. Accessed September 21, 2024. https://www.fda.gov/consumers/consumer-updates/skip-antibacterial-soap-use-plain-soap-and-water
181. Bunse, M., Daniels, R., Gründemann, C., et al. (2022). Essential oils as multicomponent mixtures and their potential for human health and Well-being. *Frontiers in Pharmacology, 13*, 13. https://doi.org/10.3389/fphar.2022.956541
182. de Groot, A. C. (2019). *Myroxylon pereirae* resin (balsam of Peru)—A critical review of the literature and assessment of the significance of positive patch test reactions and the usefulness of restrictive diets. *Contact Dermatitis, 80*(6), 335–353. https://doi.org/10.1111/cod.13263
183. Braakhuis, A. (2019). Evidence on the health benefits of supplemental propolis. *Nutrients, 11*(11), 2705. https://doi.org/10.3390/nu11112705
184. Rybczyńska-Tkaczyk, K., Grenda, A., Jakubczyk, A., Kiersnowska, K., & Bik-Małodzińska, M. (2023). Natural compounds with antimicrobial properties in cosmetics. *Pathogens., 12*(2), 320. https://doi.org/10.3390/pathogens12020320
185. Sławińska, N., Prochoń, K., & Olas, B. (2023). A review on berry seeds—A special emphasis on their chemical content and health-promoting properties. *Nutrients, 15*(6), 1422. https://doi.org/10.3390/nu15061422
186. Prosche, S., & Stappen, I. (2024). Flower power: An overview on chemistry and biological impact of selected essential oils from blossoms. *Planta Medica, 90*(07/08), 595–626. https://doi.org/10.1055/a-2215-2791
187. Burnett, C. L., Bergfeld, W. F., Belsito, D. V., et al. (2021). Safety assessment of *citrus* peel-derived ingredients as used in cosmetics. *International Journal of Toxicology, 40*(3 suppl), 77S–99S. https://doi.org/10.1177/10915818211032617
188. Friebe, K., Effendy, I., & Loffler, H. (2003). Effects of skin occlusion in patch testing with sodium lauryl sulphate. *The British Journal of Dermatology, 148*(1), 65–69. https://doi.org/10.1046/j.1365-2133.2003.04851.x
189. Presley, C. L., Militello, M., Barber, C., et al. (2021). The history of surfactants and review of their allergic and irritant properties. *Dermatitis, 32*(5), 289–297. https://doi.org/10.1097/DER.0000000000000730
190. de Wildt, S. N., Kearns, G. L., Leeder, J. S., & van den Anker, J. N. (1999). Glucuronidation in humans. *Clinical Pharmacokinetics, 36*(6), 439–452. https://doi.org/10.2165/00003088-199936060-00005
191. Richard, K., Hume, R., Kaptein, E., Stanley, E. L., Visser, T. J., & Coughtrie, M. W. H. (2001). Sulfation of thyroid hormone and dopamine during human development: Ontogeny of phenol sulfotransferases and arylsulfatase in liver, lung, and brain. *The Journal of Clinical Endocrinology and Metabolism, 86*(6), 2734–2742. https://doi.org/10.1210/jcem.86.6.7569
192. Blume-Peytavi, U., Hauser, M., Lünnemann, L., Stamatas, G. N., Kottner, J., & Garcia, B. N. (2014). Prevention of diaper dermatitis in infants—A literature review. *Pediatric Dermatology, 31*(4), 413–429. https://doi.org/10.1111/pde.12348
193. Dyer, O. (2022). Johnson and Johnson will discontinue talc based baby powder worldwide. *BMJ, 378*, o2046. https://doi.org/10.1136/bmj.o2046
194. Johnson & Johnson Consumer Health. (2022). *Johnson & Johnson consumer health to transition global baby powder portfolio to cornstarch*. Accessed September 29, 2024. https://www.jnj.com/media-center/press-releases/johnson-johnson-consumer-health-to-transition-global-baby-powder-portfolio-to-cornstarch
195. Kay, J. (2016). Johnson & Johnson removes some chemicals from baby shampoo, other products. *Scientific American*. Accessed September 29, 2024. https://www.scientificamerican.com/article/johnson-and-johnson-removes-some-chemicals-from-baby-shampoo-other-products/
196. Vukmanović, S., & Sadrieh, N. (2017). Skin sensitizers in cosmetics and beyond: Potential multiple mechanisms of action and importance of T-cell assays for *in vitro* screening. *Critical Reviews in Toxicology, 47*(5), 422–439. https://doi.org/10.1080/10408444.2017.1288025
197. Šikić Pogačar, M., Maver, U., Marčun Varda, N., & Mičetić-Turk, D. (2018). Diagnosis and management of diaper dermatitis in infants with emphasis on skin microbiota in the

diaper area. International Journal of Dermatology, 57(3), 265–275. https://doi.org/10.1111/ijd.13748

198. Hertiš Petek, T., Petek, M., Petek, T., & Marčun, V. N. (2022). Emerging links between microbiome composition and skin immunology in diaper dermatitis: A narrative review. *Children, 9*(1), 112. https://doi.org/10.3390/children9010112
199. Jenkins, B. A., & Belsito, D. V. (2023). Lanolin. *Dermatitis, 34*(1), 4–12. https://doi.org/10.1089/derm.2022.0002
200. Johnson, H. (2023). Lanolin: The 2023 American Contact Dermatitis Society allergen of the year. *Cutis, 112*(2). https://doi.org/10.12788/cutis.0825
201. Russell, M. F., Sandhu, M., Vail, M., Haran, C., Batool, U., & Leo, J. (2024). Tallow, rendered animal fat, and its biocompatibility with skin: A scoping review. *Cureus*. https://doi.org/10.7759/cureus.60981
202. Fiume, M. M., Bergfeld, W. F., Belsito, D. V., et al. (2013). Safety assessment of bis-diglyceryl polyacyladipate-2 and bis-diglyceryl polyacyladipate-1 as used in cosmetics. *International Journal of Toxicology, 32*(5 suppl), 56S–64S. https://doi.org/10.1177/1091581813504596
203. Norberg, S., Alander, J., & Jacobsson, R. (2005). *Lanolin substitute based on shea*. Accessed September 8, 2024. https://patentimages.storage.googleapis.com/6f/8d/29/197c794d0c53cf/US20050233015A1.pdf
204. Kunyk, O., Leal Filho, W., & Pasichnyi, V. (2024). Study of the equivalence of substituting lanolin with a plant-based alternative in cosmetic products. *Sustainable Chemistry and Pharmacy, 39*, 101555. https://doi.org/10.1016/j.scp.2024.101555
205. Naangmenyele, Z., Banye, E. Z., Bekoe, E. M. O., Zakaria, I., & Amuah, E. E. Y. (2023). Fundamentals of shea butter production; input–output analyses and profit maximization in northern Ghana. *Research in Globalization., 6*, 100113. https://doi.org/10.1016/j.resglo.2023.100113
206. Burnett, C. L. (2024). *Amended safety assessment of lanolin-derived ingredients as used in cosmetics*. Accessed September 24, 2024. https://cir-reports.cir-safety.org/view-attachment/?id=fd11d33d-1cf7-f5c2-b131-d6a2e397b0c4
207. Jackson, K. T., & Dennis, C. (2017). Lanolin for the treatment of nipple pain in breastfeeding women: A randomized controlled trial. *Maternal & Child Nutrition, 13*(3). https://doi.org/10.1111/mcn.12357
208. Matthieu, L., & Dockx, P. (1997). Discrepancy in patch test results with wool wax alcohols and Amerchol® L-101. *Contact Dermatitis, 36*(3), 150–151. https://doi.org/10.1111/j.1600-0536.1997.tb00398.x
209. Silverberg, J. I., Patel, N., Warshaw, E. M., et al. (2022). Lanolin allergic reactions: North American contact dermatitis group experience, 2001 to 2018. *Dermatitis, 33*(3), 193–199. https://doi.org/10.1097/DER.0000000000000871
210. Sugai, T., & Higashi, J. (1975). Hypersensitivity to hydrogenated lanolin. *Contact Dermatitis, 1*(3), 146–157. https://doi.org/10.1111/j.1600-0536.1975.tb05356.x
211. Bourdillon, K., McCausland, T., & McCabe, M. (2023). Multi-residue analysis of certain lanolin nipple care products for trace contaminants. *BMC Chemistry, 17*(1), 8. https://doi.org/10.1186/s13065-023-00919-0
212. Perez-Parada, A., Colazzo, M., Besil, N., et al. (2011). Pesticide residues in natural products with pharmaceutical use: Occurrence, analytical advances and perspectives. In *Pesticides in the modern world - Trends in pesticides analysis* (pp. 357–390). InTech. https://doi.org/10.5772/16489
213. Andersen, K. E., & Mose, K. F. (2016). Preservatives. In *Quick guide to contact dermatitis* (pp. 147–157). Springer. https://doi.org/10.1007/978-3-662-47714-4_13
214. Schaefer, K. (2008). Heating up the preservative debate. *Cosmetics & Toiletries*.
215. Nowak, K., Jabłońska, E., & Ratajczak-Wrona, W. (2021). Controversy around parabens: Alternative strategies for preservative use in cosmetics and personal care products. *Environmental Research, 198*, 110488. https://doi.org/10.1016/j.envres.2020.110488

216. Rau, A. H. (2001). Multi-sensory technologies for today's effervescent bath and shower products. *Cosmetics & Toiletries., 116*(10), 49–56.
217. Kabara, J. J., & Brady, M. B. (1984). Contamination of bar soaps under "in-use" conditions. *Journal of Environmental Pathology, Toxicology and Oncology, 5*(4–5), 1–14.
218. McBride, M. E. (1984). Microbial flora of in-use soap products. *Applied and Environmental Microbiology, 48*(2), 338–341. https://doi.org/10.1128/aem.48.2.338-341.1984
219. Heinze, J. E., & Yackovich, F. (1988). Washing with contaminated bar soap is unlikely to transfer bacteria. *Epidemiology and Infection, 101*(1), 135–142. https://doi.org/10.1017/S0950268800029290
220. Blaser, M. J. (2014). *Missing microbes: How the overuse of antibiotics is fueling our modern plagues*. Henry Holt and Co.
221. Herman, A. (2019). Antimicrobial ingredients as preservative booster and components of self-preserving cosmetic products. *Current Microbiology, 76*(6), 744–754. https://doi.org/10.1007/s00284-018-1492-2
222. Bilal, M., & Iqbal, H. M. N. (2019). An insight into toxicity and human-health-related adverse consequences of cosmeceuticals - A review. *Science of The Total Environment., 670*, 555–568. https://doi.org/10.1016/j.scitotenv.2019.03.261
223. Aerts, O., Verhulst, L., & Goossens, A. (2016). Ethylhexylglycerin: A low-risk, but highly relevant, sensitizer in 'hypo-allergenic' cosmetics. *Contact Dermatitis, 74*(5), 281–288. https://doi.org/10.1111/cod.12546
224. Steinberg, D. C. (2010). Frequency of preservative use. *Cosmetics & Toiletries., 125*(11), 46–51.
225. Dréno, B., Zuberbier, T., Gelmetti, C., Gontijo, G., & Marinovich, M. (2019). Safety review of phenoxyethanol when used as a preservative in cosmetics. *Journal of the European Academy of Dermatology and Venereology, 33*(S7), 15–24. https://doi.org/10.1111/jdv.15944
226. Lilienblum, W. (2016). Opinion of the scientific committee on consumer safety (SCCS)—Final version of the opinion on phenoxyethanol in cosmetic products. *Regulatory Toxicology and Pharmacology, 82*, 156. https://doi.org/10.1016/j.yrtph.2016.11.007
227. Kwa, M., Welty, L. J., & Xu, S. (2017). Adverse events reported to the US Food and Drug Administration for cosmetics and personal care products. *JAMA Internal Medicine, 177*(8), 1202. https://doi.org/10.1001/jamainternmed.2017.2762
228. Janetos, T. M., Akintilo, L., & Xu, S. (2019). Overview of high-risk Food and Drug Administration recalls for cosmetics and personal care products from 2002 to 2016. *Journal of Cosmetic Dermatology, 18*(5), 1361–1365. https://doi.org/10.1111/jocd.12824
229. Akgul, Ö., & Bakan, K. (2021). Aerobic bacteria isolated from used cosmetic products and evaluation of antibiotic resistance. *Ankara Universitesi Eczacilik Fakultesi Dergisi*. https://doi.org/10.33483/jfpau.850561
230. Werner, Y., & Lindberg, M. (1985). Transepidermal water loss in dry and clinically normal skin in patients with atopic dermatitis. *Acta Dermato-Venereologica, 65*(2), 102–105. https://doi.org/10.2340/0001555565102105
231. Agner, T., & Serup, J. (1990). Sodium lauryl sulphate for irritant patch testing – A dose-response study using bioengineering methods for determination of skin irritation. *The Journal of Investigative Dermatology, 95*(5), 543–547. https://doi.org/10.1111/1523-1747.ep12504896
232. de Jongh, C. M., Jakasa, I., Verberk, M. M., & Kezic, S. (2006). Variation in barrier impairment and inflammation of human skin as determined by sodium lauryl sulphate penetration rate. *The British Journal of Dermatology, 154*(4), 651–657. https://doi.org/10.1111/j.1365-2133.2005.06989.x
233. Jakasa, I., De Jongh, C. M., Verberk, M. M., Bos, J. D., & Kežić, S. (2006). Percutaneous penetration of sodium lauryl sulphate is increased in uninvolved skin of patients with atopic dermatitis compared with control subjects. *The British Journal of Dermatology, 155*(1), 104–109. https://doi.org/10.1111/j.1365-2133.2006.07319.x

234. McFadden, J. P., Holloway, D. B., Whittle, E. G., & Basketter, D. A. (2000). Benzalkonium chloride neutralizes the irritant effect of sodium dodecyl sulfate. *Contact Dermatitis, 43*(5), 264–266. https://doi.org/10.1034/j.1600-0536.2000.043005264.x
235. Ananthapadmanabhan, K. P., Moore, D. J., Subramanyan, K., Misra, M., & Meyer, F. (2004). Cleansing without compromise: The impact of cleansers on the skin barrier and the technology of mild cleansing. *Dermatologic Therapy, 17*(s1), 16–25. https://doi.org/10.1111/j.1396-0296.2004.04S1002.x
236. Fiume, M. M., Heldreth, B., Bergfeld, W. F., et al. (2013). Safety assessment of decyl glucoside and other alkyl glucosides as used in cosmetics. *International Journal of Toxicology, 32*(5 suppl), 22S–48S. https://doi.org/10.1177/1091581813497764
237. Burnett CL, Bergfeld WF, Belsito D V., et al. Safety assessment of alkyl betaines as used in cosmetics. International Journal of Toxicology 2018;37(1 suppl):28S-46S. doi:https://doi.org/10.1177/1091581818773354.
238. Abrutyn, E. S. (2020, November 1). Formulas deciphered: Facial wash. *Cosmetics & Toiletries*. Accessed August 27, 2024. https://www.cosmeticsandtoiletries.com/formulas-products/skin-care/article/21834620/formulas-deciphered-facial-wash
239. Huanbutta, K., Sripirom, P., Phetthong, P., et al. (2023). Dissolvable shower gel tablets with enhanced skin benefits. *International Journal of Cosmetic Science, 45*(6), 739–748. https://doi.org/10.1111/ics.12882
240. Yorke, K., Potanin, A., Jogun, S., Morgan, A., Shen, H., & Amin, S. (2021). High-performance sulphate-free cleansers: Surface activity, foaming and rheology. *International Journal of Cosmetic Science, 43*(6), 636–652. https://doi.org/10.1111/ics.12740
241. Maxwell, R., Costache, M. C., Giarrosso, A., Bosques, C., & Amin, S. (2021). Optimizing interactions between soluble silk fibroin and capryl glucoside for design of a natural and high-performance co-surfactant system. *International Journal of Cosmetic Science, 43*(1), 68–77. https://doi.org/10.1111/ics.12676
242. Mactaggart, E., Orchard, D., & Mui, T. M. (2021). Baby wipes and nappy rash—What is the relationship? A review. *The Australasian Journal of Dermatology, 62*(4), 470–477. https://doi.org/10.1111/ajd.13715
243. Yu, J., Treat, J., Chaney, K., & Brod, B. (2016). Potential allergens in disposable diaper wipes, topical diaper preparations, and disposable diapers. *Dermatitis, 27*(3), 110–118. https://doi.org/10.1097/DER.0000000000000177
244. Becker, L. C. (2015). *Safety assessment of polysorbates as used in cosmetics*. Accessed September 16, 2024. https://cir-reports.cir-safety.org/view-attachment/?id=657faea7-8c74-ec11-8943-0022482f06a6
245. Rodriguez, K. J., Cunningham, C., Foxenberg, R., Hoffman, D., & Vongsa, R. (2020). The science behind wet wipes for infant skin: Ingredient review, safety, and efficacy. *Pediatric Dermatology, 37*(3), 447–454. https://doi.org/10.1111/pde.14112
246. Aschenbeck, K. A., & Warshaw, E. M. (2017). Allergenic ingredients in facial wet wipes. *Dermatitis, 28*(6), 353–359. https://doi.org/10.1097/DER.0000000000000268
247. Chang, M. W., & Nakrani, R. (2014). Six children with allergic contact dermatitis to Methylisothiazolinone in wet wipes (Baby wipes). *Pediatrics, 133*(2), e434–e438. https://doi.org/10.1542/peds.2013-1453
248. Guin, J., Kincannon, J., & Church, F. (2001). Baby-wipe dermatitis: Preservative-induced hand eczema in parents and persons using moist towelettes. *American Journal of Contact Dermatitis, 12*(4), 189–192. https://doi.org/10.1053/ajcd.2001.28052
249. Nelson, S. A., & Yiannias, J. A. (2009). Relevance and avoidance of skin-care product allergens: Pearls and pitfalls. *Dermatologic Clinics, 27*(3), 329–336. https://doi.org/10.1016/j.det.2009.05.009
250. Wallen-Russell, C. (2018). The role of every-day cosmetics in altering the skin microbiome: A study using biodiversity. *Cosmetics, 6*(1), 2. https://doi.org/10.3390/cosmetics6010002
251. Jolly, A., Kim, H., Moon, J. Y., Mohan, A., & Lee, Y. C. (2023). Exploring the imminent trends of saponins in personal care product development: A review. *Industrial Crops and Products, 205*, 117489. https://doi.org/10.1016/j.indcrop.2023.117489

252. Hellweg, T., Sottmann, T., & Oberdisse, J. (2023). Recent advances in biosurfactant-based association colloids—Self-assembly in water. *Frontiers in Soft Matter, 2*, 2. https://doi.org/10.3389/frsfm.2022.1081877
253. Karnwal, A., Shrivastava, S., Al-Tawaha, A. R. M. S., et al. (2023). Microbial biosurfactant as an alternate to chemical surfactants for application in cosmetics Industries in Personal and Skin Care Products: A critical review. *BioMed Research International, 2023*(1). https://doi.org/10.1155/2023/2375223
254. Myles, I. A., Castillo, C. R., Barbian, K. D., et al. (2020). Therapeutic responses to *Roseomonas mucosa* in atopic dermatitis may involve lipid-mediated TNF-related epithelial repair. *Science Translational Medicine, 12*(560). https://doi.org/10.1126/scitranslmed.aaz8631
255. Capone, K. A., Dowd, S. E., Stamatas, G. N., & Nikolovski, J. (2011). Diversity of the human skin microbiome early in life. *The Journal of Investigative Dermatology, 131*(10), 2026–2032. https://doi.org/10.1038/jid.2011.168
256. Fiume, M. M., Heldreth, B. A., Bergfeld, W. F., et al. (2015). Safety assessment of alkyl esters as used in cosmetics. *International Journal of Toxicology, 34*(2 suppl), 5S–69S. https://doi.org/10.1177/1091581815594027
257. Khan, B. A., Mahmood, T., Menaa, F., et al. (2019). New perspectives on the efficacy of gallic acid in cosmetics & nanocosmeceuticals. *Current Pharmaceutical Design, 24*(43), 5181–5187. https://doi.org/10.2174/1381612825666190118150614
258. Rizwan, K., Zubair, M., Rasool, N., Riaz, M., Zia-Ul-Haq, M., & De Feo, V. (2012). Phytochemical and biological studies of *Agave attenuata. International Journal of Molecular Sciences, 13*(5), 6440–6451. https://doi.org/10.3390/ijms13056440
259. Duda-Madej, A., Stecko, J., Sobieraj, J., Szymańska, N., & Kozłowska, J. (2022). Naringenin and its derivatives—Health-promoting phytobiotic against resistant bacteria and fungi in humans. *Antibiotics, 11*(11), 1628. https://doi.org/10.3390/antibiotics11111628
260. Sidana, J., Singh, B., & Sharma, O. P. (2016). Saponins of agave: Chemistry and bioactivity. *Phytochemistry, 130*, 22–46. https://doi.org/10.1016/j.phytochem.2016.06.010
261. Zhu, Z., Wen, Y., Yi, J., Cao, Y., Liu, F., & McClements, D. J. (2019). Comparison of natural and synthetic surfactants at forming and stabilizing nanoemulsions: Tea saponin, Quillaja saponin, and Tween 80. *Journal of Colloid and Interface Science, 536*, 80–87. https://doi.org/10.1016/j.jcis.2018.10.024
262. Kareru, P., Keriko, J., Gachanja, A., & Kenji, G. (2008). Direct detection of triterpenoid Saponins in medicinal plants. *African Journal of Traditional, Complementary, and Alternative Medicines, 5*(1). https://doi.org/10.4314/ajtcam.v5i1.31257
263. Österlund, C., Hrapovic, N., Lafon-Kolb, V., Amini, N., Smiljanic, S., & Visdal-Johnsen, L. (2023). Protective effects of naringenin against UVB irradiation and air pollution-induced skin aging and pigmentation. *Cosmetics, 10*(3), 88. https://doi.org/10.3390/cosmetics10030088
264. Byrd, A. L., Belkaid, Y., & Segre, J. A. (2018). The human skin microbiome. *Nature Reviews. Microbiology, 16*(3), 143–155. https://doi.org/10.1038/nrmicro.2017.157
265. Lousada, M. B., Lachnit, T., Edelkamp, J., et al. (2021). Exploring the human hair follicle microbiome. *The British Journal of Dermatology, 184*(5), 802–815. https://doi.org/10.1111/bjd.19461
266. Kong, H. H., Oh, J., Deming, C., et al. (2012). Temporal shifts in the skin microbiome associated with disease flares and treatment in children with atopic dermatitis. *Genome Research, 22*(5), 850–859. https://doi.org/10.1101/gr.131029.111
267. Nakatsuji, T., Chen, T. H., Narala, S., et al. (2017). Antimicrobials from human skin commensal bacteria protect against *Staphylococcus aureus* and are deficient in atopic dermatitis. *Science Translational Medicine, 9*(378). https://doi.org/10.1126/scitranslmed.aah4680
268. Boyle, R. J., Bath-Hextall, F. J., Leonardi-Bee, J., Murrell, D. F., & Tang, M. L. (2009). Probiotics for the treatment of eczema: A systematic review. *Clinical and Experimental Allergy, 39*(8), 1117–1127. https://doi.org/10.1111/j.1365-2222.2009.03305.x

269. Zuccotti, G., Meneghin, F., Aceti, A., et al. (2015). Probiotics for prevention of atopic diseases in infants: Systematic review and meta-analysis. *Allergy, 70*(11), 1356–1371. https://doi.org/10.1111/all.12700
270. Yu, J., Ma, X., Wang, X., et al. (2022). Application and mechanism of probiotics in skin care: A review. *Journal of Cosmetic Dermatology, 21*(3), 886–894. https://doi.org/10.1111/jocd.14734
271. Theodorou, I. M., Kapoukranidou, D., Theodorou, M., et al. (2024). Cosmeceuticals: A review of clinical studies claiming to contain specific, well-characterized strains of probiotics or Postbiotics. *Nutrients, 16*(15), 2526. https://doi.org/10.3390/nu16152526
272. Foster, K. R., Schluter, J., Coyte, K. Z., & Rakoff-Nahoum, S. (2017). The evolution of the host microbiome as an ecosystem on a leash. *Nature, 548*(7665), 43–51. https://doi.org/10.1038/nature23292
273. Akdis, C. A. (2021). Does the epithelial barrier hypothesis explain the increase in allergy, autoimmunity and other chronic conditions? *Nature Reviews Immunology, 21*(11), 739–751. https://doi.org/10.1038/s41577-021-00538-7
274. Clemente, J. C., Pehrsson, E. C., Blaser, M. J., et al. (2015). The microbiome of uncontacted Amerindians. *Science Advances, 1*(3). https://doi.org/10.1126/sciadv.1500183
275. Oh, J., Byrd, A. L., Deming, C., Conlan, S., Kong, H. H., & Segre, J. A. (2014). Biogeography and individuality shape function in the human skin metagenome. *Nature, 514*(7520), 59–64. https://doi.org/10.1038/nature13786
276. Blaser, M. J., & Falkow, S. (2009). What are the consequences of the disappearing human microbiota? *Nature Reviews. Microbiology, 7*(12), 887–894. https://doi.org/10.1038/nrmicro2245
277. Wallen-Russell, C., & Wallen-Russell, S. (2017). Meta analysis of skin microbiome: New link between skin microbiota diversity and skin health with proposal to use this as a future mechanism to determine whether cosmetic products damage the skin. *Cosmetics, 4*(2), 14. https://doi.org/10.3390/cosmetics4020014
278. Whitlock, D. R., & Feelisch, M. (2009). Soil bacteria, nitrite and the skin. In *The hygiene hypothesis and Darwinian medicine* (pp. 103–115). Birkhäuser Basel. https://doi.org/10.1007/978-3-7643-8903-1_6
279. Cavagnaro, T. R., Jackson, L. E., Hristova, K., & Scow, K. M. (2008). Short-term population dynamics of ammonia oxidizing bacteria in an agricultural soil. *Applied Soil Ecology, 40*(1), 13–18. https://doi.org/10.1016/j.apsoil.2008.02.006
280. Kikuchi, S., Fujitani, H., Ishii, K., Isshiki, R., Sekiguchi, Y., & Tsuneda, S. (2023). Characterisation of bacteria representing a novel *Nitrosomonas* clade: Physiology, genomics and distribution of missing ammonia oxidizer. *Environmental Microbiology Reports, 15*(5), 404–416. https://doi.org/10.1111/1758-2229.13158
281. Arp, D., Sayavedra-Soto, L., & Hommes, N. (2002). Molecular biology and biochemistry of ammonia oxidation by *Nitrosomonas europaea*. *Archives of Microbiology, 178*(4), 250–255. https://doi.org/10.1007/s00203-002-0452-0
282. Whitlock, D. R., Jamas, S., & Weiss, L. (2018, July 10). *Ammonia oxidizing bacteria for treatment of eczema*. Accessed October 21, 2024. https://patentimages.storage.googleapis.com/51/93/08/1884f4a48d0b91/US10017731.pdf
283. Whitlock, D. R., Jamas, S., Weiss, L., & Gryllos, L. (2024, September 17). *Ammonia-oxidizing Nitrosomonas eutropha strain D23*. Accessed October 21, 2024. https://patents.google.com/patent/US12091652B2/en?oq=12%2c091%2c652%2c
284. Maura, D., Elmekki, N., & Goddard, C. A. (2021). The ammonia oxidizing bacterium *Nitrosomonas eutropha* blocks T helper 2 cell polarization via the anti-inflammatory cytokine IL-10. *Scientific Reports, 11*(1), 14162. https://doi.org/10.1038/s41598-021-93299-1
285. Silverberg, J. I., Lio, P. A., Simpson, E. L., et al. (2023). Efficacy and safety of topically applied therapeutic ammonia oxidising bacteria in adults with mild-to-moderate atopic dermatitis and moderate-to-severe pruritus: A randomised, double-blind, placebo-controlled,

dose-ranging, phase 2b trial. *EClinicalMedicine., 60*, 102002. https://doi.org/10.1016/j.eclinm.2023.102002
286. Notay, M., Saric-Bosanac, S., Vaughn, A. R., et al. (2020). The use of topical *Nitrosomonas eutropha* for cosmetic improvement of facial wrinkles. *Journal of Cosmetic Dermatology, 19*(3), 689–693. https://doi.org/10.1111/jocd.13060
287. AOBiome Therapeutics. *AOBiome cosmetics FAQ*. Accessed October 21, 2024. https://aobiomecosmetics.com/
288. AOBiome Cosmetics. *AOBiome cosmetics products D23*. Accessed October 21, 2024. https://aobiomecosmetics.com/products/d23
289. Bozzo, P., Chua-Gocheco, A., & Einarson, A. (2011). Safety of skin care products during pregnancy. *Canadian Family Physician, 57*(6), 665–667. Accessed September 18, 2024. https://www.ncbi.nlm.nih.gov/pmc/articles/PMC3114665/
290. Mutanda, T., Mokoena, M. P., Olaniran, A. O., Wilhelmi, B. S., & Whiteley, C. G. (2014). Microbial enzymatic production and applications of short-chain fructooligosaccharides and inulooligosaccharides: Recent advances and current perspectives. *Journal of Industrial Microbiology & Biotechnology, 41*(6), 893–906. https://doi.org/10.1007/s10295-014-1452-1
291. Fiume, M. M., Heldreth, B., Bergfeld, W. F., et al. (2016). Safety assessment of microbial polysaccharide gums as used in cosmetics. *International Journal of Toxicology, 35*(1 suppl), 5S–49S. https://doi.org/10.1177/1091581816651606
292. Guéniche, A., Bastien, P., Ovigne, J. M., et al. (2010). *Bifidobacterium longum* lysate, a new ingredient for reactive skin. *Experimental Dermatology, 19*(8). https://doi.org/10.1111/j.1600-0625.2009.00932.x
293. Jung, Y. O., Jeong, H., Cho, Y., et al. (2019). Lysates of a probiotic, *lactobacillus rhamnosus*, can improve skin barrier function in a reconstructed human epidermis model. *International Journal of Molecular Sciences, 20*(17), 4289. https://doi.org/10.3390/ijms20174289
294. Cherian P. (2024). *Safety assessment of yeast-derived ingredients as used in cosmetics*. Accessed October 18, 2024. https://cir-reports.cir-safety.org/view-attachment/?id=5cf8f722-d0c7-ee72-9e81-4ddc60723a28
295. Raj P. (2022). *Safety assessment of radish root-derived ingredients as used in cosmetics*. Accessed October 18, 2024. https://cir-reports.cir-safety.org/view-attachment/?id=6557fa47-aa02-9a9d-a8c5-c63ae9d92683
296. Yan, X., Tsuji, G., Hashimoto-Hachiya, A., & Furue, M. (2022). *Galactomyces* ferment filtrate potentiates an anti-Inflammaging system in keratinocytes. *Journal of Clinical Medicine, 11*(21), 6338. https://doi.org/10.3390/jcm11216338
297. Lee, W. A., Moreira Keupp, G., Brieva, H., & Warren, M. R. (2010, May 27). *Compositions containing extracts from radish*. Accessed October 18, 2024. https://patents.google.com/patent/US20100129305A1/en?oq=US+20100129305+A1
298. Scholz, D., & Segura, E. (2020, August 11). *Method for the treatment of acne*. Accessed August 27, 2024. https://patentimages.storage.googleapis.com/b6/42/2f/8fe232c0387d87/US10736938.pdf
299. Li, J., Chaytor, J. L., Findlay, B., McMullen, L. M., Smith, D. C., & Vederas, J. C. (2015). Identification of didecyldimethylammonium salts and salicylic acid as antimicrobial compounds in commercial fermented radish kimchi. *Journal of Agricultural and Food Chemistry, 63*(11), 3053–3058. https://doi.org/10.1021/jf5063588
300. Okeke, C. A., Khanna, R., & Ehrlich, A. (2023). Quaternary ammonium compounds and contact dermatitis: A review and considerations during the COVID-19 pandemic. *Clinical, Cosmetic and Investigational Dermatology, 16*, 1721–1728. https://doi.org/10.2147/CCID.S410910
301. Swanson, K. S., Gibson, G. R., Hutkins, R., et al. (2020). The international scientific Association for Probiotics and Prebiotics (ISAPP) consensus statement on the definition and scope of synbiotics. *Nature Reviews. Gastroenterology & Hepatology, 17*(11), 687–701. https://doi.org/10.1038/s41575-020-0344-2

302. Ma'or, Z., Temmerman, R., & Zhang, X. (2023). Topical application of Synbiotic *bacillus* preparations positively affects skin (micro) biology. *Journal of Cosmetics, Dermatological Sciences and Applications., 13*(02), 107–123. https://doi.org/10.4236/jcdsa.2023.132011
303. Tong, Y., Guo, H., Abbas, Z., et al. (2023). Optimizing postbiotic production through solid-state fermentation with *bacillus amyloliquefaciens* J and *Lactiplantibacillus plantarum* SN4 enhances antibacterial, antioxidant, and anti-inflammatory activities. *Frontiers in Microbiology, 14*, 14. https://doi.org/10.3389/fmicb.2023.1229952
304. Nakatsuji, T., Hata, T. R., Tong, Y., et al. (2021). Development of a human skin commensal microbe for bacteriotherapy of atopic dermatitis and use in a phase 1 randomized clinical trial. *Nature Medicine, 27*(4), 700–709. https://doi.org/10.1038/s41591-021-01256-2
305. Nizioł-Łukaszewska, Z., Bujak, T., Wasilewski, T., & Szmuc, E. (2019). Inulin as an effectiveness and safe ingredient in cosmetics. *Polish Journal of Chemical Technology, 21*(1), 44–49. Accessed September 16, 2024. https://intapi.sciendo.com/pdf/10.2478/pjct-2019-0008
306. Taylor, E. & Hanzel, D. (2021, December 18). *Compositions and methods for treatment of skin disorders*. Accessed October 18, 2024. https://patentimages.storage.googleapis.com/22/d0/ae/711a61a5aed463/US11207357.pdf
307. Bouslimani, A., da Silva, R., Kosciolek, T., et al. (2019). The impact of skin care products on skin chemistry and microbiome dynamics. *BMC Biology, 17*(1), 47. https://doi.org/10.1186/s12915-019-0660-6
308. Bouslimani, A., Porto, C., Rath, C. M., et al. (2015). Molecular cartography of the human skin surface in 3D. *Proceedings of the National Academy of Sciences, 112*(17). https://doi.org/10.1073/pnas.1424409112
309. HelloBiome. *HelloBiome*. Accessed October 15, 2024. https://hellobio.me/
310. Sequential. *Test your microbiome*. Accessed October 2, 2024. https://www.sequential.bio/test-your-microbiome
311. Gallinee. *Skin health testing kits*. Accessed October 2, 2024. https://www.gallinee.com/product/skin-health-testing-kit/
312. Sequential. *Our testing platform*. Accessed October 15, 2024. https://www.sequential.bio/ourtestingplatform
313. MyMicrobiome AG. *Cosmetic standards*. Accessed October 2, 2024. https://microbiome--friendly.com/en/cosmetic-standards
314. MyMicrobiome AG. *MyMicrobiome worldwide*. Accessed October 2, 2024. https://microbiome-friendly.com/en/mymicrobiome-worldwide
315. MyMicrobiome AG. *Certification/seals*. Accessed October 15, 2024. https://microbiome--friendly.com/en/certification/seals
316. MyMicrobiome AG. *Microbiome-friendly*. Accessed October 15, 2024. https://microbiome--friendly.com/en/
317. MyMicrobiome AG. *Tested and certified microbiome-friendly products*. Accessed October 2, 2024. https://www.mymicrobiome.info/en/tested-and-certified-microbiome-friendly-products
318. Schulte to Brinke, A., Mehlich, A., Doberenz, C., & Janssens-Böcker, C. (2021). Acidification of the skin and maintenance of the physiological skin pH value by buffered skin care products formulated around pH 4. *Journal of Cosmetics, Dermatological Sciences and Applications., 11*(01), 44–57. https://doi.org/10.4236/jcdsa.2021.111005
319. Krutmann, J., Liu, W., Li, L., et al. (2014). Pollution and skin: From epidemiological and mechanistic studies to clinical implications. *Journal of Dermatological Science, 76*(3), 163–168. https://doi.org/10.1016/j.jdermsci.2014.08.008
320. Mim, M. F., Sikder, M. H., Chowdhury, M. Z. H., Bhuiyan, A. U. A., Zinan, N., & Islam, S. M. N. (2024). The dynamic relationship between skin microbiomes and personal care products: A comprehensive review. *Heliyon., 10*(14), e34549. https://doi.org/10.1016/j.heliyon.2024.e34549

321. Chang, Y. S., Trivedi, M. K., Jha, A., Lin, Y. F., Dimaano, L., & García-Romero, M. T. (2016). Synbiotics for prevention and treatment of atopic dermatitis. *JAMA Pediatrics, 170*(3), 236. https://doi.org/10.1001/jamapediatrics.2015.3943
322. Skjerven, H. O., Lie, A., Vettukattil, R., et al. (2022). Early food intervention and skin emollients to prevent food allergy in young children (PreventADALL): A factorial, multicentre, cluster-randomised trial. *The Lancet., 399*(10344), 2398–2411. https://doi.org/10.1016/S0140-6736(22)00687-0

Hair

Valerie George

Abstract This chapter explores the complex biology and chemistry of human hair, offering a comprehensive examination of its structure, composition, and classification. It begins by contextualizing the cultural and historical significance of hair, before delving into its three major anatomical regions: the cuticle, cortex, and medulla. Emphasis is placed on keratin composition and bonding mechanisms, elucidating how these underpin hair strength, texture, and behavior. This chapter also examines melanin's role in hair color and the biochemical processes behind graying. A detailed discussion follows on hair classification systems, including morphological and mechanical characteristics, fiber diameter, porosity, and density. The physiology of the scalp and its microbiome are analyzed in relation to hair health, with particular attention to common scalp disorders and aging-related changes. This chapter concludes with an overview of the hair growth cycle and the formulation science behind shampoos, conditioners, styling products, and colorants, linking cosmetic chemistry to hair's unique biology.

Keywords Hair · Keratin · Scalp · Hair growth · Shampoo · Hair color

Humans have been changing the way their hair looks for thousands of years, with coiffures on record dating back to the ancient Egyptians and Romans. Hair did not just serve a physiological, evolutionary purpose. Whether it was braiding, curling, incorporating accessories, or even wearing a wig, how hair appeared was often linked to societal status, wealth, or religious beliefs.

Our ancestors did not have the knowledge we do about hair today, but some traditions—such as changing hair styles to reflect what is *en vogue*—remain an important aspect of human culture today.

V. George (✉)
CEO, Simply Formulas, Dallas, TX, USA
e-mail: valerie@simplyformulas.com

K. Alexander, P. Romanowski (eds.), *Cosmetic Science*, AAPS Introductions in the Pharmaceutical Sciences 15, https://doi.org/10.1007/978-3-032-03596-7_4

Hair is one of the most intriguing biological materials in the human body because it is one of the only ones that is not composed of living tissue. While hair is biologically dead, it is chemically alive, and the chemistry of the hair fiber serves as the fundamental basis for how our hair looks and feels, as well as determines how cosmetic formulators select ingredients to care for and style tresses.

To understand hair in its totality, it is essential to explore its structure, composition, and growth cycle to uncover how its unique physical properties influence everything from its health to how we change its appearance.

Hair Structure

Human hair, whether on our head or body, appears as a shaft-like structure protruding from a little hole in our skin, the follicle. It has three general regions: the cuticle, cortex, and medulla (Fig. 1).

Cuticle

The cuticle is the outermost region of the hair shaft, formed from overlapping keratinized cells that are large and flat like shingles on a roof. Much like shingles on a roof protect the living quarters of a house, the cuticle is the protective barrier for the

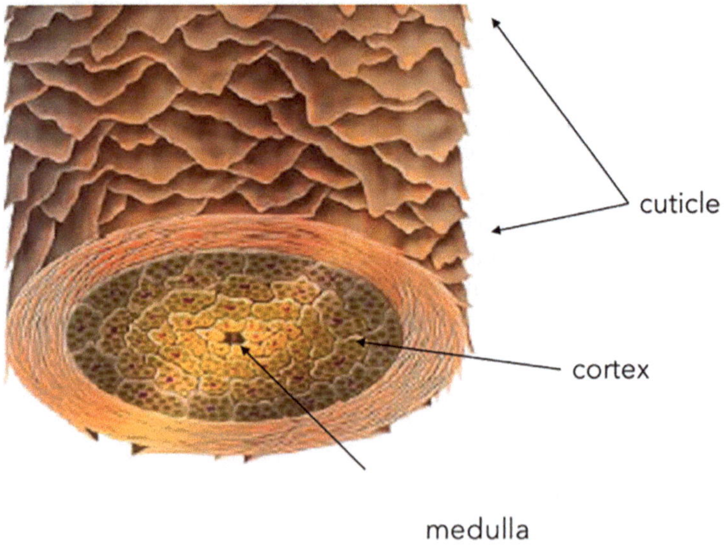

Fig. 1 A simplistic cartoon demonstrating the three regions of the hair shaft. (Source: Ashland Presentation, 2014)

soft, inner structure of the hair fiber. The cuticles are oriented such that they appear to be starting from the scalp follicle and pointing toward the tip of the hair fiber.

A head hair (as opposed to body hair) has approximately 5–10 overlapping layers of cuticle cells, with each cuticular layer approximately measuring 0.4–0.5 μm thick [1, 2]. Asian hair tends to have more cuticle layers than Caucasian, while African hair has fewer cuticle layers than both Asian and Caucasian fibers. It has also been found that coarse hair is enwrapped by more cuticle layers, and finer hair less. The impact of the number of cuticle layers on various facets of hair mechanics and behavior is an ongoing area of interest in hair research.

The cuticle layers are not only coated in lipids, primarily 18-methyl eicosanoic acid (18-MEA) in its outermost layer, but the cuticle cells are held together by fatty acids [3]. This lipid-cuticle glue structure is referred to as the cuticular cell membrane complex (CMC). This lipid structure is crucial to supporting the cuticle in mitigating water exposure to the internal hair fiber. Studies report varying levels of overall lipids across varying hair ethnicities, but the consensus is African hair contains the highest amount of lipids over Asian and Caucasian hair types, which are relatively the same [4]. When the fiber's lipids are lost, hair is more hygroscopic and susceptible to environmental assault. This dilapidated condition is often called porosity by consumers.

Since the cuticle is the outermost layer of the hair fiber, it is the first line of defense and is relatively chemically resistant. Thanks to a high quantity of stable sulfur proteins and surface lipids, hair is naturally protected from damage. The cuticle also helps regulate the moisture content of the hair fiber.

The cuticle's health and physical condition are responsible for how the rest of the world sees our hair. It contributes to our luster, combability, hygroscopity, and electrostatic charging. When consumers talk about hair being damaged, tangled, or porous, they are really speaking about the state and health of their hair's cuticle.

Cortex

Moving inward from the cuticle is the cortex. The cortex forms 80% of the bulk of the hair fiber, and is composed primarily of protein and lipids, which we call the cortex Cell Membrane Complex. The cellular glue in the CMC does not just contain fatty acids, but additionally free fatty acids, ceramides, and cholesterol [5]. This mortar holds together intermediate filaments, which comprise the bulk of our hair fiber, and their chemistry will be discussed extensively in section "Keratin".

The cortex is not (or at least should not be) readily visible to others, so it is not directly responsible for hair's appearance. However, the cortex is partially responsible for determining the shape hair has, and it is responsible for hair's strength. An intact cortex means strong hair.

Another important structure found in the cortex is melanin. Melanin is an insoluble, polymeric pigment produced by melanocytes in the scalp. While there are differing schools of thought on the exact mechanism of melanin formation in hair,

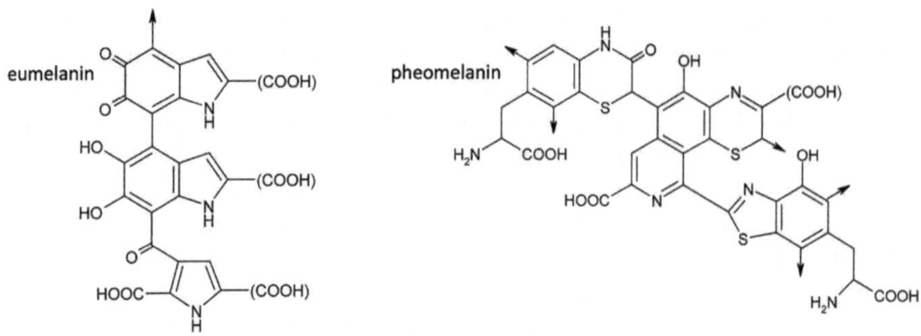

Fig. 2 The structures of eumelanin and pheomelanin. (Source: Author, 2025)

it is generally agreed that melanin is formed via conversion of the amino acid tyrosine to dopaquinone, by via the enzyme tyrosinase in the melanocyte. The dopaquinone is further reacted, oxidized, and polymerized to form the melanin polymer. Once produced, melanin is stored in an organelle called the melanosome until it is transferred to neighboring keratin produced in the hair follicle. Although there are many types of melanin, there are two predominant types found in human hair–eumelanin and pheomelanin [6] (Fig. 2).

Eumelanin is a very granular pigment and varies in size. It exists in between fibrils formed from cortical cells in clumps that appear as granules. These granules are covered in a soft protein skin, holding all the eumelanin molecules together. The size and distribution of the granules contributes to how dark hair appears. Eumelanin is responsible for black to very dark red colors.

Pheomelanin yields yellowish and reddish colors in hair and similarly exists in granules, although they are less consistent in shape and size compared to eumelanin, as it is diffused throughout the hair fiber. Pheomelanin is a polymer like eumelanin, although it has a sulfur component present, which is thought to come from cysteine being integrated in its synthesis. The sulfur atom lends a bit of stability to pheomelanin, compared to eumelanin. This stability is best visualized by thinking of someone who has lightened their dark hair with a chemical lightener or spent a significant amount of time in the sun. The brown-black pigment fades quickly, leaving behind the hallmark color of pheomelanin, a yellow to orange to red color.

In terms of the eumelanin to pheomelanin distribution in the fiber and how it relates to perceived hair color, redheads tend to have more pheomelanin than eumelanin, brunettes have more eumelanin than pheomelanin, and blondes have very little of both eumelanin and pheomelanin. The intensity of the color is contributed by the quantity and distribution of melanin present. Individuals with albinism are devoid of all melanins.

Gray or white hair is also absent of melanin, although it is different from albinic hair in that the hair fiber once contained melanin but no longer does. The exact mechanisms of hair graying are not yet understood but are generally attributed to decreased melanin production in the melanosomes, meaning less melanin is transferred to the keratin fiber in the follicle. This is a naturally occurring phenomenon as we age and can largely be attributed to genetics. Oxidative stress is another critical pathway contributing to decreased production of melanin.

Free radicals, also known as reactive oxygen species (ROS), are highly reactive molecules naturally occurring as a byproduct of cellular metabolism. In our scalp, the primary ROS is hydrogen peroxide [7]. In addition to being naturally occurring, ROS can form from external factors, such as UV light exposure, smoking, pollution and internal factors such as stress. Although our bodies form ROS, they are typically eliminated in our cells through enzymatic activity. As we age, the ability to recycle ROS into functional molecules is reduced, causing an accumulation of ROS in our hair follicles, and thus melanocytes. This build of up of hydrogen peroxide leads to DNA and protein damage in our melanocytes, as well as inactivates tyrosinase. Over time, the melanocyte's ability to produce melanin is decreased, resulting as a gradual loss of color in the hair fiber.

From an evolutionary perspective, melanin's role in the hair is to provide protection from ultraviolet (UV) light. As a result, people with darker hair and skin are primarily found in regions with intense sunlight. Conversely, in regions with less sunlight (e.g., northern latitudes) lighter hair, which has less melanin, is more prevalent.

It is a common misconception that ultraviolet (UV) light from the sun is directly responsible for causing hair color to fade [6]. This has always been a perplexing consumer notion as melanin is composed of a highly conjugated, electron-rich structure that renders it stable to ultraviolet light. Melanin is also the photoprotective element in human skin, protecting our cellular DNA from damage, so the thought that UV rays can destroy melanin found in keratin does not make sense. Melanin's electron pi system can both absorb and scatter UV radiation throughout the fiber, leaving the melanin pigment relatively unscathed.

It has recently been discovered that the lesser-thought-of sun emission, visible light, can cause photodegradation of melanin, as well as oxidation and degradation of hair's internal lipids. Despite the energy of a photon of UV light being higher than that of visible light, the number of photons of visible light far surpasses the number of photons of UV emitted from sunlight, making the assault on melanin increased. For every 1 photon of ultraviolet light, there are roughly 10 photons of visible light [8].

When visible light hits hair, it does not absorb into the keratin protein. Instead, it passes through the hair, directly interacting with and decomposing the melanin pigments, drastically destroying their structure. Once the conjugated electron system is destroyed, color is no longer visible. This effect is more demonstrative if the hair has been chemically lightened or processed, presumably from the removal of the photoprotective melanin from the fiber. Darker, virgin (undamaged) hair fares better.

There is one caveat about UV light's role in hair. While melanin does offer photoprotective effects to hair, exposure to UV light can cause significant changes to the protein comprising the internal structure of the hair fiber. When UV scatters throughout the hair, it interacts with the protein chains within keratin. Ultimately, protein degradation occurs, leaving the hair more porous. The increase in hair porosity causes the melanin to essentially fall out of the hair during washing, indirectly contributing to hair fade.

UV light can also change the color of graying hair. When UV light interacts with the keratin protein, it degrades the amino acid tryptophan through an electron cascade [9]. Specifically, chromophores such as N-formyl kynurenine, kynurenine, and 3-hydroxykynurenine are formed from the degradation, leaving gray hair with a yellow tinge.

Trace minerals are also found in the cortex. Overall mineral content in hair is quite low, but it does exist. Minerals can enter the hair fiber from sweat deposits, medicine and diet. The environment can also deposit minerals into the hair. For example, air pollution contains heavy metals, and shower or bath water can leave behind calcium, magnesium, copper, and iron, which can get deposited into the hair. Swimming pools can deposit chlorine and copper into the hair. The former oxidizes the hair protein, leaving it damaged, while the latter is responsible for the greenish hue of blonde swimmers [10]. Even shampoos can leave behind minerals, if the shampoo is an anti-dandruff shampoo with zinc pyrithione or selenium sulfide.

Minerals generally have little impact on virgin hair, although it has been demonstrated that chemically treated or bleached hair is highly susceptible to mineral uptake, presumably from an increase in negatively charged sites on hair. Minerals like calcium and magnesium can impact the fiber mechanics of hair, leaving it feeling dry and stiff. Copper and iron continually cause damage within the cortex of the hair, although they go relatively unnoticed until a hair coloring service is performed with hydrogen peroxide. Hydrogen peroxide is rapidly decomposed into water and oxygen, releasing a significant amount of heat in the process, potentially leading to third-degree burns.

Medulla

The medulla is not often discussed in hair literature. Its evolutionary origin is unknown, and its function is uncertain. It generally exists in the center of the hair fiber, following the direction of the hair shaft, and was once thought to be an empty tube filled with nitrogen gas. Recent research has demonstrated that the medulla is a loose-knit network of proteins and lipids [11]. African hair has the highest quantity of lipid distribution in the medulla over Caucasian hair. While African hair almost always has a medulla, Caucasian hair sometimes does not have a medulla at all, or the medulla is interrupted and not continuous. Asian hair does not have a high lipid content in its medulla, unlike African hair, and it generally has a continuous medulla throughout each strand.

It is not yet understood how the medulla contributes to hair's structure and behavior and the medulla is historically not an area of focus in hair research. However, the medulla is important in distinguishing human hair from other mammalian hair in forensic science, as animal hair or fur has their own distinct medulla pattern, depending on the animal.

Keratin

If melanin is responsible for hair's color, keratin is responsible for hair's structure and strength. Keratin is the primary complex protein that composes 65–95% of the hair fiber, depending on the moisture content of hair, and the terms protein and keratin are often used interchangeably.

Although keratin is a generic term, it is actually a broadly used term to describe the entire keratin family that comprises hair's structure. There are over 50 types of keratins classified in hair, each contributing to hair's structural integrity, strength, and mechanical properties. These keratins can also be placed into two generic categories, acidic (type I) and basic (type II) keratins [12]. It is thought that one basic and one acidic keratin protein will combine to form the intermediate filaments that run parallel along the hair shaft, giving the hair its shape and strength. There are also different keratin proteins found in the hair follicle's basal layer that contribute to the differentiation and keratinization processes as the hair fiber grows from the follicle [13].

Although keratin is a complex protein, its origins are simple, as with any protein, whose basic monomer is the amino acid. There are 18–20 amino acids found in hair, with cysteine and glutamic acid serving as the primary amino acids in hair, totaling 28–32% of its amino acid content [14] (Table 1).

Table 1 The typical amino acid composition of human hair keratin, expressed as a percentage range. Cysteine and glutamic acid are the most abundant, playing a vital role in the formation of structural bonds within the hair fiber

Amino acid	Composition
Alanine	2.8–3.5%
Arginine	8.8–9.6%
Aspartic acid	5.6–6.5%
Cysteine	14.0–16.5%
Glutamic acid	14.3–15.5%
Glycine	4.5–5.2%
Histidine	0.8–1.1%
Isoleucine	2.3–2.5%
Leucine	6.4–6.9%
Lysine	2.6–3.1%
Methionine	0.5–0.9%
Phenylalanine	2.2–2.8%
Proline	7.0–7.8%
Serine	9.6–10.8%
Threonine	6.5–7.5%
Tryptophan	0.8–1.2%
Tyrosine	2.1–2.7%
Valine	5.0–5.8%

Source 15: Robbins

The amino acid composition of hair does depend on genetics, diet, and hair weathering, which is the concept that the chemistry and physics of the hair fiber changes as it is exposed to external elements, such as water, ambient oxygen, UV light, pollution, and chemical treatments. The amino acid composition of the hair towards the root of the fiber, versus the tip (which is considerably older), has shown that older, more weathered hair contains fewer cysteine, tyrosine, and histidine amino acid residues [15].

Amino acids link together to form a primary structure, akin to a pearl necklace, by joining at the amino and carboxyl groups. The bond that joins them side-by-side is called the peptide bond, and a series of amino acids linked together is called a peptide. This primary structure is not enough to form a complete, functional protein. Fortunately, amino acids have acidic, basic, and ionizable side groups that allow the formation of various cross-linked bonds. This interaction of side groups with other amino acids allows the transition of a protein from its primary structure to a secondary structure, as the peptide chain folds on itself. This secondary structure is often referred to as an alpha-helix or beta-sheet. The former is a coil-like structure, the latter a pleated sheet (Fig. 3).

It is the network of different bonds formed from side group interactions that form this secondary, alpha-helix structure: covalent, ionic, and hydrogen. There is also an extensive opportunity for Van der Waals interactions.

Cysteine is the most crucial amino acid, not just because it is found in the highest magnitude in hair, but because it is the only amino acid that can covalently cross-link in the hair fiber to form a disulfide bond, or a cystine residue, (Fig. 4). Since covalent bonds are the strongest types of bonds, so are disulfide bonds. There is an estimated one cystine linkage per four turns of the alpha-helix, which is about 1 linkage every 14–16 amino acids. While disulfide bonds provide most of the strength in hair, they are extremely susceptible to oxidation or reduction reactions, such as hydrogen peroxide in hair color or thioglycolic acid in permanent waves.

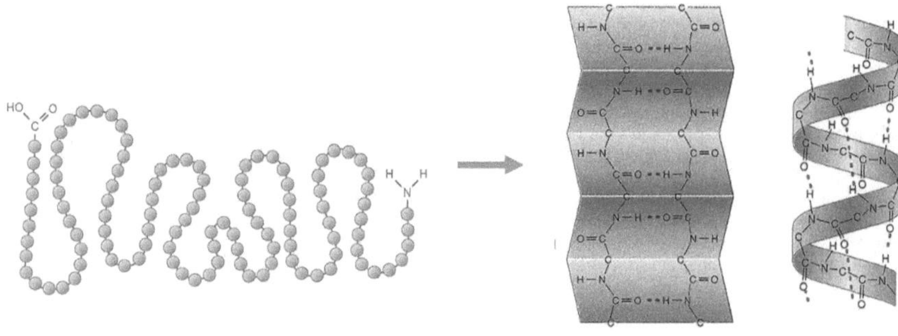

Fig. 3 The primary structure of a protein, which progresses into its secondary structure, an alpha helix or beta sheet. (Source: The McGraw-Hill Companies, Inc., University of Pennsylvania, 2006)

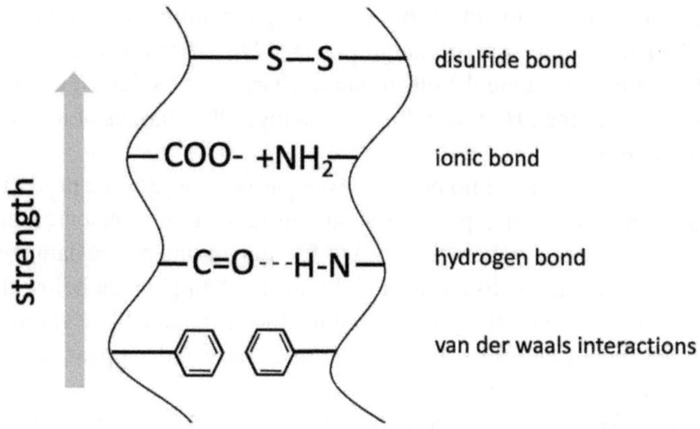

Fig. 4 A representation of bond types found in human keratin. (Source: Author, 2025)

Ionic bonds are also known as electrostatic links or salt bridges in hair. These result from the formation of positively and negatively charged amino acid (ionizable) side groups. Oppositely charged groups are attracted to each other in chemistry; therefore, these diametrically opposing groups combine.

The strength of these ionic bonds in hair is strongest when the number of positive and negative charges is maximal, at hair's isoelectric point. Much like salt is an ionic bond, and can be disrupted when dissolved in a dielectric medium like water, so can ionic bonds in hair. The COO- group of side groups becomes temporarily protonated, while the ammonium ion of side groups is deprotonated. This also happens when the hair is exposed to acidic or alkaline media. However, once this media is removed from the hair, such as drying the hair to remove water, the electrostatic linkages can reform. This means that ionic bonds give hair its strength when dry, but not wet.

Hydrogen bonding is the most frequent type of bond within the hair, as it occurs between oxygen or hydrogen atoms and C-O and N-H groups in neighboring amino acids. Much like salt bridges, hydrogen bonds can be broken by water that disrupts the hydrogen bonding network. Thus, when hair is wet, hydrogen bonds contribute no strength. When hair returns to a dry state, hair strength and stability is improved as hydrogen bonds are restored.

Van der Waals interactions are primarily employed between hydrophobic side groups of amino acids, such as phenylalanine, valine, and leucine. These interactions do not necessarily contribute to hair's strength but help stabilize the internal structure of hair. Even though Van der Waals interactions primarily result from hydrophobic groups, they can still be impacted by water solvation of the hair as it changes the overall charge density of the keratin protein.

As illustrated, hair's strength comes from a combination of permanent and temporary bonds. The ease with which noncovalent bonds dissipate illustrates the importance of maintaining the integrity of covalent bonds, so that when hair is wet, there is still a support network in place.

A beneficial aspect of the ionizable side groups in amino acids comprising keratin is that hair has some sort of charge present. Different chemical groups on the surface of the hair determine the electrostatic charges of the hair. This is influenced by overall pH; when the pH of hair shifts, so changes the charges of ionizable groups on the hair's cuticle [16].

The pH at which there are no positive or negative charges on a protein is considered the protein's isoelectric point. For human hair keratin, research has demonstrated this value to be anywhere from 3.3 to 5.1, depending on the damage sustained to the hair, but this value is likely around 3.6 [17]. At a pH value below the isoelectric point of 3.6, the keratin protein's ionizable groups will develop a positive charge. As pH increases above 3.6, the cationic charges disappear, and hair is left with an excess of negative charges.

Since hair's pH is much higher, between 4.5 and 5.5, hair almost always has an excess of negative charges present. The carboxyl functional groups from the side groups of aspartic acid and glutamine, as well as sulfonic acid groups formed from broken disulfide bonds (also called cysteic acid) are responsible for this property. Contributing factors to the excess of negative charges on hair include weathering, such as exposure to UV light, ambient oxygen, water, removal of lipids from the hair, and intentional chemical processing. In fact, these factors lead to an increase in cysteic acid on the surface of hair, a hallmark of damaged keratin.

Turning back to hair's internal structure, once the alpha-helix is formed, two helices come together to form a dimer, or coiled coil, through electrostatic and hydrogen bonding interactions. Van der Waals interactions help stabilize the formation. Two dimers coil with each other to form protofibrils, which then subsequently bundle together to form a microfibril. These microfibrils form intermediate filament proteins which further organize and pack together into macrofibrils, which are arranged in a spiral formation in cortical cells [18]. These bundles form part of the cortex's CMC, running parallel to the length of the hair shaft, the *ortho-* and *para-* nature of these cortical cells that contribute to hair's shape (Fig. 5).

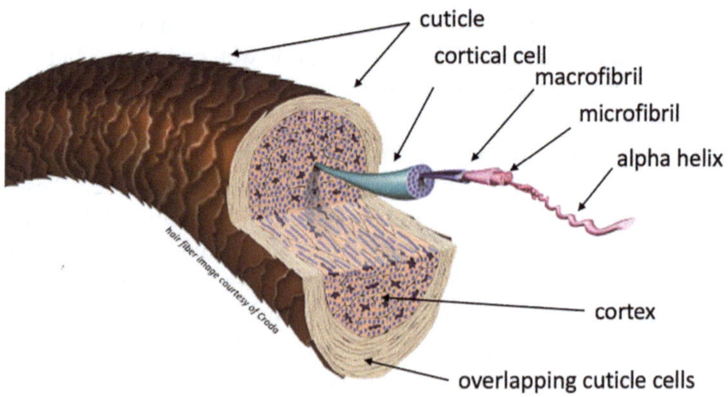

Fig. 5 An assembly model of keratin protein in the hair fiber to form the cell membrane complex (CMC). (Source: Croda, The Science of Hair Strength [White Paper], 2017)

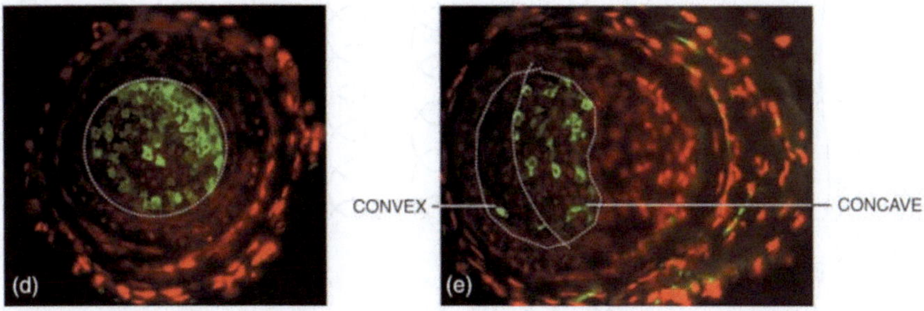

Fig. 6 Distribution of ortho- and para- cortical cells in cross sections of a straight hair fiber (**a**) and a curly hair fiber (**b**). (Source: Thibaut et al. [20])

The keratin proteins formed in hair do not exist in one long, continuous length down the hair fiber. Rather, they are organized in fragmented cortical cells with either *ortho-* or *para-* configurations, ranging from 1 to 6 μm thick and 50–200 μm long [19]. It used to be thought that the presence of these cortical cells was determined by race, as some studies showed Asian hair had no orthocortical cells while African hair contained predominantly orthocortical cells, but research has demonstrated that it is a factor of hair fiber shape [20].

Orthocortical cells are generally found in the exterior of the CMC, closer to the cuticle of the hair, and tend to be more tightly packed and organized into straight lines. This helps provide the hair strength and rigidity. In straight hair, paracortical cells are generally found interior to orthocortical cells. They are less organized than their *ortho-* counterpart and are stacked more haphazardly. This lends to their flexibility. This is advantageous for wavy or curly hair types, where it is believed that there is an asymmetric distribution of *para-* to *ortho-* cortical cells, where orthocortical cells dominate the convex side of curvature, while paracortical cells lie to the concave side. This dual arrangement of rigid to flexible keratin protein arrangement allows for bending in the hair (Fig. 6).

Hair Classification Systems

Older hair research historically classifies hair by ethnicity, into generic groups denoted as Asian, Caucasian, and African. From this research, hair fiber morphology has been dissected into three shapes: Asian hair is round, Caucasian hair is oval, and African hair is elliptical. This generality is rather accepted, but recent research has shown that hair's shape is considerably less simplistic. Hair is rarely perfectly round, and on one head of hair, multiple shapes can be encountered. In the case of African hair, where an elliptical nature is observed, the degree of ellipticity can drastically change from one segment of a strand to another [21].

While there is an abundance of hair knowledge from the 1960s (the golden era of hair research), the world's population is becoming increasingly blended, requiring a

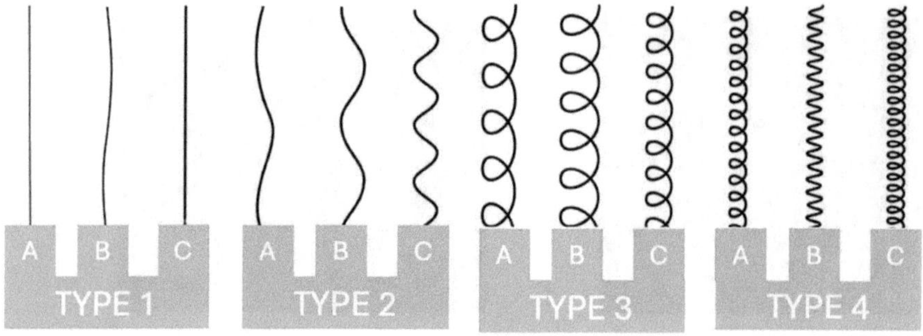

Fig. 7 One variation of Andre Walker Hair Typing System. (Source: Author, 2025)

different way to look at hair's shape. In 1997, celebrity hair stylist Andre Walker developed a hair typing system to help consumers select products from his line to better care for their hair based on its shape and texture. The system originally classified hair into four types, with each type containing additional subtypes, A, B, or C. The four numerations considered hair's straightness, waviness, curliness, or coiliness. The original system considered A, B, or C subtypes for types I and II, whereas types III and IV only contained subtypes of A and B (Fig. 7).

Over the years, the hair typing system has evolved to include additional subtypes for each parent typing, and variances, as well as differing opinions about the structure of this classification system. It depends on the user or community relying on the typing system, as the typing requirements are quite subjective and based on self-assessment.

Since its rise in popularity, there has been interest in whether the typing system can be made objective and without reference to racial origin. In 2007, researchers at L'Oréal attempted to identify simple measurements that represent each facet of the Andre Walker typing system, by evaluating hair morphology parameters from over 2000 study participants all over the world [22]. They collected data by measuring hair's curve diameter, measuring hair's curl index, the ratio of stretched hair length to the length at rest, number of waves, and number of twists. Their results indicated 8 hair types, but did not include aspects such as hair density, hair fiber shape [roundness] or growth rate.

Researchers at Spelman College measured the geometric properties of various hair types under the Andre Walker and L'Oréal typing systems, using scanning electron microscopy and photogrammetry [23]. They then studied the fiber's corresponding mechanical properties to generate correlations. As consumers maintain interest in a typing system, more quantitative research will come.

A new aspect of the hair typing system is porosity; while hair researchers do not generally use the term porosity, salon professionals typically associate porosity with the level of damage sustained from chemical treatments. The more damaged the hair, the less hydrophobic and more porous it is to water. In hair typing, porosity is often referred to as the innate condition of the cuticle and is often associated with a specific hair type. For example, hair with low porosity has a tightly packed, flat cuticle, making it resistant to moisture. Hair that is highly porous has a lifted cuticle, causing hair to absorb moisture quickly, but is also purported to lose it quickly, leading to the perception of dryness.

Type 4 hair is often associated with being highly porous, and this consumer may often complain of dryness. It is often stated the curl structure does not allow for the scalp's natural oils to travel down the hair shaft and lubricate the hair shaft, leading to more water loss. However, dryness is a consumer perception where hair that feels rough to the touch inherently feels drier than hair that feels lubricated, and it often has very little to do with the water content of hair. Consumers are experiencing a physics issue, where there is more resistance along the hair fiber due to the shape of the hair fiber, where twists, turns, coils, and a roughened cuticle lead to hair that does not feel smooth on the fingers. The increased propensity to breakage that African hair experiences is also likely due to fiber physics, where stress and strain impact hair that is curled differently over hair that is wavy or straight.

Two of the aspects overlooked by L'Oréal researchers in their 2007 typing work were hair fiber diameter and hair density. The salon professional community, as well as consumers, typically identify with hair that is labeled as fine, medium or coarse, referring to their subjective assessment of the hair fiber's diameter. Fine hair has different requirements for care from coarse hair and may be more susceptible to breakage compared to larger diameter fibers with the same force placed on it [24]. While finer fibers may stretch more before breaking, they have less load-bearing capacity.

There is very little published research conducted on how fiber diameter changes over the course of one's life, as the length of time for such a required study is prohibitive. One study reported that mean fiber diameter increases as humans age, peaking in the fourth decade of life [25]. Hair fiber diameter begins to decrease in the 50s, which aligns with the timing in which the hair follicle miniaturizes.

Hair density is also an important factor, especially for those experiencing thinning hair. Hair density is the number of hairs per square area on a scalp and helps determine how thick or thin a person's hair appears. It is segmented into low, medium, and high density. While it does not necessarily play a role into how hair care products are developed for consumers, it can play a role into perceived hair volume or thickness.

It is completely possible for an individual with fine hair to have high fiber density on the scalp, just like it is possible for someone with a large, coarse fiber diameter to have low fiber density, appearing as if their hair was thinning. Hair fiber density does decrease with age, as hair follicles enter dormancy [25].

Another aspect for classifying hair, although not in any formal typing system, is the type of damage hair has. As hair undergoes significant weathering over the course of its life cycle, it loses the protective 18-MEA lipid layer on the outer cuticle. This leaves the cuticle more exposed to environmental assault [3]. Eventually, microscopic cracks begin to form, creating weak spots prone to physical stress from brushing, combing, or manipulating the hair too aggressively. This mechanical damage to the hair eventually leads to breakage in the middle of the hair shaft or splitting at the ends of the hair.

If an individual applies chemical treatments to their hair, they likely have an abundance of cysteic acid on their cuticle and a reduced number of disulfide bonds in the cortex of the hair. Their reliance on hair strength comes from alternative bonds in the hair, such as ionic or hydrogen bonding. When chemically damaged hair is wet, very few disulfide bonds are in place to keep the hair together, so the wet hair significantly stretches.

Another indication that hair is severely damaged from chemical treatments is the reduced surface tension of water droplets on their hair. Or rather, the complete lack of surface tension. One cannot wet a healthy hair—meaning, hair that is healthy is relatively hydrophobic and water beads up readily on it. Chemically processed hair lacks this attribute and takes on more water thanks to its porosity. It therefore also takes longer for severely damaged hair to dry.

Heat styling is extremely detrimental to the hair fiber. Heat from a blow dryer or hot styling tool temporarily dissociates hydrogen bonds within hair, allowing the user to temporarily realign the hydrogen bonds in hair once it cools in the new position, granting the hair a new conformation. Over time, due to exposure to ambient humidity or when the hair is rewet with water, the hydrogen bonds once again dissociate, returning to their preferred place.

Human keratin begins to denature around 150–180°C (320–356°F) [26], yet most thermal tools heat well beyond, up to 200°C or 450°F. Prolonged exposure to heat permanently denatures the keratin's confirmation and eventually melts the hair. Routine heat exposure to thermal tools leaves an abundance of stress fractures in the cuticle, leading to mechanical breakage. Furthermore, hair contains free and bound water within its structure. Free water begins to vaporize and steam within the hair's cortex, building up pressure. Eventually, this water explodes through the hair fiber, leaving a gaping hole in the cuticle with the cortex exposed. Hair that has been previously lightened with thermal damage appears even more catastrophic; thermal damage on the cuticle exacerbates the damage in the cortex from chemical treatments, leading to exponential breakage.

Scalp

The human scalp is often overlooked in the hair world, but it is the foundation for very strand of hair on our head. The scalp is a highly complex and specialized structure, home to hair follicles, sebaceous glands, nerve endings, and blood vessels (Fig. 8). It is not just a support structure, but rather a dynamic ecosystem that intersects biologically dead hair and biologically living tissue, or what some researchers have coined "zombie hair." [27]

Although not its original etymology, an acronym has been developed for the word scalp to help those studying its physiology remember its structure—*S*kin of scalp, *C*onnective tissue, *A*poneurosis, *L*oose areolar tissue, *P*eriosteum, bone.

The skin of the scalp is quite thick, with a 30-year-old having a mean scalp thickness of 8 mm [28]. As a comparison, facial skin is 1.51 mm at the nose and 1.97 mm

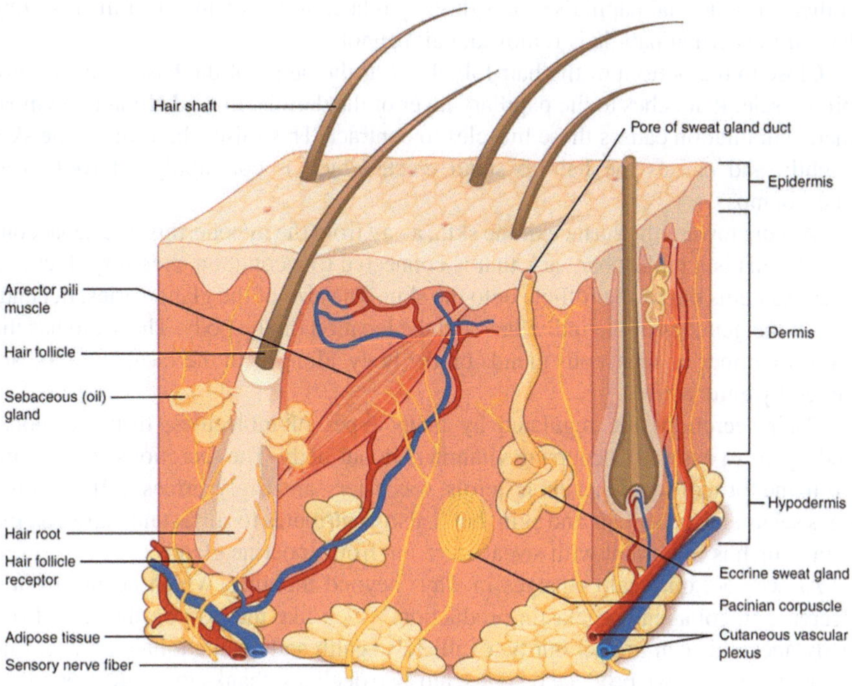

Fig. 8 A basic physiological view of the skin of the scalp. (Source: 5.1 Layers of the Skin." Anatomy & Physiology, eCampusOntario, https://ecampusontario.pressbooks.pub/hhnp4/chapter/5-1-layers-of-the-skin/)

at the infraorbital region. Total scalp thickness does vary with gender and age; a 90-year-old only has about 5 mm of scalp thickness. Females tend to have a total thicker dermis than males, reaching maximum thickness at 35 years versus 55 years, respectively. The epidermis is about the same thickness in both genders.

Connective tissue, particularly the aponeurosis, is a thin sheath of connective tissue that helps connect muscles to bones, like tendons. Loose areolar tissue is a thin connective tissue layer that separates the periosteum of the skull from the aponeurosis. It contains numerous blood vessels. The periosteum is a dense layer of vascular connective tissue that envelops bones.

The skin of the scalp is where hair originates, from pore-like structures in the skin called hair follicles. The follicle is shaped like an upside-down bulb, extending anywhere from 0.25 to 0.5″ below the skin's surface, deep into the hypodermis. This depth allows the follicle to receive a blood supply, which nourishes a structure at the base of the follicle called the dermal papilla. The dermal papilla is the only dermal element in the hair follicle, and it is composed of specialized cells that play a crucial role in the hair growth cycle. The extensive vascularization of the dermal papilla the necessary nutrients and hormones required for stimulating hair growth within the follicle. The dermal papilla's size is directly related to the diameter of the resulting hair. If the dermal papilla is removed, hair cannot grow.

Close to the bottom of the hair follicle, near the bulge of the base is an arrector pili muscle. It attaches to the papillary layer of the dermis. In cold climates, sympathetic stimulation causes these muscles to contract. This raises the level of the skin slightly and causes the hair to stand erect, which is commonly referred to as goosebumps.

Moving toward the surface of the skin, away from the hypodermis, is a sebaceous gland. This is a multilobar sack that is connected to a common excretory duct that excretes items out of the follicle onto the skin. The size of the gland varies, depending on the genetics of the individual and the location on the body. The scalp has the largest number of sebaceous glands on the body, along with the face, and they are generally quite active.

Their secretions are regulated by many types of molecules, from hormonal androgens to naturally occurring vitamin A in our body. The secretions contain lipids in the form of squalene, cholesterols, wax esters, and triglycerides [29]. Not only does sebum lubricate hair and skin, but it also contributes to surfactant action on the skin, which is important with sweat secretion from apocrine glands.

As we age, our scalp changes in ways beyond thinning, which occurs from a decrease in collagen and elastin production. Much like the skin on the rest of our body ages, this can impact our hair follicle's health and resultant hair growth. The amount of sebum we produce on our scalp also declines, thanks to a reduction in the number, size and productivity of sebaceous glands [30]. This may be a welcome phenomenon for those with oily hair. Increased reactive oxygen species introduce chronic (low-grade) inflammation in the hair follicle, causing the sheath lining the hair follicle to disappear. The hypodermis additionally shrinks, hindering the ability of the hair follicle to receive adequate nutrition. This creates an unfavorable environment for hair growth, leading to a decrease in hair fiber density and diameter.

Microbiome

The scalp skin has another important element—the microbiome. This is a commensal microbial population that exists on the skin to maintain its health and offer immune support. Their diversity is critical to balancing the scalp environment. They are responsible for the metabolism of amino acids, B-vitamins, and biotins—important elements for healthy hair growth.

While there are many bacteria and yeast species on the scalp, *Cutibacterium acnes* and *Staphylococcus epidermidis* are the core bacterial species. The two predominant yeast species are *Malassezia furfur* and *Malassezia restricta*, but there are many *Malassezia* yeast present [31].

In a normal, healthy scalp, *C. acnes* and *M. restricta* are the predominant organisms identified. One study quantified *C. acnes* at 74%, and *M. restricta* at 97% of the microbial population. In a dandruff scalp, *M. restricta* is reduced to 84%, with an increase of other yeast organisms to 16%. This yeast imbalance causes a bacterial imbalance; *C. acnes* is reduced to an astonishing 15% of the population, and *S. epidermidis* increases to a whopping 60%. Additionally, other bacteria account for 25% of the remaining bacterial population [32].

This characterization allows identification of potential biomarkers for various scalp disorders as it relates to the scalp microbiome. By qualifying and quantifying the major bacterial and fungal species present in the scalps of healthy and dandruff individuals, we can target treatments or avoid ingredients that contribute to a shift in the commensal population. For example, there is a proper ratio of sebum, moisture content and pH value of the scalp for a harmonious environment. By honoring the ideal scalp environment, organism populations will be kept in check. Additionally, topical formulations should avoid ingredients that cause opportunistic yeast to thrive, like triglycerides greater than C8 in fatty acid chain length.

Scalp Disorders

True scalp disorders require medical intervention, where a trained physician or dermatologist can establish a clinical diagnosis and suggest a treatment plan that may or may not include prescriptive therapies. When it comes to cosmetics, consumers can often treat the concerns themselves, either through over-the-counter drug products, or cosmetic products.

The most prevalent consumer concern related to scalp health is the presence of dry, flaky skin, which is frequently misattributed to dandruff. An in-house study was conducted by this chapter's author utilizing dermatological imaging that assessed over 5000 scalps. The study found that most individuals did not exhibit true dandruff. Dandruff is clinically characterized by excessive skin desquamation, manifesting as white to yellowish flakes or crust, often accompanied by overproduction of sebum [33]. Most subjects who self-reported dandruff as a concern were found to have only isolated, mild flakes, consistent with dry skin, rather than the shedding observed in dandruff. This suggests that many individuals are mistaken in their self-diagnosis of dandruff, with dry skin being the more likely condition.

Dandruff can be treated using over-the-counter drugs, typically in the form of a shampoo. The FDA permits a few anti-dandruff actives, such as:

- Coal tar, 0.5–5%
- Salicylic acid, 1.8–3%
- Selenium sulfide, 1% (non-micronized)
- Sulfur, 2–5%
- Zinc pyrithione, 0.3–2% (rinse off)

For dry skin, consumers tend to reach for oils, as oils are beneficial for reducing the appearance of dry skin in moisturizers. Given opportunistic yeast feast on triglycerides, this is not recommended. Physical exfoliants do not work well to slough off dead skin because they do not have the room to physically abrade skin in the hair. Some products employ exfoliating acids, like glycolic acid, which chemically remove dead skin cells. The author of this chapter recommends using a clarifying shampoo and manipulating the fingers on the scalp to loosen dry skin.

Some consumers do not have a dry skin issue, rather they have challenges with overproduction of oil on the scalp, which leads to greasy-looking hair. Oil production is largely influenced by genetics and hormones but can also be influenced by diet and environmental factors. This condition is hard to treat because it is an internal imbalance within the follicle. Consumers with oily scalp often wash more frequently or employ the use of an absorbent leave-on product, such as dry shampoo, to the symptoms.

Hair Growth Cycle

The largest concern about scalp and hair, overall, is hair thinning. While all hair thins with age, there are genetic and environmental factors that contribute to premature scalp thinning in some individuals. Stress, rapid changes in hormone levels, side effects from prescription medicine, autoimmune disorders, nutritional deficiencies, or chemical treatments are some causes. However, androgenetic alopecia, also known as hereditary baldness, is the leading cause of hair loss [34].

Hair grows outward from the hair follicle in a particular cadence (Fig. 9) [35]. The first phase is called the anagen phase. This is a proliferation phase where the dermal papilla signals to epithelial stem cells in the bulge of the hair bulb. Once these stem cells are stimulated, the inferior segment of the hair follicle grows downward, forming a bulb around the dermal papilla. The dermal papilla can then signal dermal matrix cells in the bulb to proliferate and grow upward, forming a new hair.

The length of the anagen phase varies; for hair on the scalp, the anagen phase can last between 2 and 6 years. However, for eyebrows and eyelashes, this cycle lasts only a few months. Hairs in the anagen phase grow approximately a half inch per month, which averages to be 6 inches per year. If you grew your hair out for 3 years, you would have 18″ of hair! On average, approximately 85–95% of head hairs at any given time are in the anagen phase.

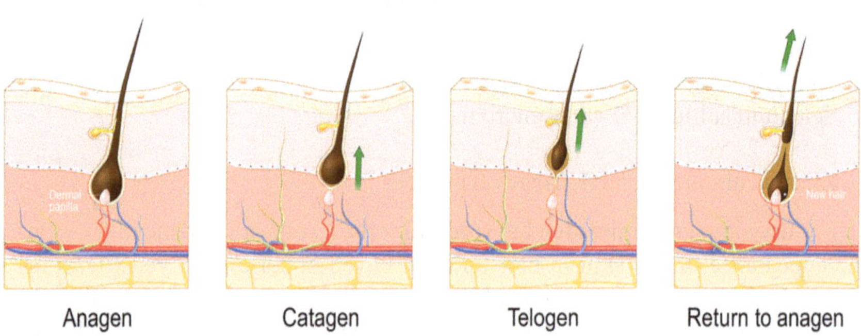

Fig. 9 The hair growth cycle. (Source: iStock, 2025)

Once the hair has completed its growth trajectory, the hair enters a transition phase called the catagen phase. This phase only lasts a few weeks. Cell division in the dermal matrix stops, meaning hair stops growing. The inferior segment of the hair bulb regresses, and the hair forms a white, hard node on the end called a club hair. This node is evident with the naked eye when examining hairs that have fallen out while brushing hair.

These club hairs no longer have a living portion in the follicle and are just resting until they fall out of the follicle. This resting phase lasts about 100 days, and this time period is called the telogen phase. After this time, the club hair falls out and the hair follicle can reenter the anagen phase.

As we age, the nutrient supply to the hair follicle diminishes, and the hair follicle can miniaturize. Hairs grow shorter and become smaller, finer, and lighter. This phenomenon can also occur with chronic exposure to an androgen called dihydrotestosterone (DHT) [36]. DHT is synthesized from testosterone via an enzyme called 5-alpha-reductase. DHT can activate the *Wnt* signaling pathway, which produces growth factors that trigger the hair growth cycle. However, too much DHT for too long can downregulate (decrease) this pathway.

Higher levels of DHT in hair follicles can lead to a shorter cycle of hair growth, resulting in shorter and thinner hair. It can also extend how long the hair stays in the telogen phase, ultimately preventing hair from reverting back to the anagen phase. This overexposure to DHT must occur over a long enough period before the overstimulated androgen receptors start shutting down. Before they fully shut down, the growth cycle gets shorter and shorter. If your hair was in anagen for 4 years, it might shorten to 2 years. Hair does not grow as long before it falls out. Eventually, once the hair falls out, it is not replaced.

Hair thinning is a medical disorder that can only be treated by drug actives in the United States. The FDA has monographed Minoxidil as an over-the-counter drug treatment. Its mechanism is not fully understood, but it is thought that it protects androgen receptors from DHT [37]. It is currently available in several solution percentages, such as 2% and 5%, depending on which gender the Minoxidil is targeted toward, and how severe the hair loss is. It is also available orally.

Finasteride is a prescription-only oral drug treatment that works by inhibiting 5-alpha-reductase. By blocking the conversion of testosterone to DHT, the androgen receptors would not be overstimulated. While testosterone does bind to androgen receptors, it does not have the affinity that DHT does.

Laser therapy devices require FDA clearance as medical devices, but they are not monographed for the purpose of hair regrowth. Surgical treatments are also available but fall outside of the scope of this text.

Cosmetic ingredients are also purported to support hair regrowth, even though this is a drug claim. These actives can be peptides or plant extracts standardized in specific flavonoids. The two mechanisms by which ingredient manufacturers report these "actives" act are either stimulation of the *Wnt* pathway or by blocking the enzymatic conversion of testosterone to DHT. The evidence to support their efficacy is typically conducted via in vitro testing, which can—at first glance—look promising. However, the in vivo testing of these same ingredients is often lackluster. The studies are not robust, and the user groups are not large enough for statistical significance.

The real issue is that if a brand wants to make a hair regrowth claim, at least in the United States, they cannot use cosmetic ingredients not monographed by the FDA for this purpose. Nonetheless, there are a plethora of anti-thinning brands on the market promising fuller-looking hair, although most of them utilize fine print on their literature that states the fuller look comes from the reduction of hair breakage.

Types of Hair Treatment Products

Shampoos

Shampoos are cleansing products designed to remove dirt, oil, debris or previously applied styling products from the scalp and hair. The level of detergency varies depending on the formulation, as does the level of conditioning a shampoo affords. Some shampoos are positioned as clarifying, meaning they are intended to deep clean the scalp and hair while depositing very little conditioning ingredients back onto the hair. Shampoos can offer nourishing characteristics, depositing conditioning agents onto the hair fiber during the shampooing process. Other shampoos may have a specific function, such as removing metal built up in hair (i.e., chelating shampoo), or are intended to deposit color back onto the hair (i.e., toning shampoo). Shampoos may also address common concerns consumers have, such as color protection, strength, damage repair, lack of volume or dandruff.

Regardless of the function, there are common ingredient classes every shampoo uses, aside from water.

Surfactants

A surfactant is a surface-active agent that features a hydrophilic (water-loving) head and a hydrophobic (water-hating) tail. Surfactants reduce the cohesive forces between water, effectively reducing surface tension. This allows water to spread more easily onto the hair, solubilizing any oils, dirt or debris it encounters. The mixture is then able to be rinsed away thanks to the surfactant's amphiphilic nature.

The surfactant's structure also allows for a special arrangement in aqueous solutions called micelles. As the hydrophobic tail does not want to interact with water upon incorporation into the formula, the surfactant rearranges itself such that the hydrophilic head aggregates with other surfactant molecules to face the water, hiding the hydrophobic tail on the interior (Fig. 10). The micelles can vary in size depending on the surfactants used and the charge density of the anionic surfactant.

Anionic surfactants are the most common class used in shampoos as they are quite effective detergents. They bear a negative charge on the hydrophilic head,

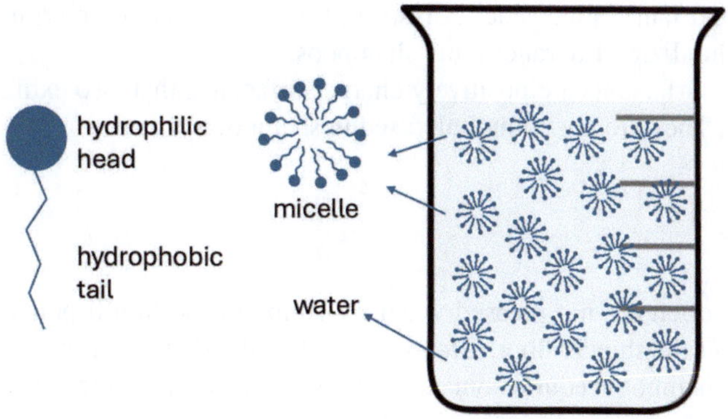

Fig. 10 A surfactant on its own and rearranged into a micelle once incorporated into water. (Source: Author, 2025)

which makes it useful for thickening shampoo formulations. The most commonly used anionic surfactants are the alkyl and ether sulfates, sodium lauryl sulfate, and sodium laureth sulfate. Although they are effective and safe for use in shampoo formulations, consumers often look for "less harsh" alternatives that are designated sulfate free. Sodium C14–16 olefin sulfonate, sodium lauroyl methyl isethionate, and sodium cocoyl isethionate are common choices.

Anionic surfactants are often the foam generators in shampoo formulations, with varying contributions to foam height and foam density. Some produce large bubbles spaced far apart, while others produce tight, creamy bubbles packed closely together. Due to their high use levels, anionic surfactants are often considered the primary surfactants in shampoo formulations.

Amphoteric surfactants are almost always used in conjunction with anionic surfactants as their natural salt content interacts with the anionic surfactant head, generating a thickening effect in formulations. Amphoteric surfactants can either be negatively or positively charged depending on the pH of the formulation; they are resultantly called zwitterionic due to their dually charged nature. The most common amphoteric surfactants are cocamidopropyl betaine, cocamidopropyl hydroxysultaine, and sodium cocoamphoacetate. Not only do amphoteric surfactants create synergistic thickening with anionic surfactants, but they also often generate foam very quickly in formulations, a phenomenon coined "flash foam." This immediate foaming effect is designed to please the consumer while the rest of the surfactants in the shampoo solubilize in the shower water and generate a slower-forming foam.

Non-ionic surfactants have no charge association, and they often do not generate a significant quantity of foam. While non-ionic surfactants like decyl glucoside can be considered good cleansers, they are generally mild compared to their anionic counterparts. They are most often used for their ability to solubilize and incorporate non-aqueous ingredients into a formulation, such as an essential oil or fragrance. Due to their lack of ionicity, they tend not to rinse well from the hair and should be used sparingly. The most popular nonionic surfactants for cleansing are decyl glucoside and lauryl glucoside. Polysorbate 20 is a common non-ionic surfactant used for solubilizing fragrances into shampoos.

Cationic surfactants are positively charge surfactants that also exhibit emulsifying activity. They are not commonly used in shampoo systems.

Thickeners

Consumers often are not pleased when a shampoo is so thin it pours out of their hand, all over the shower floor. Even worse, when the shampoo runs into their eyes! Therefore, shampoos require some sort of viscosity to stay in the hand for dispensing on the scalp, without dripping into the eyes.

Sodium chloride is the most efficient thickener when an anionic surfactant is present. The non-iodized version of table salt is usually incorporated into the end of the production process to attain the desired viscosity.

Other times there is a desire to thicken the water phase of a shampoo formulation. Hydroxypropyl methylcellulose helps build viscosity while adding a richer feel to the shampoo. It also helps stabilize foam once it is formed. It may be advantageous to add a thickener that is also conditioning, such as polyquaternium-10 or guar hydroxypropyltrimonium chloride. This is a guar starch typically comprised of two mannose sugars to one galactose sugar, which is then derivatized to contain a quaternary ammonium ion. This allows the guar to bear a positive charge, thus sticking to negative charged sites on hair. This improves the wet comb and feel of the hair once dry, in addition to thickening the shampoo.

Another thickener option is an acrylates copolymer, which uncoils and swells to both thicken the surfactant system and offer suspension of actives, such as coal tar or zinc pyrithione. One can also incorporate ethoxylated thickeners, such as PEG-150 distearate. This is produced via ethoxylation of a fatty acid with ethylene oxide. This not only drastically increases the size of the molecule but also increases the water solubility. This allows the PEG-150 distearate to create steric hindrance within the shampoo's micelle network and minimize the movement of water through the formulation. The result is an increase in the thickness of the shampoo.

Performance Enhancers

A performance enhancer's role is to improve the overall user experience of the shampoo. It may support in-use benefits, such as lightweight detangling while the hair is wet, or even lightweight conditioning. Perhaps the foam quality needs to be improved so there is not only more foam, but the foam lasts considerably longer while the consumer is washing their hair. Whatever the reason, this ingredient is there to help the consumer enjoy how the shampoo works for their hair.

Functionalized silicones like amodimethicone or polysilicone-29 are great ways to enhance the performance of a shampoo. Amodimethicone is a positively charged silicone that temporarily adheres to hair. It not only makes hair soft and reduces frizz, it protects artificial color from washing out of the hair. Polysilicone-29 is also functionalized to stick to hair, with the benefit of a cross-linking moiety to ensure prolonged effects through multiple washes. This silicone restores hydrophobicity to hair, which is beneficial for damaged hair missing its protective layer of 18-MEA. Thickeners that simultaneously condition are also a simple, effective way to improve the performance of the shampoo for the consumer.

Claims Ingredients

A product is not a sellable product without claims, and in a world where consumers shop for hair products by problem-solution, claims ingredients are a must. The opportunities for claims ingredients are endless, although most formulating chemists reach for hydrolyzed proteins that form a film on the hair's surface. This affords various attributes, such as increased strength, perception of moisture, and shine. Other claims ingredients may include color protectants, volumizers, UV protectants, and more.

Fragrance

A shower is not the same without the pleasing bloom of a fragrance as the shampoo starts to lather in the hair. Fragrance can not only make the cleansing process more enjoyable, but it can also evoke emotions such as relaxation or invigoration, depending on the scent. Some fragrances in shampoo can linger on the hair long after the shower, leaving hair feeling fresh and clean for days. Fragrances are also part of a brand or product's identity, such as the minty scent of Paul Mitchell's Tea Tree Special Shampoo.

Fragrances are often not able to be added directly to shampoo systems, as they are intrinsically oily in nature and will disrupt the micelle network, causing a viscosity crash. This is especially true of sulfate-free surfactant systems. A solubilizer premixed with the fragrance, such as polysorbate 20, PEG-40 hydrogenated castor oil, or glycolipids is recommended.

Preservative

Where there is water, there is the opportunity to support microbial life. As cosmetic products are not required to be sterile, and they are often used in wet environments such as a shower, proper preservation is critical. Shampoos often require a shampoo-specific preservation system, as they are, in the author's experience, highly susceptible to mold growth. Additionally, some preservatives are inactivated by ethoxylated surfactants, so those would need to be avoided should your shampoo contain them.

It is also important to select preservatives that allow the clarity or viscosity of the formulation to be maintained. All too often a preservative, such as caprylyl glycol, will crash the viscosity.

Conditioners

After using shampoo, it is often helpful to use a conditioner on strands to restore oils and lubricity. A hair conditioner is an emulsion that contains conditioning ingredients to help make hair feel softer, look smoother, and is easier to manage. A conditioner is the final step in which a consumer will judge whether the brand's products were effective for their hair type, and this subconscious evaluation occurs in two stages: the wet stage and dry stage.

The wet stage of a conditioner is when the conditioner is applied to wet hair, and again when the product is rinsed from the hair. The consumer assesses the ability of the conditioner to adequately coat strands by combing the conditioner through the hair, and by assessing hair feels during the rinsing stage. During application, the consumer dispenses the conditioner into their palm, noting the viscosity. A consumer with coarse, dense hair would be pleased to see a rich cream; a consumer with fine, silky strands would not.

The dry stage of evaluation occurs when the hair is dry. The consumer should be able to brush or comb their hair with relative ease. Hair should appear shiny and—if desired—feel soft and silky. There should not be a significant amount of static, and the hair should not be overtly frizzy. This is the final stage in which a consumer evaluates the efficacy of a conditioner.

Regardless of the function, there are common ingredient classes every conditioner has, aside from water.

Conditioning Agents (and Emulsifiers)

Conditioners function by lubricating the hair with emollients and conditioning agents that contain a quaternary ammonium compound, monikered quats within the cosmetics industry. Quats are positively charged molecules which ionically stick to negatively charged sites on the hair fiber. This allows varying alkyl chains attached to the quaternary ammonium group to coat the hair fiber, forming a smooth, conditioning layer. By forming this conditioned layer, quats reduce the force required for combing hair, which is perceived by the consumer as a detangling effect. Because the force required is reduced, breakage is reduced and thus allows conditioning agents to afford reparative claims. Quats also act as anti-static agents, which reduces the appearance of frizz and improves manageability of hair.

Common quaternary ammonium compounds found in conditioners are behentrimonium chloride, behentrimonium methosulfate, cetrimonium chloride and stearamidopropyl dimethylamine. Each compound offers benefits in the wet stage and dry stage of hair conditioning and they are often used in tandem to offer a comprehensive conditioning approach for the consumer.

Another benefit of long-chain alkyl quats is their ability to act as emulsifiers in conditioning products, and they are classified as cationic surfactants. As such, they are often the sole emulsifiers used in traditional hair conditioners. The type of cationic surfactant used also helps determine the viscosity, in conjunction with the use of fatty alcohols. Thin, more fluid conditioners are preferred by consumers with thin or fine hair, while thicker, richer conditioners are preferred by coarser-haired consumers. The thinness or thickness of the conditioner also plays a role in how conditioning the consumer perceives the product to be.

Emollients

Conditioners also employ emollients as lubricants. Alongside quats, emollients such as plant oils or esters can coat the hair fiber, eradicating the rough feel of the cuticle layer. This lubrication not only improves wet and dry combing of hair, particularly in the wet stages, but it gives the perception of conditioned and repaired hair. There are special formulation techniques in which these lubricants can have longer substantivity to hair, such as through coacervation with other quaternized materials that are attracted to the hair, such as guar hydroxypropyltrimonium chloride. Silicones, particularly functionalized silicones, not only serve as emollients, but also deliver specialized attributes to the hair.

Performance Enhancers

Aside from the conditioner-emulsifier playing a role in conditioning hair, it may be necessary to enhance the performance of a conditioner, from its spreadability, to hair softening ability, to frizz reduction. Conditioners may also help impart strength, reduce breakage, prevent color fading, and more. Various ingredients can function as performance enhancers, although functionalized silicones (that may also be quaternized) play an important role. Hydrolyzed proteins also offer numerous benefits from modifying hair feel, to delivering on the claim the conditioner is purported to have.

Humectants

While conventional humectants do not have any function on the hair in a rinse-off product, they are helpful in stabilizing the emulsion. Humectants are most particularly helpful in freeze-thaw conditions, and glycerin is a popular humectant for this purpose.

When a conditioner contains a significant quantity of emollients, conditioning agents, and has a rich enough viscosity for the target hair type, it is considered a treatment. These treatment products bear specific claims related to treating the hair for a specific malady, such as damage, frizz or breakage.

Leave-In Conditioners

Conditioners are traditionally applied to the hair while hair is wet, remain on the hair for a brief period, then are rinsed from the hair. However, some conditioners are designed to be left on the hair instead of being rinsed out and are classified as leave-in conditioners. These products generally contain a lower percentage of conditioning agents because the product will not be rinsed off. Additionally, quats are highly regulated for safety in leave-on products and thus are quite restricted in the amounts that can be used in this category.

Styling Products

Styling products do not have conditioning qualities, but rather are applied to wet, damp, or dry hair to aid in defining, shaping, lifting, styling, or sculpting the hair. While the most popular format is a styling gel, pomades and mousses are also popular modalities for reshaping coiffures. The product format primarily plays a role in the consumer's preference for residue on the hair. Pomades tend to be the heaviest on the hair as they contain high quantities of waxes and oils. Mousses and gels are generally lightweight, leaving little tangible residue behind.

Styling products are classified by both the strength of hold and duration of hold. High-hold styling products often leave the hair quite stiff, as they utilize high levels of polymers like polyvinylpyrrolidone (PVP) or VP/VA copolymer. Reducing the polymer level can decrease the strength of the hold, to medium hold or low hold. This may also reduce the duration of hold. Alternatively, using more flexible polymers, such as polyurethanes or polyquaterniums, will give a softer hold, and the quantity used will dictate the duration.

Polymers are quite susceptible to plasticizers, and using humectants like glycerin or other glycols can decrease the strength of hold the polymer can offer. While this can be detrimental if attempting to achieve a high-hold pomade, it can often work to benefit if looking for a more lived-in, less stiff look on the hair.

Hair sprays are not considered styling products but are rather considered finishing products. This has less to do with function (both hold hair in its new conformation) but rather is a regulatory discernment. Hair sprays are designed to be applied to the hair in a pump or aerosol format and contain volatile organic compounds (VOCs) like denatured alcohol or propellants. This allows the medium in which the styling polymer is dispersed to evaporate from the hair, leaving only the polymer behind. Due to environmental concerns, VOCs are highly regulated by various states in the US, particularly California. Finishing products like shine sprays or hair sprays (which simply add shine or hold hair in its place) are allowed to contain significantly higher VOC content than styling products. Hair sprays therefore should not make styling claims, but rather only claims for finishing a style.

Hair Coloring and Lightening

Hair Lightening

Consumers want to lighten their hair for numerous reasons; perhaps they desire to appear blonder, or they are preparing their hair to be a completely different color, such as purple, and need to erase their existing color. Regardless of the reason, the first step in lightening hair is to degrade existing melanin.

If we took a cross section of a brown hair fiber, and stained it, we would notice dark granules in the cross section, like chocolate chips in a cookie. These granules are deposits of eumelanin in the hair. If we started lightening the hair with a conventional hair bleach, the granules would start disappearing, leaving a reddish, diffused pigment. This reddish, diffused pigment is pheomelanin, which degrades much more slowly than the eumelanin granules. Eventually, the pheomelanin will degrade too. This degradation is a result of melanin's electron-rich polymeric structure undergoing oxidative attack and disintegrating. Hair lightening preparations employ an oxygen-rich molecule for such purposes called hydrogen peroxide.

Hydrogen peroxide did not come into use for hair lightening until around 1867 [38]. A French chemist and a British chemist demonstrated the use of hydrogen peroxide for bleaching hair at the Paris Exhibition. They created a lotion containing a 3% solution of hydrogen peroxide and found that it lightened the hair well compared to historical methods. Their preparation was extremely popular and is still the basis for lightening existing pigment in hair today.

The hydrogen peroxide solutions used initially were not very fast and did not lighten hair very well. Hydrogen peroxide is best able to depolymerize melanin at an alkaline pH [39]. Today, it is used in combination with alkaline materials and persulfate salts. The persulfate salts are not strong oxidizers on their own, and in fact are pretty much ineffective by themselves. However, in combination with hydrogen peroxide, persulfates exhibit a synergy that rapidly decolorizes hair over either alone.

The most common persulfates used in conventional hair lighteners are sodium persulfate, potassium persulfate, and ammonium persulfate. The latter liberates ammonia upon dissolution with water, which helps increase the alkalinity of the system for greater efficacy. Persulfates exist as powders and must be formulated in an anhydrous environment to avoid premature dissolution. The most typical format for a hair lightener containing persulfates is a powder format, although they can be incorporated into an anhydrous cream. Other ingredients in the lightening powder are stabilizers, which help maintain a high pH environment over the duration of the lightening process and consistency modifiers, such as starches. These help the product form a creamy paste that does not drip onto the consumer.

The lightening powder is mixed with a predetermined quantity of hydrogen peroxide solution at the point of application to the consumer, and the product is left on the consumers hair for up to 50 minutes. The final pH of the lightener ranges from 11 to 12. The hydrogen peroxide solution is expressed either by percentage of hydrogen peroxide contained within, or by volume, which is expressed as the number of liters of oxygen in gaseous form released by the decomposition of one liter of hydrogen peroxide solution. Table 2 expresses the relationship between percentage and volume, in common denominations available to the salon professional or consumer.

Although we want hydrogen peroxide to react quickly on hair, we do not want it to react in the packaging on shelf before it gets to the consumer. Since hydrogen peroxide decomposes most readily in alkaline pH, it needs to be stored at a low pH. This is done by having the final formula be an acidic solution. Since metals like copper and iron also catalyze hydrogen peroxide, it may be prudent to formulate with chelating agents to mitigate its decomposition.

In terms of evaluating the efficacy of a lightener, the universal salon professional measurement for measuring lightness is called level. The hair level system ranges from 1 to 12; the lower the number, the darker the hair is. Level 1 hair is generally defined as jet black, while levels 10–12 are designated for platinum blonde, depending on the brand. Level 6 is a dark blonde. Lighteners will generally advertise the magnitude of lift to be anywhere from 5 to 9 levels of lift. Meaning, in one application, a lightener with 7 levels of lift will take a level 1 hair to a level 8, which is very

Table 2 Common concentrations of hydrogen peroxide solutions used in hair lightening preparations and their corresponding volume equivalents, as expressed in salon professional terminology

% Hydrogen peroxide	Volume equivalent
1.5%	5 Volume
3.0%	10 Volume
6.0%	20 Volume
9.0%	30 Volume

Source: Author, 2025

light blonde. The fiber diameter, coarseness, and hair damage level play a role in the ability of the lightener to achieve the advertised lift.

There are a few downsides to lightening hair through conventional lightening practices. When we apply bleach mixtures onto the hair, the molecules are not selective for melanin. In fact, they interact with all molecules in the hair, such as hydrogen bonds, salt bridges, disulfide bonds, amino acids with amino side groups, and more. Hair lightening is thus extremely damaging to keratin. Hair is often left with extensive quantities of cysteic acid, as well as open spaces in the hair where melanin granules once resided. As hair is weathered differently depending on how long it has been growing from the head, and what other chemical processes have been done, the damage may also not be uniform. This makes future coloring services and treatments challenging.

Hair Coloring

Now that hair has been lightened, it is time to deposit color onto the hair. Hair color is generally separated into four categories and is usually designated by the permanence, or longevity, of the color result. When formulating each product category, the hair fiber's chemistry is taken into consideration, employing various dye molecules or ingredients to achieve the intended effect.

Temporary Color

Temporary color is just that—temporary. It is only intended to remain on the hair for a single event, until the hair is washed, and the colorant is removed. Temporary hair color is ready to use, meaning the effect is immediate and obvious. There is no chemical processing required for the end color result, and the color that you see in the mass of the product is the color result on the hair fiber.

Because the effect is temporary, there can be no absorption of the colorant into the cuticle or cortex of the hair fiber. To achieve this, pigments or pearls, which are physical, insoluble particles, are employed. Examples include iron oxides or lakes. In some instances, acid dyes can be used. These are large, anionic, water-soluble colorants that are too large to penetrate the hair due to their high molecular weight and are not formulated in such a way that allows for partial penetration.

Since pigments are physically unable to penetrate hair, the formula must be created in such a way to temporarily stick the colored molecules to the hair. This is achieved through use of polymers, such as PVP or VP/VA copolymer.

Examples of product formats for this color system are hair chalks, root touch-up sprays, and temporary color sprays. As the colorants are not embedded into the hair fiber, any shampoo system can remove the product (and color) from the hair fiber. This is a great option for young children or those looking for a quick change of color. The color result achieved from this type of color application can either be natural looking, in the case or root touch up sprays, or fun and bright.

Semi Permanent Color

This color category has more longevity than temporary color, and lasts up to 4–12 washes, depending on the color and formulation. Like temporary color, semi-permanent color is ready to use, requires no chemical processing, and the color of the neat product is generally the color result on the hair.

Semi-permanent color needs both adsorption onto and absorption into the hair fiber, so it employs acid, basic and nitro dyes to achieve this effect. Acid dyes yield extremely bright colors. They are able to show color due to their conjugated electron system centering around an azo group. While they are large and anionic, they can be used in low pH formulations that generate a positive charge on the hair, allowing the negatively charged dyes to stick to the positively charged sites. It is more common to formulate using solvents that can aid in partial penetration of the acid dyes into the fiber.

On the contrary, basic dyes are positively charged dye molecules that readily stick to negatively charged sites on hair. Their cationicity allows for instantaneous color effects. Like acid dyes, basic dyes have a high molecular weight and primarily rely on their cationic character to stick to hair, although they can partially penetrate with the right formulation. While basic dyes also have a conjugated electron system, they tend to be less vibrant in color than acid dyes. The optimal pH for basic dyes lies between 4 and 9, allowing for maximum negative charges on the hair fiber, allowing the basic dyes to stick to the hair.

Nitro dyes are a lesser used category of colorants. They are non-ionic, low molecular weight molecules that readily penetrate the hair due to their small size. They are required to have a nitro group present on a benzene ring, which provides the required conjugated electron system and thus color. The substitutions present on the benzene ring yield a variety of colors, from yellows to oranges to reds to blues. While acid dyes heavily stain skin (which is basic), nitro dyes do not stain skin. However, they have poor water solubility so appropriate formulation techniques must be employed to ensure their incorporation. Since they are non-ionic and small, their retention in hair comes from Van der Waals interactions. Unfortunately, because they are small and can penetrate the hair easily, they can also leave the hair easily.

The product format for semi-permanent color is usually a consumer-friendly format, such as a shampoo or cream that can be applied for the hair quickly and rinsed off. In some cases where extreme color intensity is desired, a conditioning cream can be left on the hair up to 30 minutes before rinsing from the hair. The longevity of the color result depends on the affinity of the dye for the hair fiber, the health and porosity of the consumer's hair, and the overall formulation (Fig. 11).

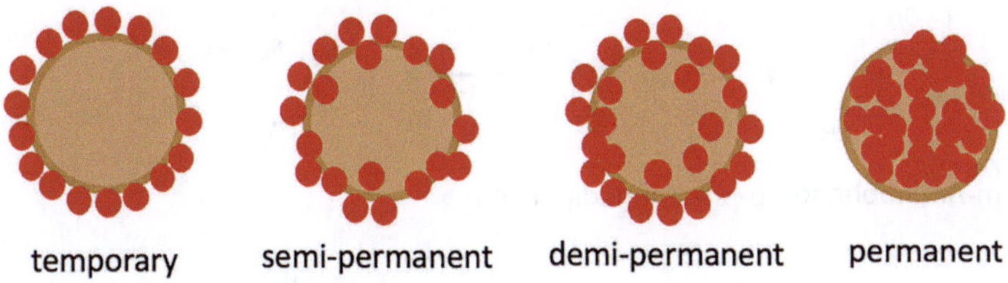

Fig. 11 How dyes interact with the hair fiber, for each colorant category. The circle represents the cross section of hair, while the red dots represent the dye molecules and how they interact with the fiber. (Source: Author, 2025)

Demi-Permanent Color

Demi-permanent color uses a different chemistry from semipermanent and temporary color, in that demi-permanent color requires chemical processing. The color seen in the neat product is not the end color result on hair, as oxidation must occur for color to be seen. The dyes contained within are called oxidative dyes, as they require a source of oxygen. This is provided via combination at the time of use with a low-dose hydrogen peroxide developer.

Demi-permanent color lasts up to 4–6 weeks and is historically a North American salon professional product format which has just gained popularity in Europe. This color format is considered the "stained glass of hair," with just enough color deposit to enhance the natural hair color of the consumer. To achieve longevity, demi-permanent color requires a minimally alkaline environment that swells the hair fiber to allow the penetration of undeveloped oxidative dye molecules. Once inside the hair, the dye molecules develop with 5 or 10 volume hydrogen peroxide to form a final color. As the pH is not quite alkaline and the dose of hydrogen peroxide is low, the environment does not allow for any lightening of naturally occurring melanin pigment. This process takes anywhere from 20 to 30 minutes.

Due to the increased longevity expectation over semi-permanent color, dyes must both absorb into and adsorb onto the hair, and certain nitro and basic dyes may be employed in this format. Oxidative dyes do not contain conjugated electron systems and are not chromophores on their own. When they are joined together through oxidation, their electrons become conjugated, and the molecules form a color.

There are two categories of oxidative dyes, primaries and secondaries. To form color through an oxidation pathway, one molecule from each category is required in equimolar amounts. Once hydrogen peroxide is introduced, the molecules undergo an SN1 reaction to form the chromophore [40]. The color result depends on the combination of primary and secondary in the formulation.

For example, p-phenylenediamine (primary) and m-aminophenol (secondary) create a reddish-brown color, while 1-hydroxyethyl-4,5-diamino pyrazole sulfate (primary) and m-aminophenol (secondary) form an orange color (Fig. 12). Because

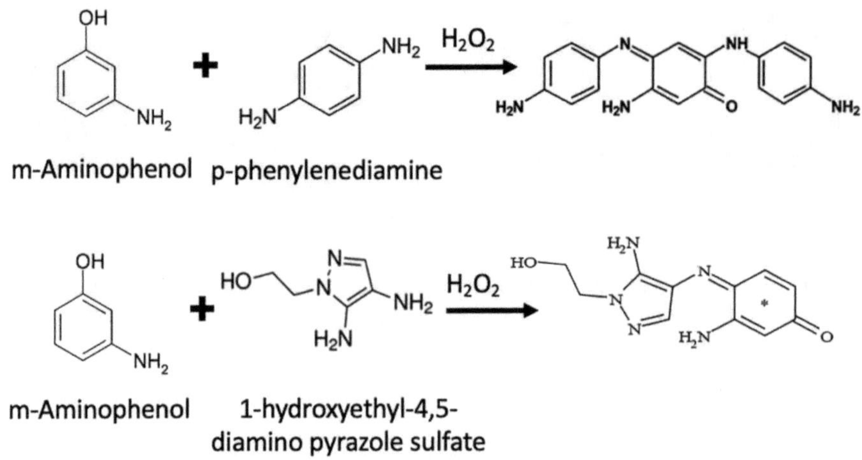

Fig. 12 Oxidation reaction between an uncolored secondary and uncolored primary, forming a polymerized color molecule; the final result can be reddish brown or bright orange depending on the combination. (Source: Author, 2025)

the uncolored dye molecules' size effectively doubles after penetrating the hair, the molecules are trapped, contributing to the product's longevity.

As this is a reactive product, it is necessary to prevent the product from reacting before it reaches the consumer. A demi-permanent color formulation would thus also contain antioxidants, such as ascorbic acid or sodium bisulfite. It also contains a chelating agent to trap any metal ions which may later get introduced to the hydrogen peroxide mixture. As we do not want a highly alkaline system, a demi-permanent formulation may contain a very low level of alkalizer, such as monoethanolamine.

Permanent Color

From a chemistry viewpoint, there are similarities between demi-permanent and permanent color, with the exception that permanent color is just that—permanent. The consumer must recolor their hair due to regrowth of their original hair color from the root, not because the previously dyed permanent color has faded or shifted so radically.

The chemical principles of demi-permanent hair color are still employed, but the ingredient choices are slightly different to accommodate for the desired level of permanency and color change required by consumers. While demi-permanent color is meant to complement one's natural hair color, permanent color is meant to replace one's natural color. Therefore, a higher volume of hydrogen peroxide is used over demi-permanent color, such as 20 volume. The pH of the permanent color system is also considerably higher, 10–11. This allows for hydrogen peroxide to work effectively to remove the natural pigment in the hair, making way for the new color that will deposit.

As permanent color relies completely on the absorption of the colorants into the hair fiber, strictly oxidative dyes are employed. To swell the hair to allow for maximum penetration, a strong alkalizer is used, such as ammonia. Monoethanolamine is used when the hair color is designated as ammonia-free, but it does not have the same synergy with hydrogen peroxide that ammonia does to remove existing pigment from the hair. It is often thought that monoethanolamine is considered less damaging than ammonia because it does not have the strong, pervasive odor that ammonia has. At equimolar concentrations, monoethanolamine has been demonstrated to strip the cuticle layer more than ammonia and contribute to 40% more protein loss in hair [41]. Because monoethanolamine is less efficient than ammonia, higher quantities often must be used, making the end results even more catastrophic to the hair fiber. Nonetheless, consumers continue to demand ammonia-free color.

The chemistry employed in permanent hair color today has been around since the 1880s and is still the basis of chemistry we use today. While some may view the hair color industry as antiquated and lacking innovation, hair color research groups at L'Oréal, Wella and Henkel, have significantly invested in the development of new dye molecules, as well as explored alternative chemistries to color hair. Hair dyes are extensively studied for toxicology and are one of the most studied classes of ingredients in personal care today for safety.

There has been a rise, however, in the propensity of allergies to hair color over the last handful of decades. While there is an increase in the use of hair colorants, p-phenylenediamine has been used commercially for over 140 years to color hair. Some sources point to the rise in tattoos, which involves the injection of ink into the skin's dermis. These inks use a variety of potentially sensitizing colorants. Another theory is that the allergenicity rate of p-phenylenediamine coincides with the rise in tourism along coastal communities where black henna tattoos are popular [42]. Black henna tattoos contain p-phenylenediamine combined with henna. Since the p-phenylenediamine has no secondary to react with, it eventually auto-oxidizes to react with itself, forming an extreme sensitizing compound called Bandrowski's Base [40]. Although black henna tattoos are banned virtually everywhere, they are still pervasive in beach communities.

On its own, henna is not black. It is a light reddish-brown color that has been used for millennia to color both skin and hair. The active coloring component in henna is lawsone, which is a highly conjugated chromophore that can penetrate the hair fiber, albeit slowly [43]. The henna is supplied in a powder form where it is mixed with hot water into a paste before it is applied onto the hair. The only mechanism by which henna can penetrate is through diffusion, which takes a considerable amount of time. The result can look quite natural, especially when henna is combined with other plant extracts such as indigo or cassia.

While henna is considered the only natural form of color, there are safety considerations to take into place. The first is the henna plant scavenges heavy metals from the soil, and it is very difficult to find reputable suppliers that can control the heavy metal content. The heavy metals inevitably end up penetrating into the hair fiber. When the consumer wants to return to conventional color or lightener with hydrogen peroxide, the metals catalyze the decomposition of hydrogen peroxide, emitting extreme heat. This causes the potential to injure the consumer.

It is also worth noting that the genotoxicity of lawsone is inconclusive as mixed results have been reported in its ability to mutate chromosomes [44]. Scientific bodies such as the SCCS in Europe have placed restrictions on its use, as well as requested further genotoxicity studies.

Bond Builders

The historical conversation around hair's bonds was that bonds were broken during hair color or lightening services, leading to damaged hair. The only remedy was to condition and treat the hair after. That is, until a new product category touting to rebuild bonds from breaking disrupted the market in 2014—bond builders.

Olaplex was the first bond-building brand to enter the market with a proprietary technology, bis-aminopropyl diglycol dimaleate, a maleic acid derivative. Their bond building solution is added directly to the hair color or hair lightener mixture at time of application to the consumer and is followed by a conditioning treatment. The theory is that by reducing the number of broken disulfide bonds during a professional coloring or lightening service, the hair is less prone to breakage after.

The technology was originally touted to help repair and strengthen the hair by restoring the disulfide bonds that are broken during chemical processing like coloring or lightening, specifically referring to thiol groups within the hair [45]. While bismaleimide compounds can cross-link to thiol groups, very few thiol groups exist in damaged hair. Most broken disulfide bonds are rapidly oxidized to sulfonic acid groups, which are no longer reactive. Cross-linking activity of bismaleimide compounds also occurs at a neutral pH, whereas hair color is alkaline. However, bismaleimide compounds *can* react with amines under alkaline conditions, of which there are plenty in the hair [46].

Since the launch of Olaplex, a plethora of brands have launched their own bond builders. While few make claims of rebuilding disulfide bonds, consumers nonetheless recognize the ability of bond-building products to improve the mechanical properties of their hair, restoring its strength. Many companies use maleic acid or malic acid as bond-building agents in their formulations, and for good reason. Research shows that low-pH molecules containing carboxylic acid can improve hair's fiber mechanics [47]. For example, a 5% citric acid solution can displace water from hair, improving hair's overall strength.

Glossary

18-MEA (18-Methyl Eicosanoic Acid) A lipid that contributes to hair's natural hydrophobicity and smoothness.

Androgenetic alopecia A common genetic form of hair loss due to hormonal influence.

Cell membrane complex (CMC) A lipid-protein network that holds cuticle and cortex cells together, influencing moisture retention.

Cortex The central portion of the hair shaft, responsible for strength and elasticity.

Cuticle The outermost protective layer of hair made of overlapping keratinized cells.

Dermal papilla A structure at the base of the follicle that regulates hair growth.

Disulfide bond A strong covalent bond between sulfur atoms in cysteine amino acids that determines hair strength and shape.

Eumelanin The dark pigment responsible for brown and black hair colors.

Hydrogen bond A temporary bond that gives hair structure, disrupted by water.

Hydrogen peroxide A chemical used in hair bleaching and oxidative dye reactions.

Keratin A structural protein that makes up the majority of the hair fiber.

Medulla The innermost, often fragmented, part of the hair shaft with an unclear biological function.

Melanosome An organelle in melanocytes that produces and transfers melanin into hair.

Pheomelanin The pigment responsible for red and yellow hues in hair.

Photodegradation The breakdown of hair components, including melanin and proteins, due to light exposure.

Porosity The hair's ability to absorb and retain moisture, influenced by cuticle condition.

Quaternary ammonium compound (Quat) A positively charged conditioning agent that adheres to hair to reduce static and improve manageability.

Salt bridge An ionic bond between positively and negatively charged amino acids, contributing to hair strength.

Sebaceous gland A gland in the scalp that secretes sebum to lubricate hair and skin.

References

1. Takahashi, T., Hayashi, R., et al. (2006). Morphology and properties of Asian and Caucasian hair. *Journal of Cosmetic Science, 57*, 327–338.
2. Hardy, D. (1973). Quantitative hair form variation in seven populations. *American Journal of Physical Anthropology, 39*, 7–18.
3. Tanamachi, H., et al. (2010). 18-MEA and hair appearance. *Journal of Cosmetic Science, 61*(2), 147–160.
4. Dekio, S., & Jidoi, J. (1990). Amounts of fibrous and matrix substances in hairs of different races. *The Journal of Dermatology, 17*, 62–64.
5. Csuka, D. A., et al. (2023). A systematic review on the lipid composition of human hair. *International Journal of Dermatology, 62*(3), 404–415.
6. Takahashi, T., & Nakamura, K. (2004). A study of the photolightening mechanism of blond hair with visible and ultraviolet light. *Journal of Cosmetic Science, 55*(3), 291–305.
7. Seiberg, M. (2013). Age-induced hair greying—The multiple effects of oxidative stress. *International Journal of Cosmetic Science, 35*(6), 532–538.
8. Pinker, R. T., et al. (2005). The solar spectrum and its variation: A literature review. *Geophysical Research Letters, 32*(6).

9. Daly, S., et al. (2009). Fluorescence and coloration of grey hair. *International Journal of Cosmetic Science, 31*(5), 347–359.
10. Fair, N. B., & Gupta, B. S. (1987). The chlorine-hair interaction. I. Review of mechanisms and changes in properties of keratin fibers. *Journal of Cosmetic Science, 38*(6), 359–370.
11. McMullen, R. L., Gillece, T., & Schiess, T. (2022). Physicochemical properties of textured hair. *Journal of Cosmetic Science, 72*(6), 711–731.
12. Rogers, M. A., et al. (2006). Human Hair Keratin-Associated Proteins (KAPs). *International Review of Cytology, 251*, 209–263. Academic Press.
13. Moll, R., Divo, M., & Langbein, L. (2008). The human keratins: Biology and pathology. *Histochemistry and Cell Biology, 129*(6), 465–484.
14. Zviak, C., et al. (1986). *The science of hair care* (1st ed.). CRC Press. Chapter 1.
15. Robbins, C. R., et al. (1970). Amino acid composition of human hair. *Textile Research Journal, 40*(10), 891–896.
16. Malinauskyte, E., et al. (2020). Effect of equilibrium pH on the structure and properties of bleach-damaged human hair fibers. *Biopolymers, 111*, 11.
17. Wilkerson, V. (1935–1936). The chemistry of human epidermis. *The Journal of Biological Chemistry, 112*, 329–335.
18. Jones, L. N., & Pope, F. M. (1985). Isolation of intermediate filament assemblies from human hair follicles. *Journal of Cell Biology, 101*(4), 1569–1577.
19. Randebrock, R. (1964). Neue Erkenntnisse über den Morphologischen Aufbau des Menschlichen Hares. *Journal of the Society of Cosmetic Chemists, 15*, 691–706.
20. Thibaut, S., et al. (2007). Human hair keratin network and curvature. *International Journal of Dermatology, 46*(suppl. 1), 7–10.
21. Syed, A. (2013, October 28) Hair ethnicity and ellipticity: A preliminary study. *Cosmetics & Toiletries.* www.cosmeticsandtoiletries.com/testing/method-process/article/21835140/hair-ethnicity-and-ellipticity-a-preliminary-study
22. Loussouarn, G., et al. (2007). Worldwide diversity of hair curliness: A new method of assessment. *International Journal of Dermatology, 46*(Suppl 1), 2–6.
23. Gaines, M., et al. (2023). Reimagining hair science: A new approach to classify curly hair phenotypes via new quantitative geometric and structural mechanical parameters. *Accounts of Chemical Research, 56*(11), 1330.
24. Wortmann, F.-J., & Zahn, H. (1994). The stress/strain curve of α-keratin fibers and the structure of the intermediate filament. *Textile Research Journal, 64*(12), 737–743.
25. Tajima, M., et al. (2007). Characteristic features of Japanese women's hair with aging and with progressing hair loss. *Journal of Dermatological Science, 45*(2), 93–103.
26. Wortmann, F. (2012). Thermal denaturation and structural changes of α-helical proteins in keratins. *Journal of Structural Biology, 177*(2), 553–560.
27. Bertolini, M., et al. (2018). Bridging the "dead hair"-"live follicle" divide in applied hair research. *Journal of Cosmetic Science, 69*(5), 315–322.
28. Hori, H., et al. (1972). The thickness of human scalp: Normal and bald. *The Journal of Investigative Dermatology, 58*(6), 396–399. https://doi.org/10.1111/1523-1747.ep12540633
29. Hoover, E., Aslam, S., & Krishnamurthy, K. (2025). *Physiology, sebaceous glands*. [Updated 2022 Oct 10]. In: StatPearls [Internet]. Treasure Island: StatPearls Publishing. Available from: https://www.ncbi.nlm.nih.gov/books/NBK499819/
30. Pochi, P. E., Strauss, J. S., & Downing, D. T. (1979). Age-related changes in sebaceous gland activity. *Journal of Investigative Dermatology, 73*(1), 108–111.
31. Lin, Q., et al. (2021). Malassezia and staphylococcus dominate scalp microbiome for seborrheic dermatitis. *Bioprocess and Biosystems Engineering, 44*(5), 965–975. https://doi.org/10.1007/s00449-020-02333-5
32. Clavaud, C., et al. (2013). Dandruff is associated with disequilibrium in the proportion of the major bacterial and fungal populations colonizing the scalp. *PLoS ONE, 8*(3), e58203. https://doi.org/10.1371/journal.pone.0058203

33. Pople, J. E., et al. (2019). Changes in epidermal morphology associated with dandruff. *International Journal of Cosmetic Science, 41*(4), 357–363.
34. Walter, K. (2022). Common causes of hair loss. *Journal of the American Medical Association, 328*(7), 686.
35. Messenger, A. G. (1993). The control of hair growth: An overview. *Journal of Investigative Dermatology, 101, no. 1, suppl*, S4–S9.
36. Randall, V. A. (2008). Androgens and hair growth. *Dermatologic Therapy, 21*, 314–328.
37. Messenger, A. G., & Rundegren, J. (2004). Minoxidil: mechanisms of action on hair growth. *British Journal of Dermatology, 150*(2), 186–194.
38. IARC Working Group on the Evaluation of Carcinogenic Risks to Humans. Occupational exposures of hairdressers and barbers and personal use of hair colourants; some hair dyes, cosmetic colourants, industrial dyestuffs and aromatic amines. Lyon (FR): International Agency for Research on Cancer; 1993. (IARC Monographs on the Evaluation of Carcinogenic Risks to Humans, No. 57.) Occupational Exposures of Hairdressers and Barbers and Personal use of Hair Colourants. Available from: https://www.ncbi.nlm.nih.gov/books/NBK493259/
39. Wolfram, L., & Hui, I. (1970). The mechanism of hair bleaching. *Journal of Cosmetic Science, 21*(9), 875–900.
40. Corbett, J. F. (1972). The autooxidation of P-Phenylenediamine. *Journal of Cosmetic Science, 23*(6), 683–693.
41. Bailey, A. D., et al. (2014). Comparison of damage to human hair fibers caused by monoethanolamine- and ammonia-based hair colorants. *Journal of Cosmetic Science, 65*(1), 1–9.
42. Cartwright-Jones, C. (2015). *'Black henna' and the epidemic of para-phenylenediamine sensitization: Mapping the potential for extreme sensitization to oxidative hair dye*. For presentation at the Society of Cosmetic Chemists' 70th annual scientific meeting, New York.
43. Cui, H., et al. (2022). Recent advancements in natural plant colorants used for hair dye applications: A review. *Molecules (Basel, Switzerland), 27*(22), 8062.
44. SCCS (Scientific Committee on Consumer Safety). (2013, 19 September). *Opinion on Lawsonia inermis (henna).*
45. Pressly, E. D., & Hawker, C. J. (2014, April 21). *Methods for fixing hair and skin*. Patent No. US9095518B2, issued by the United States Patent and Trademark Office.
46. Wong, S. S. (1991). *Chemistry of protein conjugation and cross-linking* (p. 34). CRC Press.
47. Evans, T. (2018, September 28). Your hair on acid: The influence of carboxylic acids. *Cosmetics & Toiletries*. www.cosmeticsandtoiletries.com/testing/efficacy/article/21837243/your-hair-on-acid-the-influence-of-carboxylic-acids

Color Cosmetics

Kathy Cruz

Abstract This chapter focuses on eye, face, and lip color cosmetic technologies. Color cosmetic product forms are explained in detail along with the myriad of systems in which they are created. Many examples of color formulations are provided with ingredient options and recommended use levels. This chapter also includes current and future technology-based ingredient trends in addition to safety, regulatory, social and environmentally responsible ingredient upgrades that have occurred over the years.

Keywords Alternatives · Anhydrous · Aqueous · Biobased · Biodiversity · Blush · Bronzer · Brow · Circular beauty · Clean beauty · Color correctors · Compostable · Concealer · Conscious beauty · Cruelty free · Degradability · Eco-conscious · Ecofriendly · Emulsions · Environmentally friendly · Eye · Eyeliner · Eyeshadow · Face · Face powder · Foundation · Gels · Green · Greenwashing · Heavy metals · Highlighter · Hot pours · Ingredient trends · Lip · Lip balm · Lip gloss · Lip liner · Lip oil · Lipstick · Liquid · Liquid lipsticks · Loose powder · Low waste · Mascara · Mica · Microplastics · Nanomaterials · Non-toxic · Organic · Parabens · Pigments · Preservatives · Pressed powder · Primer · Recyclable beauty · Roundtable on sustainable palm oil · RSPO · Silicone · Suspensions · Talc · Tinted moisturizers · Treated pigments · Upcycling · Vegan · Volatile · Zero waste

Part 1a: Cosmetic Overview

Color Cosmetics is typically broken down into four categories: Eye, Lip, Face and Nail. Nail will be addressed in chapter "Innovations in Nail Beauty: Past, Present, and Future". This chapter will focus on the many color cosmetic products that fall within the eye, lip and face categories.

K. Cruz (✉)
Director of Applications Laboratory, Kobo Products, Inc., South Plainfield, NJ, USA

K. Alexander, P. Romanowski (eds.), *Cosmetic Science*, AAPS Introductions in the Pharmaceutical Sciences 15, https://doi.org/10.1007/978-3-032-03596-7_5

Eye Category

Eye makeup consists of a variety of products including mascara, eyeshadow, eyeliner and brow formulas all designed to enhance and beautify the eye area.

Mascara

Mascara's primary function is to adorn the eyelashes by darkening, lengthening, curling and/or thickening. These developments are complex as they are a synergistic combination of three separate technologies: formula, brush and wiper. Within each of these technologies, there are many facets.

Mascara Formula technology: While most mascaras are emulsion systems, the formula types vary based on the intended performance goals. For example, waterproof mascaras are typically anhydrous, solvent-based formulations while washable mascaras are water-based emulsions.

Mascara wiper technology: The size of the wiper's orifice determines the amount of product that is deposited on the brush and its' bristles as the brush is pulled out of the package. This, of course, also determines the amount of mascara bulk that is added to the lashes.

Mascara Brush technology: Bristles on a mascara brush are made from different materials. Wire and plastic are the most prevalent. The shape and spacing of the bristles determine how much product is deposited on lashes, and ultimately, the performance of the mascara product.

Eyeshadow

Eyeshadow is another way to enhance the eyes. Available in many colors and finishes, eyeshadow formulas are also quite versatile. Swiped across the eyelid, and sometimes applied as a thin line beneath the lower lid, various color and finish combinations yield different eye effects. Eyeshadows appear commercially in many forms including pressed powder, stick, cream, liquid, gels and pencils.

Eyeliner

Lining the eyes, whether along the lash lid or on the waterline, is a technique that defines the eyes and attracts attention. Anhydrous pencils and crayons, anhydrous volatile gel pots and liquid felt-tip flow-through systems are among the many forms of eyeliner products.

Brow Products

Eyebrow products have many functions including shaping the brows, filling in sparse brows and coloring the brows.

Face Category

Face makeup consists of a variety of products including foundation, tinted moisturizers, concealer, color correctors, blush, bronzers, highlighters, primers and face powder.

Foundation

Much like the foundation of a house, makeup foundation is the base from which all other makeup is built upon. Foundations even out skin tone and create a smooth, level canvas.

Consumers have much to consider when selecting foundation. Each selection factor is highly personalized. First is form and application type. There are various forms to choose from such as emulsions, anhydrous sticks, loose powder or pressed powder. Coverage is the next layer of decision-making. Depending on the consumer's skin condition, he/she may need anywhere between sheer and full coverage. Finish is another selection criteria for foundation. Does the consumer prefer matte, natural, dewy? Other performance attributes such as SPF, long wear, sweatproof and transferproof are also considered. Finally, color selection completes the foundation decision-making process. Selecting one's personal shade can be quite challenging. Undertone should always be considered. It is much easier for the consumer when in-store sampling is available, so that they leave the store with a perfect skin match.

Concealer

Concealers are similar to foundations in form availability but contain a higher concentration of pigments. This elevated level of coverage is meant to cover up blemishes and various skin imperfections as well as lighten up undereye circles.

Color Correctors

Color Correctors are intended to camouflage skin imperfections, undereye circles and uneven skin tones. Complementary color technology is utilized for these products.

Tinted Moisturizers

Creating a shade palette to match all skin tones is no easy task, as not every company can afford a large size of line. This is where tinted moisturizers have an advantage over foundation. These products contain a smaller concentration of pigment which enables each sku to cover a range of skin tones versus highly pigmented foundations which are more skin tone specific.

Blush/Bronzer/Highlighter

Whether the consumer wants to add a healthy glow, build definition or contour the shape of the face, this category is quite versatile. Blush should be buildable so that depending on the consumer's desired look, he or she can apply as a subtle healthy glow or a more dramatic pop of color.

Primer

The main function of a primer is to extend the wear of makeup as well as to create a smooth and even face finish, often with blurring, soft focus properties.

Face Powder

Face powder, whether in loose or pressed form, is intended to set foundation in place. It will typically have a mattifying effect. However, face powder can also be used as a sort of foundation. Consumers who are not in need of a heavily pigmented face product will use face powder alone as a simple foundation. It is also key to note that in very hot climates, powders are commonly used as foundation.

Lip Category

There are many types of lip products including lipstick, lip gloss, lip oils, lip balms, lip liners, and liquid lipsticks. Many of these developments can be formulated to be long-wearing and transfer-proof.

Lipstick

Lipstick is an anhydrous stick that is structured with waxes, colored with pigments and enriched with emollients for payoff.

Lip Gloss/Lip Oils

These fluid, anhydrous systems have a high shine and tend to provide sheer pigment coverage.

Lip Balm

Often in a supported stick format, lip balms are typically formulated with waxes for structure, emollients for slip and cushiony semi-solid hydrocarbons for lubricity and moisturization purposes. This ingredient mix produces an occlusive layer on the lips which helps prevent trans-epidermal water loss (TEWL) for moisturization properties.

Lip Liner

Mostly in pencil form, lip liners are formulated to provide an occlusive line at the edge of the lips with the intention of preventing color bleeding. These products come in many colors so that they can be matched to the multitude of lipstick shades on the market.

Liquid Lipsticks

Known for being highly pigmented, liquid lipsticks are also typically designed to be long-wearing. Highly effective film formers, intense pigment loads and pigment suspension agents are customary ingredients in these products.

Part 1b: Color Cosmetic Systems

Almost all cosmetic products are available in the same variety of systems which include pressed and loose powders, anhydrous hot pours, aqueous gels, emulsions and volatile liquid suspensions. This section will cover a selection of systems for the delivery of color cosmetics.

Pressed and Loose Powders

The most prevalent cosmetic products developed in powder form are eyeshadow, face powder and blush. These products appear on the market in both loose and pressed (compact) forms. Powders consist of a main base powder ingredient, binding agents (dry and wet), pigments, fillers, aesthetic modifiers and preservatives.

Typical powders are commonly formulated with a main powder base ingredient. Traditionally, talc has been the basis for these formulas. Known for its creamy texture and providing opacity behind pigments, talc is the softest mineral on earth. Due to current health concerns that will be addressed later in this chapter, many powder products are now reformulated with talc alternatives.

Formulating pressed powders can be quite tricky. To formulate a stable pressed powder that withstands harsh transit conditions in addition to normal consumer habits, a pressed powder cake should stay intact throughout the life of the product. The product should not shatter even if dropped or distressed in a consumer's purse. The formulator must delicately balance the ratios between the liquid binder, dry binder, fill weight and pressing parameters.

To ensure the integrity of the cake, pressed powders must undergo a severe drop test. There are several variations of this test that are company-specific. A common drop test is dropping the filled powder pan from 12 inches three times onto a hard surface. Another severe test is dropping the pan from 36 inches and then again from 12 inches. To pass this test, there must be minimal to no cracking of the cake. Failures include large cracks or complete shattering of the cake. The pass and fail drop test examples are shown in Images 1 and 2. The failed glaze test is shown in Image 3.

Image 1 Pressed powder: passed drop test example. (Pictures taken at Kobo Products, Inc. US Application Lab in South Plainfield, NJ 2024)

Image 2 Pressed powder: failed drop test example. (Pictures taken at Kobo Products, Inc. US Application Lab in South Plainfield, NJ 2024)

Image 3 Failed glaze test example. (Pictures taken at Kobo Products, Inc. US Application Lab in South Plainfield, NJ 2024)

Glazing is a pressed powder phenomenon that chemists should be aware of. Glazing means that with use, the product will eventually show a glazed surface that no longer releases product (see Image 3). This essentially means that the product is too hard. There is either too much binder in the formula or the pressing parameters are inaccurate. Each formula is unique to its' optimal combination of binders and pressing factors therefore glazing can be remedied by adjusting these aspects.

Pressed powder formula breakdown:

Formula functional agents	Raw ingredient possibilities	Typical use level
Main system carrier/ filler	Talc; talc alternatives; treated mica, silica	10–70%
Wet binder	Dimethicone, oils (i.e., caprylic/capric triglyceride), esters (i.e., isopropyl palmitate)	2–15%
Dry binder	Zinc myristate, magnesium myristate	4–6%
Aesthetic modifiers	Boron nitride, nylon, silica, dimethicone	5–15%
Colorants	Pigments, pearlescent pigments	10–40%
Preservatives	Propylparaben, phenoxyethanol, glyceryl caprylate (and) glyceryl undecylenate	0.8–1%

A Talc-Free Bronzer formulation example is given in Table 1

Pressed Powder Formula Example:
Talc-free Bronzer

Anhydrous Hot Pours

Anhydrous hot pour formulations are the basis for many cosmetic products including lip gloss, lipstick, eyeshadow, eyeliner pot, and cream to powder compact makeup. These systems require high temperature filling and specific cooling rates which are unique to each formulation. Cooling rate is key as it will determine the softness of the stick or pot.

Formula functional agents	Raw ingredient possibilities	Typical use level
Waxes	Ozokerite, beeswax, carnauba, microcrystalline, polyethylene	7–25%
Shine agents	Polybutene, silicone	5–20%
Emollients	Oils, esters, petrolatum	10–20%
Stabilizers	Silica	0.2–3%
Fillers	Mica	2–10%
Aesthetic modifiers	Boron nitride, nylon, silica	2–15%
Film formers (if applicable)	Glyceryl hydrogenated rosinate (and) caprylic/capric trigliceride	1–7%
Sunscreen actives (optional)	Mineral actives: titanium dioxide, zinc oxide; organic actives	Follow current FDA guidelines.
Colorants	Pigments, pearlescent pigments	5–40%
Preservatives	Phenoxyethanol; 1,2 hexanediol (and) caprylyl glycol	0.6–0.8%
Flavor (if applicable)		0.1–1%
Sweetener (if applicable)	Sodium saccharin, sucralose	0.2–0.5%
Promotional ingredients (if applicable)	Vitamins (i.e., tocopherol)	0.01–0.50%
Active ingredients (if applicable)	Hyaluronic acid	Varies per active

Table 1 Pressed powder formula example

Phase	Percentage (%)	Ingredient INCI	Ingredient function
1	22.50	Mica (and) Stearoyl glutamic acid	Main system carrier (talc replacement)
	12.95	Boron nitride	Aesthetic modifier
	10.00	Silica	Bulking agent/filler
	5.00	Zinc Myristate	Dry binding agent
	1.00	Titanium dioxide (and) Stearoyl glutamic acid (and) Polyhydroxystearic acid	Treated pigment
	0.80	Iron Oxides (CI 77492) (and) Stearoyl Glutamic Acid (and) Polyhydroxystearic Acid	Treated pigment
	0.50	Iron Oxides (CI 77491) (and) Stearoyl Glutamic Acid (and) Polyhydroxystearic Acid	Treated pigment
	0.25	Iron Oxides (CI 77499) (and) Stearoyl Glutamic Acid (and) Polyhydroxystearic Acid	Treated pigment
	1.00	Glyceryl Caprylate (and) glyceryl Undecylenate	Preservative
2	15.00	Caprylic/Capric triglyceride	Wet binder/aesthetic modifier
3	20.00	Synthetic Fluorphlogopite (and) Iron Oxides (CI 77491) (and) Silica	Pearlescent pigment
	9.00	Mica (And) Titanium Dioxide (and) Iron Oxides (CI 77491)	Pearlescent pigment
	1.00	Iron Oxides (CI 77499) (and) Mica	Pearlescent pigment
	100.00%		

https://www.koboproductsinc.com/formulations/KPP-134A.pdf

An anhydrous stick formulation example is given in Table 2.

Anhydrous Formulation Examples:
Peachy Glow Highlighter Stick

Aqueous Gels

Aqueous gels have a light feel due to the high level of water in these formulas. These systems work well for low coverage blushers, cool feeling eyeshadows and liquid eyeliners.

Formula functional agents	Raw ingredient possibilities	Typical use level
Main carrier	Water	60% +
Thickener(s)	Xanthan gum, carbomer, hydroxyethylcellulose, magnesium aluminum silicate	0.5–1%
Film former	Polyvinylpyrrolidone, acrylates copolymer	0.5–7%
Colorants	Dyes, pigments, pearlescent pigments	0–25%
Preservative	Phenoxyethanol, ethylhexylglycerin	1.00%
Chelating agent	Disodium EDTA	0.1%–0.5%
Promotional ingredients (if applicable)	Vitamins, extracts	0.001–0.5%
Active ingredients (if applicable)	Hyaluronic acid	Varies per active

An aqueous gel formulation example is given in Table 3.

Aqueous Gel System Example:
The Simplest Eyeliner in the World

Emulsions

There are multiple types of emulsions used in color cosmetics. The most common are oil-in-water, water-in-oil, and water-in-silicone emulsions. Due to the growing prevalence of treated pigments, water-in-oil systems are becoming increasingly pervasive for highly pigmented liquid foundations since the treated pigments are compatible with the external oil phase.

Foundation Emulsion Basic Structure—W/O; W/Si

Formula structural/ functional agents	Raw ingredient possibilities	Typical use level
Main silicone/ volatile carriers	Dimethicone, isododecane, C13–15 alkane	40–70%

Table 2 Highlighter stick formula example

Phase	Percentage (%)	Ingredient INCI	Ingredient function
1	40.80	Tripelargonin	Emollient
	17.00	Isocetyl Stearoyl stearate	Ester
	7.14	Glyceryl hydrogenated Rosinate (and) Caprylic/Capric triglyceride	Film former
	2.86	Caprylic/Capric triglyceride	Emollient
	2.00	Polyglyceryl-3 Diisostearate	Aesthetic modifier
	0.60	1,2-Hexanediol (and) Caprylyl glycol	Preservative
	0.60	Iron Oxides (CI 77492) (and) Stearoyl Glutamic Acid (And) Polyhydroxystearic Acid	Treated pigment
	0.30	Iron Oxides (CI 77491) (and) Stearoyl Glutamic Acid (and) Polyhydroxystearic Acid	Treated pigment
2	6.00	Helianthus Annuus seed wax	Wax, structure agent
	4.00	Pentaerythrityl Tetrabehenate	Ester-based wax, structuring agent
	2.00	C20–40 alkyl stearate	Ester-based wax, structuring agent
3	5.00	Synthetic Fluorphlogopite (and) Silica (and) Titanium Dioxide (and) Iron Oxides (CI 77491)	Pearlescent pigment
	3.75	Mica (and) Titanium Dioxide (and) Iron Oxides (CI 77491)	Pearlescent pigment
	3.75	Synthetic Fluorphlogopite (and) Titanium Dioxide (and) Iron Oxides (CI 77491)	Pearlescent pigment
	2.00	Silica	Aesthetic modifier, stabilizer
	1.20	Silica	Aesthetic modifier, stabilizer
	1.00	Cellulose (and) hydrogenated lecithin	Aesthetic modifier
	100%		

https://www.koboproductsinc.com/formulations/KMU-057-EU.pdf

Table 3 Eyeliner formula example

Phase	Percentage (%)	Ingredient INCI	Ingredient Function
1	73.00	Water	Main system carrier
	1.00	Phenoxyethanol (and) Ethylhexylglycerin	Preservative
2	1.00	Hydroxyethylcellulose	Thickener
3	25.00	Black 2 (CI 77266) (and) PEG-40 Hydrogenated Castor Oil (and) Water (Aqua) (and) Acrylates/Ethylhexyl Acrylate Copolymer (and) Aminomethyl Propanol (and) Sodium Dehydroacetate	Pigment dispersion
	100%		

https://www.koboproductsinc.com/formulations/KEL-030-EU.pdf

Formula structural/ functional agents	Raw ingredient possibilities	Typical use level
Emollients	Dicaprylyl carbonate	5–20%
Oil phase thickeners	Disteardimonium hectorite	5–10%
Water	Water	5–20%
Humectants	Glycerin	1–5%
Chelating agent	Citric acid, disodium EDTA, tetrasodium EDTA	0.1–0.5%
Emulsifiers	Polyglyceryl-4 isostearate (and) cetyl PEG/PPG-10/1 dimethicone (and) hexyl laurate; bis-PEG/PPG-14/14 dimethicone (and) dimethicone; cetyl PEG/PPG-10/1 dimethicone	4–10%
Aesthetic modifiers	Silica, cellulose	0.5–3%
Film formers	Trimethylsiloxysilicate, hydrogenated glyceryl rosinate	Varies per material, level of wear
Sunscreens (optional)	Zinc oxide, TiO2, organic filters	Follow current FDA guidelines
Pigments	TiO2, iron oxides	2–15% + (varies per coverage level)
Preservative	Hydroxyacetophenone, 1,2-hexanediol (and) caprylyl glycol; glyceryl caprylate (and) glyceryl undecylenate	0.5% - 1%
Fragrance (if applicable)		Varies
Promotional ingredients (if applicable)	Vitamins, extracts	0.001–0.5%
Active ingredients (if applicable)	Hyaluronic acid	Varies per active

A foundation emulsion formulation example is given in Table 4.

Foundation Emulsion Formulation Examples:

O/W Natural Medium Toffee Foundation

A matte cream foundation with exceptional full coverage and comfort on skin.

A W/Si high coverage mineral only SPF foundation formulation example is given in Table 5.

W/Si High Coverage Mineral Only SPF Foundation

Mineral-based Sunscreen Foundation

Table 4 Foundation emulsion formula example

Phase	Percentage (%)	Ingredient INCI	Ingredient function
1	59.99	Water	Main carrier
	2.00	PEG-20 methyl glucose Sesquistearate	Emulsifier
	1.00	Gluconolactone (and) sodium benzoate	Preservative
	1.00	Potassium Cetyl phosphate	Emulsifier
2	3.00	Montmorillonite (and) sucrose Distearate (and) sucrose stearate	Emulsifier
	3.00	Glycerin	Humectant
3	5.00	Hydrolyzed Jojoba esters (and) water	Emollient
4	11.25	Shea butter ethyl esters	Ester
	9.00	Titanium dioxide (and) Octyldodecanol (and) jojoba esters (and) Trihydroxystearin	Treated pigment dispersion
	1.50	Iron Oxides (CI 77492) (and) Octyldodecanol (and) Jojoba Esters (and) Trihydroxystearin	Treated pigment dispersion
	1.00	Dimer Dilinoleyl dimer Dilinoleate	Ester
	0.20	Iron Oxides (CI 77491) (and) Octyldodecanol (and) Jojoba Esters (and) Trihydroxystearin	Treated pigment dispersion
	0.05	Iron Oxides (CI 77499) (and) Octyldodecanol (and) Jojoba Esters (and) Trihydroxystearin	Treated pigment dispersion
5	2.00	Cellulose	Aesthetic modifier
	0.01	Phenethylalcohol (and) Caprylhydroxamic acid (and) glycerin	Preservative
	100%		

https://www.koboproductsinc.com/formulations/KLF-318.pdf

Volatile Liquid Suspensions

Longwear eyeliner is a good example of volatile liquid suspension systems. These products need to evaporate quickly upon application to prevent smudging.

Formula functional agents	Raw ingredient possibilities	Typical use level
Volatile carriers	Water, volatile silicone, isododecanol	10–70%
Thickeners	Hydroxyethylcellulose, xanthan gum	0.5–1%
Aesthetic modifiers	Mica, cellulose, silica	0.5–10%
Film former	Trimethylsiloxysilicate	10–30%
Colorants	Pigments, pearlescent pigments	5–30%
Preservative	Phenoxyethanol and ethylhexylglycerin (for water-based formula)	0.80%

A Volatile Liquid Suspension formulation example is given in Table 6.

Volatile Liquid Suspension Formula Example:
Violet Sparkle Liquid Eyeliner

Table 5 W/Si mineral only SPF foundation formula example

Phase	Percentage (%)	Ingredient INCI	Ingredient function
1	16.00	Zinc oxide (and) Dimethicone (and) Isononyl Isononanoate (and) Polyglyceryl-6 Polyricinoleate (and) PEG-10 Dimethicone (and) hydrogen Dimethicone	Sunscreen active dispersion
	15.00	Butyloctyl salicylate (and) titanium dioxide (and) alumina (and) hydrogen Dimethicone (and) Polyhydroxystearic acid	Sunscreen booster dispersion
	9.00	Dimethicone	Emollient
	7.00	Titanium dioxide (and) Polyglyceryl-4 Isostearate (and) Cetyl PEG/PPG-10/1 Dimethicone (and) hexyl Laurate (and) isopropyl titanium Triisostearate	Treated pigment dispersion
	4.00	Lauryl Peg-9 Polydimethylsiloxyethyl Dimethicone	Emulsifier
	4.00	Argania Spinosa Kernel oil (and) Tocopheryl acetate (and) Bisabolol	Sunscreen booster
	4.00	Trimethylsiloxysilicate (and) Dimethicone	Film former
	3.20	Iron Oxides (CI 77492) (and) Polyglyceryl-4 Isostearate (and) Cetyl PEG/PPG-10/1 Dimethicone (and) Hexyl Laurate (and) Isopropyl Titanium Triisostearate	Treated pigment dispersion
	2.00	Peg-15 Lauryl Dimethicone Crosspolymer (and) Isododecane	Emulsifier
	0.80	Quaternium-90 Sepiolite (and) Quaternium-90 montmorillonite	Rheology modifier
	0.65	Iron Oxides (CI 77491) (and) Polyglyceryl-4 Isostearate (and) Cetyl PEG/PPG-10/1 Dimethicone (and) Hexyl Laurate (and) Isopropyl Titanium Triisostearate	Treated pigment dispersion
	0.35	Iron Oxides (CI 77499) (and) Polyglyceryl-4 Isostearate (and) Cetyl PEG/PPG-10/1 Dimethicone (and) Hexyl Laurate (and) Isopropyl Titanium Triisostearate	Treated pigment dispersion
2	25.60	Water	Water phase carrier
	4.00	Butylene glycol	Humectant, solvent
	1.50	Magnesium Sulphate Heptahydrate	Viscosity agent
	0.60	1,2-Hexanediol (and) Caprylyl glycol	Preservative
	0.50	Phenylpropanol (and) Propanediol (and) Caprylyl glycol (and) tocopherol	Preservative
3	1.00	Silica	Aesthetic modifier
	0.80	Vinyl Dimethicone/Methicone Silsesquioxane Crosspolymer	Aesthetic modifier
	100%		

https://www.koboproductsinc.com/Formulas.aspx?c=Foundation-Concealer

Table 6 Liquid eyeliner formula example

Phase	Percentage (%)	Ingredient INCI	Ingredient function
1	66.00	Water	Main system carrier
	3.00	Glycerin	Humectant
	1.00	Hydroxyethylcellulose	Thickener
2	12.00	Black 2 (CI 77266) (and) PEG-40 Hydrogenated Castor Oil (and) Water (Aqua) (and) Acrylates/Ethylhexyl Acrylate Copolymer (and) Aminomethyl Propanol (and) Sodium Dehydroacetate	Pigment dispersion
3	15.00	Calcium aluminum borosilicate (and) titanium dioxide	Pearlescent pigment
4	2.00	Polyethylene	Thickener
	1.00	Propylene glycol (and) Diazolidinyl urea (and) Iodopropynol Butylcarbonate	Preservative
	100%		

https://www.koboproductsinc.com/formulations/KEL-019-BR.pdf

Part 2: Current Ingredient Trends

Technology-Based Ingredient Upgrades

Treated Pigments

Treated pigments entered the world of cosmetics in Japan at the end of the 1970s. It was not until the 1980s that this innovation materialized in the USA, thanks to three companies: Miyoshi, Clark Colors and Kobo Products. Initial treatment launches were Isopropyl Titanium Triisostearate (ITT) and Silicone. In less than a decade, a wider assortment of treatments became available to cosmetics including, but not limited to, methicone, dimethicone, triethoxycaprylylsilane, lecithin, magnesium myristate, jojoba esters, and silica.

The use of this new pigment innovation grows more and more each year as companies begin to understand and leverage the myriad of benefits. Most treatments will render pigments hydrophobic and thereby enhance dispersibility properties in an oil phase. Treatments can also enhance adhesion and wear properties in addition to improving physical and chemical stability of formulations.

How Are Pigments Treated?

The most utilized surface modification methods include physisorption or chemisorption processes. Physisorption depends on Van der Waals or electrostatic forces while chemisorption chemically bonds treatments to pigment surfaces. Surface treatment bonding examples are shown in Image 4.

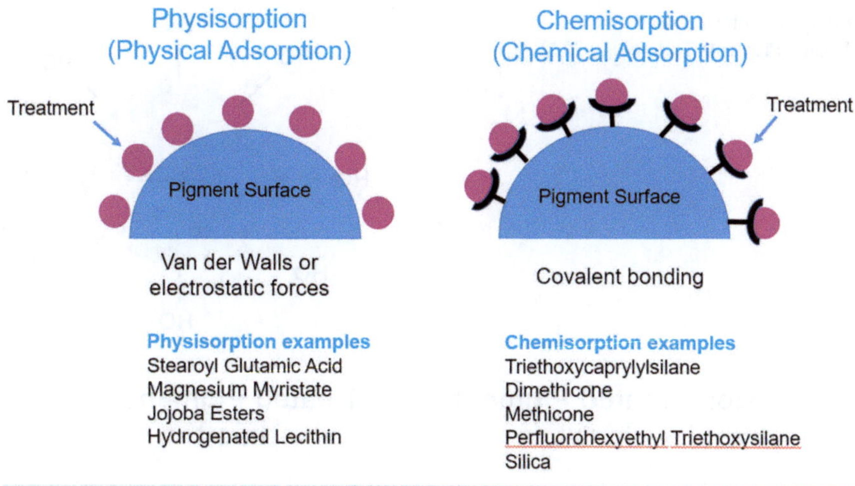

Image 4 Kobo Webinar: Treated Pigments Technology 2022

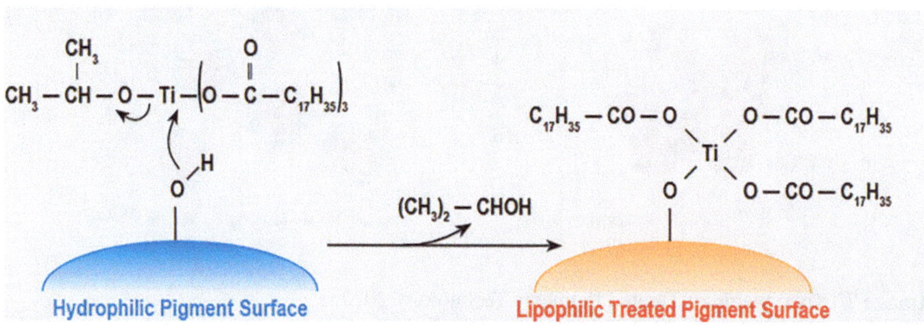

Image 5 Chemical reaction of ITT and Pigment surface. (Image reference: https://www.koboproductsinc.com/Downloads/Kobo-TitanateITT.pdf)

The chemical reaction for ITT surface treatment is shown in Image 5.

Treated V. Untreated Pigments

Untreated pigments have hydrophilic surfaces that display hydroxyl groups, thereby adsorb moisture. This leads to an array of negative issues including pigment agglomeration, insufficient wettability and dispersibility within formulations, poor formula stability, inadequate pressability of pressed powders and poor chemical stability of metal oxides. The structure of an untreated pigment is shown in Image 6.

The comparison of non-treated and treated pigment behavior is shown in Image 7.

Treated pigment surfaces, with an appropriate coating, provide the steric hinderance and electrostatic repulsion that is necessary to yield a uniform distribution. While untreated pigment agglomeration ultimately leads to an undesirable feeling

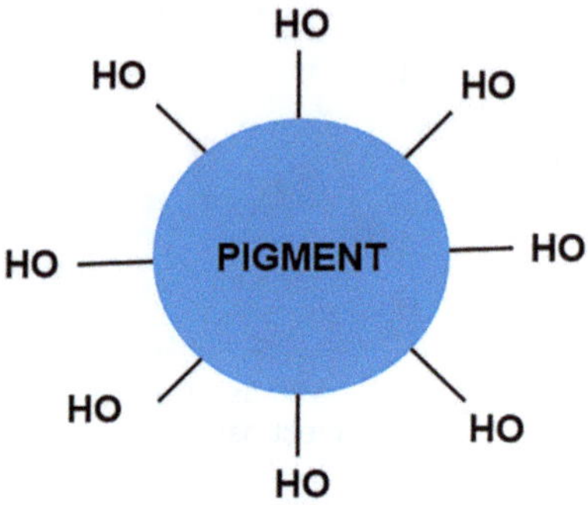

Image 6 Kobo webinar: Treated Pigments Technology 2022

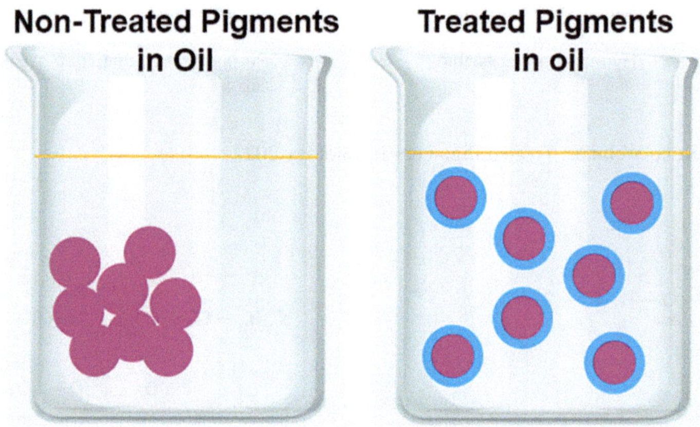

Image 7 Kobo webinar: Treated Pigments Technology 2022

on skin, the uniformity of treated pigments provides a smoother skin feel. In addition, treated pigments enhance pigment wetting, render hydrophobic properties, thereby providing water resistance, full color development, and overall better stabilization in the vehicle.

The Dispersibility Difference of Treated Versus Non-treated TiO2 is Shown in Image 8.

For example, Titanium Dioxide, due to its naturally hydrophilic surface chemistry, disperses well in aqueous media but not in organic media. Once treated with ITT, the surface becomes organophilic due to the fatty acid groups aligning with the surrounding oil and esters.

The difference in treated versus non-treated pigments in a liquid lip system is shown in Image 9.

Image 9 confirms that treated pigments develop more fully versus untreated pigments. Untreated pigments need more processing time under the homogenizer versus treated pigments. The treated pigment phase appears completely homogenous without agglomerates and has full color development.

Image 8 Kobo paper, Fig. 2: https://www.koboproductsinc.com/Downloads/Kobo-Surface-Treated-Pigments-and-Powders-Combined-June2015.pdf

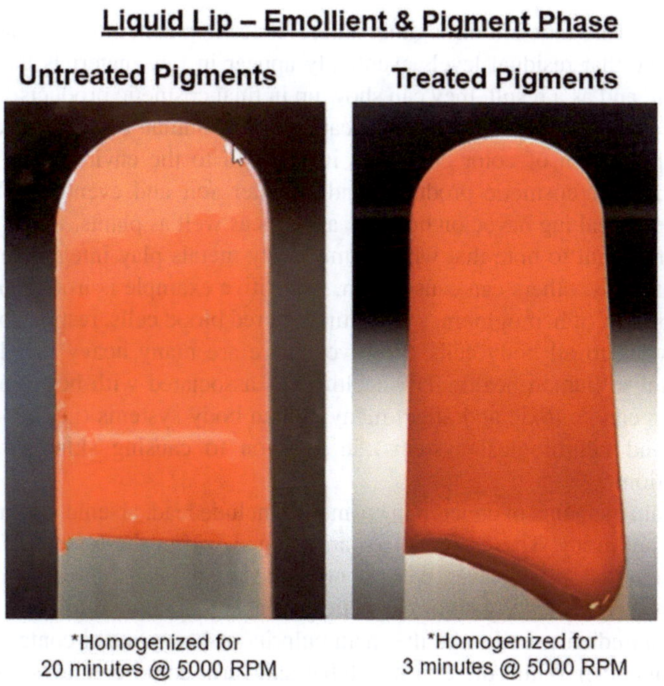

Image 9 Kobo Webinar: Treated Pigments Technology 2022

Safety and Regulatory-Based Ingredient Upgrades

The most impactful changes to color cosmetics have been the outcome of regulatory changes. Governing bodies such as REACH (European Regulatory Union—Registration, Evaluation, Authorization and restriction of Chemicals), FDA (USA's Food and Drug Administration) and PCPC (USA Trade Association—Personal Care Products Council) assess and prohibit the use of certain materials based on their research. REACH is highly aggressive and while the USA has recently followed their main principles, California has been commanding the lead with regards to safety of raw materials in the USA. Below are some of the most pervasive changes that have recently impacted the cosmetics industry.

Safety and Health Ingredient Upgrades

Pigments—Heavy Metals

Heavy metals are inherent throughout nature. Present in soil, minerals, and water, it is inevitable that residual levels eventually appear in raw materials through their cultivation, and as a result, they can show up in final cosmetic products.

There are many ways heavy metals can enter the human body. This can happen via the application of color cosmetics in addition to the environment influence. Discarded color cosmetic products tend to enter soil and eventually leach into waterways, wreaking havoc on humans, animals as well as plants.

It is important to note that while some heavy metals play integral roles within the human body, others can cause harm. A positive example is iron (Fe). Iron is a key component of hemoglobin, a constituent of red blood cells, responsible for carrying oxygen to all body cells. However, there are many heavy metals that are detrimental to human health. The health risks associated with heavy metal contamination can be toxic and affect many human body systems (i.e., reproductive, nervous and neurological systems) in addition to causing skin irritation and discoloration.

The main elements of concern in cosmetics include lead, arsenic, cadmium, mercury and antimony. These substances are banned within the bounds of cosmetic products regulation. However, these contaminants have been detected in personal care products, particularly color cosmetic products that contain mineral pigments. (Natural ingredients tend to be the main culprits of heavy metal contamination in color cosmetics.) While there is no global standard, some countries have issued contamination limits for specific elements. For example, the US FDA "set a maximum content limit for *mercury* (1 mg/kg) in cosmetic products as well as regulatory limits for color additives used in cosmetics including *arsenic, lead* and *mercury*. In addition, a guidance limit for *lead* (10 mg/kg) in cosmetic lip products has been advised."[1]

Toxic heavy metals should be eliminated whenever possible in cosmetic formulation and manufacturing. When it is not possible to remove within GMP and the inevitability of contamination exists, safety evaluations should be completed on final products to guarantee product safety for the public.

Silicone Alternatives

Silicones have played a major role in the cosmetic industries since the 1950s. Applied in basically all product types (cosmetics, personal care, skin care, hair care), this category of materials contributes many functions to formulations. Not only are these materials known for exceptional sensorial aesthetics, but they also provide film-forming, antifoaming, emulsifying and viscosity controlling properties. Volatile silicones typically yield a light, silky, powdery (non-greasy) feel with rapid spreading and quick dry time on the skin while non-volatile silicones provide longer playtime and brilliant slip while also contributing non-greasy aesthetics and conditioning properties.

The molecular structure variations of silicones are vast, resulting in each of them possessing unique properties. Structurally, silicones are polymers of siloxane groups. These distinct silicone and oxygen bonds are the basis for many structures that are available in numerous molecular weights and shapes. Silicones are categorized as either cyclic or linear. The most prevalent cyclic silicones (known as cyclomethicones) used are octamethylcyclotetrasiloxane (D4) and decamethylcyclopentasiloxane (D5). Linear silicones are polydimethylsiloxanes, also known as PDMS or dimethicone.

With the increase of consumer awareness of raw ingredient safety to humans and the environment, silicone concerns include poor biodegradability, damage to the environment, and negative health effects on all forms of life. As a result, in 2020, the European Union banned the use of D5 in wash-off products "in concentration equal or greater than 0.1% by weight."[2]

Creating alternative ingredients that match the elegant tactile perception of silicone while maintaining safety and environmental guidelines is not an easy task. Nevertheless, many replacements have emerged and are continually being developed. For example, while carbon-based emollients do not compare in feel, blending alkanes with esters seems to be a creative solution to mimic the velvety, dry tactile sensations of silicones. There is no doubt this research will continue as we move towards healthier and safer ingredients for the future.

---

[1] https://www.cosmedesk.com/blog/posts/heavy-metals-impurities-in-cosmetics/

[2] https://joanmorais.com/silicone-alternatives/

Nanomaterials

Nanotechnology began with positive intentions for many different industries. In addition to cosmetics, nanotechnology is used in the food and pharmaceutical businesses. This technology is not specific to any one type of material. It refers to the size of the particles. Nanoparticles are very tiny. They are categorized as less than 100 nanometers in diameter. Essentially invisible to the naked eye, nanoparticles can exhibit properties that are vastly different from their larger counterpart of the same material. For example, when zinc oxide and titanium dioxide are severely reduced in particle size, their UV protection properties increase, thereby yielding a higher SPF value when tested in a sunscreen formula. These extra-fine particle sizes also feel nicer and yield more transparency on skin versus the same material in larger sizes.

Nanomaterials enter the human body via skin permeation or inhalation. Since nanoparticles are so tiny, they tend to become airborne easily and thereby inhaled into the lungs. Their small size also enables them to easily penetrate the skin. Once in the dermis, these particles enter body cells and the bloodstream where they can spread to all parts of the body causing negative health effects. Aside from causing inflammation within the lining of the lungs, evidence suggests that inhaled nanoparticles enter the bloodstream thereby having the capability to travel to all other organs, causing health concerns throughout the human body.

Talc

Talc was once a ubiquitous ingredient throughout color cosmetics due to its many performance characteristics. First and foremost, talc is the softest mineral on earth. It ranks #1 on the Mohs Scale of Hardness. Therefore, this material was used to add supreme softness to cosmetic products, particularly in pressed powders. Talc also provides high absorption properties, hence its use in formulas to absorb excess oil on the skin. In addition, talc offers opacifying properties for cosmetic color products. The Mohs Hardness Scale is shown in Image 10.

Unfortunately, talc has been associated with serious health concerns. When contaminated with asbestos, talc poses a risk of cancer. In addition, talc in its powder form is associated with respiratory illness such as Mesothelioma. Consequently, many cosmetic and personal care companies are being legally pursued for their use of talc.

As a result of these negative health effects, many cosmetic companies are avoiding talc in new launches and reformulating classic products with talc replacements. Research is on the rise and new replacements are emerging every day. While there doesn't seem to be an equivalent supremely soft mineral, blends of minerals and treatments of powders seem to be vital options for alternative materials.

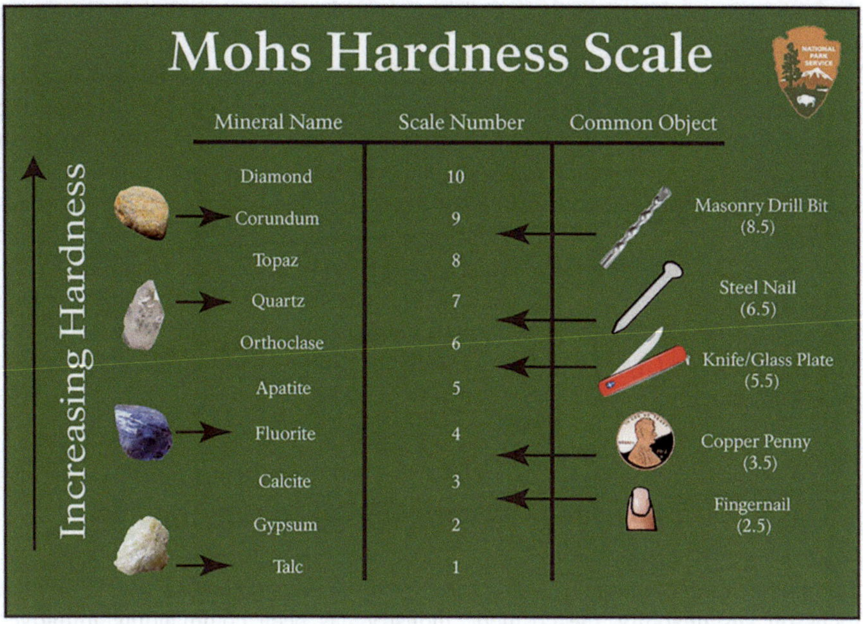

Image 10 Reference: https://www.nps.gov/articles/mohs-hardness-scale.htm

Preservatives

Due to health concerns, many regulatory bodies have restricted or prohibited the use of many traditional preservatives in cosmetics. Some classis examples include formaldehyde, methylene glycol, quaternium-15 and specific parabens.

Formaldehyde is a known carcinogen. Both quaternium-15 and methylene glycol are related to formaldehyde. Quaternium-15 is a formaldehyde releasing agent and is also known as a skin irritant and allergen. Methylene Glycol is a hydrated reaction product of formaldehyde and through specific processes, could potentially turn into formaldehyde gas, making it just as toxic as formaldehyde.

Parabens are esters based on parahydroxybenzoic acid. The paraben family is comprised of many structures including long and short chains. There are six kinds of parabens which include methylparaben, ethylparaben, propylparaben, isopropylparaben and isobutyl paraben.

Highly effective against gram-positive bacteria, gram-negative bacteria, mold and yeast at very low levels, parabens have been in existence for about 100 years. These are preservatives that are also inexpensive, therefore at one time, they were vastly prevalent in cosmetics.

Unfortunately, many paraben health concerns have surfaced over the years. Parabens can cause skin irritation, but more egregiously, parabens are linked to reproductive toxicity as they can damage reproductive organs in both men and women. In addition, longer chain and branched parabens can simulate estrogen, therefore are considered endocrine disruptors which links their activity in cancer cell production.

As a result of these health concerns, many government agencies and retailers have banned the use of long-chain and/or branched parabens. Even though the shorter chain parabens don't appear to have the health concerns of the long chain and branched parabens, many finished good companies have since dropped the use of parabens altogether. Paraben-free has become a very common claim for product safety.

In the endeavor to have safe product preservation systems, multifunctional ingredients are serving as a new approach. These materials accomplish more than one function within a formulation, with preservation being a secondary property. For example, Caprylyl Glycol is a skin conditioning agent that also functions as an antimicrobial agent.

Social & Environmental Responsibility Ingredient Upgrades

The preservation of natural resources is currently a strong goal worldwide. With a focus on sustainability, the cosmetic industry has been commanding initiatives toward eliminating materials that damage the earth and/or are not renewable. The below topics are examples of how the industry is reacting to environmentally destructive materials and practices within cosmetics.

Microplastics

Historical microplastic materials used in cosmetics include Polymethylsilsesquioxane, Polymethyl Methacrylate, Polybutyl Methacrylate, Nylon, and Polyethylene. These materials are defined as plastic particles that are less than 5 millimeters in size. They have contributed to many cosmetic products, providing properties such as silky textures, enhanced slip and feel, soft focus effect and optical blurring, sebum absorption, enhanced blendability of formulas on skin and mascara volumizing.

Despite the many positive performance characteristics, microplastics are threatening the globe in many ways. They drift into oceans where they float on the surface as well as sink deep into the sea. These materials then get ingested and/or breathed in by marine life. Microplastics also pollute beaches where they become intermingled with sand and show up in the food chain. For example, these particles have been spotted in beer, milk, and tap water. The discovery of these materials in human organs, defecation, tumors and bloodstream proves the traction from the environment to the food chain and ultimately into human and animal bodies.

The best solution to end the negative impact of microplastics on the earth is to eliminate all future use of these materials, thereby preventing any future particles from being introduced into the environment. Many cosmetic companies are setting microplastic-free goals for the near future. Much research has already emerged and is ongoing for microplastic replacements. Alternatives are of mineral, plant and marine origins. Derivatives of silica, cellulose and algae are among the many options that chemists have today. These replacements absolutely fulfill many properties offered by traditional polymer microspheres.

Mica

The sourcing of raw materials is becoming increasingly pervasive. Many finished good companies are requesting sourcing data from their raw material suppliers. Sheet mica, which is prevalent in cosmetics, has been one of the first materials to become scrutinized for multiple reasons.

While sheet mica mines exist in many countries, many of these mines are illegally operated. Mining often has small tunnels and as such, children have been exploited in these dangerous and severely laborious environments. Mining sheet mica is also impacting the environment including issues such as deforestation, water contamination and dust emissions.

Two resolutions have arisen to the mica situation: synthetic mica and the creation of the RMI (Responsible Mica Initiative). Synthetic mica is an alternative solution since it is created in a lab. Formed by the processing of manganese, metal and aluminum into magnesium aluminum silicate sheets which are crystal, these sheets are then processed into a powder. This synthetic form of mica can be easily identified in an ingredient listing as Synthetic Fluorphlogopite or simply, Fluorphlogopite. While synthetic mica is not as cost effective as mined mica, this alternative solves the child labor issue as well as environmental impact caused by sheet mined mica. Another solution that keeps naturally mined mica on the table, while eliminating child labor concerns, has been the creation of the RMI. This organization "works to create responsible and traceable mica chains. Its goals are to end child labor and create work environments that follow legal mandates."[3]

RSPO (Roundtable on Sustainable Palm Oil)

Palm oil is one of the globe's most utilized plant-based raw materials. A highly cost-effective ingredient, its ubiquity in many industries (including food, health, household and beauty) have caused increasing demands for this material.

The RSPO was created to ensure the sustainability of palm oil. Members of this organization have committed to specific principles such as the prevention of deforestation for the purpose of new oil palm production, elimination of child labor, inclusion of smaller farmers, assurance of legal regulations, and proper inspection of plantations.

[3] https://www.treehugger.com/what-is-mica-powder-5216422#:~:text=With%20child%20exploitation%20being%20common,for%20families%20in%20production%20areas._

Clean Beauty

Clean Beauty refers to a movement in the cosmetics and personal care industry indicating the formulas are created without harmful ingredients. Although not officially defined by one organization, many companies have devised their own definition for this initiative. Main tenets of this movement resonate between the larger companies with minor differences.

It is important to note that clean beauty does not necessarily mean organic, natural or green. It simply means that the ingredients used are not toxic, nor do they compromise human health. Many ingredients restricted within the clean beauty movement include parabens, sulfates, phthalates, fragrance, talc, formaldehyde, toluene, cyclic silicones, and nanomaterials.

Vegan and Cruelty Free

Vegan and Cruelty free claims have also surfaced in recent years. Vegan means the ingredients are not animal derived while Cruelty Free ensures the product has not been tested on animals. Vegan ingredients prohibit ingredients such as carmine, honey, beeswax, lanolin, and collagen.

Part 3: Future Ingredient Trends

The future technology trends are undoubtedly based on the health of the world. Consumers are becoming increasingly more aware of the environmental consequences of specific ingredients. Ingredient labels are being scrutinized and consumer purchasing habits are becoming progressively more decisive, geared toward the longevity and health of the planet and all its' inhabitants. A study from 2021 proved that more than half of the global population has altered their purchasing habits and that over one-third of the population will spend more on sustainable goods.

In this quest to establish an eco-friendlier and more sustainable world, a multitude of new ingredient trends are emerging. Below are new emerging terms and trends that will shape the future of cosmetic formulation.

Awareness Terms

Green, Eco-conscious and Conscious Beauty have become catchall phrases for products that are developed while keeping the health of the globe at the forefront of mind.

Ecofriendly and Environmentally Friendly indicates that the product, both formula and package, does not cause damage to the environment in any way. This includes manufacturing and disposal processes.

Biodiversity is the vast array of animals, plants and all other living beings on earth.

Principles and Ethics Term

Greenwashing refers to the practice of companies making inflated and/or dishonest claims which deceptively draws in ecologically aware consumers.

Formula/Ingredient Origins

Biobased

Since consumers are becoming more educated and in-tune with sustainable ingredient origins, the future of cosmetic raw materials will be focused on biomass. *Biomass* refers to organic materials that are renewable. As such, *biobased* formulas are the way of the future. This means that formulas will be completely based on renewable materials such as plants, marine life, and agriculture leaving non-renewable resources such as petrochemical-based materials in the past.

Upcycling

New raw ingredients will also arise from the process of upcycling. *Upcycled* ingredients signify the creative usage of food waste. By recreating food waste into newly useful, functional materials, we decrease the enormous amount of planetary food waste as well as unearth new supply chain sources. Most of this source comes from commodity crop farms. Examples include fruit seeds (plums, avocado), coffee grinds, and barley.

Safety Trends

Non-Toxic, Organic

Ensured formula safety will be the forefront of all future cosmetic products. Formulas will be *non-toxic* which means that these products will not contain ingredients that are known, or speculated, to cause toxic effects on consumers. There will also be a push for more organic claims. *Organic* claims refer to both the naturality as well as the processing of ingredients. In the USA, the USDA (United States Department of Agriculture) requires organic ingredients to meet specific requirements including non-GMO, contain 95% organic material, minimization of synthetics use in production, etc.

Return to Earth

Circular Beauty, Recyclable Beauty

Closely tying into the sustainable and renewable trend, there will be a strong focus on less or zero product waste. Products will fit the circular beauty model which is a circular business model where there is no waste or chemical pollution. This model focuses on using the concepts of reuse, reduce, recycle and upcycle as much as possible. This will apply to the entire product, including both formula and packaging.

Recyclable Beauty is a term that refers to this focus of reducing waste and environmental pollution via the creation of products that can be reused or recycled. The difference between a linear versus a circular economy is shown in Image 11.

Degradability

In the near future, products will be more biodegradable versus degradable. Degradable materials break down with time and is not a natural process. Chemical reactions are needed to break down these materials. On the other hand, biodegradable products essentially involve the natural, biological breakdown of products by microorganisms, which results in the formation of natural compounds, carbon

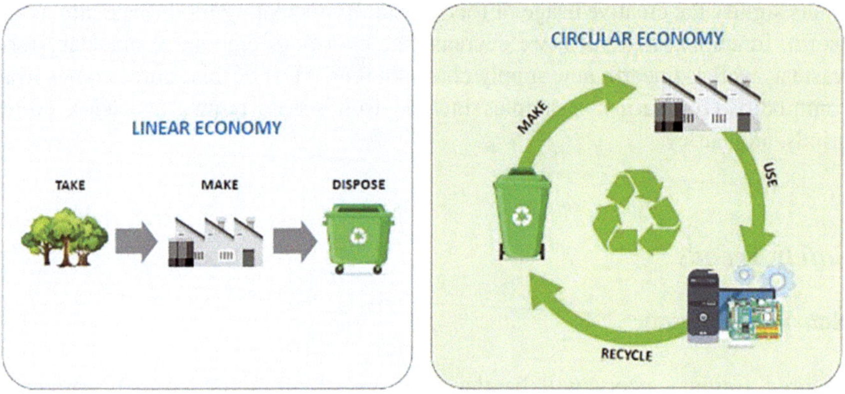

Image 11 Ref.:https://www.linkedin.com/pulse/relevance-circular-economy-approach-speciality-chemical-raman/

dioxide and water vapor. This leaves no toxic pollution to the environment. Ingredients used in these products are restored to nature. Many things are biodegradable, but some may take thousands of years.

Compostable

Compostable materials will also be a future trend. These materials disintegrate via microorganisms, humidity and heat into non-toxic, natural components. This is a biodegradable process but happens at a speed proportional to other organic materials, such as those found in a compost pile (leaves, grass, food scraps). Timing is typically less than 12 weeks. While personal composts are not highly prevalent, local based commercial composting will become more accessible.

Zero Waste, Low Waste

By implementing the circular beauty model with biodegradable and compostable products, Zero and Low Waste claims will prevail. This essentially means that no or very little waste is created by a product, or at the very least, there is a substantial reduction in the waste produced.

Glossary

Biobased Term used to indicate that the product is based on renewable materials such as plants, marine life, and agriculture (biomass).

Biodegradable Term that describes the natural, biological breakdown of products by microorganisms, which results in the formation of natural compounds, carbon dioxide and water vapor. This leaves no toxic pollution to the environment.

Biodiversity The vast array of animals, plants and all other living beings on earth.

Biomass Organic materials that are renewable such as plants, marine life, and agriculture.

Carcinogen Cancer-causing agent.

Chemisorption Surface modification method involving chemical bonding.

Circular Beauty A circular business model where there is no waste or chemical pollution. This model focuses on using the concepts of reuse, reduce, recycle and upcycle as much as possible.

Clean Beauty Movement in the cosmetics and personal care industry indicating that formulas are created without harmful ingredients.

Compostable Term that describes the natural disintegration of materials via microorganisms, humidity and heat into non-toxic, natural components. This is a biodegradable process but happens at a speed proportional to other organic materials, such as those found in a compost pile (leaves, grass, food scraps). Timing is typically less than 12 weeks.

Conscious Beauty Term used to indicate that a product was developed while keeping the health of the globe in mind.

Cruelty Free Claim that indicates product has not been tested on animals.

Degradable Term that indicates the non-natural breakdown of products by use of chemical reactions.

Eco-conscious Term used to indicate that a product was developed while keeping the health of the globe in mind.

Ecofriendly Term used to indicate that the product, both formula and package, does not cause damage to the environment in any way. This includes manufacturing and disposal processes.

Environmentally Friendly Term used to indicate that the product, both formula and package, does not cause damage to the environment in any way. This includes manufacturing and disposal processes.

FDA Governing body in the USA. Food and Drug Administration.

Fluorphlogopite Lab created mica.

Glazing Pressed powder phenomenon in which the pressed cake exhibits a glazed surface where product no longer releases. This essentially means that the powder is too hard. Binder concentrations and/or pressing parameters need to be adjusted.

Green Term used to indicate that a product was developed while keeping the health of the globe in mind.

Heavy Metals Naturally occurring metallic elements of high molecular weight and density with the capacity to be toxic at very low levels.

Low Waste Term that indicates the final breakdown of product produces a substantial reduction in waste.

Mesothelioma An aggressive form of cancer that develops in the mesothelium (tissue layer that encompasses most internal organs).

Microplastics Very small plastic particles (less than five millimeters in length) that are causing damage to the environment and all forms of life.

Mohs Scale of Hardness Classification system of mineral hardness. The scale is comprised of minerals of varying hardness levels (10—hardest (diamond), 1—softest (talc)). Hardness is dependent on crystal structure and the mineral's resistance to scratching.

Multifunctional Ingredients Ingredients that contribute a variety of properties to a formula. These are becoming highly pervasive in the product preservation category.

Nanomaterials Materials of very small size (less than 100 nanometers).

Non-Toxic Term that indicates that a product will not contain ingredients that are known, or speculated, to cause toxic effects on consumers.

Organic Claim that refers to both the naturality as well as the processing of ingredients.

Palm Oil An edible vegetable oil from the oil palm tree. Palm oil is one of the globe's most utilized plant-based raw materials.

PCPC USA Trade Association—Personal Care Products Council.

Physisorption Surface modification method involving Van der Walls or electrostatic forces.

REACH Governing body in Europe. European Regulatory Union—Registration, Evaluation, Authorization and restriction of Chemicals.

Recyclable Beauty Term that refers to the focus of reducing waste and environmental pollution via the creation of products that can be reused or recycled.

RMI Responsible Mica Initiative: Organization created to ensure the responsible cultivation of natural mica. RMI addresses both child labor as well as environmental impact of mica mining.

RSPO Roundtable on Sustainable Palm Oil. Non-profit organization dedicated to making palm oil a sustainable resource.

Sheet mica Slim layers of silicate minerals. This naturally mined material is widely used in cosmetics due to its reflective and refractive properties.

Surface Treatment Modification of a pigment or powder surface that ultimately changes its normal properties. Depending on the treatment, surfaces can become hydrophobic, hydrophilic, lipophilic, or both lipophobic and hydrophobic. Additional benefits include the reduction of surface activity, improvement of skin compatibility, physical and chemical stability and increased pigment loading capabilities.

Synthetic mica Lab-created mineral (Fluorphlogopite).

Upcycled Signifies the creative usage of food waste.

Vegan Claim that indicates formulas are not animal derived.

Zero waste Term that indicates the final breakdown of product produces no waste at all.

References

Face Section

1. https://www.aunaturalecosmetics.com/blogs/blog/blushing-beauty-the-benefits-of-using-blush-in-your-daily-makeup-look-1#:~:text=Blush%20is%20a%20cosmetic%20product,for%20a%20more%20dramatic%20effect.
2. https://www.lorealparisusa.com/beauty-magazine/makeup/face-makeup/bronzer-vs-contour-difference#:~:text=Bronzer%20is%20applied%20to%20the,eyelids%20as%20an%20eyeshadow%20alternative.
3. https://cosmetics.specialchem.com/tech-library/article/anatomy-of-a-pressed-eyeshadow

Silicone Replacements

4. https://joanmorais.com/silicone-alternatives/
5. https://cosmethicallyactive.com/silicones-in-cosmetics-and-their-impact-on-the-environment/
6. https://www.sciencedirect.com/science/article/abs/pii/S0001868622000811

Heavy Metals

7. https://www.cosmedesk.com/blog/posts/heavy-metals-impurities-in-cosmetics/
8. https://www.safecosmetics.org/chemicals/lead-and-other-heavy-metals/
9. https://www.sciencedirect.com/science/article/pii/S2772416623001614
10. https://www.cosmedesk.com/blog/posts/heavy-metals-impurities-in-cosmetics/).

Talc

11. https://www.nps.gov/articles/mohs-hardness-scale.htm
12. https://www.safecosmetics.org/chemicals/talc/#:~:text=Some%20talc%20may%20contain%20the,avoided%20in%20the%20pelvic%20areas.
13. https://www.byrdie.com/is-talc-in-makeup-bad
14. https://www.ewg.org/skindeep/ingredients/706427-TALC/
15. https://www.rockngem.com/what-is-mohs-scale-of-hardness/
16. https://rocksminerals.flexiblelearning.auckland.ac.nz/minerals/talc.html#:~:text=Talc%20is%20the%20softest%20mineral,thermal%20metamorphism%20of%20siliceous%20dolomites.

Preservatives

17. https://www.safecosmetics.org/chemicals/parabens/
18. https://www.ewg.org/what-are-parabens
19. https://www.ewg.org/news-insights/news/2022/11/how-multitasking-preservatives-might-help-make-cosmetics-safer
20. https://www.ewg.org/skindeep/ingredients/701067-CAPRYLYL_GLYCOL/

RSPO

21. https://rspo.org/as-an-organisation/our-standards/
22. https://www.dnv.us/services/round-table-on-sustainable-palm-oil-rspo-standard-182924/

Biodegradability/Green/Future Trends

23. https://www.cossma.com/ingredients/article/biobased-ingredients-36781.html
24. https://edis.ifas.ufl.edu/publication/AE483
25. https://www.nivea.co.uk/about-us/sustainability-at-nivea/biodegradable-skincare-at-nivea
26. https://www.goodstartpackaging.com/biodegradable-vs-compostable-what-is-the-difference/
27. https://nyscc.org/blog/biodegradability-considerations-for-cosmetic-ingredients/
28. http://goodstartpackaging.com

Safety/Health and Social Responsibility

29. https://www.seventhgeneration.com/blog/biobased-plant-based-what-do-they-mean
30. https://thegoodfaceproject.com/articles/what-is-clean-beauty
31. https://cosmeticsbusiness.com/what-is-upcycled-beauty-the-latest-in-beauty#:~:text=Ingredient%20upcycling&text=The%20beauty%20industry%20has%20found,sustainability%20that%20is%20often%20overlooked.
32. https://www.cosmeticsdesign-europe.com/Article/2024/04/24/The-rise-and-rise-of-upcycled-ingredients-for-cosmetics
33. https://emilieheathe.com/blogs/editorial/green-beauty-glossary
34. https://www.scratchmagazine.co.uk/feature/decoded-conscious-beauty-terms-you-need-to-note/
35. https://www.cossma.com/ingredients/article/biobased-ingredients-36781.html
36. https://www.perfectcorp.com/business/blog/general/the-complete-guide-to-beauty-tech?utm_id=15691952531&utm_source=google&utm_medium=cpc&utm_campaign=PFU_Search_Content_CS_NA&utm_term=cosmetic%20news&utm_content=CS1&gad_source=1&gclid=CjwKCAjwydSzBhBOEiwAj0XN4BzNeCNst4FGRfiIDYgUZrIF2oDdCqdSliOSx2vDbIJHmyxFTUTXcRoCwdoQAvD_BwE
37. https://medium.com/@vijay.mishra_4096/why-are-bio-based-cosmetics-and-personal-care-ingredients-increasingly-becoming-popular-89d3dab41dac
38. https://www.acme-hardesty.com/the-5-elements-of-sustainable-beauty-products/

Microplastics

39. https://www.oceancare.org/en/stories_and_news/microplastics/?utm_campaign=plastik&utm_source=gad-g&utm_medium=sea&utm_content=plastik-en&gad_source=1
40. https://koboproductsinc.com/Downloads/Kobo-Microspheres.pdf
41. https://www.treehugger.com/what-is-mica-powder-5216422#:~:text=With%20child%20exploitation%20being%20common,for%20families%20in%20production%20areas.

Clean Beauty

42. https://www.foreo.com/mysa/what-is-clean-beauty-and-why-does-it-matter
43. https://thegoodfaceproject.com/articles/what-is-clean-beauty

Innovations in Nail Beauty: Past, Present, and Future

Doug Schoon

Abstract This chapter explores the evolution, formulation, and technological advancements in nail beauty products. Beginning with a historical overview, it discusses the finger nail anatomy/structure before examining the chemical composition and formulation of traditional nail polish, including ingredients such as film formers, resins, solvents, plasticizers, and pigments. This chapter provides a detailed analysis of UV-cured nail systems, discussing photopolymerization chemistry, ingredients, safety considerations, and proper curing techniques. Additionally, it covers dip powder systems based on cyanoacrylate chemistry, liquid and powder acrylic systems, and various nail care products like lotions, creams, oils, and strengtheners. Throughout, this chapter emphasizes the importance of proper formulation and important safety considerations, highlighting how innovations in polymer chemistry and pigment technology continue to drive industry advancements while addressing potential health concerns.

Keywords Nail plate anatomy · Nail polish formulation · UV gel nails · Photopolymerization · Cyanoacrylate · Dip powder nails · Liquid and powder acrylic · Nail care products · Nail strengtheners · Film formers · Plasticizers · Pigments · Colorants · Polymerization · Crosslinking · Nail lamp safety · Adhesion · Viscosity · Nail plate hydration · Allergic reactions · Polymer chemistry

Products designed to enhance fingernail beauty and improve their properties have undergone remarkable evolution over the past century, shaped by cultural influences, fashion trends, and technological advancements. The landscape of nail care was forever changed in 1916 with the launch of the first modern liquid nail polish by Cutex, replacing antiquated methods of nail coloring with a convenient and durable alternative. The 1970s heralded a new era in nail care with the advent of acrylic

D. Schoon (✉)
Schoon Scientific + Regulatory Consulting, LLC, Dana Point, CA, USA
e-mail: DSchoon@schoonscientific.com

K. Alexander, P. Romanowski (eds.), *Cosmetic Science*, AAPS Introductions in the Pharmaceutical Sciences 15, https://doi.org/10.1007/978-3-032-03596-7_6

Fig. 1 The original 1912 Cutex advertisement for their paste polish

nails (also known as liquid and powder), offering durability and versatility for intricate designs. In the 1980s, UV gel technology revolutionized nail care, providing a quicker, easier solution for nail coatings that could withstand daily wear and tear and provide a more durable and chip-resistant alternative to traditional nail polish. This was followed by cyanoacrylates impregnated into originally paper and linen, and finally settling on fiberglass or silk to create "wrap nails." (Fig. 1).

Fashion demands led to UV-curing nail coatings, mimicking traditional solvent-based nail polish by offering a palette of vibrant shades and finishes. More recently, marketers have created a demand for safer, eco-friendly nail products that claim to be free from harmful chemicals, to appeal to health-conscious consumers. Additionally, the rise of social media platforms has transformed nail art into a global phenomenon, with artists showcasing their creativity. Today, cosmetic scientists are challenged to innovate nail products using safer materials, formulations, processes, and techniques, reflecting the dynamic interplay of science, culture, and creativity.

Natural Nail Anatomy (Fig. 2)

Medical literature and the Internet are filled with misinformation about nail anatomy. Most experts agree that the proximal nail fold (PNF) is the entire flap of skin covering the nail matrix, extending from the edge of the visible nail plate to the first joint of the finger. However, there is some disagreement about the eponychium.

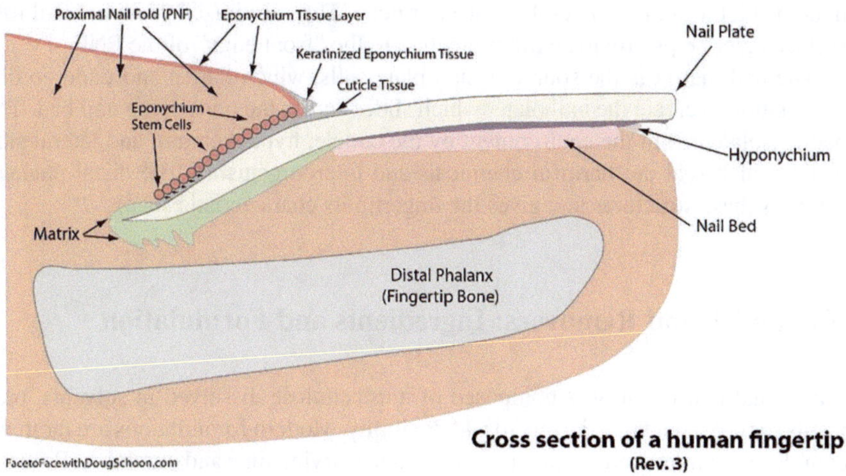

Fig. 2 Cross-section of the fingertip

Eponychium literally means "*upon the nail.*" It is the tissue that covers the new growth of the nail plate. Therefore, the eponychium is the underside portion of the proximal nail fold that covers the new growth of nail plate and creates the cuticle tissue. The eponychium is surprisingly thin- approximately 0.1–0.15 mm thick (0.004–006 inches)—and consists of only one layer of stem cells. The eponychium extends from the nail matrix to the front edge of the proximal nail fold. The nail matrix also contains stem cells, but their job is to make the nail plate. The term cuticle is widely misunderstood. There is only one type of cuticle tissue—the thin, colorless layer of dead tissue that adheres to the nail plate. This tissue comes from the eponychium. The cuticle becomes strongly adhered to the nail plate while it is still located underneath the proximal nail fold. The cuticle then pulls away and continues to ride on the upper surface of the nail plate as it grows outward.

The outer layer of the proximal nail fold is covered with tissue called stratum corneum, which is typical of other parts of the body. The area where this skin meets the nail plate has a different appearance because its surface has become keratinized. In other words, just the outer surface layers undergo a process to produce more keratin. This keratinization process hardens the PNF (proximal nail fold) surface and gives it a slightly different appearance. Like protective armor, this hardening surface layer creates a protective shield that can resist attack by infectious organisms or potentially harmful chemicals. Normally, the PNF has a smooth face layer until it is chemically attacked, physically abraded/cut or injured. Then the surface layer becomes thicker and harder to provide greater protection against any further abuse or injury.

The PNF hardened surface skin is NOT cuticle tissue. The cuticle is softer and much stickier, having superior nail plate adhesion. The cuticle forms an important seal between the PNF and nail plate, preventing infectious organisms from getting underneath the skin to infect the matrix or bone. The keratinized proximal nail fold surface layer helps provide extra protection to the "front edge" of the PNF.

The nail matrix is the source of nail plate cells, which flatten and undergo differentiation to create the nail plate, which slides across the underlying nail bed. The bed is isolated from the environment by the cuticle, hyponychium, and lateral side walls, which seal out harmful chemicals and microorganisms. The distal phalanx supports these structures and gives the fingertip its characteristic shape.

Nail Polish and Removers: Ingredients and Formulation

Traditional nail polish was composed of nitrocellulose dissolved in solvents, pigments for color, and plasticizers to add flexibility. Modern formulations are far more complex with improved application, appearance, drying time and durability. Primary components include film formers, resins, plasticizers, solvents, pigments, and various functional additives. Formulation of nail polish involves an understanding of polymer chemistry, solvent interactions, pigment dispersion, and the rheology of the final product. Formulators must consider the volatilization rates of solvents, the compatibility of film formers with plasticizers, and the stability of pigments, viscosity, application ease, film uniformity, and the selection of appropriate thickeners and rheology modifiers. The coating must provide a balance between flexibility and strength to create toughness is crucial for wear resistance.

In the past, nitrocellulose provided a shiny, durable coat; however, newer formulations utilize alternatives like polyurethane, acrylates copolymer, adipic acid/neopentyl glycol/trimellitic anhydride, as well as cellulose acetate butyrate or polyvinyl butyral, which enhance flexibility, durability, and reduce yellowing. Resins such as tosylamide/formaldehyde resin and polyester resins such as hexanediol/adipate/citrate crosspolymers are used to increase the adhesion of the film to the nail surface and improve gloss. Marketing demands have led to the use of resins not associated with formaldehyde release. Plasticizers like camphor, acetyl tributyl citrate, neopentyl glycol, sucrose acetate isobutyrate, and triphenyl phosphate (TPP) make the film flexible and resistant to cracking. The choice of plasticizer affects the surface hardness, gloss, and drying time. Solvents such as ethyl acetate, butyl acetate, ethanol, diacetone alcohol, isopropanol, and butanol are critical for dissolving the film formers, adjusting the viscosity, and controlling the evaporation rate of solvents to control drying time and film formation.

Solid pigments, lakes, and liquid dyes are used to achieve the desired color and opacity. Dyes can stain the nail plate and are not light stable, but they are used in lower concentrations, if at all. Lake pigments, such as D&C Red 27 AL lake, are made by precipitating a liquid dye with an inert binder, typically a metallic salt. Only cosmetic colorants with FD&C, D&C and CI designations should be used. Some thermochromic, photochromic and fluorescent pigments are allowed, but there may be restrictions in some regions depending on their composition. Pearlescent pigments are flat plates of mica coated with titanium dioxide, iron oxide, or other metal oxides, and glitters are used in nail polish for special effects. Due to environmental concerns over microplastics, the use of cellulose-based bio-glitters is on the increase. Surface treatment of pigments with silanes or other coupling agents can improve pigment dispersion and stability within the polish, preventing color change and settling.

Efficient pigment dispersion is critical. Pigments must be evenly dispersed throughout the solvent matrix to prevent settling or agglomeration, which can cause streakiness and uneven color. The dispersion process involves breaking down pigment agglomerates into primary particles and stabilizing them in the solvent to prevent reagglomeration by dispersing agents, which adsorb onto the pigment surface to enhance steric or electrostatic stabilization. Pigment concentrates, also known as color chips, are pre-dispersed pigment mixtures that offer more consistent coloration, easier handling of pigments, and reduced exposure to pigment dust. These concentrates can be mixed directly with the nail polish base to save time and ensure a higher degree of reproducibility and color consistency between batches.

Nail plate staining is a common complaint and is caused by colorant migration into the nail plate. Certain colorants are more likely to stain the nail plate than others. D&C Red No. 6 (CI 15850) and Red No. 34 (CI 15880) can leave a pink or orange stain on the nails, while Yellow 5 Lake (CI 19140:1) yellows the nail plate. Phthalocyanine blue (CI 74160), chromium oxide (CI 77288) and carbon black (CI 77266) can cause staining due to their strong pigmentation, which can seep into the upper layers of the nail plate and leave a residue that is difficult to remove, especially on nail plates with surface damage or cracks. Ferric Ferrocyanide (CI 77510) can cause a blue or greenish stain [1, 2].

Solvents such as ethyl acetate, butyl acetate, ethanol, diacetone alcohol, isopropanol, and butanol are critical for dissolving film formers, adjusting the viscosity, and controlling the evaporation rate of solvents to manage drying time and film formation. Naturally sourced solvents allow for "natural" claims and include bio-ethanol, ethyl acetate, butyl acetate, and ethyl lactate, although being bio-derived does not improve their safety profile for users.

Thickeners and rheology modifiers control the viscosity and flow characteristics of nail polish, making it easy to apply and providing a smooth, even finish. For example, silica, clays, and stearalkonium bentonite are added to improve the suspension of pigments and control flow properties. Achieving an ideal viscosity is challenging due to the complex, non-Newtonian behavior of nail polish, which exhibits both shear thinning and thixotropic properties, leading to viscosity decreases with increased shear rate and helping to prevent pigment settling. This ensures the polish does not become too fluid after application or shaking. Bentonite and hectorite facilitate these desired behaviors, allowing the polish to develop a lower viscosity when brushed, mixed, or shaken, but returning to a higher viscosity when such stress is removed.

Other functional additives include UV stabilizers such as benzophenone and octocrylene, which absorb UV light, protecting pigments from degradation/fading. Antioxidants like ascorbic acid (vitamin C) and tocopherol (vitamin E) are widely used to prevent oxidation and are substantially more effective than the lower-cost tocopherol linoleate, or the almost completely ineffective tocopherol acetate [3–6]. Adhesion promoters improve the compatibility of the nail polish film with the nail surface, often by forming a chemical or physical bridge that binds the polish firmly to the nail. Low concentrations of adhesion promoters such as acrylic oligomers, phosphoric acid esters, trimethylsiloxysilicate, or silanes like triethoxycaprylylsilane or 3-methacryloxypropyltrimethoxysilane are used. Dimethicone, cyclomethicone, or phenyl trimethicone enhances shine and improves water resistance but must be used judiciously to prevent loss of adhesion. Marketing additives also encompass "nourishing" ingredients like vitamins and oils to promote nail health; however, cosmetics are not allowed to imply nutritional or medical drug-like effects, and the INCI name should be used to avoid confusion. Often, vitamins, essential oils, and botanical extracts are used solely for marketing purposes, providing no functional value to nail polishes and can adversely affect adhesion and lower the dry film's cohesion.

Topcoats applied over the colored layer provide a protective seal that enhances durability, adds gloss, and can speed up drying time. Their formulation is like that of base coats but optimized for surface hardness/scratch resistance, wear resistance, gloss, and clarity. UV stabilizers prevent discoloration/fading, while plasticizers ensure flexibility and adhesion with the underlying color layer. Fast-drying topcoats utilize quickly evaporating solvents, and sometimes glitters or other special effect colorants are added to modify the middle color layer.

The middle color layer consists of pigments, lakes, dyes, and sometimes pearlescent particles or glitters suspended in a blend of film-forming polymers, plasticizers, and solvents, all balanced to ensure the polish is easy to apply, durable, and dries quickly. Pigments can be pretreated to ensure even dispersion and prevent rapid settling or clumping.

Base coats are applied directly to the nail plate to vastly improve adhesion to the plate and colored layers, as well as to protect the plate from staining. Certain base coats also include marketing ingredients like keratin, which offers no benefits to the nail plate but appealing to consumers. Many consumers and marketers consider base coats to be unnecessary and time-consuming, but salon professionals understand they can increase wear time from one or two days to up to 14 days and help avoid nail staining from the color layer.

Water-based polishes can significantly reduce Volatile Organic Compound (VOC) emissions by utilizing water-dispersible film formers such as polyurethane dispersions or acrylic emulsions and water-soluble plasticizers, e.g., glycerin, however, preservatives like phenoxyethanol or sodium benzoate are essential to prevent microbial growth, which is a non-issue for traditional nail polish formulations. Water-based polishes are perceived as safer for consumers/salon workers, minimizing the risk of respiratory issues, skin irritation, and other solvent-related health concerns, but have poor adhesion, quickly peel, have slow dry times, poor durability, and struggle with pigment dispersion and stability, as well as lower gloss finishes with reduced color vibrancy, despite 40 years of attempts to develop more competitive formulations.

Acetone removers work quickly to dissolve nail polish, making them a popular choice. Besides being highly flammable, undiluted acetone creates a dry appearance on the surface of the nail plate and surrounding skin and has a strong odor. The non-acetone removers typically contain ethyl acetate, isopropyl alcohol, or propylene carbonate and can be less drying and have a milder odor compared to acetone. However, they require more time and effort to remove polish effectively, especially for polishes containing glitters. Prolonged and repeated use of nail polish solvents can lead to chronic skin conditions and nail plate surface damage; consequently, many removers now include conditioning agents like glycerin, panthenol or natural oils to mitigate these effects.

Nail Polish Testing/Evaluation

An array of test methods is used for quality control and research/development, enabling cosmetic chemists to tune polish properties, ensuring the final product meets quality and performance standards. Dry time methods are performed on uniform thin films made using a drawdown blade. Flash point testing determines flammability, while rheometers and viscometers measure flow and viscosity under different shear rates and conditions to simulate application. Adhesion tests, such as tensile testing, involve the polish being applied to a substrate and subjected to forces that try to detach it, providing quantitative data, while quickly removing adhesive tape applied to the dried surface estimates adhesion quality. Scratch tests and cross-hatch tests are also useful methods.

Water immersion tests for a specified period simulate exposure to moisture to determine its effect on adhesion and appearance. Chemical resistance tests expose dry polish to common substances such as detergent, soap, or alcohol hand sanitizers to demonstrate the ability to resist exposure. Stability testing, including exposure to light, heat, and air, assesses whether the color changes over time or under various environmental conditions. Gloss meters provide quantitative data on the reflectivity and shine of the nail polish surface and can guide/optimize the selection of shine-enhancing ingredients or those that provide surface textures such as silicas, micronized waxes, or polyethylene terephthalate. Colorimeters make color measurements for consistency and quality control. Hardness is measured by rocking hardness testers and pencil hardness testers to determine the resistance to surface scratching. Flexibility and film integrity tests involve bending and flexing the substrate to observe the polish's behavior under strain, identifying formulations that maintain their integrity without cracking or flaking. However, the most important and useful tests are real-world wear tests where the nail polish is applied to nails, and the wear is observed over a period of days. These tests can include normal daily activities, including hand washing, to assess durability and adhesion over time.

Understanding the extent to which ingredients in nail polish can penetrate the nail plate is crucial for ensuring product safety, especially for formulations claiming strengthening. An intact, healthy human nail plate is an excellent protective barrier that shields the nail bed from environmental damage, microbial infections, UV light, and chemical penetration. Not many, but certain substances can penetrate this barrier. The degree of penetration depends on molecular size (<500 Daltons) and polarity: smaller, linear, and more hydrophilic molecules can penetrate the nail more easily than hydrophobic. Surface-damaged, cracked, or diseased nails are more permeable, potentially allowing greater penetration of chemicals, especially for hydrophobic molecules.

Advanced imaging, such as scanning electron microscopy and digital color microscopy, aids in understanding the microstructure of nail polish films, assessing pigment dispersion, and evaluating the interaction between the polish and the nail plate.

Emerging technologies and materials, such as bio-based plasticizers, advanced polymer systems, new pigment stabilization techniques, and crosslinking agents in topcoats designed to increase scratch resistance, offer promising avenues for future advancements. Nanotechnology offers exciting possibilities for nail polish, including the creation of novel color effects, enhanced mechanical properties, and improved adhesion [7].

UV-Cure Nail Coatings

UV-cured nail coatings, including UV gel manicure products, have changed the nail industry by providing a long-lasting, durable, and shiny alternative to regular nail polish and liquid/powder artificial nails. This technology originally came from the dental industry, where similar materials and curing methods were used for dental fillings. In the 1980s, this technology was adapted for nails, and the first UV-cured nail coatings were introduced in Europe. A key part of the UV-curing process is the UV lamp. UV-emitting LED nail lamps are popular because they reduce manicure times, use less power and create fairly consistent levels of UV intensity. All nail lamps emit UVA to cure UV nail coating products. There are no exceptions to this rule. UV-curing nail lamps produce UV energy with wavelengths between 350 and 400 nanometers. The UVA emitted by a nail lamp is a range of wavelengths, not just one wavelength. For example, nail lamps claiming to emit 405 nm actually emit a range of wavelengths, and 405 nm is just the middle of the range of 390 nm to 420 nm. The wavelengths responsible for curing are in the range of 390 nm to 400 nm, which is invisible to the eye. The intensity of the UV is even more important than the wavelength range.

Intensity is a crucial factor in the UV curing of nail coating products that determines how quickly and effectively the UV gel or UV polish cures. A lamp with higher UV intensity will deliver more energy to the nail coating in a shorter amount of time, causing it to cure faster. The opposite is true: UV intensity that is too low will not properly cure UV nail products. The unit used to measure UV intensity is "milliwatts per square centimeter" (mW/cm^2). Nail lamps on the market range from 4 mW/cm^2 to 55 mW/cm^2, so there is a tremendously wide range of intensities being sold. Undercuring results in soft, sticky, or easily damaged surfaces or bubble-like spaces under the nail coating as a result of adhesion loss. If the intensity is too high for the formulation, this can cause the nail coating to cure too quickly, leading to issues like excess heat, shrinkage, brittleness, and even under-curing. How can excessively high UV intensities cause under-curing? When the intensity is too high, this can cause the upper surface of some darker colored UV gels to cure too quickly, which can then block UV from penetrating to cure the lower layers.

Wattage does NOT measure UV strength. This common myth is at the heart of the confusion about nail lamps. Wattage indicates how much electricity the bulb uses. Lumens are a measure of how well the human eye will perceive the brightness of a light. Since UV is invisible, it cannot be measured in lumens. UV is measured in milliwatts per centimeter squared (mW/cm^2). Do not be fooled by this common marketing tactic; higher wattage does not mean a UV nail lamp is better than one with a lower wattage.

It is important that UV energy penetrate deep enough to cure the product properly, which is called the "depth of cure." UV light penetrates poorly and is easily absorbed by ingredients in the formula, thus converting it into heat and visible light. Only a tiny fraction of UV from the sun can penetrate the natural nail plate; most is blocked by the plate and will not reach the nail bed. The same is true when UV nail lamps are used.

UV energy struggles to reach deep into thick layers or through dark colors. Thick layers (3 mm or 0.12 inches or more) are much harder to properly cure than thin layers (about 1 mm or 0.04 inches). A thick layer may harden on the top surface, leaving the bottom side softer and not fully cured. For this reason, it is wise to apply UV gel products in several thin layers and completely cure each layer before adding the next. Colorless gels allow UV light to pass through them more easily. However, dark colors (like black, brown, red, and glitter) can block much of the UV light due to the "umbrella effect," which occurs when pigment particles cast a shadow below them, blocking most of the UV light from reaching deeper layers. This results in under-curing, and over time can lead to adverse skin reactions. To help prevent this, use thin layers of colored gels, especially for dark colors, and avoid skin contact with uncured or partially cured UV gel formulations of any kind, since prolonged and/or repeated contact can lead to the development of skin allergies.

Both the wavelength range and the wave intensity influence how deeply the UV can penetrate into the UV gel. The 405 nm range LED lamps (390-400 nm) do not penetrate as well as shorter wavelengths and tend to focus more on curing the surface. The 395 nm range LED lamps (380–410 nm) have better depth curing compared to the 405 nm range, but the 365 nm range LED lamps (350 nm to 380 nm) can penetrate the deepest into the UV gel and are better for curing thicker layers or darker colors. Matching the lamp's intensity to the gel's requirements is crucial for proper curing. Dual-cure nail lamps use two different types of LEDs that emit in a broader range, for example, 395 nm and 365 nm LEDs, which would emit in the range of 410 nm to 350 nm. Clearly, there is no such thing as a universal UV nail lamp that properly cures any UV nail product; nor are there UV-curing nail products that properly cure with any nail lamp. Those making these erroneous claims are contributing to the growing number of consumers and professionals developing skin irritation and persistent or permanent skin allergies that worsen over time.

The sellers of UV nail coating products are directly responsible for product safety, which includes ensuring that users receive all the information needed to work safely with the products. For example, manufacturers of UV-curing products are responsible for testing their products to ensure they can be used safely, including providing clear instructions about how to safely and properly cure the products for the benefit of both nail technicians and their clients. They are responsible for determining which wavelength ranges and intensities work best for safe curing of their specific formulation. Also, they are responsible for passing this information to users in the form of directions for proper use.

Nail Lamp Safety

The safety of all types of UV nail lamps, including LED lamps, has been established and confirmed by numerous safety tests. Both traditional fluorescent UV nail lamps and LED nail lamps were tested using the appropriate international standards [8]. The study was performed by Dr. Dowdy and Dr. Sayre; both are world-renowned experts in measuring and understanding the effects of UV on skin [9].

Their results have been supported by additional studies. These studies demonstrate the safety of UV nail lamps and show they are well within accepted safe levels. In fact, these scientists found they were even safer than expected. These scientists normally studied the effects of tanning beds on skin, but they concluded UV nail lamps are NOT like tanning beds and concluded, "*UV nail lamps are vastly less hazardous.*"

Their study, which followed international protocols, demonstrates UV exposure from nail lamps is safer than sunlight exposure and determined the skin cancer risks are 11–46 times lower than natural noonday sunlight. These scientists concluded that the risks presented by UV nail lamps are *"comparatively trivial"* and there is virtually no risk to the nail bed because the nail plate's natural UV resistance is comparable to the UV resistance provided by an SPF 40 sunscreen. Also, research demonstrated that the hand's backside is 3 1/2 times more resistant than a person's back, making it the most UV-resistant part of the body.

A second independent study was conducted by Brown University that compared UV nail lamp exposure to exposure from UV medical treatments for psoriasis. Dermatologists have used UVA lamps since 1974 to treat those with psoriasis, and medical research has determined that such treatments are safe. Researchers at Brown University compared these safe levels of exposure for psoriasis to nail client exposure from UV nail lamps and concluded nail clients would need to receive 250 years of "weekly" UV gel manicures to receive the same level of exposure as patients undergoing these types of psoriasis treatments, which are considered safe by medical doctors worldwide [10]. Unfortunately, some raise unfair suspicions about nail lamps by designing unfair and inappropriate studies with the intent of making nail lamps look bad. Such studies are flawed and biased, often designed to support specific agendas, misguided beliefs, or to manipulate the news media and public opinion. By presenting false/distorted information, the authors of these types of junk science studies hope to create confusion and doubt.

It is important to note that accurately measuring the intensity and wavelengths of a nail lamp requires a properly calibrated high-end spectrophotometry and following internationally accepted testing procedures. Inexpensive devices purchased on the internet yield inaccurate and unreliable results.

Chemistry and Ingredients

Ultraviolet (UV) cured nail coatings, commonly known as gel nail polishes, have revolutionized the nail care industry due to their superior durability, high-gloss finish, and rapid curing. This technical overview explores the chemistry of UV-cured nail coatings; providing a detailed examination of the key ingredients and processes involved in their formulation and application, with a focus on safety and recent developments in the field.

UV nail coatings undergo photopolymerization, specifically free radical polymerization initiated by UV light. This process involves several distinct stages outlined below:

1. Initiation: Upon exposure to the correct range of UV wavelength, typically between 350 and 400 m), and sufficient UV intensity, the selected photoinitiator(s) absorb these specific photons and decompose into unstable free radical species. This can occur through two primary mechanisms: α-cleavage (type I photoinitiators), for example, benzoin ethers, benzil ketals; and hydrogen abstraction (type II photoinitiators), for example, benzophenone/amine systems.
2. Propagation: Free radicals attack the carbon-carbon double bonds (C=C) of monomers and oligomers, creating a new radical.
3. Chain Transfer: The growing polymer radical may transfer its activity to another molecule, which can lead to branching or termination.
4. Termination: The polymerization process ends when two radicals react with each other, either by combination or disproportionation.

Acrylic oligomers form the backbone of the polymer network and largely determine the final properties of the cured film. Proper selection of the oligomer is key to a successful formulation. There are hundreds of different oligomers. They come from these general categories:

Urethane Acrylates: Offer excellent flexibility and adhesion, due to this general structure: R-NH-COO-R′-OOC-CH=CH2, where R and R' are various organic groups. By varying the side groups, oligomers can be specifically tailored to have enhanced properties, such as improved crack resistance, better adhesion, smoother surface texture or quicker removal.

Epoxy Acrylates: Provide hardness and chemical resistance. Typically synthesized by reacting acrylic acid with epoxy resins: CH2-CH-R + CH2 = CH-COOH → R-CH(OH)-CH2-O-CO-CH=CH2.

- Reactive monomers are added to lower the viscosity of the formulation, increase curing and form covalent bond crosslinks between the growing polymer chains, for example, polyester acrylates offer a wide range of hardnesses and flexibilities and are formed by esterification of polyols with acrylic acid and dicarboxylic acids. Some examples are:
  - Isobornyl methacrylate aka IBOMA (CAS# 7534-94-3): A cyclic monomer with a unique molecular structure which helps to improve hardness and reduce film shrinkage upon polymerization.
  - Triethylene glycol dimethacrylate aka TEGDMA (CAS# 109–16-0): CH2 = C(CH3)-COO-(CH2CH2O)3-CO-C(CH3) = CH2. Increases cross-linking density of the polymer, which increases surface hardness/scratch resistance, increases solvent resistance, and increases toughness/durability when used judiciously.
  - Trimethylolpropane triacrylate aka TMPTA (CAS# 15625–89-5): A trifunctional monomer creates an even higher density of crosslinks around it, linking many polymer chains into a dense network, when used judiciously. Excessive crosslinking increases shrinkage and creates cured coatings with built-in internal stresses, which can lead to cracking and brittleness.
  - 2-Hydroxyethyl methacrylate aka HEMA (CAS# 868–77-9), and 2-hydroxyethyl Methacrylate phosphate aka HEMA-phosphate (CAS# 52628–03-2) are being phased out in many formulations due to increased potential for allergic reaction and should be avoided. HEMA named Allergen of the Month, by the British Association of Dermatologist in June 2018. A clinical review of this potent allergen was published in 2023 [1, 11].

Structure/Activity relationships are shown when a specific physical property is the result of the chemical composition and molecular structure of the substance. As an example, varying the length of the chains between the reactive groups that create the crosslinks and active groups is a useful way to enhance flexibility and toughness. For example, triethylene glycol dimethacrylate with three ethylene glycol units creates less flexible crosslinks than tetraethylene glycol dimethacrylate with four ethylene glycol units. Another often seen example is demonstrated with (meth)acrylate oligomers having large numbers of hydroxyl substituent groups (R = OH) that are more likely to have significant adhesion to the nail plate when cured as part of a film-forming blend, largely due to their high polarity and increased hydrogen bonding with the nail plate surface. Understanding how to use structure/activity relationships to enhance the performance of the nail coating is an important tool that should not be overlooked.

The Function and Structures of Photoinitiators

Photoinitiators are crucial for initiating the polymerization process. Below are examples of commonly used photoinitiators in UV nail coatings, along with their peak absorption wavelengths. It is important to remember that the "peak absorption" represents a wider range of wavelengths. More detailed information about how sensitive a photoinitiator will be to other wavelengths can be determined by viewing the UV absorption spectrum for the photoinitiator, which usually can be found on the Internet.

- Bis(2,4,6-trimethylbenzoyl)-phenylphosphineoxide (Irgacure 819): Peak absorption at 370 nm. Type I photoinitiator is effective for pigmented systems and deeper curing.
- 2-Benzyl-2-(dimethylamino)-1-[4-(4-morpholinyl)phenyl]-1-butanone (Irgacure 369): Peak absorption at 366 nm. Type I photoinitiator, suitable for highly pigmented formulations.
- Diphenyl(2,4,6-trimethylbenzoyl)phosphine oxide (TPO): Peak absorption at 380 nm. Type I photoinitiator is excellent for fast curing and color stability. However, regulatory issues with this ingredient in the EU have many looking for an alternative.
- Ethyl (2,4,6-trimethylbenzoyl) phenyl phosphinate (Lucirin TPO-L): Peak absorption at 380 nm. Type I photoinitiator, liquid form, allows easy incorporation into formulations.
- 2-Hydroxy-1-[4-(2-hydroxyethoxy)phenyl]-2-methyl-1-propanone (Irgacure 2959): Peak absorption at 370 nm. Type I photoinitiator, known for its low odor and low yellowing properties.

- The choice of a photoinitiator and its concentration is completely dependent on the UV wavelengths and intensity of the UV nail lamp. Often, a combination of photoinitiators is used to ensure efficient curing throughout the coating. Photoinitiators that are particularly well-suited for use with LED UV lamps have good absorption in the 365–405 nm range, which will help to ensure more efficient curing of the nail coating formulation.

Pigments and Colorants

Inorganic and organic colorants are used and selected for their stability under UV light and compatibility with the polymer system. Examples include iron oxides (red, yellow, and black), titanium dioxide (white), and organic lakes and toners (various colors).

Typically, iron oxides have high UV absorption and high photostability, meaning they can absorb and dissipate UVA without undergoing any photodegradation, unlike the organic lakes which are much more susceptible. The ability to absorb UV leads to poor curing for UV formulations containing high concentrations of iron oxides, especially black.

UV Stabilizers

Both the polymers and colorants can discolor over time, and the nail coating can become brittle if not stabilized. This is a main reason for the premature aging of a nail coating. Long-term UV exposure is a typical reason for this. To prevent this, UV absorbing ingredients are utilized, typically in very low concentrations, for example, 0.1% to 1%. Some common examples are: benzophenone-1 (BP-1), benzophenone-3 (BP-3 aka Oxybenzone), ethylhexyl methoxycinnamate (octinoxate), butyl methoxydibenzoylmethane (avobenzone), octocrylene, and methylene bis-benzotriazolyl tetramethylbutylphenol (MBBT). These absorb UV energy and convert it to visible blue light and heat.

Adhesion Promoters

Adhesion to the surface of the nail plate can be improved using adhesion promoters, which can create either or both hydrogen and longer lasting covalent bonds. Typically, these are 0.5% or less of the total formula. Examples are silane coupling agents such as 3-(trimethoxysilyl)propyl methacrylate, or (3-glycidoxypropyl) trimethoxy silane, which can form covalent bonds with the nail surface. Phosphate esters such as ethoxylated trimethylolpropane triacrylate phosphate can improve adhesion without the skin irritation potential of methacrylic acid and acrylic acid, which should be avoided due to their potential to cause skin corrosion and allergic reactions.

Solvents, Diluents and Plasticizers

Solvents lower the viscosity of UV gel nail coatings by dissolving or dispersing ingredients such as high molecular weight oligomers, making it easier for them to flow. Some commonly used solvents are as follows:

- Isobornyl Methacrylate aka IBOMA (CAS# 7534-94-3): A reactive monomer that acts as a solvent, reduces the viscosity, and improves adhesion.
- Butyl Acetate (CAS# 123-86-4): Used to decrease viscosity and improve application. It has good solvency for many oligomers.

- Ethyl Acetate (CAS# 141-78-6): Used to reduce viscosity and has the advantage of evaporating quickly.
- Propylene Carbonate (CAS# 108-32-7): Used to adjust viscosity and improve the flow of the gel.
- Isopropyl Alcohol (CAS# 67-63-0): Used to adjust consistency and improve clarity.
- Diacetone Alcohol (CAS# 123-42-2): Acts as a solvent to control viscosity and flow.

Careful use of solvents to balance viscosity and application properties can ensure a smooth and even finish on the nail plate, but long-term stability of the solvent blend should also be a consideration. Incompatibilities can develop over time, resulting in any of the following: discoloration, formation of a solid precipitate, phase separation, gel formation, or mass polymerization.

Diluents also reduce gel viscosity and aid in application and are better able to maintain the integrity of the formulation. For example, reactive diluent monomers participate in the curing process, unlike the traditionally used volatile solvents. Here are some common examples:

- Ethyl Methacrylate (EMA): Can be used in small amounts to reduce viscosity in UV gel formulations. This reactive monomer polymerizes and contributes to the final hardness and durability of the cured gel. Typical use level is 2–10%.
- Hydroxypropyl Methacrylate (HPMA): Lowers viscosity and improves nail plate adhesion. It can also improve crack resistance. Typical use level is 5–30%.
- Isobornyl Methacrylate (IBOMA): Reduces viscosity, enhances adhesion and improves surface scratch resistance. Typical use levels are 3–20%.
- Triethylene glycol dimethacrylate (TEGDMA): A difunctional methacrylate used to build crosslinks in the polymer network and thereby improve durability, when used judiciously. Excess concentrations can increase shrinkage and lower adhesion. The typical use level is 2–10%.
- Trimethylolpropane Triacrylate (TMPTA): A trifunctional low viscosity acrylate monomer that reduces viscosity while increasing surface hardness and chemical resistance. Increases crosslink density; use judiciously. Excess amounts can cause brittleness. The typical use level is 1–3%.

Diluent choice is driven by compatibility with the main "workhorse" gel components and their ability to participate in the curing process and polymerize into the polymer matrix to the benefit/improvement of the coating's final physical properties.

Plasticizers like camphor, acetyl tributyl citrate, neopentyl glycol, sucrose acetate isobutyrate and triphenyl phosphate (TPP), isopropyl myristate, polypropylene glycol (PPG) and diisobutyl adipate make the film flexible and resistant to cracking. The choice of plasticizer and its concentration can lower surface hardness while increasing toughness and scratch resistance.

Formulation Considerations

Several factors must be considered to achieve a carefully balanced formulation and to achieve optimal performance and safety:

- Viscosity/Flow: Typically ranges from 200 to 2000 cP, adjusted using reactive diluents or non-reactive plasticizers. Temperature-dependent, so formulations should be tested at various temperatures. Note: Samples should be conditioned for 24 hours at a constant temperature before measuring the viscosity. The product's viscosity is dependent on its recent thermal history, not just its current temperature.
- Reactivity: Influenced by photoinitiator concentration (usually 1–5%), monomer/oligomer ratio, number of functional reactive sites, and the intensity of the UV wavelengths that can activate the photoinitiator. Excessive amounts of photoinitiator can be detrimental to the system and lead to a lowering of physical properties and an increase in allergic reactions. UV gel nails will harden when they are only 50% cured, so hardening is not a sign of proper cure. To minimize the potential for adverse skin reactions, UV-curing nail products should be cured to a minimum of 70% for basecoats and color coats, while the topcoat should be cured to a minimum of 90%. Methods for testing this are described below.
- Shrinkage: Higher reactivity leads to faster cure times but may increase shrinkage and exothermic heat generation. Exotherms must be controlled to prevent discomfort or damage to the nail bed tissue and can be controlled by using lower concentrations of photoinitiators or applying and curing thinner layers. Lower reactivity may improve adhesion but can increase exposure to unreacted ingredients when the coating is filed or removed by soaking in solvents. Shrinkage can be minimized by using high molecular weight oligomers, cyclic monomers, and lower overall concentrations of cross-linker or using longer chain cross-linkers. Typical shrinkage ranges from 8% to 12%. Excessive shrinkage (>12%) can lead to poor adhesion and stress cracking.
- Flexibility: Balanced to prevent cracking or chipping by varying the ratio of flexible to rigid oligomers and monomers or by incorporating a suitable plasticizer.
- Adhesion: Critical for long-lasting wear. Influenced by the choice of oligomers, adhesion promoters, and surface preparation. Under curing can lead to separation between the nail plate and product, which can have a "bubble-like" appearance.
- Stability: Formulations must be stable during storage and application. UV stabilizers and antioxidants can improve shelf life. Some examples are as follows:
  - Hydroquinone aka HQ (CAS# 123-31-9) typically used at 0.01% to 0.1%. It acts as a radical scavenger to prevent premature polymerization during storage.
  - Butylated Hydroxytoluene aka BHT (CAS# 128-37-0) at concentrations of 0.01% to 0.1%. It is effective in preventing oxidative degradation and polymerization of the monomers in the formula.

  - Benzophenone (CAS# 119-61-9) typically 0.05% to 0.5%. It is used primarily as a UV absorber but also contributes to the stabilization of the formula by capturing free radicals.
  - 4-Methoxyphenol aka MEHQ (CAS# 150-76-5) is usually in the range of 0.02% to 0.2%. Effective as an inhibitor for acrylic monomers and prevents premature polymerization. Packaging in opaque containers helps prevent premature curing.
- Wise Formulation Practices: Increasing focus on developing formulations with reduced allergenic potential, avoiding known sensitizers like HEMA, acrylic acid, and certain low molecular weight acrylates. Larger molecular sizes and bulky structures reduce skin penetration.
- Removal Properties: While durability is desirable, the coating should be removable without excessive damage to the natural nail. Balancing crosslink density to allow for acetone penetration during removal, as well as the addition of non-reactive polymers that are soluble in organic solvents.
- Pigment Compatibility: Most pigments will block UV from penetrating into the coating, so it is critical to test the depth of cure for heavily pigmented formulations to ensure they do not interfere with the curing process.
- Gloss and Surface Properties: Achieving a high gloss is often desired in nail coatings. Surface smoothness is affected by pigment dispersion and oligomer selection. Also, silicone-based wetting agents can help to ensure smooth, reflective surfaces, but must be used judiciously to prevent them from interfering with adhesion.
- Regulatory Compliance: Ensuring all ingredients comply with relevant cosmetic regulations (e.g., EU Cosmetics Regulation, FDA guidelines). Adhering to VOC (Volatile Organic Compound) limits where applicable.

Product Testing

Testing UV-cured nail coatings is crucial for several reasons, primarily to ensure that the coatings cure properly and to minimize the risk of adverse skin reactions. Incomplete curing can leave unreacted materials in the coating that can later irritate the skin or cause allergic reactions, for example, exposure to partially cured dusts. Through rigorous testing, manufacturers can determine the optimal UV exposure time and intensity to ensure complete curing, thereby reducing potential health risks as well as preventing excess heat generation, usually during the first 15 seconds of the curing process. Durability and shine are also improved since properly cured coatings are more resistant to chipping, peeling and scratching.

Differential Calorimetry

Allergic skin reactions to under-cured UV gels, and nail bed burns from over-cured UV gels are leading causes of injury to clients, often due to a mismatch between the UV gel and nail lamp [12]. Under-curing is a common problem that occurs because UV-cured products harden and can be filed when they are only 50% cured. The product looks and feels cured even when it has not polymerized to a safe level. Differential calorimetry methods precisely measure the heat emitted by the UV nail lamp separately from the exotherm created each time a molecule of UV gel joins a growing polymer chain. This allows for the determination of the total heat released by a nail product during UV curing, which is directly proportional to the degree of curing. Besides calculating the percentage of reactive species curing, this testing can also be used to determine maximum exotherm, activation energy, and polymerization rate, allowing product developers to measure the effects of photoinitiators, colorants, and other additives on polymerization, as well as to check the quality of their incoming products and determine which commercially available nail lamps work well with the formulations. The image below is an example of such a device. Three high-precision thermocouples are placed into the chambers at the end of the device. The chambered end is then inserted into a UV nail lamp, and the exotherms are recorded during cure. The UV gel sample is placed in the center chamber, while the outer two chambers are empty and used to record the lamp's emitted heat, which is subtracted from the sample's recorded temperature to reveal the heat of polymerization which is used to calculate the curing properties of the sample (Fig. 3).

Crosshatch Adhesion Testing

Crosshatch testing is an effective method for evaluating adhesion. It involves scoring the coated surface in a grid or crosshatch pattern using a specialized tool/blade to create small, equally spaced cuts through the coating. The UV gel coating can be applied to different substrates for this testing, for example, glass, plastic or hoof/horn samples, and cured under the nail lamp. A grid pattern is made by cutting through the cured gel in a crisscross manner, ensuring the cuts go down to the

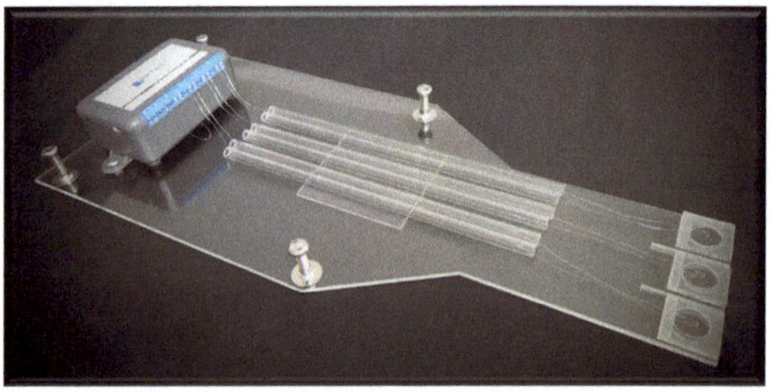

Fig. 3 Differential calorimeter used for determining the percent conversion of reactive monomers and oligomers into polymers

substrate without damaging it. After creating the grid, an adhesive tape with a consistent tack level is applied over the scored area, pressed down to ensure good contact, and quickly pulled off at a consistent angle to check if any sections of the coating are lifted or removed. A high level of coating retention indicates strong adhesion, while significant peeling or flaking indicates weak adhesion.

Surface Hardness

Pencil hardness testing is used to evaluate the hardness and surface durability of UV gel nail coatings. Resistance to scratching gives insight into the quality of the gel's finish. A series of pencils with different hardness levels, ranging from soft (6B) to hard (9H), are used. The hardest pencil that doesn't leave a scratch on the gel surface is recorded as the coating's hardness level. This method is simple, cost-effective, and provides a clear, standardized measure of the gel coating's hardness.

Flexibility

Mandrel bend testing assesses flexibility and adhesion by evaluating the ability of the cured gel to withstand bending or deformation without cracking, flaking, or losing adhesion to the nail plate. The UV gel coating is applied and cured on a flexible substrate, for example, thin plastic sheets. A set of mandrels (cylindrical rods) of varying diameters is used, ranging from large to small. The smaller the mandrel, the more severe the bend that will be applied to the coating. After each bend, the UV gel coating is visually inspected for signs of cracking, flaking, or loss of adhesion (delamination). The diameter of the smallest mandrel around which the coating can be bent without showing any damage or failure is recorded as the result of the test. Cracking or delamination with larger mandrels indicates lower flexibility or weaker adhesion. The ability to bend without cracking is critical for maintaining the appearance and longevity of the gel coating [13].

Safety Considerations

While UV nail coatings are generally considered safe when properly applied, there are some concerns that formulating chemists should address. Dust and filings from under-cured UV products contain significant amounts of uncured monomers, oligomers or other substances. Filing or soaking off the nail coatings can release unreacted ingredients and increase skin exposure risks. Over time, repeated exposure to these uncured substances can cause adverse skin reactions or trigger existing skin problems. Typical symptoms are redness, itching, swelling, water blisters, onycholysis, numbness, an unusual sensitivity to heat and pressure or, in extreme conditions, a total loss of the nail plate. Continued use causes the symptoms to progress and worsen. Curing to greater than 75%, as determined by differential calorimetry, and avoidance of certain monomers, for example, hydroxyethyl methacrylate and acrylic acid, can minimize the potential for such problems.

UV-cured nail coatings are a sophisticated application of polymer chemistry in the cosmetics industry. Due to the unique combination of durability, rapid curing, and aesthetic appeal, they have become a staple in professional and retail nail care. As research continues, we can expect further innovations in this field, particularly in developing better curing formulations and improved test methods.

Dip Powder Nail Coatings

Dip powder nails are a popular nail enhancement technique. This technique combines cyanoacrylate monomers blended with fine colored powders to create a strong, chip-resistant finish. A layer of low-viscosity cyanoacrylate monomer is applied to the precleaned natural nail plate. Then each nail is either dipped into the colored powder or, conversely, the powder is sprinkled onto the nail plate. This step is typically repeated three times to build up layers of the desired thickness and color intensity, after which an activator liquid is used to trigger polymerization, creating a durable coating. Once hardening occurs, usually within several minutes, the resulting coating is filed, shaped, and buffed to ensure a smooth surface. A final topcoat of cyanoacrylate is applied to seal and protect the powder layers, adding shine and further enhancing the durability of the manicure.

Cyanoacrylate monomers cure via anionic polymerization, which is initiated by moisture (water or hydroxide ions) on the nail plate surface or in the environment. Cyanoacrylate monomers contain a highly reactive carbon-carbon double bond (C=C) flanked by a cyano group (-CN) and an ester group (-COOR), which are electron-withdrawing groups. This makes the monomer highly susceptible to nucleophilic attack, particularly by hydroxide ions or moisture. The nucleophiles attack the electrophilic carbon of the monomer's double bond, opening the double bond to create a reactive carbanion intermediate. Once initiated, the carbanion formed from one monomer attacks the next cyanoacrylate monomer, leading to a chain reaction. The polymerization proceeds rapidly as monomers continue to link together, forming long polymer chains. The speed of this reaction is a key feature, resulting in a fast-setting process. Polymerization generally continues until all available monomers have reacted, thus ensuring complete cure. Inhibitors (such as HQ or MEHQ) are often included to stabilize the monomer during storage by preventing premature polymerization. Cyanoacrylate monomers are cured into a hard, durable polymer film that adheres strongly to the nail plate to form a protective, glossy layer. Many of the plasticizers or thickeners previously described in this chapter can be included in the formulation to modify properties, such as improved flow/viscosity or greater flexibility of the cured coating.

Due to their extreme sensitivity to moisture, cyanoacrylate monomers must be carefully manufactured under controlled conditions. They are manufactured by reacting cyanoacetic acid with an alcohol to yield cyanoacetate ester, which is subsequently reacted with formaldehyde and heated to yield the final monomer. The monomer is distilled under vacuum in the presence of a stabilizer such as hydroquinone (100–1500 ppm). Weak acids such as sulfur dioxide are also utilized. The viscosity is typically adjusted to 400–600 centipoise (cP) using methacrylate polymers, e.g. polymethyl methacrylate, and then packaged in an air-tight container.

The powdered "dip" component is typically an acrylic polymer such as polymethyl methacrylate or a copolymer of polymethyl/ethyl methacrylate, usually between 10 and 30 microns, blended with various cosmetic pigments/lakes to provide the desired color and to create an even dispersion throughout. Special effect glitter or mica particles may be included to add a shimmer or glittery effect. These colorants should be cosmetic grade to ensure they are safe for use on nails.

These cyanoacrylate-based nail coatings are less likely to cause allergic skin reactions because the polymerization process is highly efficient and typically goes to completion, meaning that there is little to no unreacted monomer left in the final cured product. Once the polymerization process is complete, the risk of skin sensitivity caused by the monomer itself is minimal. However, there is still a potential for allergic reactions or skin sensitivity that arises from the inhibitors added to prevent premature polymerization during storage, since they remain in the cured coating and can sometimes cause irritation or allergic reactions in sensitive individuals, especially with prolonged or repeated exposure. Thus, while the polymerized cyanoacrylate itself is generally inert and safe, individuals prone to skin sensitivities should be mindful of the inhibitors in the formulation, as these are the more likely culprits for allergic reactions. While dip powder systems are generally considered safe, some people are sensitive to the adhesive or activator liquids used in the process. It is essential to ensure proper ventilation and avoid skin contact with the chemicals. Additionally, using clean, uncontaminated powder is crucial, as dipping multiple clients' nails into the same jar could potentially spread bacteria or infection.

Brush-on activators used in dip powder nail systems are typically composed of amines or amine derivatives, which catalyze polymerization. The most common activator ingredient is the slightly basic aromatic tertiary amine dimethyl-p-toluidine, aka DMPT (CAS# 99-97-8), typically in 1–2% concentration. DMPT has a long history of safe use in this application, likely because it is brushed onto the nail coating. Spraying this substance should be avoided to prevent excessive inhalation or direct skin contact. Usually, the DMPT is dissolved in water-free solvent carriers such as ethyl acetate. Other stabilizers or mild acids may be included to balance reactivity. The pH of the substrate and formulation is an important consideration since polymerization is highly sensitive to pH and the presence of basic (alkaline) conditions can greatly accelerate polymerization. This explains why pH values greater than 8 can cause the reaction to proceed almost instantaneously, while under acidic conditions, polymerization is slowed or inhibited.

Liquid and Powder Nail Coatings

Acrylic monomer liquid and polymer powder nail coating systems are widely used to create durable and strong artificial nails. Their name incorrectly implies that these are the only systems based on "acrylics," when all curable nail ingredients are primarily from the acrylic family. These types of nail products involve a chemical reaction between a liquid monomer and a powdered polymer, resulting in the formation of a hard coating that can be sculpted into various nail shapes. This method is popular due to its strength, longevity, and versatility in creating custom nail designs. This method also allows the coating to be easily sculpted and to extend beyond the free edge of the nail plate, which is much more difficult to do with other types of nail coatings. The liquid component is composed mainly of monofunctional methacrylate monomers, most commonly ethyl methacrylate aka EMA (CAS# 97-63-2), as well as di or tri-functional methacrylates, which link together the growing strands of EMA polymer with crosslinks. Some examples of crosslinking monomers are ethylene glycol dimethacrylate aka EGDMA (CAS# 97-90-5), 1,4-butanediol dimethacrylate aka BDDMA (CAS# 2082-81-7), diethylene glycol dimethacrylate aka DEGDMA (CAS# 2358-84-1), and trimethylolpropane triacrylate aka TMPTA (CAS# 15625-89-5). Multi-functional monomers increase the crosslink density of the polymer network, resulting in enhanced hardness and mechanical strength, which are useful for high-performance coatings and durable artificial nails. These monomers are used in varying concentrations to fine-tune the physical properties of the cured acrylic, such as flexibility, strength, and resistance to wear or cracking, typically 10% or less of the monomer blend.

The polymer powder is usually a high molecular weight copolymer of both methyl and ethyl methacrylate (MMA/EMA copolymer) which contribute specific properties to the final product. The MMA provides rigidity and strength, while the EMA imparts flexibility to increase overall durability and crack resistance to the nail coatings. The size of the copolymer particles typically varies between 30 and 80 microns and are often coated with pigments/lakes and other additives to modify the color, strength, and performance of the final product.

To initiate the polymerization process when the powder and liquid monomer are mixed, the MMA/EMA copolymer powder is coated with benzoyl peroxide aka BPO (CAS# 94-36-0) which serves as the polymerization initiator. BPO molecules (typically used at concentrations of 0.5–2%) break in half to create two identical free radicals, when exposed to the liquid monomer. The primary catalyst in the liquid is typically a tertiary amine, typically used at 0.5–2%. The amine catalyst accelerates the decomposition of benzoyl peroxide at room temperature. Commonly used tertiary amines are N,N-dimethyl-p-toluidine aka DMPT (CAS# 99-97-8) or N,N-dimethylaminoethyl methacrylate aka DMAEMA (CAS# 2867-47-2). This reaction occurs via a redox mechanism, where the amine donates an electron to the benzoyl peroxide, breaking the peroxide bond and releasing phenyl radicals. The free radicals formed initiate the polymerization of the monomers in the liquid, leading to the hardening of the acrylic. Upon mixing with the liquid monomer, benzoyl peroxide decomposes to form free radicals. These free radicals then attack the double bonds in the monomers, initiating the polymerization process and causing the liquid to harden into a solid acrylic structure.

After the nail plates are cleaned, shaped, and lightly filed to increase the surface area for better adhesion, a primer or bonding agent may be applied to enhance adhesion between the natural nail and the acrylic. In the past, methacrylic acid aka MA (CAS# 79-41-4) was the most used primer. Contrary to popular belief, it does not etch the nail surface to improve adhesion. In fact, nail clippings can be soaked in methacrylic acid for long periods without creating any surface damage. Instead, the MA-based primers dissolve tightly bonded oils from "nooks and crannies" on the surface of the nail plate to increase surface area for improved adhesion. MA covalently bonds with the growing polymer chain and increases hydrogen bonding between the nail plate surface and the nail coating. MA is extremely corrosive and can chemically burn soft tissues; therefore, alternative adhesion promoters have gained in popularity. Other nail plate adhesion promoters mentioned in this chapter are incorporated directly into the monomer blend, which eliminates the need for such a primer.

Incorporating 2-Hydroxyethyl Methacrylate aka HEMA (CAS# 868-77-9), Hydroxypropyl Methacrylate aka HPMA (CAS# 27813-02-1), or Acetoacetoxyethyl Methacrylate aka AAEMA (CAS# 21282-97-3) into the monomer liquid improves adhesion of the monomer blend to the natural nail. These functionalized methacrylate monomers have hydroxyl or acetoacetoxy groups that enhance adhesion. The first two function by increasing hydrogen bonds with the nail surface while AAEMA creates a covalent bond, which is significantly stronger. These monomers also better "wet" the surface of the nail plate to promote stronger adhesion. Typically, these are used in concentrations of 5–10%. As mentioned previously HEMA has fallen out of favor due to its tendency to cause allergic reactions with prolonged/repeated skin exposure.

Liquid and powder systems are used in a very specific fashion. A small amount of liquid monomer is dispensed into a dappen dish. A brush is then dipped into the monomer and used to pick up a bead of acrylic powder to form a small ball. The liquid monomer immediately begins dissolving BPO to initiate polymerization. The bead is then placed onto the nail or a nail form, and the technician sculpts the material into the desired shape. The mixture remains moldable for a short period (3–4 minutes) allowing the technician to shape and smooth the nail before it hardens. Once the nail coating has hardened, it is filed, shaped, and buffed to smooth out the surface. Finally, a nail polish color layer and topcoat are applied for shine and protection.

Hydroquinone aka HQ (CAS# 123-31-9) is often found in these monomers at 0.01-0.05% to prevent premature polymerization of the monomer during storage. The powder contains various approved cosmetic organic and inorganic pigments such as titanium dioxide and iron oxides, which are added to give the acrylic powder its color. These pigments are typically finely ground to ensure smooth and even application. Silica (CAS# 7631-86-9) is often added to the powder in 0.5–3% concentration as a flow/anti-caking agent to keep the powder free-flowing and to prevent clumping. The liquid monomer, especially EMA, has a strong odor that some clients may find unpleasant. Proper ventilation is essential to minimize exposure to vapors. The so-called "Odorless" products are better described as having lower odor, since they still have a characteristic odor. These products are based on hydroxyethyl methacrylate aka HEMA. Although such products have good adhesion to the nail plate, as previously noted, HEMA is a likely allergen that should be avoided since skin contact with such products is difficult to avoid. Proper handling, such as using disposable nitrile gloves (8 mils) and good ventilation, is recommended, but often not a part of a normal nail salon's standard operating procedures. Some individuals may be sensitive to certain ingredients, like benzoyl peroxide or hydroquinone, and may develop allergic reactions or contact dermatitis.

Lotions, Creams, and Oils for Natural Nail Care

Lotions, creams, and oils are common products used to treat and maintain the health of natural nails. These products primarily aim to moisturize, strengthen, and protect nails and surrounding skin. Each product type has unique characteristics based on its formulation and the balance of water, oils, and active ingredients.

Lotions are light emulsions of oil and water, with a higher water content, making them less greasy and quickly absorbing. They are often used for daily moisturization of both nails and surrounding skin. Water aka aqua (CAS# 7732-18-5) is the primary solvent in lotions that hydrates and delivers other ingredients to the skin and nails. Glycerin (CAS# 56-81-5) is a powerful humectant that attracts water to the nails and skin, as is panthenol (CAS# 81-13-0), which also improves nail flexibility and prevents/reverse brittleness. These are typically used in 1–5% concentration.

Urea (CAS# 57-13-6) is well known for both its hydrating properties as well as its ability to soften and smooth the skin around the nails. It is hygroscopic and naturally found in the skin's outermost layers (stratum corneum), where it serves as a key component of the natural moisturizing factor (NMF). At concentrations below 8%, urea primarily acts as a humectant and provides mild hydration to improve skin elasticity, while keeping the nail plate moisturized. In this range, it helps regulate the water balance within the skin and nail plate by breaking hydrogen bonds in the keratin structures of the skin, making it more permeable to water. This increases the moisture content of the skin and nails without disrupting their structural integrity. However, when used above certain concentrations, particularly above 8%, urea begins to exhibit stronger keratolytic properties that can dissolve the skin and damage the nail plate. Its ability to break down the intercellular matrix holding together corneocytes (dead skin cells) becomes significantly more pronounced, leading to dissolution of the stratum corneum and potential damage to the nail plate. Higher concentrations are often utilized in callous softening/removal products. Urea interferes with the hydrogen bonding in keratin, the primary protein found in skin and nails. By breaking these bonds, urea loosens the structural cohesion of the corneocytes, promoting desquamation (shedding) of the outermost skin cells. This is desirable in cases of hyperkeratosis, where there is excessive skin buildup, such as in calluses or psoriasis, but it can also lead to unwanted skin thinning if used inappropriately. High concentrations of urea can cause excessive swelling of the nail plate by pulling water into its structure. While moderate hydration can improve flexibility, overhydration can weaken the structural integrity of the nail plate. When nail plates are repeatedly exposed to high concentrations of urea, they may become soft and prone to damage. Understanding the dual nature of urea is critical in cosmetic formulations, particularly for products intended for nail and skin care.

Avocado oil aka Persea Gratissima (Avocado) Oil (CAS# 8024-32-6) is a light vegetable oil that penetrates the skin quickly and helps seal moisture into the nails. It contains vitamins and essential fatty acids that increase nail plate flexibility. Alpha-hydroxy acids (AHAs), such as lactic acid (CAS# 50-21-5) can help to gently exfoliate dry skin and improve the absorption of other ingredients; however, AHAs in concentrations above 10% pose a significant risk to the nail plate. At these levels, these acids can begin to break down the keratin structure, leading to thinning, weakening, and softening of the nail. Prolonged use at high concentrations can result in nails that are more prone to splitting, peeling, and brittleness.

Other useful ingredients include squalane (CAS# 111-01-3), which is a highly stable emollient that mimics the skin's natural oils and supports the skin barrier. Sodium PCA (CAS# 28874–51-3) is a naturally occurring humectant found in the skin. Sorbitol (CAS# 50-70-4) is a sugar alcohol that functions as a humectant and adds a smooth texture to formulations. Allantoin (CAS# 97-59-6) is a skin-soothing ingredient that helps to treat dry or irritated skin and promotes the healing of minor nail and skin injuries.

Nail creams are thicker than lotions due to a higher oil content. They create a barrier that locks in moisture, making them ideal for nighttime use or in dry climates. They often include additional emollients to support nail flexibility and strength. Some common ingredients include shea butter aka Butyrospermum Parkii (CAS# 91080-23-8), beeswax (CAS# 8012-89-3) which forms a protective barrier on the nail plate and cuticle, preventing moisture loss, and vitamin E aka tocopherol (CAS# 59-02-9) is an antioxidant that helps repair damaged nails and promotes healthy growth. It is useful to note that vitamin E acetate is not a useful or effective substitute for tocopherol. It is used only because it has a much lower cost. For example, to make tocopheryl acetate the only active site on the molecule tocopherol that is capable to serve as a proton donor and quench free radicals is modified to a form that is unreactive and must first undergo bioconversion to the tocopherol form before it can serve as a proton donor. Tocopheryl acetate claims to be more stable toward oxygen and heat and oxygen, but this further demonstrates its inability ability to serve as a proton donor and functioning as an antioxidant. Clearly, modifying tocopherol to tocopheryl acetate or linoleate drastically reduces tocopherol's ability to be an antioxidant. [3–6]

Hydrolyzed keratin (CAS# 69430-36-0) is another example of a relatively useless ingredient. This large protein is unlikely to penetrate the nail plate or skin and therefore unlikely to provide any substantial benefits.

Nail oils are often blends of pure oils and active ingredients designed to be massaged into the nails and skin primarily to increase moisturization of brittle nails and dry skin. These oils are extremely slow to absorb into the nail plate and can take many weeks of daily use to accumulate in useful quantities. Their primary route is through existing damage such as small cracks. Once these gain access to the nail plate, they can act as plasticizers. Plasticizers are critical components in cosmetic formulations, particularly in products intended to improve the flexibility and pliability of skin and nails. Their primary function is to modify the mechanical properties of the keratin structure in nails and skin, making them more flexible, resilient, and less prone to brittleness or cracking. They function by inserting themselves between the layers of nail cells in the plate (typically 30–40 layers) reducing the intermolecular forces and reducing internal friction when the plate is bent. This disruption of intermolecular interactions lowers the glass transition temperature (Tg) of keratin, making it less rigid and more pliable. This decreases the stiffness and increases the flexibility of the nail plate.

Some common ingredients in nail oils are sweet almond oil aka Prunus Amygdalus Dulcis (CAS# 8007-69-0), olive oil aka Olea Europaea (CAS# 8001-25-0), jojoba oil aka Simmondsia Chinensis (Jojoba) Seed Oil (CAS# 61789-91-1], and squalane (CAS# 111-01-3). It is important to understand that natural oils can slowly penetrate over time into damage or cracked nails and do not easily penetrate healthy, intact nail plates. Studies performed by the author using Photo-induced force microscopy, which combines atomic force microscopy (AFM) and infrared spectroscopy to acquire both topography and IR spectral information with ~5 nm spatial resolution confirmed the eventual penetration of some oils which daily application over a one-month period [14]. Micro-FTIR has also been widely used in the analysis of cosmetic ingredients, especially in studying their absorption and penetration in formulations. The technique utilizes infrared light to identify chemical bonds in materials by measuring the absorption and transmission at various wavelengths, making it a powerful tool for assessing the composition, purity, and structural changes in cosmetics.

In the context of cosmetic ingredients, micro-FTIR is particularly useful for verifying the identity of raw materials and detecting impurities or adulterants in formulations. For example, it can analyze emollients, surfactants, preservatives, and pigments, helping to ensure that these ingredients meet quality and regulatory standards. Additionally, FTIR plays a role in evaluating the stability of cosmetic products over time, ensuring that they maintain their efficacy and safety during storage and use [15].

Lotions, creams, and oils formulated for nail care are designed to deliver moisture and improve the flexibility/appearance of the nail plate and surrounding skin. Choosing the right formulation—whether it is a light lotion for daily use or a rich cream or oil for intensive treatment—depends on the condition of the nails and the specific needs of the user. For instance, those with brittle nails will require different treatment than those with thin, flexible nails.

Nail Strengthener

Nail strengtheners are formulated to improve the durability of the nail plate. These products address brittle, weak, peeling or overly flexible nail plates. They are available in various forms, including base coats, serums, and creams, and contain a combination of film-forming agents, plasticizers, proteins, and conditioners previously described in this chapter. In addition to these, crosslinking agents reinforce the existing keratin structure by forming covalent bonds between keratin chains to result in a more rigid and durable nail plate. These should not be used on brittle nails, since crosslinking will exacerbate these issues. Cross-linkers should be targeted for weak, thin nail plates. The two most widely used ingredients are methylene glycol aka formaldehyde (CAS# 50-00-0). Methylene glycol is a liquid that slowly releases formaldehyde (FA). Since formaldehyde is a dry anhydrous gas and cannot be directly incorporated into cosmetic formulations, substances that release low ppm concentrations of FA over time are utilized [16]. The released FA leads to the creation of crosslinks in the keratin structure, significantly increasing nail rigidity and surface hardness. However, overuse leads to over crosslinking, resulting in the nail plate becoming overly brittle. Due to this and the potential for allergic reactions, its use has become less common.

Dimethyl Urea aka DMU (CAS# 96-31-1) is a safer alternative to formaldehyde. It forms crosslinks between keratin molecules, improving nail strength without the potential for irritation that formaldehyde can cause. Due to its size and bulkier shape, DMU (MW 88) is much less likely to penetrate deep into the nail plate to overcross-link and cause excessive brittleness. DMU does not react as quickly when compared to formaldehyde (MW 30), but this makes it less likely to cause brittleness, and it is not considered to be an allergen, which makes it preferable for nail plate hardening [17].

Conclusion

The science and technology of nail products have advanced significantly, offering a wide range of useful formulations and unique applications. From traditional nail polishes to UV-cured gels, acrylic systems, and specialized treatments, the nail care industry continues to evolve, driven by innovations in polymer chemistry, pigment technology, and a deeper understanding of nail anatomy and biology.

The importance of proper formulation cannot be overstated, as it directly impacts not only the aesthetic qualities of the product but also its safety for nails and skin. Cosmetic scientists must balance many factors, including color, shelf-life, viscosity, curing mechanisms, adhesion, and wear resistance, while also considering potential allergenic properties and environmental impacts. The future of nail care products lies in continued research and innovation, with a focus on enhancing nail health alongside appearance. As our understanding of the natural nail and polymer chemistry deepens, we can expect to see even more advanced formulations that offer improved durability, easier application, and better compatibility with natural nails.

References

1. Harleen, A., & Antonella, T. (2017). Safety and efficacy of nail polish products. *Cosmetics, 4*(3), 24. https://doi.org/10.3390/cosmetics4030024
2. Schoon, D., & Baran, R. (2019). Cosmetics: The care and adornment of the nail. In R. Baran (Ed.), *Disease of the nail and their management* (5th ed., p. 646). Wiley-Blackwell.
3. Draelos, Z. Ed. Cosmeceuticals, 2009, 2, Saunders Elsevier, pg. 44
4. Mambro, V., Azzolini, A, et al, Comparison of antioxidant activities of tocopherols alone and in pharmaceutical formulations, International Journal of Pharmaceuticals, 2003, 262(1–2) pg. 93–99, DOI: https://doi.org/10.1016/S0378-5173(03)00333-8.
5. Linus Pauling Institute, Micronutrient Information Center, Oregon State University. https://lpi.oregonstate.edu/mic/vitamins/vitamin-E
6. Charolotte, C., Bordat, C., et al. (2021). Comparison of α-tocopherol, α-tocopherol acetate, and α-Tocopheryl polyethylene glycol succinate 1000 absorption by Caco-2 TC7 intestinal cells. *Nutrients, 13*(1), 129. https://doi.org/10.3390/nu13010129
7. Lin, L. L., Nufer, K. L., Tomihara, S., & Prow, T. W. (2015). Non-invasive nanoparticle imaging technologies for cosmetic and skin care products. *Cosmetics, 2*(3), 196–210. https://doi.org/10.3390/cosmetics2030196
8. ANSI/IESNA RP-27.3-96, Recommended Practices for Photobiological Safety for Lamps-Risk Group Classification & Labeling.
9. Dowdy, J. C., & Sayre, R. M. (2013). Photobiological safety evaluation of UV nail lamps. *Photochemistry and Photobiology, 89*, 961–967. https://doi.org/10.1111/php.12075
10. UV nail lamp use is not associated with skin cancer risk, 2012, SH News 1, 1818. https://doi.org/10.1007/s40014-012-1818-0.
11. de Groot, A., Rustemeyer, T., 2-hydroxyethyl methacrylate (HEMA): A clinical review of contact allergy and allergic contact dermatitis. Part 2. Cross- and co-sensitization, other skin reactions to HEMA, position of HEMA among (meth)acrylates, sensitivity as screening agent, presence of HEMA in commercial products and practical information on patch test procedures, Contact Dermatitis, 2024; 90(1), pg. 1–16, DOI: https://doi.org/10.1111/cod.14430.
12. Schoon, D. The latest facts explaining the importance of using the correct nail lamp to ensure proper curing of UV nail products. https://schoonscientific.com/resource_type/technical-articles/
13. ASTM D522-93a. (2008). Standard test methods for mandrel bend test of attached organic coatings. *ASTM International, 2008*. https://doi.org/10.1520/D0522-93AR08
14. Molecular Vista. https://molecularvista.com/applications/nanoscale-chemical-imaging-soft-biological-tissues/
15. Hicks, A, Role of imaging techniques in analyzing cosmetic formulations: From microscopy to spectroscopy. Research & Reviews: Journal of Pharmaceutical Analysis, 2024, 13:2, pg. 5, DOI: https://doi.org/10.4172/2320-0812.13.2.005.
16. Formaldehyde, International Agency for Research on Cancer (IARC) Monographs on the Evaluation of Carcinogenic Risks to Humans, 2006, Vol. 88, 39–325.
17. Nail Strengthening Compositions and a Method for Strengthening Nails, US Patent # 5785959.

The Science of Beauty in Personal Care Products

Paolo Giacomoni

Abstract By Science of Beauty, we generally understand the scientific activities aimed at learning about the physiology and the biochemistry of skin. This chapter will discuss healthy skin and the factors of skin damage and of skin aging, describing topical treatments aimed at maintaining skin homeostasis and at promoting repair. This chapter will not deal with invasive treatments, reconstructive surgery, and laser peeling.

Since aging is the accumulation of damage, the factors of skin aging that have been identified as of today will be listed and their mode of action will be summarized in the description of the microinflammatory theory of skin aging.

Topical treatments able either to prevent damage or to remove damage after environmental aggressions will be presented.

Healthy epidermis will be described and some of the factors necessary for its normal functioning will be quoted.

A second goal of the Science of Beauty is to prepare products able to transiently conceal surface imperfections such as dark spots, fine lines and wrinkles, and sagging skin. Ways to prolong the duration of the cosmetic effect (the so-called Cinderella effect) will be described.

Keywords Skin aging · Skin homeostasis · Antioxidants · Energetic balance · Repair · Surface modification

P. Giacomoni (✉)
Insight Analysis Consulting, Brownsboro, AL and L-Raphael Beauty Institutes, Geneva, CH, Switzerland

K. Alexander, P. Romanowski (eds.), *Cosmetic Science*, AAPS Introductions in the Pharmaceutical Sciences 15, https://doi.org/10.1007/978-3-032-03596-7_7

Introduction

Brands in the skin care industry have generated mottos to be recognized in every market. Among the most successful ones I remember: *Every woman can be beautiful*; *Because I am worth it* and *Great skin can be created.*

Since the canons of beauty change with culture and ethnicity and since I am not an expert in the psychology of self-reward, this chapter will deal with healthy skin and with the factors of skin damage and of skin aging. This chapter will describe topical treatments aimed at maintaining skin homeostasis, at efficiently protecting the skin and promoting its repair, as well as at transiently concealing surface imperfections such as dark spots, fine lines and wrinkles, and sagging skin. Ways to prolong the duration of the cosmetic effect will also be described.

Healthy Skin

The individual skin variability is so large, that it would be a formidable task to try and propose reasonably narrow ranges of biomedical parameters so that, when one skin's measurements fall within those ranges, one can safely say that the skin is "healthy." A skin is often defined "healthy" when in the absence of signs of disease or discomfort: healthy skin does not present with erythema, discoloration, itch, bacterial or fungal infection, acanthosis, hyperkeratosis, parakeratosis, dryness, acne, rosacea, atopic dermatitis, and so forth.

Healthy skin is resilient, elastic, moist, lubricated, smooth, radiant, glowing, and luminous. The loss of one or more of these characteristics, per se, does not make the skin unhealthy; it makes it just appear to be in bad shape.

Avoiding the skin to lose its good shape and restoring the good shape of skin in bad shape is one of the goals of the so-called science of beauty.

Damaging Healthy Skin

Professor Albert Kligman famously said: "The skin has memory." And indeed, an aggression on the skin, suffered without immediate consequences at a given moment in life, will reveal its effects years later. The most remarkable example of this is the onset of age spots, which appear on anatomical zones that were exposed to solar radiation 20 or 30 years earlier. The same is true for all the phenomena that appear because of a slow, chronic aggressions such as the ones suffered by exposure to ultraviolet radiation, gravitational forces, wounds, air pollution, infections, traumas, cold, anoxia, psychological stress, cigarette smoke, hormonal imbalance, protein glycation, ethanol consumption, sleep deprivation and even excessive physical exercise.

Aging is not a disease, albeit there are aging-associated diseases. Aging, and in particular skin aging, is a process that leads to the loss of "good shape." If we understand the mechanisms of aging, we might be able to delay the onset of aging-associated diseases and of aging-associated skin bad shape.

Aging has been defined as the accumulation of damage [1–4] and all the skin-damaging factors evoked above are indeed factors of skin aging. They share the capability to trigger an inflammatory response and the analysis of their mode of action has allowed me and my coworker Patrizia D'Alessio to propose the microinflammatory model of skin aging [3]. The microinflammatory model was later generalized into the microinflammatory hypothesis of skin aging, thanks to the work performed with my collaborator Glen Rein [5–8].

Skin Aging

Aggressions due to the lifestyle or to the environment are capable of accelerating skin aging. While working with the European Network for the Biology of Aging I pointed out in 1996 that several factors of skin aging share as a common feature, the capability to trigger an inflammatory response by inducing the synthesis and the mobilization of intercellular adhesion molecule-1 (ICAM-1) in the endothelium [3]. The first factors of aging to be identified were ultraviolet radiation, tractions, wounds, infections, trauma, anoxia, cigarette smoke, and specific hormonal imbalances. Protein glycation, stretching, electromagnetic fields, psychological stressors, and neuropeptides were later recognized to be able to induce ICAM-1 synthesis and mobilization and [5].

These phenomena trigger a cycle of self-maintained inflammatory responses, which comprises the induction of mobilization and neosynthesis of ICAM-1 and the onset of oxidative steps, and are described in [3, 5–7]. Reactive Oxygen Species released during this process are displayed in Table 1.

The microinflammatory hypothesis of skin aging has been described [7] and is summarized below. One of the innovative aspects of the microinflammatory hypothesis is that skin aging can be accelerated also by factors that do not damage skin cells, that is to say that pro-inflammatory aggressions can exert their action also in ways that were previously neglected. The aging of skin results from three oxidative steps that are the consequence of the synthesis and the mobilization of intercellular

Table 1 Reactive oxygen species released during the microinflammatory process

ROS	Biochemical/biophysical origin
Hydrogen peroxide	Rolling over of immune cells: first oxidative step
Singlet oxygen	Released by myeloperoxidase: second oxidative step
Hydrogen peroxide	Oxidative burst: third oxidative step

adhesion molecule 1 (ICAM-1) within the endothelium of cutaneous vessels. Agents able to provoke this synthesis and mobilization contribute to skin aging [3].

The first oxidative step takes place when vascular cell-adhesion molecule-1 (VCAM-1) activates endothelial NADPH oxidase, an enzyme committed to the production of reactive oxygen species (ROS). This activity is required for VCAM-1-dependent lymphocyte migration [9]. Circulating inflammatory cells bind ICAM-1 and roll over and release hydrogen peroxide [10], endothelial cells lose intercellular contact, round up, and allow monocytes and macrophages to cross the vascular wall.

The second oxidative step occurs in the extracellular matrix of the dermis, when inflammatory cells can either follow chemotactical signals to reach damaged somatic cells or agents of infection, or just exert their lytic functions randomly. In all cases, reactive oxygen species are released together with specific proteases. In particular, singlet Oxygen is released by resident myeloperoxidase.

The third oxidative step takes place when the immune cells release H_2O_2 in the presence of cells to destroy, engulf, and remove (such as damaged somatic cells, foreign bacteria, or molds), or even when, as in the case of anoxia, there is no clearly defined target and smooth muscle cells surrounding the veins are destroyed. In this situation, nearby resident cells can be directly affected by this oxidative burst. In these serendipitously damaged cells, the arachidonic acid cascade is triggered, and the cells release prostaglandins and leukotrienes which will be sensed by resident mast cells. Mast cells will release histamine and TNF-α, and they will induce the synthesis and the mobilization of ICAM-1, thus perpetuating the inflammatory process.

Skin aging is the result of the accumulation of oxidative and proteolytic damage experienced over time by the extracellular matrix of the dermis, together with the remodeling of fibers in a disorganized mode. The microinflammatory hypothesis emphasizes the aging of the cutaneous connective tissue and of the extracellular matrix.

Macroscopic consequences of this hypothesis have been experimentally verified. For instance, the fact that post-UV-repair and wound-healing share extracellular matrix (ECM) remodeling as a common feature offers an understanding of why blood vessels are deeper down in aged skin than in young skin. The sagging of the dermis is the consequence of a modified ECM, too, and because of the chronic action of gravitational forces, it provokes an overall increase of the cutaneous surface of the skin, particularly of the face. Nerves and muscles pull the skin, and wrinkles form where the skin is attached to muscles. Wrinkles have a neuromuscular cause as it can be inferred by their absence in hemiplegics, or when the individual is under general anesthesia or when is injected with botox. The increase of skin surface area takes into account the fact that, notwithstanding a nearly constant rate of turnover of the keratinocytes through the life span, the thickness of the epidermis is diminished with aging. This is more the consequence of the stretching of the skin than of a modification of the turnover rate of the keratinocytes. Indeed, the turnover of the keratinocytes does not change with age, and this is confirmed by the fact that the thickness of the *stratum corneum* is, in fact, constant with age [11].

Factors of Aging That Do Not Damage Cells

Some of the factors of skin aging damage cells, and we will discuss them in detail below. What about other factors of aging that are not directly damaging agents?

Traction and gravitational forces provoke the activation of phospholipase A2, an enzyme involved in the arachidonic acid cascade and therefore in the onset of inflammation.

Anoxia induces ICAM-1 synthesis and the diapedesis (crossing of the endothelial wall) of macrophages. When in the dermis, macrophages digest the ECM around veins or other blood vessels and attack smooth muscle cells so that the vessels collapse and turn into a "spider vein," and the veins collapse and turn varicose.

Glucose binds to proteins in a nonenzymatic glycation process, and glycated proteins are inducers of ICAM-1 synthesis.

Electromagnetic fields associated with computers provoke the release of histamine, IL-1 and IL-6, and these cytokines trigger the inflammatory response.

Neuropeptides regulate the expression of cell-adhesion molecules on both leukocytes and endothelial cells in a coordinated effort to control neurogenic inflammation.

Similar effects are achieved by sleep deprivation, exposure to cold, consumption of ethanol, neuro-mediators and excessive physical exercise [5–8].

In the frame of the microinflammatory hypothesis of skin aging, one can easily understand the relevance of using antioxidants in the attempt of reducing the rate of accumulation of damage and therefore of aging. When avoiding the damaging factors is impossible, one has to design effective anti-aging strategies in addition to using appropriate antioxidants. Different factors of aging call for different cares. For instance, one could fight the negative effects of anoxia by performing appropriate physical exercises; the consequences of the unwanted release of neuropeptides could be counteracted by careful stress management; age accelerating adventitious infections and traumas should be taken care by prompt medical treatments; the pro-inflammatory effects of sleep deprivation can be modulated by accurate management of one's leisure and protein glycation can be avoided by nutritional strategies, whereas the effects of smoking tobacco can only be treated cosmetically and the only possible way to avoid its effects is to quit smoking. These six strategies, together with the protection against cell-damaging factors, constitute the seven strategies (aka the Seven Pillars of Healthy Aging) proposed by the L-Raphael Institutes and are summarized in Table 2.

Table 2 Different factors of aging call for different cares

Anoxia	Physical exercise
Infections, traumas	Medical care
Protein glycation	Nutritional strategies
Neuropeptides	Stress management
Sleep deprivation, cold exposure	Leisure programs
Tobacco smoking, alcohol consumption	Aesthetic treatments (or quitting)
UV radiation	Protection, repair, Age management

Cell-Damaging Factors

When a cell is damaged, the arachidonic acid cascade is triggered, and prostaglandins and leukotrienes are released. Mast cells sense these signaling molecules and in turn release histamine and TNF-α. Vasodilation follows, ICAM-1 is synthesized and mobilized, and immune cells are recruited in the dermis. When in the dermis, immune cells release ROS and lytic enzymes that damage the extracellular matrix (ECM). We are going to describe in detail some of the most relevant cell-damaging factors, analyze their mode of action, and suggest interventions to limit their damaging potential.

Ultraviolet Radiation

Ultraviolet radiation (UVR) is known to provoke erythema, pigmentation, mutations, immune depression, skin aging, and cancer. Its mechanisms of action have been studied over the years. When UVR is absorbed by DNA, it induces the formation of DNA damage in the form, among others, of cyclobutane pyrimidine dimers (CPD).

The DNA repair enzymatic panoply of the marsupial *Monodelphis domestica* contains photolyase, an enzyme able to bind to CPDs and to absorb blue light. Photolyase can use the energy of absorbed blue light to undo the dimers and restore the DNA in its pristine state. Ronald Ley was able to show that no erythema and no CPD are found in the skin of the marsupial when it is exposed to blue light after exposure to UVR [12]. Ley's experiment did show that CPDs trigger the reactions leading to skin redness (the so-called solar erythema).

In mammalian skin, there is no photolyase. Upon formation of a CPD, the cellular enzymatic machinery starts the process of removing the damage (DNA repair). During this process, the DNA is nicked on the right and left side of the damage and free DNA ends are generated. Free DNA ends activate poly ADP-ribose polymerase (PARP), an enzyme that splits nicotinamide adenine dinucleotide (NAD) to nicotinamide and ADP-ribose and then polymerizes ADP-ribose in a poly ADP-ribose chain (a poly-anion) bound to the histones. Because of electrostatic repulsion and competition for binding histones, (DNA itself is a poly-anion) the damage is made accessible to DNA repair enzymes.

This process facilitates repairing DNA at the cost of NAD [13]. NAD is a key electron transporter in glycolysis. Without NAD, energy production is impaired, cells undergo blebbing and necrosis, DNA repair might not be completely achieved, contact hypersensitivity and delayed type hypersensitivity are suppressed, and an inflammatory reaction is set on. Administering NAD in liposomes post-UVR remarkably reduces the intensity of the consequent erythema [13]. NAD carries two negative charges and cannot penetrate cells unless it is formulated in liposomes, but its precursor, vitamin B3, does easily cross the stratum corneum and the topical application of vitamin B3 restores energy production. It also restores the immune response in humans (when topically applied at 0.2%–5% after UVR), and boosts DNA repair (in skin explants exposed to UVR and dipped in 50 μM nicotinamide-containing solution) [14–16].

Nucleic acids (DNA and RNA) are not the only molecules in the skin that can absorb UVR. Proteins absorb UVR. The molar extinction coefficient of amino acids in the UVB is about 100 times smaller than the extinction coefficient of nucleotides, but cells contain 100 times more proteins than DNA, and the extracellular matrix (ECM) contains lots of proteins, so proteins are a relevant player in the absorption of UVR. Cells contain lipids, too, but lipids do not absorb UVR.

In addition to proteins, the skin contains molecules called photosensitizers that, upon absorption of a UV photon, can transfer an electron to molecular oxygen, thus forming the superoxide ion O_2*-, or transfer the absorbed energy to molecular oxygen, thus generating the highly reactive singlet oxygen $^1O_2$. Photosensitizers can be endogenous, like some porphyrins, tryptophan and tyrosine, or exogenous, like benaxoprofen, furocoumarins and tetracycline. Tyrosine and tryptophan exposed to UV generate hydrogen peroxide [17] that, in the presence of transition metals, is converted in the hydroxyl radical. A few sun-filters, like ZnO or TiO_2 and terephtalylidenedicamphor sulphonic acid (Mexoryl SX) do generate singlet Oxygen when they absorb UV photons.

Oxidative Stress by UVR

Direct absorption of UVB by proteins does not induce stereochemical modifications. Protein modifications as well as lipid modifications are mainly the consequences of photo-sensitized oxidation [18].

Reactive Oxygen Species relevant to skin damage are displayed in Table 3.

In cells, the vast majority of lipids are phospholipids, confined to cell membranes. The phospholipids are so oriented that the charged phosphate heads are in contact with water molecules in the cytoplasm and in the external environment, while the lipid tails are associated in hydrophobic bonds at the interior of the cell wall. The access of foreign molecules to the cytoplasm is regulated either by

Table 3 Reactive oxygen species relevant to skin damage

ROS	Biochemical/biophysical origin
Hydrogen peroxide	Photosensitization of amino acids (Tyr, Trp) oxidative burst of inflammatory cells
Hydroxyl radical	Fenton Reaction
Superoxide	Charge transfer to molecular oxygen upon absorption of a UV photon by a photosensitizer
Singlet Oxygen	Energy transfer to O_2 upon absorption of a UV photon by a photosensitizer or by specific commercial sun-filters. Oxidative burst of inflammatory cells

appropriate “pumps” or by the complex mechanisms of endocytosis or pinocytosis. Not many molecules can cross the membrane freely: one of them is the protonated superoxide radical HO_2* that does not carry a charge and can create havoc inside a cell, penetrating down to the nucleus.

When ROS are within a cell membrane, the lipid bilayer is their preferred target. Upon peroxidation, phospholipids bind molecular Oxygen, acquire a defined electric polarity, and find a potential minimum by moving toward water, thus disrupting the hydrophobic structure and disorganizing the cellular membrane.

A most remarkable phenomenon that can be observed in cultured cells exposed to UV is the onset of zeiosis, also called blebbing [19]. Blebbing consists of the appearance of bubbles on the surface of the cells, hence the name zeiosis (in Greek: boiling) reminiscent of the bubbles generated in boiling water. It is a cytoskeleton-linked phenomenon [20] that has been shown to occur in human epidermis in vivo [21]. Histologically, zeiosis gives rise to the phenomenon of spongiosis, the appearance of isolated granules in the interstitial space that is generated after the rounding up of the cells as a consequence of UV stress [21].

Blebbing occurs both in culture and in vivo. It is therefore reasonable to use it as a meaningful end point to determine the efficacy of stressing or of anti-stress treatments in experiments with cultured cells. Another consequence of UV exposure in cultured cells is the fragmentation of chromatin, a specific indicator of apoptosis [22]. These two phenomena allow one to quantify overall cellular damage by counting cells that have undergone blebbing, and to quantify apoptosis by counting cells with fragmented chromatin.

UVR and ROS: Why Antioxidants?

In addition to the three oxidative steps in the microinflammatory reaction to UVR, ROS can be generated by the photosensitizer-induced charge transfer that produces the superoxide anion. In this process, many steps are necessary to generate one lipid peroxide as follows. O_2^{*-} can reduce Iron or Copper. Via the Fenton reaction, reduced transition metals split H_2O_2 into OH- and OH*. The hydroxyl free radical can strip a hydrogen atom from a lipid molecule to generate water and a lipid free radical

$$\mathrm{L-H+OH^{*} >>> L^{*} + H_2O}$$

In the presence of molecular Oxygen, a Lipid peroxide radical is formed.

$$\mathrm{L^{*} + O_2 >>> L-O-O^{*}}$$

that can strip hydrogen atoms from nearby lipids thus propagating the peroxidative cascade.

Another reactive oxygen species generated via photosensitization is singlet Oxygen because of energy transfer from an excited photosensitizer to molecular

Oxygen. Singlet Oxygen, too, induces Peroxy Radicals. Preferred targets of singlet Oxygen are the Carbon atoms linked by a double bond in alkenes so that, by binding $^1O_2$, an unsaturated lipid molecule turns into a lipid peroxide in one single step

$$L-H+{}^1O_2 >>> L-O-O-H$$

In the presence of reduced transition metals, the lipid peroxide will produce a lipid oxide and the Hydroxyl anion while oxidizing the transition metal. The lipid oxide.

can strip a Hydrogen atom from a nearby lipid, generating a lipid free radical.

$$L-O^* + L-H >>> L-O-H+L^*$$

In the presence of molecular Oxygen, the lipid free radical will generate a lipid peroxide L-O-O* able to strip a hydrogen atom from a lipid and trigger the peroxidative cascade.

Alternatively, L-O* can react with molecular Oxygen to give O-L-O-O*, a molecule able to strip an hydrogen atom from a nearby lipid molecule, resulting in one lipid peroxide and one lipid radical O-L-O-O-H + L* and the lipid radical L* can in turn react with molecular Oxygen to generate L-O-O* that can strip a hydrogen atom from nearby lipids (see for instance [35]).

Lipid peroxides can be scavenged by α-tocopherol, acetyl zingerone or bakuchiol. $^1O_2$ can also be scavenged by carotenoids or tryptophan. Its production can be hindered by appropriate molecules, such as Micah and acetyl zingerone.

Another ROS, peroxynitrite, is produced by the reaction between nitric oxide (NO) and superoxide radicals (O_2^{*-}) that is effective in inducing lipid peroxidation, damaging DNA and oxidizing sulfhydryl groups in proteins. As such, peroxynitrites could be damaging to the cytoskeleton, one of the indirect targets of the nefarious action of UVR on cells [20, 22]. Often neglected by the skin care industry, nitric oxide is a vaso-dilating agent generated by inducible and constitutive nitric oxide synthase, two enzymes often involved in unwanted phenomena such as cutaneous itch. Since peroxy-nitrite formation is dependent on the presence of the superoxide anion, we will keep in mind that the scavenging of peroxynitrite will be a windfall profit when using scavengers of superoxide. Experiments were performed to learn how to counteract the nefarious action of UV-induced ROS.

UVR and α-Tocopherol—Cultured Cells

Cultured normal human keratinocytes were exposed to either 1200 J/m2 UVB or 200,000 J/m2 UVA (with a filter system that reduced to practically nil the contamination by UVC). In the presence of α-tocopherol or of its solvent, non-irradiated cells did not differ from the untreated control.

24 hours after UVA irradiation, 75% of the cells were damaged (blebbing) and 60% of the cells were apoptotic (chromatin fragmentation).

24 hours after UVB irradiation, 75% of the cells were damaged and 95% were apoptotic. The addition of α-tocopherol either before or before-and-after UVA irradiation totally abolishes cell blebbing but only partially reduces apoptosis, if at all. The addition of α-tocopherol either before or before-and-after UVB irradiation reduces cell blebbing by about 60% and apoptosis by about 30% [23].

Similar results are observed when α-tocopherol is added only after the irradiation, irrespective of the time of addition, up to at least 7 h after irradiation [24]. These results seem to indicate that the kinetics of formation of the oxidative damage leading to morphological modifications and cell death is relatively slow.

UVR and α-Tocopherol—Clinical Study

The epidermis of 12 human volunteers was either untreated or treated with a topical formulation containing α-tocopherol acetate in nano-capsules and exposed to two individual MED from a solar simulator. Biopsies were taken before and 2, 4 and 7 h after exposure to UV. The analysis of biopsies allowed the morphological changes induced by UV to be detected. Modifications of intercellular interactions and changes in nuclear morphology were apparent 2 and 4 h after irradiation. By 7 h after exposure, the epithelium seemed to have partially recovered its structure and morphology. Treating the epidermis before the exposure to UV with α-tocopherol acetate in nano-capsules abolished the early morphological modifications induced by UVR [21].

UV and Singlet Oxygen

Perhaps the most reactive of all ROS, singlet Oxygen plays a major role in the generation of UV-induced damage and in accelerating the rate of skin aging.

One of the ultrastructural signs of photoaging is the disorganization of the elastic fibers [6]. This is thought to be the consequence of proteolysis in the extracellular matrix followed by a slow remodeling due to the very long time of half renewal of collagen 1. Experiments performed with UVA-irradiated human fibroblasts as well as with UVA-irradiated human epidermis show that UVA induces the expression of collagenase, a protease responsible for the digestion of dermal elastic fibers [24]. Similar results were obtained when cultured cells were exposed to singlet oxygen [25]. It is therefore reasonable to surmise that singlet oxygen is the mediator between UVA irradiation and the digestion of elastic fibers.

It is therefore of interest to ascertain whether singlet oxygen scavengers such as β-carotene do indeed abrogate the effects of singlet oxygen. Experiments were performed with cultured human HaCat keratinocytes pre-treated with 1.5 μM β-carotene and exposed to UVA. In irradiated cells, expression profiles indicate that the presence of β-carotene inhibits the degradation of the extracellular matrix induced by UVA [26].

In addition to inducing interstitial collagenases, matrix metallo-proteinases and other proteases, singlet oxygen plays a role in the UV-induced oxidative damage on lipids and proteins [27] and can therefore provoke membrane damage and play a role in the onset of necrosis. Xanthine, a singlet oxygen scavenger, is effective in avoiding the formation of sunburn cells in mice skin explants when applied at concentrations between 1 mM and 10 mM during irradiation with 600–1200 J/m2 UVB [28].

Other singlet oxygen scavengers or quenchers are worthwhile being quoted. Bakuchiol has been tested for 12 weeks on 50 healthy volunteers aged 40–55, as an anti-photoaging agent when formulated at 0.5% in topical creams. It appeared that Bakuchiol significantly decreases the surface of the area covered with wrinkles, and that it reduces hyperpigmentation. The experiment was performed against 0.5% retinol, and no difference in the efficacy of the two actives was observed [29].

Another quencher of singlet oxygen is acetyl zingerone. It does

(i) Attenuate intracellular ROS in cultured keratinocytes under UVA radiation [30].
(ii) Inhibit the post-UV formation of pyrimidine dimers in melanocytes (a process that continues taking place after the end of UV irradiation), possibly by scavenging peroxynitrite as a key precursor in the process [31].
(iii) Improve the integrity of the ECM by repressing matrix metalloproteases [31].
(iv) Visibly improve the signs of facial photoaging (wrinkle, uneven pigmentation and redness) following daily application of lotion containing 1% AZ over 8 weeks compared with its placebo lotion [32].

An alternative strategy to scavenging singlet Oxygen is to hinder its formation. Porphyrins are ubiquitous in human skin [33] and generate singlet oxygen when exposed to solar radiation. An aromatic cyano-acrylate quenches the triplet state of protoporphyrin IX exposed to visible radiation and hinders the production of singlet oxygen [34]. This molecule, Bis (cyano butylacetate) anthracenediylidene (trade name: Micah), has been tested on female skin explant, and in vivo on the skin of 10 volunteers. Unpublished results indicate that female skin explants treated with Micah produced 30% less free radicals after UVA irradiation than the untreated control. After UVA, the level of 8-OH-dG (8-oxo-Guanine, an oxidative product of DNA, catalyzed by singlet oxygen) increased more than twofold in the non-treated samples, and was equal to the non-irradiated control level in Micah-treated explants.

For the in vivo test, four zones of the skin of ten volunteers were used: untreated control, UV-irradiated control, non-UV-irradiated with Micah, and UV-irradiated with Micah. Immediately and 24 h after the exposure to 20 J/cm^2 UVA, suction blisters were performed. The interstitial fluid and the roof of the blister were analyzed for IL-6, MMP-1 and 8-OH-dG. In Micah-treated zones, the levels of IL-6 and MMP-1 after UVR were remarkably reduced as compared to the controls irradiated in the absence of Micah and the level of 8-OH-dG was zero. (Hallstar

Table 4 Reactive oxygen species and appropriate antidotes

Reactive oxygen species counteracting molecules	
Hydroxyl radical OH^*	Mannitol
	Ergothioneine
	α-tocopherol–Ascorbic acid
Peroxy radical LOO *	α-tocopherol–Ascorbic acid
Superoxide ion O_2^{*-}	α-tocopherol–Ascorbic acid
Peroxynitrite $ONOO^-$	Bakuchiol
	α-tocopherol–Ascorbic acid
Singlet oxygen $^1O_2$	β-Carotene
	Xanthine, Astaxanthin, Lycopene, etc.
	Acetyl Zingerone, Bakuchiol
	Micah

Ascorbic acid reduces oxidized α-tocopherol: when added together in a topical product, it recycles α-tocopherol that is generally present at low concentrations

information via personal communication. I had the possibility of reading and discussing the reports from the laboratory that performed the analyses.) A more detailed description of the technologies against UV-induced oxidative stress can be found in [35].

The reactive oxygen species (ROS) produced by the interaction of UVR with biological matter together with some appropriate antioxidants, are reported in Table 4. Because the superoxide anion and the hydroxyl radical play together to generate Peroxy Radicals, we will consider that vitamin E (α-tocopherol) is an "antidote" against UV-generated superoxide, hydroxyl radical and peroxy radical.

Acne

Acne is characterized by excessive sebum production and by clogged pores where anaerobic bacteria proliferate and provoke inflammatory reactions. The clogging of the pores is likely due to the increased viscosity of sebum lipids that become saturated upon peroxidation by locally generated singlet oxygen.

Bakuchiol has been shown to be an effective scavenger of singlet oxygen generated using hydrogen peroxide and the molybdate anion. As such it can be surmised that it could inhibit the singlet oxygen-induced peroxidation of squalene, thus hindering the clogging of the pores and counteracting the onset of acne. In a pilot study with 54 volunteers in four groups (A: control, B; 2% salicylic acid, C: 1% bakuchiol and D:

2% salicylic acid and 1% bakuchiol) it was shown that Bakuchiol alone clears 42% of the acne in 4 weeks, whereas salicylic acid clears 34% and the combination of the two clears 48% of the acne. In the untreated control group, the clearing of acne was 5% after 4 weeks [36].

Maintaining Skin Homeostasis

Homeostasis is a complex word used to indicate the set of biochemical processes that take place in a normally functioning healthy skin. Homeostasis can be disrupted with undesired consequences, such as damage formation or synthesis and mobilization of ICAM-1. It is therefore necessary to do our best to maintain a smoothly rolling homeostasis. We do not know all the factors contributing to the maintenance of homeostasis, but we know some relevant ones.

The energetic metabolism is of primordial importance in cell and tissue physiology. All the pathways leading to energy production are dependent on the presence of the coenzyme nicotinamide adenosine dinucleotide. This molecule has been recently promoted to the rank of "longevity molecule" because in its absence, life is interrupted. The same happens, of course, with other molecules, like water or Oxygen that were never given that level of glory. Irrespective of marketing claims, energy metabolism is essential to skin homeostasis, and precursors of NAD such as nicotinamide find their place in products for topical applications.

The *stratum corneum* is constituted by corneocytes that are surrounded by a shell called cornified envelope, itself covered with ceramides. The delicate balance of lipids and ceramides is an important characteristic of healthy skin. Vitamin C plays an important role in the maturation of the stratum corneum, insofar as it is essential in maintaining the ceramide ratio [37] and henceforth the esthetic qualities of the surface such as smoothness, softness and maintenance of the moisturization balance.

In the course of a day, an epidermis can experience small traumas, injuries and other aggressions that might upset the homeostasis, for instance by triggering an unwanted, unnecessary inflammatory response. We know that melatonin, the so-called Zeitgeber because it triggers the go-to-sleep signal, is a powerful inhibitor of the transcription of 5-Lipoxygenase, an enzyme committed in the triggering of the production of pro-inflammatory signals such as leukotrienes [38]. Melatonin has therefore the capability of halting the self-maintained microinflammatory cycle that contributes so powerfully to skin aging, and therefore to help maintain a healthy cutaneous homeostasis.

Interventions to maintain skin homeostasis do not need to be performed subsequent to an aggression that impairs it. One can envision preparing the skin for future aggressions such as ultraviolet radiation. It had been shown that exposure to simulated solar UV radiation induced the synthesis and the accumulation of heat shock proteins in the irradiated epidermis, within a few hours after exposure [39, 40] and heat shock proteins exert the function of repairing or removing damaged proteins, thus clearing the cells of obstacles to their proper functioning [41]. A derivative of vitamin D, 7-dehydrocholesterol, has been shown to induce heat shock proteins in cultured human keratinocytes, and to increase the Minimal Erythemal Dose in vivo when applied before exposure to UV radiation [42] thus providing an indication that it is conceivable to prepare the skin to resist environmental aggression.

It is important to remember that one of the most important signs of healthy skin homeostasis is skin hydration and moisturization. In healthy skin, a proteolytic process degrades filaggrin, and the set of free amino acids resulting from the degradation process composes the so-called natural moisturizing factor. As a matter of fact, the composition of amino acids of filaggrin strictly mirrors the composition of free amino acids found in the epidermis, when urocanic acid, citrulline and pyrrolidone carboxylic acid (PCA) are accounted for as derivatives of histidine, arginine, and glutamic acid, respectively. This is why PCA is often incorporated in commercial moisturizers.

Another excellent moisturizer is urea. Unfortunately, when urea is incorporated in formulas, the resulting sera or emulsions are often tacky. This drawback can be overcome by adding arginine to the topical, surfactant-free formula [43]. In the absence of surfactants, arginine will be metabolized to urea in situ by resident arginases, and moisturization will be built and locked in after topical application of the product.

A clinical indicator that skin homeostasis is somehow impaired is the itch that affects the skin of elderly individuals or the skin in cold climates. Inhibitors of NO synthase can be remarkably successful in hindering the itching sensation consequent to environmental conditions or triggered by cosmetic procedures such as hair dying [44].

Another clinical indicator of poor homeostasis is a harsh and scabrous skin surface, characterized by scales and substantially dull. This was perhaps the first skin condition to be recognized by dermatologists as effectively treatable within the limits of the definition of cosmetics, and the science of exfoliation was born. Salicylic acid and many α-hydroxy acids, as well as different activators of proteases such as the resident kallikreins or *stratum corneum* trypsin and chymotrypsin, were successfully introduced to the market to achieve smooth and luminous skin [45]. Interestingly enough, some of these activators work synergistically with salicylic acid, thus decreasing the concentration of the β-hydroxy-acid needed to achieve exfoliation, thus limiting the stinging effect in the eye zone.

As of today, no topical cosmetic intervention has been found to effectively even out skin discolorations such as *lentigo senilis* (age spots) and vitiligo.

Table 5 displays possible actives for topical applications aimed at maintaining skin homeostasis.

Transient Corrections of Surface Imperfections

According to one very conservative definition, cosmetics are products aimed at modifying the appearance of the surface, without intervening at the biochemical level, let alone the clinical one. According to the FDA, if a water-containing product is advertised with the claim that water has any kind of biochemical action, water becomes *ipso facto* a drug and the product "so advertised" must abide by the law governing the preparation and the sale of pharmaceutical drugs. This legislation, by exerting an intimidating action on marketers trying to generate and advertise

Table 5 Possible actives to maintain skin homeostasis

Active	Mode of action
Vitamin B3	Energy metabolism, DNA repair, immune response
Vitamin C	Ceramide balance
Melatonin	Inhibition of transcription of 5-lipoxygenase
Vitamin D derivatives	Induction of heat shock genes
Arginine	In situ production of urea
Inhibitors of NO synthase	Anti-itch
Exfoliators	Turnover of *stratum corneum*

efficacious products, is perhaps the main cause of the proliferation and success of weird advertisements.

Products that modify the surface without biochemical action can be found in the field of the so-called color cosmetics, whose goal is to conceal surface imperfections such as uneven discolorations and local redness. Another category of surface modifiers is constituted by the so-called tensors, large polymers that bind the surface of the skin and undergo conformational changes when the water contained in the topically applied formula evaporates. The conformational change may result in a tightening effect and provoke a transient reduction of the number and of the visibility of the wrinkles. A third category is constituted by large polymers that bind water and have a transient hydration effect.

The polymers in these three categories are subject to degradation by enzymes resident in the *stratum corneum* such as hyaluronidase, glycosylases, amylases, cellulases, and proteases. As a matter of fact, these enzymes hydrolyze those polymers and accelerate the return of the surface to the status that was meant to be transiently modified. An easy and effective method to prolong the duration of these transient effects is to add to the formula an inhibitor of the resident enzymes that are susceptible to hydrolyze the active polymer.

One example of the possibility to increase the duration of the so-called Cinderella effect is given by a cosmetically acceptable composition containing a protein-based film former able to undergo a spontaneous helix-coil transition, added with an inhibitor of proteolytic enzymes. Protease inhibitors can be found in extracts of white birch, grape leaf or betel nut, for instance, or can be chemically pure such as N-α-p-tosyl-L-Lysyl-chloromethyl-chetone, leupeptin, or pepstatin. In the presence of the protease inhibitor, the duration of the tightening effect is dramatically increased when compared to the duration of the tightening effect in the absence of the protease inhibitor [46]. The duration of the effect of color cosmetics and concealers can be prolonged, too, when inhibiting the exfoliation process that detaches the corneocytes on which the cosmetic is "attached."

Another domain in the personal care industry where the modulation of the release of "actives" is of particular interest is the field of fragrances. Top notes tend to be released more quickly from the surface of the skin than the so-called base notes, so that the scent provided to the individual by an applied fragrance changes over time. To avoid this inconvenience, molecules providing top notes can be covalently linked to a polymer by bonds cleavable by *stratum corneum* enzymes (e.g., esterases). This polymer can be added to the fragrance-containing formula. Upon topical application, the enzymatic panoply of the *stratum corneum* will cleave the covalent bonds and will provoke the release of the molecules providing the top notes, thus prolonging the duration of the scent intended to be provided by the original fragrance [47].

Of particular interest, specifically because of the attention brought to consumers by the awareness of the dangers associated to solar exposure, is the increased need of self-tanning products. As of today, the best ingredient to provide self-tan is dihydroxy-acetone (DHA). At pH above 7, DHA binds to the non-protonated amino-groups of free amino acids and forms molecules of different colors, mainly light brown, gold, yellow-orange and dark brown. In some consumers, the high concentrations of DHA that are generally used generate a disagreeable odor. It is therefore necessary to find a way to achieve a tan of the same intensity, using less DHA. This is possible because DHA is a target of enzymes called kinases, enzymes that catalyze the binding of a phosphate group (PO_4^{--}) to a DHA molecule. When bound to a phosphate group, DHA cannot react with the free amino groups of the amino acids in the *stratum corneum* and cannot generate colored products. By adding kinase inhibitors to the formula, or by adding an excess of sugars that are themselves targets for the kinase, and thus allowing more DHA molecules to remain unphosphorylated, a smaller concentration of DHA (50% or less than in the original formula) will be necessary to generate a tan of the same intensity without the associate disagreeable odor [48, 49].

Conclusion

The progress of scientific research, particularly in the field of photobiology and of the interaction of solar radiation with biological matter, has allowed scientists to unravel the mechanisms of damage formation and of repair, thus providing the cosmetic chemists with information about many details that help finding ways to protect the skin against environmental aggressions, to help removing the resulting damage or to maintain a healthy homeostasis.

One of the consequences of this research is that several ingredients have been found that can be efficaciously applied to the skin. Another consequence is that the *stratum corneum* is no longer seen as a dead tissue and that its richness in enzymes has been recognized. Activating or inhibiting these enzymes can help prepare cosmetic products with enhanced efficacy and longer lasting effects.

References

1. Giacomoni, P. U. (1992). Ageing and cellular defence mechanisms. In C. Franceschi, G. Crepaldi, V. J. Cristofalo, & J. Vijg (Eds.), *Ageing and cellular defence mechanisms, Annals of the New York Academy of Sciences* (Vol. 663, pp. 1–3). New York Academy of Sciences.
2. Giacomoni, P. U. (1995). Radiazione ultravioletta, invecchiamento e blebs. *Giorn Geront, 43*, 209–210.
3. Giacomoni, P. U., & D' Alessio P. (1996). Skin ageing: The relevance of antioxidants. In S. I. S. Rattan & O. Toussaint (Eds.), *Molecular gerontology: Research status and strategies* (pp. 177–192). Plenum Press.
4. Novoltsev, V. N., Novoltseva, J., & Yashin, A. (2001). A homeostatic model of oxidative damage explains paradoxes observed in the earlier aging experiments: A fusion and extension of older theories of aging. *Biogerontology, 2*, 127–138.
5. Giacomoni, P. U., & Rein, G. (2001). Factors of skin aging share common mechanisms. *Biogerontology, 2*, 219–229.
6. Giacomoni, P. U., & Rein, G. (2004). A mechanistic model for the aging of human skin. *Micron, 35*, 179–184.
7. Giacomoni, P. U., & Rein, G. (2010). Skin aging: A generalization of the micro-inflammatory hypothesis. In M. A. Farage, K. W. Miller, & H. I. Maibach (Eds.), *Textbook of aging skin* (pp. 789–796). Springer Verlag.
8. Giacomoni, P. U., & Rein, G. (2017). Skin aging: A generalization of the micro-inflammatory hypothesis. In M. A. Farage, K. W. Miller, & H. I. Maibach (Eds.), *Textbook of aging skin* (pp. 1289–1298). Springer Verlag.
9. Tudor, K. S., Hess, K. L., & Cook-Mills, J. M. (2001). Cytokines modulate endothelial cell intracellular signal transduction required for VCAM-1-dependent lymphocyte transendothelial migration. *Cytokine, 15*, 196–211.
10. Goldman, G., Welbourn, R., Klausner, J. M., Kobzik, L., Valeri, C. R., Shepro, D., & Hechtman, H. B. (1991). Intravascular chemoattractants inhibit diapedesis by selective receptor occupancy. *The American Journal of Physiology, 260*, H465–H472.
11. Gilchrest, B. (1993). Aging of skin. In T. B. Fitzpatrick, A. Zur Hausen, K. Wolff, I. M. Freedberg, & K. F. Austen (Eds.), *Dermatology in general medicine* (pp. 150–157). McGraw-Hill.
12. Ley, R. D. (1985). Photoreactivation of UV-induced pyrimidine dimers and erythema in the marsupial Monodelphis Domestica. *Proceedings of the National Academy of Sciences of the United States of America, 82*, 2409–2411.
13. Jacobson, E. L., Giacomoni, P. U., Roberts, M. J., & Wondrak, J. T. (2001). Jacobson MK optimizing the energy status of skin cells during solar radiation. *Journal of Photochemistry and Photobiology B: Biology, 63*, 141–147.
14. Park, J., Halliday, G. M., Surjana, D., & Damian, D. L. (2010). Nicotinamide prevents ultraviolet radiation-induced cellular energy loss. *Photochemistry and Photobiology, 86*, 942–948.
15. Sivapirabu, G., Yiasemides, E., Halliday, G. M., Park, J., & Damian, D. L. (2009). Topical nicotinamide modulates cellular energy metabolisms and provides broad-spectrum protection against ultraviolet radiation-induced immunosuppression in humans. *The British Journal of Dermatology, 161*, 1357–1364.
16. Surjana, D., Halliday, G. M., & Damian, D. L. (2013). Nicotinamide enhances repair of ultraviolet radiation-induced DNA damage in human keratinocytes and in ex-vivo skin. *Carcinogenesis, 34*, 1144–1149.
17. Stoien, J. D., & Wang, R. J. (1974). Effect of near-ultraviolet and visible light on mammalian cells in culture II. Formation of toxic photoproducts in tissue culture medium by black light. *Proceedings of the National Academy of Sciences of the United States of America, 71*, 3961–3965.
18. Davies, M. J., & Truscott, R. J. W. (2001). Photo-oxidation of proteins and its consequences. In P. U. Giacomoni (Ed.), *Sun protection in man* (pp. 251–275). Amsterdam.

19. Malorni, W., Donelli, G., Straface, E., Santini, M. T., Paradisi, S., & Giacomoni, P. U. (1994). Both UVA and UVB induce cytoskeletal dependent surface blebbing in epidermoid cells. *Journal of Photochemistry and Photobiology B: Biology, 26*, 265–270.
20. Malorni, W., Iosi, F., Mirabelli, F., & Bellomo, G. (1991). Cytoskeleton as target in menadione-induced oxidative stress in cultured mammalian cells: Alterations underlying surface bleb formation. *Chemico-Biological Interactions, 80*, 217–236.
21. Giacomoni, P. U., Nadaud, J. F., Straface, E., Donelli, G., Heenen, M., & Malorni, W. (1998). Morphological alterations and cell blebbing in UV-irradiated human epidermis. *Archives of Dermatological Research, 290*, 163–166.
22. Straface, E., Falchi, M., Giacomoni, P. U., & Malorni, W. (2001). Cytotoxicity and morphological endpoints of exposure to UV: Cultured cells as a model system. In P. U. Giacomoni (Ed.), *Sun protection in man* (pp. 321–336). Amsterdam.
23. Malorni, W., Straface, E., Donelli, G., & Giacomoni, P. U. (1996). UV-induced cytoskeletal damage, surface blebbing and apoptosis are hindered by α-tocopherol in cultured human keratinocytes. *European Journal of Dermatology, 6*, 414–420.
24. Scharffetter, K., Wlaschek, M., Hogg, A., Bolsen, K., Schothorst, A., Goertz, G., Krieg, T., & Plewig, G. (1991). UVA irradiation induces collagenases in human dermal fibroblasts *in vitro* and *in vivo*. *Archives of Dermatological Research, 283*, 506–511.
25. Scharffetter-Kochanek, K., Wlaschek, M., Briviba, K., & Sies, H. (1993). Singlet oxygen induces collagenase expression in human skin fibroblasts. *FEBS Letters, 331*, 304–306.
26. Hunziker, B., Seifert, N., Riss, G., Neeb, M., Steiner, G., Hunziker, W., & Goralczyk, R. (2005). β-Carotene interferes with ultraviolet light a-induced gene expression by multiple pathways. *The Journal of Investigative Dermatology, 124*, 428–424.
27. Vile, G. F., & Tyrrell, R. M. (1995). Uva radiation-induced oxidative damage to lipids and proteins in vitro and in human skin fibroblasts is dependent on iron and singlet oxygen. *Free Radical Biology and Medicine, 18*, 721–730.
28. Miyachi, Y., Horio, T., & Imamura, S. (1983). Sunburn cell formation is prevented by scavenging oxygen intermediates. *Clinical and Experimental Dermatology, 8*, 305–310.
29. Dhaliwal, S., Rybak, I., Ellis, S. R., Notay, M., Trivedi, M., Burney, W., Vaughn, A. R., Nguyen, M., Reiter, P., Bosanac, S., Yan, H., Foolad, N., & Sivamari, R. K. (2019). Prospective, randomized, double blind assessment of topical bakuchiol and retinol for facial photoageing. *British Journal of Dermatology, 180*, 289–296.
30. Chaudhuri, R. K., Meyer, T., Premi, S., & Brash, D. (2020). Acetyl zingerone: An efficacious multifunctional ingredient for continued protection against ongoing DNA damage in melanocytes after sun exposure ends. *International Journal of Cosmetic Science, 42*, 36–4546.
31. Swindell, W. R., Bojanowski, K., & Chaudhuri, R. K. (2020). A zingerone analog, acetyl zingerone, bolsters matrisome synthesis, inhibits MMPs and represses IL-17A target gene expression. *The Journal of Investigative Dermatology, 140*, 602–614.
32. Dhaliwal, S., Rybak, I., Pourang, A., Burney, W., Haas, K., Sandhu, S., Crawford, R., & Sivamani, R. (2020 May 5). Randomized double-blind vehicle controlled study of the effects of topical acetyl Zingerone on Photoaging. *Journal of Cosmetic Dermatology, 20*, 166–173. https://doi.org/10.1111/jocd.13464
33. König, K., Meyer, H., Schneckenburger, H., & Rück, A. (1993). The study of endogenous porphyrins in human skin and their potential for photodynamic therapy by laser induced fluorescence spectroscopy. *Lasers in Medical Science, 8*, 127–132.
34. Jockush, S., Bonda, C., & Hu, S. (2014). Photostabilization of endogenous porphyrins: Excited state quenching by fused ring cyanoacrylates. *Photochemical & Photobiological Sciences, 13*, 1180–1184.
35. Giacomoni, P. U. (2020). Appropriate technologies to accompany sunscreens in the Battle against ultraviolet, superoxide, and singlet oxygen. *Antioxidants (Basel), 9*, 1091–1103.
36. Chaudhuri, R. K., & Marchio, F. (2011). Bakuchiol in the management of acne-affected skin. *Cosmetics & Toiletries, 126*, 502–510.

37. Ponec, M., Weerheim, A., Kempenaar, J., Mulder, A., Gooris, G. S., Bouwstra, J., & Mommaas, A. (1997). The formation of competent barrier lipids in reconstructed human epidermis requires the presence of vitamin C. *The Journal of Investigative Dermatology, 109*, 348–355.
38. Carlberg, C. (1995). Wiesenberg I the orphan receptor family RZR/ROR, melatonin and 5-lipoxygenase: An unexpected relationship. *Journal of Pineal Research, 18*, 171–178.
39. Brunet, S., & Giacomoni, P. U. (1989). Heat shock mRNA in mouse epidermis after UV irradiation. *Mutatation Research, 219*, 217–224.
40. Trautinger, F. (2001 October). Heat shock proteins in the photobiology of human skin. *Journal of Photochemistry and Photobiology. B, 63*(1–3), 70–77. https://doi.org/10.1016/s1011-1344(01)00203-2
41. Jonak, C., Klosner, G., & Trautinger, F. (2009 Jan 1). Significance of heat shock proteins in the skin upon UV exposure. *Frontiers in Bioscience-Landmark, 14*(12), 4758–4768. https://doi.org/10.2741/3565
42. Mammone, T., Muizzuddin, N., Goyarts, E., Gan, D., Giacomoni, P. U., Marenus, K., & Maes, D. (2004). Normal human epidermal keratinocytes treated with 7-dehydrocholesterol express increased levels of heat shock proteins. *Journal of Cosmetic Science, 55*, 149–155.
43. Wohlrab, J., Siemes, C., & Marsch, W. C. (2002 Jan-Feb). The influence of L-arginine on the regulation of epidermal arginase. *Skin Pharmacology and Applied Skin Physiology, 15*(1), 44–54. https://doi.org/10.1159/000049388
44. Giacomoni, P. (2006). Nitric oxide synthase inhibitors for topical use. EP 0812184.
45. Giacomoni, P. U., Lentini, P. J., Syed, I., & Sparacio, R. M. Cosmetic composition containing a protease activator. WO/2007/024915.
46. Giacomoni, P. U., Manirazman, A. M., Lentini, P. J., & Gedeon, H. Cosmetic composition containing a protein and an enzyme inhibitor. US 2005/0147572 A1.
47. Giacomoni, P., Gross, R., Hagver, R., & Laczynski, S. Sustained release of active molecules from polymers topically applied to skin or hair. WO 2006/047714 A3.
48. Giacomoni, P. U., & Lentini, P. J. Self-tanning cosmetic composition and methods. US2009/00925666.
49. Giacomoni, P. U., & Manirazman, A. M. Cosmetic composition and methods containing a tanning agent and liposome-encapsulated ursolic acid. US 7,794,694 B2.

Paolo Giacomoni Ph.D. is as a consultant to the skin care industry and serves as the Head of R&D of L-Raphael. He was Chief Scientific Officer of Élan Rose International and served as VP of Skin Care World Wide R&D with Herbalife. As Executive Director for R&D with Estee Lauder, he contributed to the preparation of best-selling products for Clinique and La Mer. He also served as scientific spokesperson for Clinique. In the frame of the European Society for Photo-biology, he was the Editor of two monographs on the effects of solar radiation on human skin. During his tenure at L'Oréal he served as Head of the Department of Biology and then as scientific attaché to the Director of Applied Research. Results of his research were applied in products such as Primordial (Lancôme) and Plénitude (L'Oréal). During his work in Academia, Dr. Giacomoni was Maître de Conférences at the University of Paris and was Visiting Professor at the University of Milano. Dr. Giacomoni is presently Editor-in-Chief of the Journal of Cosmetic Science. He received his Ph.D., in Biochemistry from the UNIVERSITY of PARIS, Paris France; his Masters Degree in Atomic Physics from the UNIVERSITY of MILANO, Milano, Italy and has had Post-Doctoral Training at Deutsches Krebsforschungszentrum at Heidelberg, Germany, the University of Wisconsin, Madison, WI and the University of California, San Diego, CA. Dr. Giacomoni was one of the founders of the European Society for Photobiology as well as of the European Network for the Study of the Biology of Aging. He has authored 100+ publications and patents in Biochemistry, Molecular Biology and Biology of Aging, focusing on the DNA damage and the metabolic impairment induced by UV radiation as well as on the positive effects of antioxidants.

Current Trends and Developments in Men's Personal Care Products

Sanjay Jayaram, Keerthiraj Kalmantidoddi Shivaji, Pooja G. Achar, B. R. Ganavi, Madhuchandra Kenchegowda, Riyaz Ali M. Osmani, and Hosahalli Veerabhadrappa Gangadharappa

Abstract The men's personal care industry is undergoing a transformative shift, driven by evolving societal norms, technological advancements, and increasing awareness of skin health. This chapter explores the biological differences between men's and women's skin, the unique challenges faced by men, and the diverse product categories tailored to meet their needs. It also examines the industry's efforts to overcome gender stereotypes, adopt sustainable practices, and integrate cutting-edge innovations such as smart grooming devices and eco-friendly packaging. By addressing these trends, this chapter highlights the growing importance of inclusivity, personalization, and sustainability in shaping the future of men's personal care.

Keywords Personal care · Skincare products · Haircare products · Shaving products · Cosmetic and cosmeceuticals

S. Jayaram · K. K. Shivaji · M. Kenchegowda · H. V. Gangadharappa (✉)
Department of Pharmaceutics, JSS College of Pharmacy, JSS Academy of Higher Education & Research (JSS AHER), Mysuru, Karnataka, India
e-mail: keerthirajks@jssuni.edu.in; hvgangadharappa@jssuni.edu.in

P. G. Achar · B. R. Ganavi
Department of Pharmacology, JSS College of Pharmacy, JSS Academy of Higher Education & Research (JSS AHER), Mysuru, Karnataka, India

R. A. M. Osmani
Department of Pharmaceutics, JSS College of Pharmacy, JSS Academy of Higher Education & Research (JSS AHER), Mysuru, Karnataka, India

Department of Pharmaceutics, College of Pharmacy, King Khalid University, Al-Fara, Abha, Saudi Arabia

K. Alexander, P. Romanowski (eds.), *Cosmetic Science*, AAPS Introductions in the Pharmaceutical Sciences 15, https://doi.org/10.1007/978-3-032-03596-7_8

Introduction

The skin is the body's largest organ, making up over 10% of its total mass, and the organ that allows the body to have the closest possible relationship with its surroundings. The stratum corneum (nonviable epidermis), the remaining layers of epidermis (viable epidermis), dermis, and subcutaneous tissues make up the skin's four basic layers. Nails, sweat ducts, apocrine glands, and hair follicles are some of the additional appendages that are connected. The majority of the physiological functions of mammals and humans in a generally hostile environment can be categorized as vital to their survival. This includes several functions of the skin. These roles can be broadly categorized as sensing, protecting, or preserving homeostasis [1]. The structure of the skin is shown in Fig. 1.

Because many skincare products are functional, the technology in this field is wide and distinct from many other cosmetic areas. Certain products are mainly intended to provide cosmetic impacts, while others have important pharmacological effects. All products, however, are made to work with and care for the skin, which is the body's largest organ. Understanding the business is necessary in order to comprehend the requirement for the category. With global sales of over $10 billion split between luxury and mainstream distribution, the skincare industry is very big. Prestige is about 40% and mass is about 60%. With 3.5 billion dollars, the United States has the largest market, followed by Western Europe with 3 billion dollars and

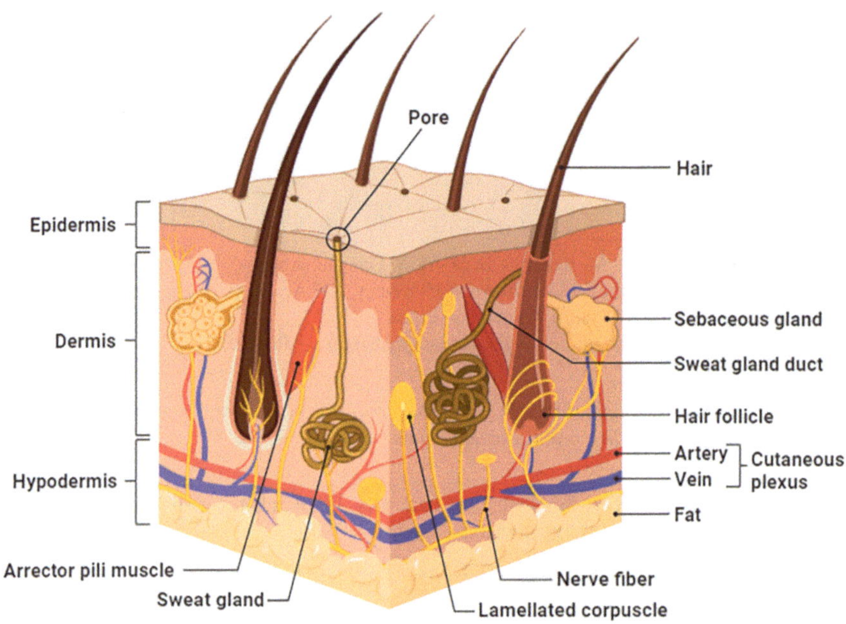

Fig. 1 Structure of the skin

Japan with 2.5 billion dollars. Additional product divisions within the company are based on skin care market subcategories. For instance, 23% of hand and body moisturizers, 34% of facial moisturizers, 13% of suncare products, and 24% of cleansers (bar soaps excluded) are sold in the United States. These ratios fluctuate between mass and prestige, and between markets, contingent upon national distribution patterns, customer consumption patterns, and the quantity of products regularly utilized. Women make up over 80% of the market's usage and likely 90% of its purchases, with the exception of hand lotions, sunscreens, and lip protection products, as well as bar soaps [2].

Although the personal care industry has traditionally catered mostly to women, males have also grown increasingly conscious of their looks in recent years. They take more time to groom themselves, which creates chances in the market for male personal care products. The contemporary man denotes a change in the social construct of what constitutes a man, expanding the parameters of acceptable male conduct and giving more options to men who are brave enough to embrace their 'feminine' sides.

Men are becoming more conscious about their appearance for a variety of reasons, including the decline of traditional masculinity, the greater success of attractive men in the workplace [3], the role of some celebrities as role models for the new man [4], etc. Over the past few years, the male personal care industry has developed steadily, and future prospects appear promising. It is anticipated that the male personal care branch will even surpass the female personal care branch in terms of growth during the next 5 years. Businesses operating in the personal care sector now have a lot of options. Large corporations are currently expanding their men's range and creating new items.

A new market has emerged as men seek to look better and take greater care of their looks. The difficulty facing this new industry is that corporations still do not fully understand men's thinking, needs, or expectations for personal care items. This ignorance results in issues with advertising campaigns and the methods used to approach males. It is crucial that businesses comprehend the needs of male customers and create promotions that are specifically suited to them. The male personal care market can only be taken over in this manner. Therefore, the challenge in the personal care sector is getting to know the male customer. Furthermore, it appears to be a problem because the makers are unable to identify which 'different kind' of males. Additionally, the makers must be able to target diverse categories of males with their marketing initiatives.

Gender-specific cosmetics are used by male consumers to accentuate their own identities [5]. Customers who are driven have confidence in the self-image products. Replicating people's fragmented self-image is one of the reasons male consumers use cosmetic items. Research indicates that males make goods purchases based on ideas and pictures that help them form their identities and self-images [6]. Self-image projective goods help customers in a representative and useful way [7]. The literature also demonstrates how different genders interpret information differently when consuming goods. The usage of these items has recognized the critical role that men play in improving the self-image of male consumers and communicating this through a symbolic connotation [8]. The formation of a man's persona is the driving force behind his use of grooming items. Self-image is a tool to improve one's self-concept and image as well as a stimulus for cosmetic consumption. Consequently, male consumers' attention to purchase cosmetic items is significantly influenced by their self-image.

Male shoppers are motivated to buy cosmetics by health concerns, which also shields them from several health issues [9]. Skincare is the main product category for men's grooming goods. Skincare products are used by 63% of Chinese male consumers. Opinion leadership and early uptake of personal care products are significantly correlated in the cosmetics industry. Research indicates that a large number of consumers use grooming items because they subtly convey a curiosity about cleanliness, well-being, and attractiveness. Male consumers are using more personal care goods as a result of growing health consciousness [10]. Consequently, the usage of grooming products by male consumers is significantly influenced by their health care concerns, which leads to the following deduction.

Physical exquisiteness that is controlled by appearances is what is meant by physical attractiveness [10]. Among its noteworthy characteristics are appearance, body size, and body form. For individuals, beauty is the key to success. Research indicates that women are drawn to attractive guys and want to maintain relationships with them. It is commonly believed that attractive people have better success rates [11]. Additionally, studies show that a person's personality has a significant impact on the brands and goods they choose to buy. Male consumers use grooming goods for a variety of reasons, including self-creation. Research indicates that men who make purchases are drawn to altering their physical appearance in order to create their identities [12]. Several studies demonstrate that societies within the same era place a high value on physical appearance and body composition. Literature has also demonstrated how outlooks, body type, and weight affect wages and job advancement in the work market. Thus, corporations favour those who are physically beautiful. Moreover, those who prioritize self-care are easily accepted by the community. Male consumers are victims of fashion and aesthetics; they use grooming items for their trendy look and body alteration [9].

Understanding Men's Skin and Hair

Biological Differences Between Men's and Women's Skin

The largest and most adaptable organ is the skin. In addition to acting as an exterior barrier, it blocks chemical and microbiological penetration and absorbs UV radiation. Apart from its immune, sensory, and autonomic roles, skin regulates the body's water and electrolyte balance and plays a crucial role in temperature regulation. Understanding the physiological, chemical, and biophysical properties of the skin allows us to create effective skincare plans. However, it is equally important to take into account how genetic and environmental variables affect the skin [13].

The most significant difference between the skin of men and women is due to male sex hormones, androgens. The main sex hormone in this category is testosterone, which is largely secreted by the testes and ovaries. Although there is some variation, adult males produce tenfold the amount of testosterone as females [14].

Researchers evaluated men's and women's skin features on different body parts. The design of studies and the creation of products tailored to an individual's gender for better dermatological or cosmetic treatment may benefit from an understanding of sex-related cutaneous features.

Anatomy, physiology, epidemiology, and illness symptoms all differ depending on gender. In terms of skin ailments, men are more likely to have infectious diseases, while women are more likely to have psychological troubles, pigmentary conditions, hair problems, autoimmune, and allergy concerns. Women are more likely to acquire sex-related dermatoses, and sex differences influence the development and prognosis of certain types of skin cancer.

The factors that govern sex differences in skin disorders are mostly unclear. Such variations may be brought about by sex hormones, behavioural characteristics, ethnicity, or environmental differences. It will be easier to prevent, diagnose, and treat skin illnesses if gender differences in human health and diseases are understood [15].

American author John Grey claims that women are from Venus and men are from Mars, outlining the fundamental differences between the two. While these differences make sense from a social standpoint, it is still unclear whether gender differences also occur in the human skin. Transepidermal water loss (TEWL), stratum corneum hydration (SC hydration), sebum concentration, and pH value are among the physiological characteristics of the largest organ in humans that seem to be gender-neutral. But as puberty sets in and hormone production ramps up, skin differences show. In addition, it is well known that the treatment requirements for male and female skin sometimes differ, as does the susceptibility to various skin conditions, including rosacea, acne, and seborrhoeic eczema, which differ between the sexes [16].

Men's and women's skin differ in a variety of biological aspects due to hormonal, anatomical, and physiological factors. Here is a thorough breakdown.

Hydration

For skin function, including control of inflammation, differentiation, and proliferation of the epidermis, the stratum corneum (SC) must be hydrated. Men and women in the same age group did not differ in their SC hydration levels, according to Rogiers et al. [17]. Man et al. found that SC hydration on the forehead was lower in men and women over the age of 70 compared to younger groups. Both men and women had similar SC hydration levels on their foreheads compared to their forearms [18]. According to a German study by Luebberding et al., young males were more hydrated in SC than women. Throughout their lives, women's SC hydration either remained constant or increased, while men's declined beyond the age of 40 [19].

Transepidermal Water Loss

Transepidermal water loss (TEWL) is used to evaluate the skin's water barrier effectiveness. TEWL is recognized as one of the most essential measures for assessing skin physiology since it indicates how much water penetrates into the protective barrier of skin. Thus, a direct measure of the skin barrier's performance in vivo is the stratum corneum's capacity to decrease evaporative water loss [14]. Men generally showed much lower TEWL than women, irrespective of age. Regardless of gender, water loss grew or stayed constant in the majority of the respondents' lifetimes [13].

Sebum

Larger pores are associated with higher sebum production, according to research by Roh et al. [20]. Before the 2012 study by Bailey et al., it was thought that sebum production was similar in males and females and that men produced more sebum overall, except on the forehead. Luebberding et al., however, demonstrated that while sebum content on the forehead decreased gradually in women as they aged, it did increase somewhat in men [21].

Skin Thickness

Owing to the distribution of terminal hair follicles throughout most of the body the skin of men is thicker than females. This difference is most noticeable on the face, where women hae just fine, colourless vellus hairs, but men have fully formed terminal hairs that are coarse and pigmented, occupying up space beneath the skin. Male facial hair contributes to mature males having a more attractive appearance than mature women. The male beard resists wrinkling because UV light triggers collagenase, which destroys cutaneous collagen [22].

Skin pH

Only men demonstrated a rise in pH value with age,Sex and sex hormones are usually considered to have no influence on the surface of pH [23]. Over the postnatal period, pH decreased rapidly in the first week, followed by a progressive drop over the next 3 weeks. However, other research was unable to show differences in neonatal sex [24].

Microcirculation

A negative, nonlinear relationship between baseline capillary density and homocysteine was found in men by Hornstra et al. The study also found that women with the highest tertile of homocysteine had less capillary density, although there was no discernible correlation between homocysteine and microvascular outcomes [25].

Skin Elasticity

Bailey et al. found that only in the skin of the abdomen did females exhibit more elastic deformation. Additionally, women felt stiffer skin; however, the difference was only noticeable on the abdomen [21].

Skin Friction

Stratum corneum moisture and skin friction coefficient were found to be significantly positively correlated by Zhu et al. on the canthus and dorsal hand skin of women and the forehead and dorsal hand skin of males [26].

Wrinkles

According to Tsukahara et al., women had significantly more wrinkles than males did among those 65–75 years of age [27].

Unique Challenges Faced by Men's Skin and Hair

Three common skin and hair challenges in men are as follows:

- Shaving-related issues
- Acne and breakouts
- Fine lines and breakouts

Shaving-related issues: For many men, shaving is an essential component of their grooming routine. However, it frequently causes its own set of issues, such as razor burns, ingrown hairs, and skin irritation. Shaving can remove the skin's natural oils, leaving it dry and prone to irritation. Preparation is essential for a smooth and irritation-free shave. Regular exfoliation: the reasons for this include our ignorance of the biology of beard hair and the underlying mechanisms of shaving-induced skin irritation, the physiology and structure of male beard hair, and the sensitivity of perifollicular facial skin to mechanical and thermal stimulation [28].

- Beard hair density: High densities of beard hair follicles can contribute to multiple stroke shaving
- Beard hair thickness
- Beard hair stiffness
- Beard hair shape
- Beard hair growth rate
- Low emerging angle
- Heterogeneity of beard hair features [29]

Acne and Breakouts

Acne is not just a concern for youngsters. Breakouts can affect men of all ages and be caused by a variety of factors such as hormonal imbalances, stress, and incorrect skincare habits. Men often have higher testosterone levels, which causes them to have oily skin. Beards and moustaches can trap oil and bacteria; therefore, facial hair is also important. Both of these variables cause clogged pores, increasing the likelihood of breaking out.

Fine Lines and Breakouts

As men and women age, their skin's collagen and elastin fibres gradually break down, resulting in the production of fine lines and deep wrinkles. Men, on average, have 20% thicker skin due to hormonal variances. Thicker skin may cause deeper creases and wrinkles, but less visible fine lines. Inconsistent skincare routines, poor sun protection, and unhealthy habits can all contribute to the formation of deep creases [30].

Impact of Lifestyle and Environmental Factors

Although everyone ages, the most visible sign of ageing is visible on the skin, which is the largest organ in the body and is most noticeable in the areas of the face, neck, hands, arms, and legs. Nonetheless, both short- and long-term health and quality of life (QOL) are significantly impacted by behaviours and habits, whether they are beneficial or negative. Numerous scientific studies and evidence-based recommendations for the prevention and/or treatment of age-related illnesses support this viewpoint. For example, numerous studies have looked into subjects like

(a) Maintaining a healthy diet
(b) Getting enough sleep to manage stress
(c) Engaging in moderate to hard workouts to improve cardiovascular health and weight management

(d) Creating healthy relationships with others to stay connected with like-minded individuals [31]

Common environmental pollutants have their greatest impact on our skin and hair. Chemicals, contaminants, mineral substances, the environment, and sun exposure are all examples of such impacts. Our bodies are damaged over time, resulting in visible dryness of the skin and lines and wrinkles, and also weak and fragile hair.

Effects Due to Pollution

Oxidative stress, another name for the harmful impacts of pollution on humans, is a persistent source of exposure and can exacerbate hair loss. One method that chemicals can harm our hair is through water. Your skin and hair are being harmed by chemicals and minerals in water and airborne particles. If left untreated, this damage will worsen and result in thinning, breaking, and dry hair follicles. Scalps that are itchy, oily, or dry can all play a major role in hair loss.

Humid or Dry Conditions

The humidity and dryness of your surroundings can significantly affect the condition of your skin and hair, depending on where you reside. If you live somewhere with a lot of heat and sunshine, UV rays can harm the proteins in your hair, and sunburns on your scalp can permanently damage your hair follicles, inhibiting future growth. Your hair may become more brittle in colder temperatures because the air is less moist. Hair strand breakage and friction can also result from wearing hats and scarves frequently.

Environmental Toxins

Our bodies absorb the toxins from the environment that we are regularly exposed to. You might be more exposed than others depending on where you work or reside (for example, chemical industries, industrial sites, warehouses, disaster, or contaminated zones). These toxins have the ability to alter hormone levels in our body, starving hair of nutrition and damaging hair follicles. Other common household toxins that may contribute to hair loss are insecticides, formaldehyde, and benzene.

Many of the negative consequences of our environment are hard to avoid on a daily basis. Numerous aspects are inherent in our surroundings as a result of our way of life, occupation, housing arrangement, and personal hygiene practices.

Impact of Lifestyle and Dietary Habits on the Prevalence of Acne Vulgaris

A common dermatological condition that is greatly influenced by lifestyle, environmental, and genetic factors is acne vulgaris. According to lifestyle assessments, people's daily water intake varied widely, with a sizable portion consuming less than 1–2 L. The majority of people slept for 4–5 h. Of the respondents, 24% frequently consumed fast food, whereas 40% exercised occasionally. Remarkably, 72% had a history of acne, which often started in the 16–20 age range. Acne mostly affected the face and was frequently mild to moderate in intensity. Of those who sought therapy, topical methods were the most popular. The causes of acne were not well understood, but there was a broad consensus about the roles that stress and diet play in the emergence of acne.

In Saudi Arabia, there is a substantial correlation between acne vulgaris and several modifiable lifestyle factors. This suggests that lifestyle modification therapies could be helpful in managing and preventing acne. This study backs the use of lifestyle recommendations in acne treatment plans, providing a comprehensive strategy for dealing with this prevalent ailment [32].

Product Categories in Men's Personal Care

Skincare Products

Skincare is essential for preserving the function, appearance, and health of the skin, the body's largest organ. Proper skincare prevents issues such as dryness, acne, and premature ageing while enhancing the skin's barrier against environmental stressors like pollution and UV rays. A consistent routine tailored to individual skin types can improve skin texture and tone. Beyond aesthetics, skincare plays a crucial role in preventing long-term damage, including skin cancer. Research shows that early and consistent use of products like moisturizers and sunscreens significantly enhances skin health. The skincare market is experiencing significant growth, driven by rising consumer interest in anti-ageing, natural, and organic products and innovations in personalized skincare. Valued at over $150 billion globally in 2023, it continues to expand due to increasing awareness of skin health and beauty routines across all demographics [33].

Types of Skincare Products

Cleansers

Cleansers are fundamental in skincare routines, designed to cleanse the skin of debris, oil, and pollutants. They support preserving the natural equilibrium of the skin without removing essential oils. Depending on skin type, individuals can choose gel, cream, or foam cleansers. Cream cleansers are better suited for dry or sensitive skin since they are mild and moisturizing, but gel cleansers are great for oily or acne-prone skin because of their deep-cleaning capabilities. The skin is kept clean and ready for additional product absorption with the use of a good cleanser [34, 35].

Exfoliants

By eliminating dead skin cells from the skin's surface, exfoliation encourages skin renewal and a smoother texture. Exfoliants come in two varieties: chemical and physical. Granules in physical exfoliants, such as scrubs, are used to manually buff away dead cells; however, excessive use might irritate them. Chemical exfoliants provide a milder yet efficient technique to exfoliate the skin by breaking the bonds that hold dead skin cells together. Examples of these acids are alpha hydroxy acids (AHAs) and beta hydroxy acids (BHAs). Frequent exfoliation can enhance skin tone, pores, and fine wrinkles by promoting cell turnover [36].

Moisturizers

Moisturizers are essential in keeping the skin hydrated, preventing dryness, and maintaining the skin's barrier function. Different moisturizers cater to various skin types. For example, water-based gels are perfect for oily or combination skin, while thick creams or oil-based moisturizers work well for dry or mature skin. Moisturizers typically contain humectants, which attract moisture to the skin, emollients that smooth skin surfaces, and occlusives that lock in moisture. A good moisturizer ensures the skin remains supple, reducing the likelihood of wrinkles or cracks [34, 37].

Serums

Serums are concentrated treatments designed to target specific skin concerns like pigmentation, wrinkles, or acne. They frequently include active components such as peptides, hyaluronic acid, vitamin C, and niacinamide. Serums are lighter than moisturizers, which enables them to reach deeper into the skin's layers. Serums are effective in delivering potent ingredients directly to where they are needed most, making them a popular choice for enhancing anti-ageing, brightening, or hydrating routines [38].

Sunscreens

Probably the most essential skincare product is sunscreen since it shields the skin from damaging ultraviolet (UV) radiation. Exposure to ultraviolet light can cause hyperpigmentation, early ageing, and a higher risk of skin cancer. Sunscreens come in two varieties: chemical and physical (mineral). Chemical sunscreens absorb UV radiation and turn it into heat, whereas physical sunscreens, like zinc oxide or titanium dioxide, sit on the skin's surface and deflect UV rays. Experts recommend applying sunscreen with at least SPF 30 daily, regardless of the weather, for optimal protection [39].

Customizing Skincare for Skin Types

Individual skin types—normal, dry, oily, combination, and sensitive skin—should be taken into consideration while choosing skincare products.

Dry Skin

Products with rich moisturizing ingredients like ceramides and glycerine are beneficial to restore hydration and prevent moisture loss.

Oily Skin

Oil-free, non-comedogenic products that control sebum production, such as salicylic acid or clay-based masks, are effective for managing shine and preventing breakouts [40].

Sensitive Skin

People who have sensitive skin should use hypoallergenic, fragrance-free cosmetics with calming components like aloe vera or chamomile to avoid irritation [41].

Haircare Products

Haircare is a crucial part of personal grooming, and the right products help maintain the health, strength, and appearance of hair. A wide range of haircare products is available today, each serving a specific purpose, from cleansing and conditioning to styling and protecting. Hair types and concerns vary widely, so it is essential to choose products that suit individual needs [42].

Types of Haircare Products

Shampoos

Shampoo is the foundation of any haircare routine, designed to cleanse the scalp and hair by removing dirt, excess oils, product buildup, and environmental pollutants. Most shampoos contain surfactants that cleanse the hair, but they can vary in composition depending on the hair type and specific needs [43].

Types of Shampoos

- Clarifying Shampoos: Deep-cleaning formulas that remove heavy buildup but may be too harsh for daily use [44] .
- Moisturizing Shampoos: Ideal for dry or damaged hair, as they cleanse while adding hydration and nourishment [45].
- Sulphate-Free Shampoos: Designed for people with sensitive scalps, sulphate-free shampoos offer gentle cleansing without stripping the hair of natural oils [46] .
- Volumizing Shampoos: Created for fine or thinning hair, these shampoos add body and lift by enhancing hair texture [47].

Conditioners

Conditioners help to seal moisture into the hair, reducing frizz and promoting elasticity, which is essential for preventing breakage. They also provide a protective layer that shields hair from damage and makes it easier to manage [48].

There are different types of conditioners to suit varying hair needs:

- Rinse-Out Conditioners: The most common form, used after shampooing to hydrate and soften hair.
- Leave-In Conditioners: Lightweight formulas that provide continuous moisture throughout the day and are ideal for dry, curly, or textured hair.
- Deep Conditioners: Intensive treatments that repair and strengthen damaged hair, often left on the hair for a longer duration.
- Hair Masks: Designed to treat specific concerns like dryness or breakage, hair masks are usually packed with nourishing oils, proteins, and vitamins to revitalize hair.

Hair Oils

Hair oils have been used for centuries to nourish and strengthen hair. They play a crucial role in sealing in moisture, smoothing hair, and preventing split ends [49, 50].

Some popular oils used in haircare include the following:

- Argan Oil: Often referred to as ‘liquid gold’, argan oil is rich in antioxidants, vitamins, and fatty acids, making it ideal for softening and hydrating hair.
- Coconut Oil: It is well known for its deep conditioning abilities, works by penetrating the hair shaft to stop protein loss and fix damage.
- Jojoba Oil: Similar to the natural oils produced by the scalp, jojoba oil is excellent for balancing oil production and soothing a dry, irritated scalp.

Heat Protectants

Heat protectants shield hair from high temperatures during styling, preventing damage, frizz, and breakage by creating a protective barrier with ingredients like silicones [51].

Hair Serums

Hair serums, often silicone-based, enhance shine, smoothness, and manageability while protecting against frizz, humidity, and environmental damage, especially for dry or treated hair [52].

Customizing Haircare for Hair Types

Like skincare, haircare should be tailored to specific hair types and concerns [51].

Dry Hair

Requires moisturizing products like hydrating shampoos, deep conditioners, and hair oils.

Oily Hair

Benefits from lightweight, clarifying shampoos that control oil production without stripping the hair of natural moisture.

Curly Hair

Needs nourishing, sulphate-free products that define curls while maintaining moisture balance.

Colour-Treated Hair

Requires products designed to protect colour, such as sulphate-free shampoos and UV protectants.

Shaving Products

Shaving is a daily or regular grooming ritual for many people, and the use of proper shaving products can significantly enhance the experience. From ensuring a smooth, close shave to preventing skin irritation, shaving products play a crucial role in achieving optimal results [53].

Types of Shaving Products

Shaving Creams [54]

Shaving creams are one of the most common and widely used shaving products. They help create a rich lather that softens the facial hair, making it easier to cut. Shaving creams also act as a lubricant between the razor and the skin, reducing the risk of cuts, nicks, and razor burns. Traditional shaving creams can be foam-based or lathered manually with a brush, depending on the preference of the user. Key ingredients in shaving creams typically include the following:

- Glycerine: A humectant that draws moisture to the skin, helping maintain hydration.
- Emollients: Softening agents that keep the skin supple during shaving.
- Fragrances: Some shaving creams are scented, although fragrance-free options are available for sensitive skin.

Shaving Gels

Shaving gels are a popular choice for people looking for a transparent or lighter lather. Unlike shaving creams, gels do not produce as much foam, allowing for better visibility of the skin, and making them ideal for precision shaving around facial hair styles, sideburns, or beard lines. Shaving gels provide excellent lubrication and are often designed for men with oily or acne-prone skin, as they feel lighter and less greasy on the face [55]. Key characteristics of shaving gels include the following:

- Transparent formula: Ideal for accurate shaving.
- Cooling ingredients: Often contain menthol or aloe vera, offering a cooling effect that soothes the skin.
- Light texture: Less dense than creams, making them suitable for people with oily skin.

Shaving Foams

Shaving foams are widely used because of their convenience. They are available in aerosol cans and are easy to apply. With a few pumps, shaving foam provides an instant lather, ready for a quick shave. However, shaving foams are generally less hydrating than creams or gels and can dry out the skin due to their high air-to-product ratio [56]. This makes them less ideal for people with dry or sensitive skin. Key benefits of shaving foams include the following:

- Quick and easy application: Convenient for those who want a fast shave.
- Lightweight texture: Feels less dense on the skin compared to creams.
- Affordable: Often the most cost-effective option for daily shaving.

Shaving Oils

Shaving oils are gaining popularity as a pre-shave or standalone product. They act as a lubricant, creating a slick surface on the skin that helps the razor glide more smoothly, minimizing friction and irritation. Shaving oils are typically applied before a cream or gel to enhance the shaving experience, but they can also be used alone for a lighter, more natural shave [57]. These oils are particularly beneficial for individuals with dry or coarse facial hair as they offer intense hydration and protection. Key features of shaving oils are as follows:

- Rich in natural oils: Often contains essential oils like jojoba, argan, or coconut oil for deep hydration.
- Reduces irritation: Helps minimize razor burn and redness.
- Pre-shave or standalone use: Can be used alone or under other shaving products to improve glide.

Aftershaves

Aftershaves are used post-shave to soothe, disinfect, and hydrate the skin. They come in various forms, including balms, lotions, and splashes, each offering unique benefits based on the user's skin type. Modern aftershaves now include soothing and hydrating ingredients to prevent irritation and promote healing [58]. Types of aftershaves are as follows:

- Alcohol-based splashes: Provide a refreshing feel but may dry out the skin.
- Balms: Non-alcoholic and hydrating, making them suitable for dry or sensitive skin.
- Lotions: Offer a balance between hydration and protection, suited for normal skin types.

Key ingredients in aftershaves include the following:

- Aloe vera: Soothes and calms irritated skin.
- Witch hazel: A natural astringent that helps to tone the skin and reduce inflammation.
- Chamomile: Known for its anti-inflammatory properties, ideal for reducing redness and irritation.

Shaving Brushes

Though not a product applied directly to the skin, shaving brushes play an essential role in the shaving routine. A good-quality shaving brush helps to lift facial hair, exfoliate the skin, and create a rich, even lather from shaving creams or soaps. Badger hair, synthetic, or boar bristle brushes are the most common types, each offering different levels of softness and water retention [55].

Fragrances and Deodorants

Fragrances

Fragrances have a wide range of uses, and even if one wants to, it may be hard to avoid contact. Fragrances can be found in a wide range of toiletries and cosmetics, air fresheners, air purifiers, toys, textiles, over-the-counter topical pharmaceuticals, and industrial settings. Many of the components used in fragrances are also utilized as culinary flavours, and some of them are found in spices naturally. Due to their antibacterial qualities, natural scent products are sometimes employed as topical medications in addition to being used in aromatherapy and herbal medicines. Unpleasant smells can be neutralized by fragrances [59]. They are added to goods in order to give them a pleasing aroma, give them a unique quality, or serve as useful additives, such as having antimicrobial properties.

Fragrance ingredients are any basic ingredient utilized for their odorous, odour-enhancing, or mixing capabilities in the creation of fragrance compounds, according to the International Fragrance Association (IFRA). A natural product or a chemically specified component can be an ingredient in a scent. Plant material that produces fragrances is processed to create natural fragrance products. Nearly every part of the plant has the scent, which can be extracted by pressing, steam distillation, or organic solvent extraction to provide essential oils, concretes, and absolutes [60]. Because natural consistency and composition vary depending on several circumstances, such as the soil and atmosphere in which they develop, it is difficult, if not impossible, to entirely standardise the contents and quality of the finished product. Usually, the plant-based volatile fragrance product is made up of several different ingredients. The unique scent of a fragrance product could be the result of the blending of multiple chemicals or a single component in the case of a complex composition. Oak moss absolute includes at least 250 components and numerous odour-determining compounds, whereas clove oil contains up to 80% eugenol, the determining odour agent [61, 62]. In the past, animal secretions—such as deer musk and whale sperm ambergris—were also utilized as the raw material for natural scent compounds. Fragrance chemical mixtures have taken their place. Prior to the scientific and technological advancements of the first half of the 1800s, natural goods made up all perfumes. However, chemists were able to discover the primary elements that determine the scent of natural fragrance materials. The industrial synthesis of synthetic scent compounds started after this discovery. Nature-identical chemicals, or synthetic versions of naturally occurring components, are frequently used in the creation of synthesized ingredients; nevertheless, completely novel chemicals are also produced. Industry data suggest that there are around 2500 distinct scent compounds in use, natural plant extracts, nature-identical substances, or wholly synthetic chemicals make up the components.

Deodorants

Deodorant is the most popular cosmetic, applied to the axilla on a daily basis by millions of individuals. Preparations for deodorants prevent bacteria from growing and acting in a way that damages the apocrine gland under the armpit. Despite having strong antibacterial qualities, typical antibacterial agents like triclosan and aluminium salts can raise the risk of Alzheimer's, prostate, and breast cancer, as well as contact dermatitis. Roll-on, creams and wipes, sticks and solids, aerosol, and wipes are the five main categories of deodorant products. By type of deodorant, the market is segmented into aerosol, roll-on, creams and wipes, sticks, and solids, among other categories [63, 64]. Odours are covered up with deodorants. There are two distinct ways that a deodorant functions [65]. The number of bacteria that create volatile, odorous chemicals is reduced by antimicrobial treatments, and any odours produced are covered up with perfume. As of late, aluminium or aluminium crystals, including zeolite minerals, have been marketed as natural deodorants [66]. These can be purchased as solid crystals. The consumer is suggested to moisten the crystals before applying the product to the underarm area in order to prevent odour. In general, deodorants are extremely safe products. Nonetheless, these products have drawn a lot of attention due to theories regarding their potential role in the rising incidence of skin issues and carcinoma, the majority of which point to the oestrogenic qualities of parabens as the primary culprit. Currently, deodorants have a variety of chemical antimicrobials added to reduce the number of germs that cause odours. In fact, a number of detrimental health outcomes, such as cancer and degenerative brain disorders, including Alzheimer's disease and encephalopathy, have been related to the common deodorant component aluminium [67, 68], despite the fact that there is still no solid evidence linking it to cancer. Fragrance-induced allergic dermatitis is significantly influenced by deodorants. It appears that deodorants include an unfavourable combination of scent chemicals that cause allergic dermatitis. It might be claimed that since deodorants are applied to sensitive parts of the

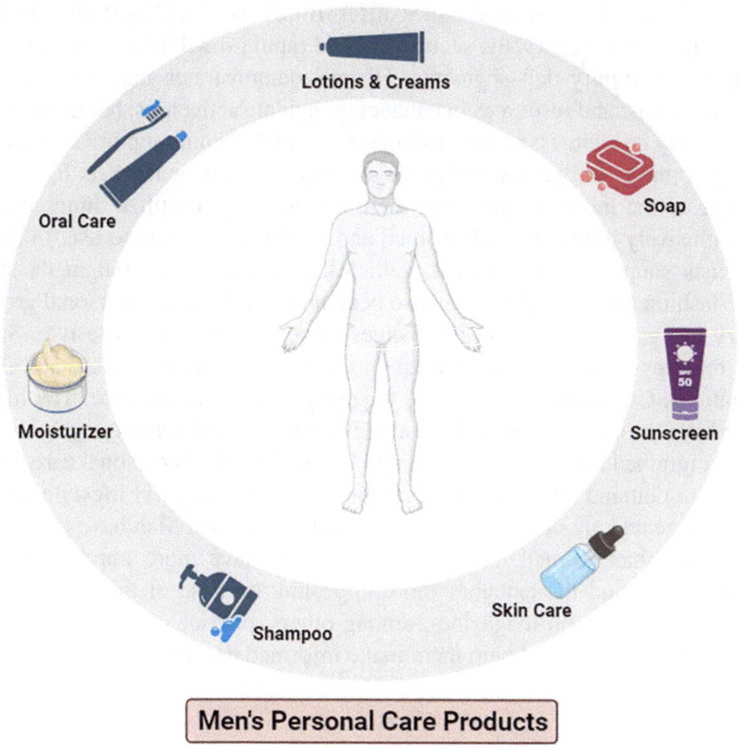

Fig. 2 Men's personal care products

body, certain aroma compounds that cause skin irritation should either be avoided entirely or used in deodorants in smaller amounts than in other products. Because of their natural origin and perceived safety, natural options that stop the growth of bacteria that cause odours are preferred and ought to be more palatable to customers [69]. A deodorant is a substance that is rubbed on the body to eliminate odours caused by germs that break down in the feet, underarms, and other places. By interfering with sweat glands, deodorants not only produce odour but also prevent perspiration. Deodorants can be sprayed on the feet and other parts of the body [70]. The different types of men's personal care products are shown in Fig. 2.

Other Emerging Categories

Because of the popularity and spread of metrosexual culture, new avenues for production have been available in the realm of male consumerism [71]. Due to its high profitability, the metrosexual market has grown to be a global phenomenon which has created new opportunities for businesses. Due to the fact that men's demands differ greatly from those of women, the market for items related to male skincare is changing. Men's skin types and issues differ from women's [72]. Within the beauty and personal care category, the section has had rapid growth and is currently one of the most opportunity-rich segments. History demonstrates that women tend to devote more time and resources in enhancing their attractiveness, but in more recent times, society has witnessed men following suit and becoming more aware of their looks, grooming, and personal cleanliness. Male consumer markets have shifted, according to the industry, and more men are choosing to utilize things that were once exclusively identified with women and forbidden for men to use. In the past, men's lists slanted towards sports, vehicles, and technology, but in the present period, fashion and cosmetics have also been included [71]. The personal grooming industry has grown as a result of changes in consumption and lifestyle. Sales of male grooming products have increased as a result of social pressure and shifting expectations. Customers are prepared to spend more to appear nice. Traditionally, women have been associated with good looks and self-care, but modern masculinity has also ingrained the expectation of good looks [73]. Men's personal care products are in more demand due to changes in purchasing patterns and lifestyle. Products that save time and are durable are highly sought after [74]. Men have more disposable income than women, and as a result, they have more purchasing power. Marketers now find the category more appealing because of this feature. Men's health, GQ, and Maxim magazines, among others, provide consumers with information about products that help them make informed decisions about what to buy.

Challenges and Opportunities

Overcoming Gender Stereotypes in Marketing

Rigid gender standards that have traditionally stressed men's grooming practices as utilitarian and minimalist, with an emphasis on deodorants, shaving, and basic hygiene items, have affected the marketing of men's personal care products [75]. Traditionally, the personal care business has categorized skincare, haircare, and other cosmetic items as belonging exclusively to women, which serves to perpetuate the idea that men's hygiene requirements are primarily utilitarian and basic [76]. Men's products were consequently marketed in a way that was typically associated with bold, minimalist packaging, sombre colour palettes, and aromas that played up masculinity, including woodsy or musky notes. But as men place more and more value on their attractiveness, mental health, and self-care, these antiquated beliefs are slowly beginning to change [77]. More than 70% of men worldwide are now open to trying skincare products that are typically sold to women, according to recent surveys, which illustrates a paradigm shift in male grooming habits and preferences [78]. The growth of social media and influencer culture, which has increased the mainstreaming, visibility, and acceptance of skincare and personal grooming for men across a range of demographics, is another factor contributing to this trend.

Men are also becoming more conscious of the long-term advantages of skincare, and the correlation between a well-groomed appearance and self-assurance or career advancement has fuelled this growing interest in personal care beyond basic hygiene [79].

In reaction to this change, some well-known companies have been instrumental in questioning conventional gender stereotypes in the promotion of personal hygiene goods. To encourage men to embrace skincare as part of a larger self-care routine while also appealing to their basic values of strength and resilience, Dove Men Care, for example, produced a series of advertisements that centred on redefined masculinity. Nivea Men, too, has taken a similar tack, fusing self-care with utility through ads that highlight real, everyday men rather than unreachable masculinity standards, thus breaking down the barriers that traditionally made skincare seem 'feminine' [80, 81]. These companies have been effective in changing the conversation so that grooming now encompasses self-expression, confidence, and general well-being in addition to cleanliness. The advertisements highlight how societal standards are shifting and that personal care and self-presentation are no longer just the purview of women. Rather, men are being urged to follow grooming regimens that improve their looks without sacrificing their sense of self. For example, social advertisements that highlight the importance of grooming in boosting well-being and self-confidence have successfully included both everyday guys and sports personalities [82]. Additionally, to reflect the expanding inclusion of male grooming, marketers have been more conscious about portraying a wider diversity of men in their commercials. This includes showcasing guys of all ages, body types, and ethnicities [83].

Notwithstanding these developments, brands that aim to break the entrenched gender norms surrounding men's care still face formidable obstacles. Men sometimes feel awkward using skincare and grooming items in public or having open discussions about them since many still identify these goods, especially those that include cosmetics, as feminine [84]. This stigma is particularly common in societies where conventional ideas of masculinity are still strongly held and where personal cleanliness and grooming beyond shaving are viewed as superfluous or even conceited. Personal care companies must therefore develop a marketing plan that not only caters to men's changing interests but also takes into account any cultural quirks that may still be resistant to these developments. The challenge for businesses is to promote male grooming in a way that balances this need without offending customers who might still adhere to these antiquated ideas of what it is to be a man [85]. To facilitate the adoption of new grooming products, brands are increasingly focusing on small package modifications, such as neutral colour schemes, minimalist designs, and the usage of words like 'all-in-one' or 'dual-purpose'. This is especially true in areas like Asia, where sales of male skincare products are expanding quickly while being positioned more as performance enhancers than as indulgences [86].

Developing marketing tactics that promote a diversity of masculinities is a more significant opportunity than simply rethinking the packaging and promotion of these items. It is crucial to launch campaigns that the mainstream skincare and personal grooming for men of all ages, including younger guys joining the workforce and older men who are athletes [87]. This calls on marketers to embrace the idea that men can be concerned about their beauty without sacrificing their masculinity and to go beyond the conventional ideas of what it means to be a man. Crucially, expanding product lines that address particular male skin issues, including oily skin, post-shave discomfort, and sensitive skin, are important in influencing consumer perceptions [88, 89]. Brands may close the divide between conventional and contemporary ideas of masculinity in grooming by showcasing how these goods solve real-world problems. Using a direct-to-consumer approach that stresses authenticity, transparency, and simplicity-appealing to men who seek simple yet effective grooming solutions—brands like California's Baxter and Harry's are likewise changing the space [90].

This societal shift has also been facilitated by the rise of influencer marketing since male influencers publicly discuss their skincare regimens and promote grooming goods on social media sites like YouTube and Instagram [91]. By putting a human face on the skincare product experience, influencers may normalize these regimens for other men and help them become more relatable to their audience. Younger generations, who are becoming more at ease with defying conventional gender norms, are especially drawn to this organic marketing strategy [92]. Further lowering the stigma associated with male grooming is the ability of firms to precisely cater to the unique demands of males through the use of data-driven marketing and customization. Men who desire discreet home delivery of customized routines have become more and more interested in personalized subscription services. The future of men's personal care marketing will depend on firms' ability to strike a balance between inclusivity, authenticity, and pragmatism while also continuing to challenge antiquated gender conventions, as these trends continue to develop [93].

Addressing Specific Skin Concerns of Men

Men's skin presents distinct challenges compared to women's, necessitating the development of gender-specific personal care products. Biologically, men's skin is approximately 20–25% thicker than women's due to higher collagen density, which makes their skin more resilient but also prone to more prominent issues like rough texture and deeper-set wrinkles over time [94]. In addition, men have more active sebaceous glands, leading to increased sebum production, which often results in oily skin, larger pores, and a higher prevalence of acne [95]. This overproduction of oil can contribute to clogged pores and frequent breakouts, requiring formulations that effectively balance oil production while maintaining skin hydration. Regular shaving, which is a significant component of most men's grooming routines, further exacerbates these challenges by irritating the skin, causing issues such as razor burn, ingrown hairs, and micro-abrasions. Shaving disrupts the skin barrier, necessitating additional care, especially in the form of moisturizers and barrier-repairing ingredients that soothe post-shave irritation [96]. As a result, the formulation of men's skincare products must account for these unique characteristics to effectively address specific concerns while providing both immediate and long-term benefits.

To address the heightened oil production and acne concerns in men, personal care products are increasingly incorporating active ingredients that specifically target these issues. For instance, niacinamide (vitamin B3) has become a favoured ingredient due to its ability to regulate sebum production, reduce the appearance of pores, and offer anti-inflammatory properties, making it highly effective for oily and acne-prone skin [97]. Moreover, salicylic acid, a beta-hydroxy acid, is frequently used in men's skincare formulations due to its ability to penetrate deeper into the pores and exfoliate from within, effectively reducing acne and preventing future breakouts [98, 99]. These ingredients are especially beneficial for men, whose skin often experiences clogged pores and persistent acne due to higher sebum levels. A study conducted by Farris found that products containing salicylic acid and niacinamide, when used in conjunction, significantly reduced oil production and improved the overall texture of men's skin. Additionally, given that men's skin tends to lose moisture rapidly after shaving, hydrating ingredients such as hyaluronic acid and glycerine are frequently included in formulations to restore hydration levels and support the skin's barrier function [100]. Hyaluronic acid, in particular, is a powerful humectant capable of holding up to 1000 times its weight in water, making it an ideal ingredient for ensuring skin stays hydrated without the heavy, greasy feel that many men prefer to avoid [101].

Another emerging trend in men's skincare is the development of multifunctional products, which are particularly appealing to men due to their simplicity and efficiency. Men tend to prefer grooming routines that are quick and effective, and as a result, companies are creating products that address multiple skin concerns within a single application. For example, after-shave balms are now being formulated with added SPF protection to reduce the need for additional sunscreen application. Sunscreen is critical, as men's skin is often less likely to receive adequate sun protection compared to women's, putting them at a higher risk for skin damage and premature ageing [102]. Moreover, many after-shave products now include ingredients like allantoin and aloe vera, which help to soothe the skin, reduce inflammation, and provide antioxidant protection to aid in post-shave recovery. This multifunctionality streamlines grooming routines, making it easier for men to incorporate comprehensive skincare into their daily habits. Additionally, the inclusion of SPF in everyday skincare products is crucial, as studies have shown that men are significantly less likely to use sunscreen regularly, contributing to higher rates of skin cancer in men compared to women [103].

Innovations in men's skincare have also extended to the use of botanical and naturally derived ingredients, responding to the growing consumer demand for clean, sustainable products. Ingredients such as tea tree oil, known for its antimicrobial and anti-inflammatory properties, have been widely adopted in acne treatments for men. Studies suggest that tea tree oil can be as effective as benzoyl peroxide in treating mild to moderate acne, without the associated irritation [104]. Charcoal, another popular natural ingredient, is incorporated into face washes and masks to draw out impurities, detoxify the skin, and help manage excess oil, making it an excellent option for men dealing with oily skin [105]. Meanwhile, plant-based antioxidants like green tea extract and vitamins C and E are included to protect against environmental stressors, such as UV radiation and pollution, which are known to accelerate skin ageing [106]. These natural ingredients not only appeal to the environmentally conscious consumer but also offer gentler alternatives to synthetic chemicals, catering to men with sensitive skin.

Men's skincare products are also increasingly adopting minimalist, eco-friendly packaging that aligns with broader trends towards sustainability in the beauty industry. According to a report by Euromonitor International, consumers are gravitating towards brands that demonstrate a commitment to environmental responsibility, and men's care is no exception. This shift has led companies to reduce plastic usage, introduce refillable containers, and use biodegradable or recyclable materials. For example, Bulldog Skincare for Men, a leader in sustainable grooming products, has transitioned to using sugarcane-based plastic packaging, reducing its reliance on fossil fuels and cutting down on carbon emissions. This sustainable approach resonates particularly with younger male consumers, who are increasingly prioritizing eco-conscious brands in their purchasing decisions. Moreover, the transparency of ingredient sourcing, cruelty-free certifications, and the use of renewable resources are becoming critical factors that influence consumer loyalty in the men's skincare market [93]. The ongoing innovation in men's care is driven not only by the unique physiological characteristics of men's skin but also by evolving consumer preferences for simpler, effective, and environmentally responsible products. As companies continue to refine their formulations and adopt cutting-edge ingredients, the future of men's skincare will likely see a stronger focus on personalization, sustainability, and multifunctionality, addressing both the biological and lifestyle-driven needs of modern men.

Sustainable Practices and Eco-Friendly Packaging

As sustainability continues to gain momentum as a pivotal concern among consumers, the men's personal care industry is increasingly adopting eco-friendly practices, reflecting a broader shift towards responsible consumption. A significant aspect of this trend involves sustainable packaging, where brands are actively utilizing recyclable materials, reducing plastic waste, and introducing refillable containers. According to a recent report, the global sustainable packaging market is projected to reach USD 500 billion by 2027, driven largely by consumer demand for environmentally friendly products [107]. Companies such as Bulldog Skincare and L'Oréal are at the forefront of this transformation, having made significant advancements in incorporating sustainability into their production processes. Bulldog Skincare, for instance, has embraced innovative packaging solutions by using sugarcane-based bioplastics, effectively reducing its reliance on fossil fuels and contributing to lower carbon emissions [108]. This shift not only aligns with the growing consumer demand for sustainable options but also showcases a commitment to environmental stewardship that resonates with increasingly eco-conscious shoppers.

In tandem with advancements in packaging, eco-friendly formulations have emerged as a critical component of sustainability within men's personal care. Brands are increasingly formulating products that utilize natural, organic, and cruelty-free ingredients, catering to a demographic that is more informed about the environmental impact of synthetic chemicals [109]. For example, the Clean Beauty movement has inspired numerous brands to reformulate their products by eliminating harmful ingredients, which is particularly relevant as consumers prioritize their health alongside environmental considerations. This alignment with consumer values presents an opportunity for brands to differentiate themselves in a saturated market by emphasizing their commitment to sustainability and transparency in product sourcing. According to a survey by Nielsen, nearly 73% of global consumers are willing to change their consumption habits to reduce environmental impact, indicating a clear preference for brands that prioritize sustainability [110].

Furthermore, transparency has become a crucial factor in building brand loyalty. Consumers are increasingly scrutinizing the sourcing and manufacturing processes of the products they purchase, leading brands to provide clearer information regarding ingredient origins and environmental impact. For instance, the implementation of certifications such as the Leaping Bunny and Ecocert has allowed brands to substantiate their claims of being cruelty-free and organic, thus fostering trust with their customer base [111]. In addition to transparency in formulations, brands are also adopting lifecycle assessments to quantify the environmental impact of their products from production to disposal. This holistic approach not only highlights a brand's commitment to sustainability but also provides valuable data that can be used to refine processes and minimize waste.

As the market landscape evolves, brands are increasingly adopting innovative strategies to enhance sustainability within their supply chains. Collaborations with sustainable ingredient suppliers, such as those utilizing regenerative agricultural practices, are becoming more common, ensuring that sourcing methods contribute positively to the environment [112]. Also, initiatives such as the circular economy, where products are designed for reuse and recycling, are being implemented by forward-thinking brands to further reduce environmental impact. By engaging consumers in these sustainable practices, companies are not only cultivating brand loyalty but also encouraging collective responsibility towards environmental conservation [113].

The shift towards sustainable practices and eco-friendly packaging within the men's personal care industry presents both challenges and opportunities for brands. As consumer awareness grows, companies that prioritize sustainability are likely to experience increased loyalty and market share. By innovating packaging solutions, adopting eco-friendly formulations, and emphasizing transparency in sourcing, brands can position themselves as leaders in a rapidly evolving market that values ethical consumption. This strategic focus on sustainability not only addresses consumer demand but also contributes to broader environmental goals, ultimately paving the way for a more responsible and conscious personal care industry.

Future Directions and Emerging Trends

Trends in Men's Personal Care (Natural and Organic Formulations, Incorporation of Advanced Technologies)

The trend towards natural and organic formulations in men's personal care products is emblematic of a broader shift within the beauty and cosmetics industry, highlighting an increasing consumer preference for sustainability and ingredient transparency. According to a market analysis by Grand View Research, the global natural and organic personal care market is projected to reach approximately USD 25.1 billion by 2025, with men's grooming products playing a significant role in this growth [114]. This shift is driven by a growing awareness among men about the potential harmful effects of synthetic chemicals, prompting brands to innovate with safer, plant-based alternatives. Ingredients such as aloe vera, known for its hydrating and soothing properties, tea tree oil, recognized for its antimicrobial benefits, and activated charcoal, acclaimed for its detoxifying capabilities, have surged in popularity. These components address common male skin concerns, including oily skin, acne, and post-shaving irritation [115, 116]. Moreover, the demand for efficacy in men's skincare has led to the incorporation of advanced technologies and biotechnology in product formulations. A notable example is the use of biosynthesized ingredients, such as lab-grown collagen and peptides, which enhance the effectiveness of anti-ageing and moisturizing products. Recent studies indicate that these ingredients can significantly improve skin hydration and elasticity, thereby providing targeted solutions that cater to men's specific skincare needs [117]. Additionally, the rise of personalized skincare, driven by advancements in technology, allows brands to offer tailored solutions that align with individual skin types and concerns. By leveraging AI and machine learning, companies can analyse consumer data to develop products that meet the unique requirements of their male clientele. This technological integration is not just limited to product formulation; it extends to packaging innovations as well. Brands are increasingly adopting eco-friendly and sustainable packaging solutions to meet the expectations of environmentally conscious consumers. For instance, biodegradable and recyclable materials are being utilized to minimize environmental impact, reinforcing the industry's commitment to sustainability [118]. Moreover, the incorporation of smart technologies in grooming devices, such as electric razors equipped with sensors to adapt to the contours of the face, is transforming the grooming experience, making it more efficient and user-friendly. These developments illustrate a holistic evolution within the men's personal care sector, wherein natural and organic formulations are not merely trends but represent a fundamental shift towards a more responsible and effective approach to grooming. The convergence of sustainability, advanced technologies, and personalized solutions is setting new standards in the industry, offering men enhanced options that prioritize both health and environmental consciousness [119]. As this market continues to expand, the opportunity for brands to innovate and adapt to these emerging trends will be crucial in capturing the attention and loyalty of a discerning male audience.

Integration of Technology: Smart Grooming Devices and Apps

The rapid evolution of technology in the personal care sector has significantly transformed grooming practices, leading to the rise of smart grooming devices and personalized skincare applications. Notably, companies like Gillette have pioneered the introduction of smart razors equipped with sensors designed to gauge skin sensitivity and adjust blade pressure accordingly, thus ensuring a more comfortable and efficient shaving experience. This innovation highlights a growing trend where consumer convenience and comfort are prioritized through the use of technology [120, 121]. Furthermore, brands are leveraging artificial intelligence (AI) to develop applications that analyse individual skin conditions and provide personalized skincare regimens. For instance, apps like 'Skin AI' allow users to upload photos of their skin, after which an AI algorithm assesses skin type, issues, and concerns, generating tailored recommendations for products and routines [122, 123]. Such personalization not only enhances user satisfaction but also fosters greater consumer trust in product efficacy, as users are more likely to engage with items specifically suited to their unique needs. The integration of technology into personal care also facilitates more informed purchasing decisions; by providing detailed insights and recommendations, these apps help demystify the vast array of available products, making it easier for men to navigate their skincare and grooming choices. The increasing availability of AI-driven diagnostics and smart devices is paving the way for a future where wearable technology plays a crucial role in monitoring skin health in real-time. Devices that measure skin hydration, sebum levels, and UV exposure are being developed, enabling users to make informed decisions about their skincare routines based on real-time data [124]. These advancements not only enhance product effectiveness but also empower men to take a proactive approach to their care, transitioning the perception of grooming from a routine task to an engaging, health-oriented activity. The potential for smart devices to provide immediate feedback on skin health represents a significant leap in personalized care, bridging the gap between technology and self-care. Moreover, this technology encourages men to adopt a more holistic approach to grooming, where insights gleaned from data can influence not only product choice but also lifestyle modifications aimed at improving overall skin health. As these technologies continue to advance, we can anticipate the emergence of integrated platforms that combine skincare management with lifestyle advice, offering comprehensive solutions that encompass diet, hydration, and environmental factors impacting skin conditions. This integration aligns well with the increasing consumer demand for products that deliver both performance and sustainability; as men become more environmentally conscious, brands that utilize smart technology to reduce waste and enhance product lifecycle will likely gain a competitive edge [125]. The future of men's personal care is being reshaped by the integration of smart grooming devices and AI-driven applications, which not only improve the efficiency and effectiveness of products but also promote greater engagement and awareness of personal grooming practices among men. As we move forward, the symbiosis between technology and personal care will continue to evolve, leading to more innovative solutions that cater to the diverse needs of today's consumers, ultimately transforming the landscape of men's grooming and personal care [126].

Innovations in Packaging and Delivery Systems

The landscape of packaging and delivery systems for personal care products is rapidly evolving, driven by technological advancements and an increasing emphasis on sustainability. Modern consumers are more discerning than ever, seeking products that not only deliver effectiveness but also align with their values, particularly regarding environmental impact. Airless pump technology, for example, has gained traction in the skincare industry as it prevents air exposure and contamination, thereby preserving the integrity and efficacy of active ingredients such as antioxidants and vitamins [127, 128]. This preservation not only enhances product performance but also extends shelf life, allowing brands to reduce waste associated with expired products. Additionally, the rise of single-use pods has emerged as a practical solution to the common issue of over-application or contamination that can occur with traditional packaging. These pods, which contain pre-measured doses of the product, help minimize waste and encourage precise usage, ultimately enhancing the consumer experience while supporting sustainability goals [129]. Biodegradable packaging is another critical innovation, with many companies shifting towards materials derived from renewable sources, such as plant-based plastics and recycled paper. These environmentally friendly alternatives are designed to decompose naturally, thereby reducing the carbon footprint of personal care products and meeting the growing demand for sustainable options [130]. Furthermore, companies are also exploring smart packaging solutions that leverage technology to enhance user engagement. For instance, some brands have developed QR codes or NFC (Near Field Communication) chips embedded in their packaging, allowing consumers to access detailed product information, usage instructions, and even tutorials through their smartphones. This interactive approach not only enriches the user experience but also fosters a deeper connection between the consumer and the brand [131]. The integration of augmented reality (AR) features into packaging is another emerging trend that enhances consumer interaction. Through AR applications, users can visualize how products will look and perform on their skin before purchasing, thus making more informed decisions [132]. In addition to these technological advancements, brands are increasingly adopting multi-step kits that simplify grooming routines. These kits often include a range of products designed to be used together, providing a cohesive skincare or grooming experience tailored to specific concerns, such as anti-ageing or hydration. Such systems not only offer convenience but also encourage consumers to adopt comprehensive skincare regimens, as they can easily follow step-by-step instructions without the confusion of mixing and matching products from different brands. By streamlining the skincare process, these kits effectively address the needs of busy individuals who may otherwise overlook daily grooming rituals. The push towards convenience is further reflected in the design of portable and travel-friendly packaging. With an increase in travel and on-the-go lifestyles, personal care brands are rethinking how products are packaged and delivered. Compact sizes, easy-to-use dispensers, and leak-proof designs are becoming standard features, appealing to consumers who prioritize practicality without sacrificing quality [133, 134]. Moreover, the trend towards refillable containers is gaining momentum, providing consumers with the option to reduce waste while still enjoying their favourite products. This not only fosters customer loyalty but also reinforces a brand's commitment to sustainability [135, 136]. The innovations in packaging and delivery systems are reshaping the personal care industry, making it more aligned with contemporary consumer expectations for sustainability, convenience, and interactivity. As brands continue to invest in these areas, they not only enhance the user experience but also contribute to a more sustainable future, responding effectively to the demands of an increasingly eco-conscious market.

Conclusion

The men's personal care industry is at a pivotal moment, characterized by a significant shift towards inclusivity, personalization, and sustainability. As societal perceptions evolve, brands are increasingly recognizing the importance of gender-neutral marketing strategies that resonate with a diverse consumer base. This chapter highlights how understanding the biological differences between men's and women's skin and hair is essential for developing targeted products that address specific challenges faced by men. The exploration of various product categories from skincare and haircare to shaving and fragrances demonstrates the breadth of options available, reflecting the industry's growing sophistication. Moreover, the challenges of overcoming entrenched gender stereotypes and the need for sustainable practices, such as eco-friendly packaging, are critical considerations for brands looking to thrive in this changing landscape. Looking ahead, emerging trends such as the rising demand for natural and organic formulations, the integration of advanced technologies in grooming devices, and innovative packaging solutions are set to redefine men's personal care. These advancements not only promise enhanced product efficacy but also align with the values of today's consumers, who are increasingly conscious of both their grooming choices and their impact on the environment. As the industry continues to innovate and adapt, it is poised to become a transformative segment of the beauty market, offering men more effective, inclusive, and sustainable personal care options than ever before.

Conflict of Interest The authors declare no conflict of interest.

Glossary

AI in skincare Technology that analyses skin conditions to recommend personalized skincare products and routines

Alpha Hydroxy Acid A chemical exfoliant commonly used in men's skincare to remove dead skin cells and improve texture and tone

Augmented reality A technology integrated into product packaging or apps, enabling men to virtually test or preview products before purchase

Beta Hydroxy Acid An oil-soluble exfoliant, like salicylic acid, is used to penetrate pores and reduce acne in oily or acne-prone skin

Biodegradable packaging Environmentally friendly packaging is designed to decompose naturally, reducing environmental impact

Collagen A structural protein in the skin that provides strength and elasticity

Exfoliants Products that help the skin regenerate by removing dead skin cells from its surface

Gentlemen's quarterly A popular men's magazine often influences trends in grooming, fashion, and lifestyle, shaping consumer preferences

Humectants Ingredients that attract water to the skin, aiding hydration

Hydration The presence of water in the skin's layers, essential for maintaining elasticity and barrier function

International Fragrance Association A regulatory body ensures the safety of fragrance ingredients used in men's grooming products like deodorants and colognes

Microcirculation The circulation of blood in the smallest blood vessels in the skin, affecting its health and appearance

Near field communication A smart packaging technology allowing men to access product details and tutorials through smartphone scanning

Occlusives Substances that form a barrier to prevent moisture loss from the skin

pH The level of acidity or alkalinity on the skin's surface, important for maintaining its health

Quality of life Refers to the overall well-being and lifestyle improvements achieved through proper grooming, skincare, and personal care routines

Sebum The natural oil produced by sebaceous glands in the skin

Skin elasticity The ability of skin to regain its natural shape after being stretched is indicative of youthfulness and health

Skin friction coefficient A measure of the resistance between the skin's surface and another object, influenced by moisture and texture

Smart grooming devices High-tech tools, like razors with sensors, are designed for efficient and adaptive personal grooming

Stratum Corneum The outermost layer of the skin, crucial for skin hydration and protection

Sun protection factor An indicator of how well sunscreen shields the skin from damaging UV radiation, which is essential for avoiding skin cancer and premature ageing

Transepidermal water loss A measure of water evaporation through the skin, indicating skin barrier effectiveness

References

1. Walters, K. A., & Roberts, M. S. (2002). The structure and function of skin. In *Dermatological and transdermal formulations* (pp. 19–58). CRC Press. [cited 2024 Sept 28].
2. Schmitt, W. H. (1992). Skin-care products. In *Chemistry and technology of the cosmetics and toiletries industry* (pp. 104–148). Springer Netherlands. [cited 2024 Sept 30].
3. Biddle, J. E., & Hamermesh, D. S. (1998). Beauty, productivity, and discrimination: Lawyers' looks and lucre. *Journal of Labor Economics, 16*(1), 172–201.
4. Pfann, G. A., Biddle, J. E., Hamermesh, D. S., & Bosman, C. M. (2000). Business success and businesses' beauty capital. *Economics Letters, 67*, 201–207. [cited 2024 Sept 19].
5. Grubb, E. L., & Grathwohl, H. L. (1967). Consumer self-concept, symbolism and market behavior: A theoretical approach. *Journal of Marketing, 31*(4), 22–27.
6. Kellner, D. (2003). *Media culture: Cultural studies, identity and politics between the modern and the post-modern*. Routledge.
7. Seitz, V. A., & Johar, J. S. (1993). Advertising practices for self-image projective products in the New Europe: A print advertising content analysis. *Journal of Consumer Marketing, 10*(4), 15–26.
8. Coley, A., & Burgess, B. (2003). Gender differences in cognitive and affective impulse buying. *Journal of Fashion Marketing and Management, 7*(3), 282–295.
9. Sturrock, F., & Pioch, E. (1998). Making himself attractive: The growing consumption of grooming products. *Marketing Intelligence & Planning, 16*(5), 337–343.
10. Souiden, N., & Diagne, M. (2009). Canadian and French men's consumption of cosmetics: A comparison of their attitudes and motivations. *Journal of Consumer Marketing, 26*(2), 97–109.
11. Mulyanegara, R. C., Tsarenko, Y., & Anderson, A. (2009). The Big Five and brand personality: Investigating the impact of consumer personality on preferences towards particular brand personality. *Journal of Brand Management* [cited 2024 Sept 23], *16*(4):234–247.
12. Härkönen, J. (2008). Labour force dynamics and the obesity gap in female unemployment in Finland. *Research on Finnish Society, 1*, 3–15. diva-portal.org, 2007 [cited 2024 Sept 23]
13. Rahrovan, S., Fanian, F., Mehryan, P., Humbert, P., & Firooz, A. (2018). Male versus female skin: What dermatologists and cosmeticians should know. *International Journal of Women's Dermatology, 4*(3), 122–130.
14. Magazine, C. A. D. (2015). Undefined. *Understanding male skin*. education.dermalogica.ca. [cited 2024 Sept 30].
15. Chen, W., Mempel, M., Traidl-Hofmann, C., Al Khusaei, S., & Ring, J. (2010). Gender aspects in skin diseases. *Journal of the European Academy of Dermatology and Venereology, 24*(12), 1378–1385.
16. Luebberding, S., Krueger, N., & Kerscher, M. (2013). Skin physiology in men and women: In vivo evaluation of 300 people including TEWL, SC hydration, sebum content and skin surface pH. *International Journal of Cosmetic Science, 35*(5), 477–483.
17. Rogiers, V., Derde, M., & Toiletries, G. V. C. (1990). Standardized conditions needed for skin surface hydration measurements. *Cosmetics and Toiletries, 37*, 73–82. researchportal.vub.beV. [cited 2024 Dec 9].

18. Man, M. Q., Xin, S. J., Song, S. P., Cho, S. Y., Zhang, X. J., Tu, C. X., et al. (2009). Variation of skin surface pH, sebum content and stratum corneum hydration with age and gender in a large Chinese population. karger.com. *Skin Pharmacology and Physiology, 22*, 190–199. [cited 2024 Oct 1].
19. Luebberding, S., Krueger, N., & Kerscher, M. (2014). Mechanical properties of human skin in vivo: A comparative evaluation in 300 men and women. *Skin Research and Technology, 20*(2), 127–135.
20. Roh, M., Han, M., Kim, D., & Dermatology, K. C. B. J. (2006). Sebum output as a factor contributing to the size of facial pores. *British Journal of Dermatology, 155*, 890–894. academic.oup.com. [cited 2024 Oct 1].
21. Bailey, S. H., Oni, G., Brown, S. A., Kashefi, N., Cheriyan, S., Maxted, M., et al. (2012). The use of non-invasive instruments in characterizing human facial and abdominal skin. *Lasers in Surgery and Medicine, 44*(2), 131–142.
22. Draelos, Z. D. (2008). Understanding male skin care needs: The challenge of an evolving market. *Cosmetics & Toiletries*. Market.img.cosmeticsandtoiletries.com, 2008 [cited 2024 Oct 1], *123*(9), 59.
23. Burry, J. S., Coulson, H. F., Esser, I., Marti, V., Melling, S. J., Rawlings, A. V., et al. (2001). Erroneous gender differences in axillary skin surface/sweat pH. *International Journal of Cosmetic Science, 23*(2), 99–107.
24. Harpin, V. (1983). Barrier properties of the newborn infant's skin. *The Journal of Pediatrics, 102*, 419. [cited 2024 Oct 1].
25. Hornstra, J. M., Hoekstra, T., Serné, E. H., Eringa, E. C., Wijnstok, N. J., Blom, H. J., et al. (2014). Homocysteine levels are inversely associated with capillary density in men, not in premenopausal women. *European Journal of Clinical Investigation, 44*(3), 333–340.
26. Zhu, Y., Song, S., Luo, W., et al. (2011). Characterization of skin friction coefficient, and relationship to stratum corneum hydration in a normal Chinese population. *Skin Pharmacology and Physiology*. population.karger.com, 2011 [cited 2024 Oct 1], *24*, 81–86.
27. Tsukahara, K., Hotta, M., Osanai, O., Kawada, H., Kitahara, T., & Takema, Y. (2013). Gender-dependent differences in degree of facial wrinkles. *Skin Research and Technology, 19*(1), e65–e71.
28. Maurer, M., Rietzler, M., Burghardt, R., & Siebenhaar, F. (2016). The male beard hair and facial skin–challenges for shaving. *International journal of cosmetic science, 38*, 3–9.
29. Maurer, M., Rietzler, M., Burghardt, R., & Siebenhaar, F. (2016). The male beard hair and facial skin – Challenges for shaving. *International Journal of Cosmetic Science, 38*, 3–9.
30. Wu, W., Yang, J., Tao, H., & Lei, M. (2022). Environmental regulation of skin pigmentation and hair regeneration. *Stem Cells and Development, 31*(5-6), 91–96.
31. Knaggs, H., & Lephart, E. D. (2023). Enhancing skin anti-aging through healthy lifestyle factors. *Cosmetics, 10*, 142. mdpi.com, 2023 [cited 2024 Oct 1].
32. Khormi, G., Aldubayyan, N., Hakami, M., & Cureus, S. D. (2024). Impact of lifestyle and dietary habits on the prevalence of acne vulgaris: A cross-sectional study from Saudi Arabia. *Cureus, 16*. ncbi.nlm.nih.gov. [cited 2024 Oct 1].
33. Emerald, M., Emerald, A., Emerald, L., & Kumar, V. (2016). Perspective of natural products in skincare. Nano ranking found for "Emerald Expert Briefings". *Pharmacy & Pharmacology International Journal, 4*, 72–75. 2016. [cited 2024 Sept 30].
34. Association KG of the AWM, 2015 undefined. Cleansers and moisturisers: The basics Nano ranking found for "Asian Libraries." search.informit.org. [cited 2024 Sept 30].
35. Griffiths-Brophy, S., Hasenoehrl, E. J., & Wei, K. (2022). Facial cleansers and cleansing cloths. In *Cosmetic dermatology: Products and procedures* (pp. 144–150). John Wiley & Sons.
36. Draelos, Z. K. (1995). Cosmetics: An overview. In *Current problems in dermatology*. Elsevier. [cited 2024 Sept 30].
37. Mezail Mawazi, S., Ann, J., Othman, N., Khan, J., Alolayan, O., Al Thagfan, S. S., et al. (2022). A review of moisturizers; history, preparation, characterization and applications. *Cosmetics, 9*, 61. mdpi.com, [cited 2024 Sept 30].

38. Kamble, R., Gamare, D., Yeole, P., & MPF. (2022). *Formulation and evaluation of herbal based anti-aging face serum*, Biomedical Chromatography; researchgate.net, 2022 [cited 2024 Sept 30].
39. Bens, G. (2020). Sunscreens. In *Sunlight, Vitamin D and Skin*. Springer. [cited 2024 Sept 30].
40. Black, D., Diridollou, S., Lagarde, J. M., & Gall, Y. (1998). Skin care products for normal, dry and greasy skin Nano ranking found for "Series in Cosmetic and Laser Therapy". In *Textbook of cosmetic dermatology*. CRC Press. books.google.com. [cited 2024 Sept 30].
41. Inamadar, A. C., & Palit, A. (2013). Sensitive skin: An overview. *Indian Journal of Dermatology, Venereology and Leprology, 79*, 9. [cited 2024 Sept 30].
42. Bolduc, C., & Shapiro, J. (2001). Hair care products: Waving, straightening, conditioning, and coloring. *Clinics in Dermatology, 19*, 431–436. cidjournal.com. [cited 2024 Sept 30].
43. Cornwell, P. A. (2017). A review of shampoo surfactant technology: Consumer benefits, raw materials and recent developments. *International Journal of Cosmetic Science, 40*(1), 16–30. [cited 2024 Sept 30].
44. Puccetti, G., & Kulcsar, L. (2020). Hair surface quality: Laser scattering as a tool for characterizing the surface condition and deposits from shampoos and conditioners. *International Journal of Cosmetic Science, 42*(1), 89–98.
45. Patil, H., & Mundecha, K. (2023). Marketing research-An examination of consumer preferences for shampoo brands Nano ranking found for shampoo brands. *Research Journal of Topical and Cosmetic Sciences, 14*, 1–6. indianjournals.com. [cited 2024 Sept 30].
46. Rahat, M., Shuja, S., Kanwal, S., Zafar, N., & Talib, S. (2024). Formulating sulfate-free shampoo: Composition, benefits, and therapeutical. *Pakistan Journal of Medical and Health Sciences, 6*(13), 4468–4483. [cited 2024 Sept 30].
47. Datta, Y. (2018). *Original paper the US shampoo market: A competitive profile*. core.ac.uk. [cited 2024 Sept 30];4(2).
48. Papageorgiou, Z., Fernandes, C., Medronho, B., Alves, L., & Rasteiro, M. G. (2023). On hair care from structure and degradation to novel biobased conditioning agents physico chemistry. *Polymers, 15*, 608. mdpi.com. [cited 2024 Sept 30].
49. Mysore, V. (2022). Indigenous knowledge on hair oils. *International Journal of Trichology, 14*, 84–90. [cited 2024 Sept 30].
50. Shahi, Z., Khajeh Mehrizi, M., Hadizadeh, M., Shahi, Z., Mehrizi, M. K., & Hadizadeh, M. (2017). A review of the natural resources used to hair color and hair care products. *Journal of Pharmaceutical Sciences and Research, 9*, 1026. [cited 2024 Sept 30].
51. Hair, W., & Unmanageable, F. *How do you treat thick wavy frizzy hair: Hair guide to healthy styling*. hairbrushy.com WAWTP Hair, FB Unmanageablehairbrushy.com. [cited 2024 Sept 30].
52. Fernandes, C., Medronho, B., Alves, L., & Rasteiro, M. G. (2023). On hair care physicochemistry: from structure and degradation to novel biobased conditioning agents. *Polymers, 15*(3), 608.
53. Cowley, K., & Dermatology, K. V. B. J. (2012). Insights into shaving and its impact on skin. *British Journal of Dermatology, 166*(SUPPL.1), 6–12. [cited 2024 Sept 30].
54. Hu, X., & He, H. (2021). A review of cosmetic skin delivery. *Journal of Cosmetic Dermatology, 20*(7), 2020–2030.
55. Trivellini, R., Perez-Meza, D., Renaud, H. J., & Gupta, A. K. (2021). Preview long hair follicular unit excision: an up-and-coming technique. *Journal of Cosmetic Dermatology, 20*(11), 3422–3426.
56. Karbelani, P., Nisa, H., Prasetyo, A., & Firdaus, F. (2013) *Chemical Product Design "Shaving Product"*. Library Hi Tech. [cited 2024 Sept 30].
57. Paniccia, P. (2000). Men's toiletries. In *Poucher's perfumes, cosmetics and soaps* (pp. 345–375). Kluwer Academic Publishers.
58. Cosmetics SB, Technology. (1972). Pre shave and aftershave preparations. books.google.com. *Cosmetics, Science and Technology*. [cited 2024 Sept 30].

59. Johansen, J. D., & Lepoittevin, J. P. (2011). Fragrances. In *Contact dermatitis* (5th ed., pp. 607–627). Springer. [cited 2024 Sept 27].
60. Müller, J. (1992). *The H&R book of perfume*. Glöss. [cited 2024 Sept 27].
61. Bernard, G., Giménez-Arnau, E., Rastogi, S. C., Heydorn, S., Johansen, J. D., Menné, T., et al. (2003). Contact allergy to oak moss: Search for sensitizing molecules using combined bioassay-guided chemical fractionation, GC-MS, and structure-activity relationship analysis. *Archives of Dermatological Research, 295*(6), 229–235.
62. Bauer, K., Garbe, D., & Surburg, H. (2008). *Common fragrance and flavor materials: Preparation, properties and uses*. John Wiley & Sons. [cited 2024 Sept 27].
63. Balsam, M., & Sagarin, E. (1972). *Cosmetics science and technology*. John Wiley & Sons. [cited 2024 Sept 27].
64. Ficheux, A., Wesolek, N., Chevillotte, G., & Roudot, A. C. (2015). Consumption of cosmetic products by the French population. First part: Frequency data. *Food and Chemical Toxicology, 78*, 159. [cited 2024 Sept 27].
65. Oliveira, E. C. D., Salvador, D. S., Holsback, V., Shultz, J. D., Michniak-Kohn, B. B., & Leonardi, G. R. (2021). Deodorants and antiperspirants: Identification of new strategies and perspectives to prevent and control malodor and sweat of the body. *International Journal of Dermatology, 60*(5), 613–619.
66. Laden, K. (1999). *Antiperspirants and deodorants* (2nd ed.). CRC Press. [cited 2024 Sept 27].
67. Shaw, C., & Tomljenovic, L. (2013). Aluminum in the central nervous system (CNS): Toxicity in humans and animals, vaccine adjuvants, and autoimmunity. *Immunologic Research*, Springer, *56*(2–3):304–316. [cited 2024 Sept 27].
68. Today, C. E. M. M. (1998). Does antiperspirant use increase the risk of aluminium-related disease, including Alzheimer's disease. *Molecular Medicine Today, 4*, 107–109. [cited 2024 Sept 27].
69. Andersen, K. E., Heisterberg, M. V., Menné, T., Menné, M., Andersen, K. E., Avnstorp, C., et al. (2011). Deodorants are the leading cause of allergic contact dermatitis to fragrance ingredients. Wiley Online Library. *Contact Dermatitis, 64*(5), 258–264. [cited 2024 Sept 27].
70. Callewaert, C., Hutapea, P., Van de Wiele, T., & Boon, N. (2014). Deodorants and antiperspirants affect the axillary bacterial community. *Archives of Dermatological Research, 306*(8), 701–710.
71. Tan, C. S. L. (2008). Make me beautiful. Japanese men's personal grooming products' consumption and brand selection. assumptionjournal.au.edu. *ABAC Journal, 28*. 2008 assumptionjournal.au.edu [Internet]. [cited 2024 Sept 27].
72. Draelos, Z. D. (2008). Understanding male skin care needs: The challenge of an evolving market. *Cosmetics and Toiletries, 123*(9) [cited 2024 Sept 27].
73. Sankaranarayanan, K. G., & Mekoth, N. (2014). Cosmetic consumption pattern among male college students: a cluster analytic segmentation approach clear *International Journal of Research in Commerce & Management, 5*(6).
74. Łopaciuk, A. (2013). *Global beauty industry trends in the 21st century*. Management, knowledge and learning international conference [cited 2024 Sept 27].
75. Ota, R., Ray, S. S., & Jena, S. (2023). A study on the growing interest of male towards grooming Industry. *International Journal of Research Publication and Reviews, 4*(3), 2386–2397.
76. Arnesdotter, E., Rogiers, V., Vanhaecke, T., & Vinken, M. (2021). An overview of current practices for regulatory risk assessment with lessons learnt from cosmetics in the European Union. *Critical Reviews in Toxicology, 51*(5), 395–417.
77. Steiner, K., & Florack, A. (2023). The influence of packaging color on consumer perceptions of healthfulness: A systematic review and theoretical framework. *Food, 12*(21), 3911.
78. Rahul, A., & Shushma, H. (2022). Analysis of factors influencing on consumer buying behavior on men's grooming products. *Neuro Quantology, 20*(19), 5105–5118.
79. Matin, A., Khoshtaria, T., & Todua, N. (2022). The impact of social media influencers on brand awareness, image and trust in their sponsored content: An empirical study from

Georgian social media users. *International Journal of Marketing, Communication and New Media, 10*(18), 88–114.

80. Yoo, H. J., & Song, E. (2021). Effects of personal hygiene habits on self-efficacy for preventing infection, infection-preventing hygiene behaviors, and product-purchasing behaviors. *Sustainability, 13*(17), 9483.
81. Amaral, N., Torelli, C., Keh, H. T., & Chiu, C. Y. (2019). Harley blow dryers and Dove Men body wash: The masculine-American stereotype and its effects on gendered marketing. *Advances in Consumer Research, 47*, 431–432.
82. Pounders, K. (2018). Are portrayals of female beauty in advertising finally changing? *Journal of Advertising Research, 58*(2), 133–137.
83. Hamshaw, R. J. T., & Gavin, J. (2022). Men's perspectives on their grooming practices and appearance concerns: A mixed methods study. *The Journal of Men's Studies, 30*(2), 251–270.
84. Hamshaw, R. J., & Gavin, J. (2022). Men's perspectives on their grooming practices and appearance concerns: A mixed methods study. *The Journal of Men's Studies, 30*(2), 251-270.
85. Duarte, P., Silva, S., & Carvalho, M. J. (2025) Is masculinity ruining the male skincare market? Comprehensive insights on consumer behavior in cosmetics. *Journal of Consumer Marketing 42*(2):159–173.
86. Khan, I., Dongping, H., Abdullah, M., Ahmad, Z., Ahmad Ghauri, T., & Ghazanfar, S. (2017). Men's attitude and motivation toward consumption of grooming products: A comparison of Chinese and Pakistani male consumers. *Cogent Business & Management, 4*(1), 1309783.
87. McNeill, L. S., & Douglas, K. (2011). Retailing masculinity: Gender expectations and social image of male grooming products in New Zealand. *Journal of retailing and Consumer Services, 18*(5), 448–454.
88. Chu, L., Yu, Y., Choi, J., Kim, Y., Xiong, M., Lee, Y., et al. (2022). Research on the skin concerns of consumers with sensitive skin and their attitude and behavior toward derma products. *Scientific and Social Research, 4*(4), 124–141.
89. Draelos, Z. D. (2012). Male skin and ingredients relevant to male skin care. *British Journal of Dermatology, 166*, 13–16.
90. Scheibling, C., & Lafrance, M. (2019). Man up but stay smooth: Hybrid masculinities in advertising for men's grooming products. *The Journal of Men's Studies, 27*(2), 222–239.
91. Kim, J., & Kim, M. (2022). Rise of social media influencers as a new marketing channel: Focusing on the roles of psychological well-being and perceived social responsibility among consumers. *International Journal of Environmental Research and Public Health, 19*(4), 2362.
92. Dwivedi, Y. K., Ismagilova, E., Hughes, D. L., Carlson, J., Filieri, R., Jacobson, J., et al. (2021). Setting the future of digital and social media marketing research: Perspectives and research propositions. *International Journal of Information Management, 59*, 102168.
93. Alenezi, S. (2024). *The impact of social media marketing on consumer buying behaviour: A case of Kuwaiti's male grooming retail market* [Doctoral dissertation, SOAS University of London].
94. Luebberding, S., Krueger, N., & Kerscher, M. (2014). Age-related changes in male skin: Quantitative evaluation of one hundred and fifty male subjects. *Skin Pharmacology and Physiology, 27*(1), 9–17.
95. Zouboulis, C. C., Baron, J. M., Böhm, M., Kippenberger, S., Kurzen, H., Reichrath, J., et al. (2008). Frontiers in sebaceous gland biology and pathology. *Experimental Dermatology, 17*(6), 542–551.
96. Fluhr, J. W., Darlenski, R., Angelova-Fischer, I., Tsankov, N., & Basketter, D. (2008). Skin irritation and sensitization: Mechanisms and new approaches for risk assessment. *Skin Pharmacology and Physiology, 21*(3), 124–135.
97. Bissett, D. L., Oblong, J. E., & Berge, C. A. (2005). Niacinamide. *Dermatologic Surgery, 31*(s1), 860–866.
98. Wiśniewska, J., Klasik-Ciszewska, S., & Duda-Grychtoł, K. (2023). Salicylic acid and its use in cosmetology. *Aesthetic Cosmetology and Medicine, 12*(3), 91–95.

99. Kumar, P., Verma, A., Ashique, S., Bhowmick, M., Mohanto, S., Singh, A., et al. (2024). Unlocking the role of herbal cosmeceutical in anti-ageing and skin ageing associated diseases. *Cutaneous and Ocular Toxicology, 43*(3), 211–226.
100. Baumann, L. (2007). Skin ageing and its treatment. *The Journal of Pathology, 211*(2), 241–251.
101. Papakonstantinou, E., Roth, M., & Karakiulakis, G. (2012). Hyaluronic acid: A key molecule in skin aging. *Dermato-Endocrinology, 4*(3), 253–258.
102. Green, A. C., Williams, G. M., Logan, V., & Strutton, G. M. (2011). Reduced melanoma after regular sunscreen use: Randomized trial follow-up. *Journal of Clinical Oncology, 29*(3), 257–263.
103. Dennis, L. K., Lowe, J. B., & Snetselaar, L. G. (2009). Tanning behaviour among young frequent tanners is related to attitudes and not lack of knowledge about the dangers. *Health Education Journal, 68*(3), 232–243.
104. Bassett, I. B., Barnetson, R. S. C., & Pannowitz, D. L. (1990). A comparative study of tea-tree oil versus benzoylperoxide in the treatment of acne. *Medical Journal of Australia, 153*(8), 455–458.
105. Budiman, I., Amirta, R., Yuliansyah, W. B. A., & Fatriasari, W. (2024). Activated carbon for cosmetics applications. In *Biomass-based cosmetics* (pp. 217–237). Springer Nature Singapore.
106. Thring, T. S., Hili, P., & Naughton, D. P. (2009). Anti-collagenase, anti-elastase and anti-oxidant activities of extracts from 21 plants. *BMC Complementary and Alternative Medicine, 9*(1), 27.
107. Morashti, J., An, Y., & Jang, H. (2022). A systematic literature review of sustainable packaging in supply chain management. *Sustainability, 14*(9), 4921.
108. Vrabič-Brodnjak, U., & Jestratijević, I. (2024). The future of baby cosmetics packaging and sustainable development: A look at sustainable materials and packaging innovations—A systematic review. *Sustainable Development, 32*(3), 2208–2222.
109. Bhardwaj, S., Nair, K., Tariq, M. U., Ahmad, A., & Chitnis, A. (2023). The state of research in green marketing: A bibliometric review from 2005 to 2022. *Sustainability, 15*(4), 2988.
110. Greener products: The making and marketing of sustainable brands – Al Iannuzzi – Google Books. [cited 2024 Sept 30].
111. Wagener, H. (2024). *Beautifully green: Beauty vloggers' assessments of sustainable cosmetic products: A qualitative content analysis* [Master's thesis, University of Twente].
112. Suphasomboon, T., & Vassanadumrongdee, S. (2023). Multi-stakeholder perspectives on sustainability transitions in the cosmetic industry. *Sustainable Production and Consumption, 38*, 225–240.
113. Suphasomboon, T., & Vassanadumrongdee, S. (2023). Multi-stakeholder perspectives on sustainability transitions in the cosmetic industry. *Sustainable Production and Consumption, 38*, 225-240.
114. Sasounian, R., Martinez, R. M., Lopes. A. M., Giarolla, J., Rosado, C., Magalhaes, W. V., Velasco, M. V., & Baby, A. R. (2024). Innovative approaches to an eco-friendly cosmetic industry: a review of sustainable ingredients. *Clean Technologies, 6*(1):176–198.
115. Kamwendo, A. R., & Maharaj, M. (2022). The preferences of consumers when selecting skin care products. *Journal of Contemporary Management, 19*(1), 82–106.
116. Guo, H. (2022). The preference for male cosmetics: An empirical analysis. *BCP Business & Management, 26*, 864–872.
117. Gomes, C., Silva, A. C., Marques, A. C., Sousa Lobo, J., & Amaral, M. H. (2020). Biotechnology applied to cosmetics and aesthetic medicines. *Cosmetics, 7*(2), 33.
118. Zhao, X., Zhang, X., & Liu, D. (2021). Collagen peptides and the related synthetic peptides: A review on improving skin health. *Journal of Functional Foods, 86*, 104680.
119. Goyal, N., & Jerold, F. (2021). Biocosmetics: Technological advances and future outlook. *Environmental Science and Pollution Research, 30*(10), 25148–25169.

120. Elsner, P. (2012). Overview and trends in male grooming. *British Journal of Dermatology, 166*, 2–5.
121. Ustymenko, R. (2023). Trends and innovations in cosmetic marketing. *Economics & Education, 8*(3), 12–17.
122. Vatiwutipong, P., Vachmanus, S., Noraset, T., & Tuarob, S. (2023). Artificial intelligence in cosmetic dermatology: A systematic literature review. *IEEE Access, 11*, 71407–71425.
123. Georgievskaya, A., Tlyachev, T., Danko, D., Chekanov, K., & Corstjens, H. (2023). How artificial intelligence adopts human biases: The case of cosmetic skincare industry. *AI and Ethics, 5*, 105.
124. Mao, P., Li, H., & Yu, Z. (2023). A review of skin-wearable sensors for non-invasive health monitoring applications. *Sensors, 23*(7), 3673.
125. Khan, N., Ray, R. L., Kassem, H. S., Hussain, S., Zhang, S., Khayyam, M., et al. (2021). Potential role of Technology innovation in transformation of sustainable food systems: A review. *Agriculture, 11*(10), 984.
126. Salvioni, L., Morelli, L., Ochoa, E., Labra, M., Fiandra, L., Palugan, L., et al. (2021). The emerging role of nanotechnology in skincare. *Advances in Colloid and Interface Science, 293*, 102437.
127. Halla, N., Fernandes, I. P., Heleno, S. A., Costa, P., Boucherit-Otmani, Z., Boucherit, K., et al. (2018). Cosmetics preservation: A review on present strategies. *Molecules, 23*(7), 1571.
128. Shivsharan, U., Raut, E. S., & Shaikh, Z. M. (2014). Packaging of cosmetics: A review. *Journal of Pharmaceutical Innovation, 3*, 286–293.
129. Chakraborty, A., Adhikary, S., Bhattacharya, S., Dutta, S., Chatterjee, S., Banerjee, D., et al. (2023). Pharmaceuticals and personal care products as emerging environmental contaminants: Prevalence, toxicity, and remedial approaches. ACS. *Chemical Health and Safety, 30*(6), 362–388. [cited 2024 Sept 30].
130. Abatan, A., Lottu, O. A., Ugwuanyi, E. D., Jacks, B. S., Sodiya, E. O., Daraojimba, A. I., et al. (2024). Sustainable packaging innovations and their impact on HSE practices in the FMCG industry. *Magna Scientia Advanced Research and Reviews, 10*(1), 379–391.
131. Thirupathi Vasuki, M., Kadirvel, V., & Pejavara, N. G. (2023). Smart packaging—An overview of concepts and applications in various food industries. *Food Bioengineering, 2*(1), 25–41.
132. Wang, Y., Ko, E., & Wang, H. (2022). Augmented reality (AR) app use in the beauty product industry and consumer purchase intention. *Asia Pacific Journal of Marketing and Logistics, 34*(1), 110–131.
133. Tarnanidis, T. (2024). Exploring the impact of mobile marketing strategies on consumer behavior: A comprehensive analysis. *International Journal of Information, Business and Management, 16*(2), 1–17. (ISSN 2218-046X). [cited 2024 Sept 30].
134. Husnain, M., & Toor, A. (2017). The impact of social network marketing on consumer purchase intention in Pakistan: Consumer engagement as a mediator. *Asian Journal of Business and Accounting, 10*(1), 167–199. [cited 2024 Sept 30].
135. Martins, A. M., & Marto, J. M. (2023). A sustainable life cycle for cosmetics: From design and development to post-use phase. *Sustainable Chemistry and Pharmacy, 35*, 101178.
136. Dube, M., & Dube, S. (2023). Towards sustainable color cosmetics packaging. *Cosmetics, 10*(5), 139. [cited 2024 Sept 30].

Fragrance: Technical and Creative Aspects

Stephen Herman

Abstract This chapter begins with the recent history of fragrance. It continues with the technical aspects of fragrance creation, including the development of synthetic aroma chemicals, the importance of natural products, and the role of fixatives. It discusses the methodologies used in fragrance formulation, such as the Jean Carles method, and explores the chemistry of aroma compounds and the role of green chemistry and upcycling in fragrance production. It addresses the stability of fragrances in diverse applications, the role of encapsulation and profragrances in controlled release, and the importance of safety evaluations. Finally, it considers the influence of digital tools and artificial intelligence on the fragrance industry, marking a significant shift in how fragrances are created and marketed.

Keywords Aroma chemicals · Essential oils · Extraction processes · Fixatives · Green chemistry · Upcycling · Controlled release · Profragrances · Safety evaluations · Digital tools

Modern History of Fragrance

The use of fragrant materials can be traced back to prehistory. Archaeological evidence is extensive, including evidence in ancient Mesopotamia, Egypt, Cyprus, the Indus Valley, and China. Balasubramanian [1] offers a good introduction to the story of ancient fragrances. In the early historic period, Greece and Rome used aromatic materials extensively for ceremonial purposes, and in personal use in a limited way for wealthy individuals.

As historically interesting, the story of fragrance is in these remote times, it was with the development of organic chemistry and the growth of an economically

S. Herman (✉)
The University of Cincinnati, College of Pharmacy, Cincinnati, OH, USA
e-mail: hermanst@ucmail.uc.edu

K. Alexander, P. Romanowski (eds.), *Cosmetic Science*, AAPS Introductions in the Pharmaceutical Sciences 15, https://doi.org/10.1007/978-3-032-03596-7_9

Table 1 Landmarks in the history of fragrance-related science

1834	Cinnamic aldehyde isolated from cinnamon oil
1837	Benzaldehyde isolated from bitter almond oil
1844	Methyl salicylate identified in wintergreen oil
1856	Cinnamic aldehyde synthesized
1863	Benzaldehyde synthesized
1868	Coumarin synthesized
1876	Vanillin synthesized
1884	D-limonene identified
1888	Discovery of nitro musks
1889	Discovery of citronellal
1919	Linalool synthesized
1923	Amyl cinnamaldehyde discovered
1934	Muscone synthesized
1939	Leopold Ruzicka wins Nobel Prize for Chemistry
1966	Calone discovered
1971	Synthesis of α and β damascene

Fig. 1 Calone

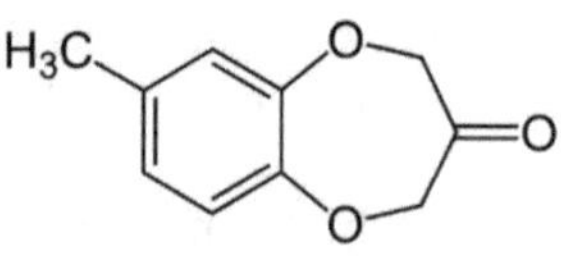

empowered middle class that the modern uses of fragrance began. Table 1 shows how the growth of science made the foundation of modern perfumery possible.

One of the last entries on the list, Calone, illustrates the impact of chemistry on the creative process of perfumery. Calone has the structure shown in Fig. 1 [2]. Calone is an unusual chemical compound that has an intense marine ozonic note with slight floral and fruit overtones. It has been used in fragrances since the 1980s, but it was not until the 1990s that it became the dominant note in several successful perfumes like Davidoff Cool Water and Giorgio Armani Aqua di Gio that firmly established the marine trend, which in turn was a key to the unisex movement.

Synthetic materials were first used in a fragrance in Fougere Royale in 1882, but it was not a particularly remarkable fragrance. The first use of aroma chemicals in a great fragrance was Jicky, created in 1889 by Aimé Guerlain. Jicky laid the foundation of great French perfumery. It marked the beginning of emotive perfumery, which no longer attempted to imitate the scent of flowers but sought instead to arouse emotions. Packaging was important for its success, and Jicky is designed in the shape of a nineteenth century medicine bottle with a stopper shaped like a champagne cork. The most remarkable tribute to Jicky is that, in updated form, it is still available today.

Around 1905, François Coty began to create the first perfume company as we now know it. His business model was to create an affordable luxury, and he formulated excellent fragrances with himself as the perfumer, used bottles from upscale sources like René Lalique Glass, and priced them within the aspirational range of a mass market. Coty was a great perfumer, and his breakthrough creation was L'Origan in 1905. But his most remarkable achievement was Chypre de Coty in 1917, where he created an entire fragrance category. Chypre was crucial to the development of the modern fragrance industry.

A pinnacle of the history of fragrance came in 1921 with Chanel No 5, after over a century, one of the best-selling fragrances in the world. Chanel No 5 was created by master perfumer Ernest Beaux and was revolutionary for its overdose of aldehydes. Its simple bottle and clean label were the opposite of the Lalique bottles favored by Coty. For a technical analysis of the fragrance beginning with its Russian origins as Rallet No 1, see Kraft et al. [3]. For a complete account of the fascinating life of Coco Chanel and her most famous fragrance, see Mazzeo [4].

France has long been the heart of the fragrance world, but World War II shifted some of the important activity to the United States, particularly the New York City area. The key locations of fragrance creation are now Grasse, Paris, and New York.

Beginning in the early 1990s, fragrance use changed from products like perfume and cologne to personal care products like bath gels and environmental applications like candles and diffusers. Fragrance types expanded in the use of food ingredients to create the gourmand category. Unisex marine and ozonic fragrances became popular, and niche brands offering personalized fragrances expanded the creative bounds of fragrance types. Individual creative geniuses like François Coty and Ernest Beaux have largely been replaced by Big Data and Algorithmic Perfumery and AI-guided scent creation [5].

Creation and Duplication

In a world where only natural products existed as fragrance raw materials, perfumers used a laborious and time-consuming trial-and-error approach to the creative process. This was fine for the formulation of old-fashioned floral and citrus fragrances but became ineffective when hundreds of new aroma chemicals became part of the palette. The new approach was codified by Jean Carles. Many years later, he explained his approach in three papers on "A Method of Creation in Perfumery" [6].

The first step for a perfumer was to develop an olfactory memory. The trainee learns contrasting odors, then materials within the same odor families. Following that are the important concepts of fragrance creation: volatility, accords, and building a fragrance pyramid from the bottom up.

The characteristic odor of a single material is termed a note. Mixtures of two or more materials, having a unified olfactory theme, are termed accords. Jean Carles started with a simple base accord. Materials of medium volatility are used to modify this base accord. Some pleasant, highly volatile notes are added to create the top note, which is the sparkle that creates the immediate impression of the fragrance.

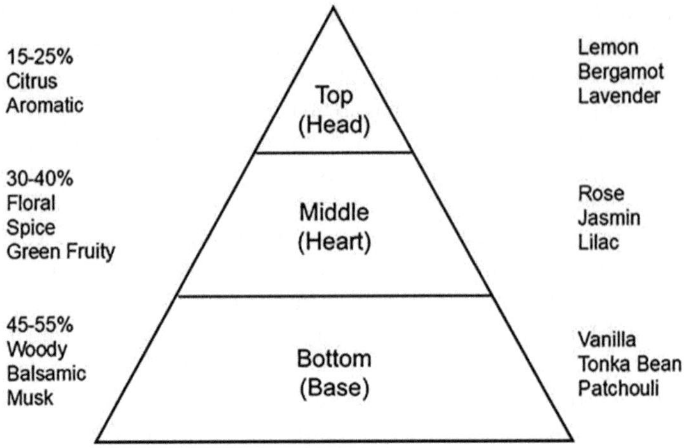

Fig. 2 Fragrance triangle

For a student perfumer, following the development of olfactory memory, the volatility of perfume materials is studied. Finally, accords are created for the top, middle, and bottom, and to this olfactive skeleton a variety of other materials are added to give roundness and complexity to the scent. This construction is commonly expressed as a fragrance triangle (Fig. 2), shown with some odor descriptors and characteristic ingredients. The fragrance out of the bottle contains all these notes: during use, the fragrance evolves as materials evaporate in order of their volatility. A key aspect of a great fragrance is that as it evolves over time, it does so in a logical manner: a citrus type does not change into a floral.

Linear fragrances can be made that do not evolve substantially over time. A linear fragrance maintains a consistent scent from the beginning to the end of its use. An important example of a classic linear fragrance is Trésor by Lancôme and Estee Lauder's White Linen. More recent examples are the Tom Ford Rose Trilogy series and APOM Pour Homme by Maison Francis Kurkdjian.

Fixatives are a vital component of fragrance, a material that reduces the volatility of other fragrance ingredients. Fixatives extend the life of the fragrance, an important property for consumer satisfaction. Fixatives typically have lower volatility than the other materials in the formula. Some common fixatives are musks, benzyl benzoate, terpinoid gums like Hercolyn D (methyl hydrogenated rosinate), and olibanum (frankincense).

Large fragrance molecules with low volatilities can be used as base notes, thus functioning as fixatives. A perfume fixative functions for all notes in a perfume but will have a greater fractional effect for middle notes rather than top notes.

Synthetics are the most common fixative materials today. Ambrox, a synthetic version of Ambergris; Benzoin, a resin that slows the dispersion of fragrance molecules and other fragrance materials in the air; and Galaxolide, a synthetic musk, are often used. Another well-known fixative is Iso E super, which dates to the 1960s and is known for its green, earthy smell. For an approach to sustainable ambrox, see Martínez-Guido et al. [7].

Aggarwal [8] established thermogravimetric analysis (TGA) to study the evaporation of methyl benzoate. Phang [9] used TGA to test benzophenone and cinnamic acid. It was determined that when an evaporation process takes place from a constant surface area, the rate of evaporation is a zero-order reaction, meaning the chemical change is independent of concentration. The coefficient of evaporation can be obtained by dividing the rate of evaporation by the cross-sectional surface area of the sample cell.

Studies of thermal analysis are helpful in determining the activation energy of evaporation values to determine thermal stability, which are critical parameters to analyze when looking at fragrance fixatives. These studies are more helpful for single substances, but for fragrance materials mixed with perfume fixatives, further study is needed. In a fragrance and fixative mixture, the situation can become more complicated as several factors come into play, such as vapor pressure, molecular weight, temperature, and surface area, as shown by Lu [10].

Fixative efficacy can be studied experimentally via thermogravimetric analysis to obtain information such as heat of vaporization and activation energy. These values are important in analyzing the strength of various fixatives and can assist a formulator in fixative choice. Complex mixtures of one or more fragrances along with fixatives can also be analyzed using these methods.

A study of evaporation via thermogravimetric analysis, concentrating on D-Limonene by Lewczyk et al. [11], found that there was no sharp boundary between the liquid and gas phase. This work provides experimental evidence that corroborates previous reports that vapor pressures determined from TGA experiments inherently contain kinetic parameters as well. Changes in the carrier gas identity and flow rate affect the thickness of the boundary layer above the evaporating surface and alter the kinetics of the process. The additional energy is hypothesized to be due to the diffusion of molecules through the boundary layer above the evaporating surface. Additional energy is hypothesized to be required due to the diffusion of molecules through the boundary layer above the evaporating surface. This effect can be seen in Fig. 3.

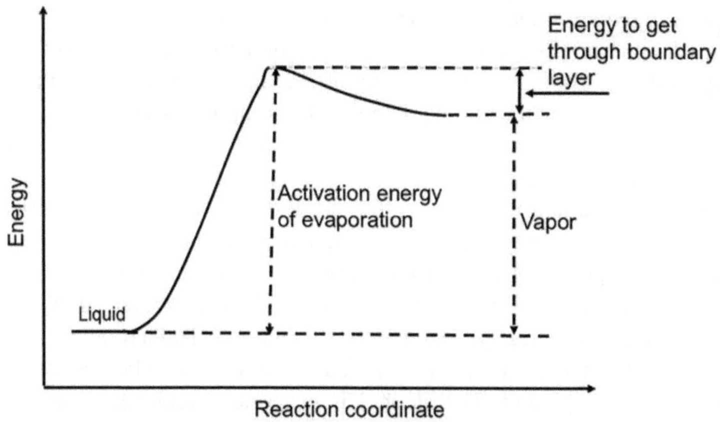

Fig. 3 Fragrance evaporation energy

Natural Products, Distillation, and Essential Oils

Before organic chemistry made synthetic aroma chemicals available, all fragrances were necessarily composed solely of natural products. After over a century of aroma chemical development, natural products are still critical for most fragrances. There are trace components of natural products that have never been exactly replicated by synthetic means. Even if it were possible to synthesize a natural component, it might be too difficult or expensive to be a practical alternative.

For a brief account of extraction methods, see Mishra [12]. Enfleurage is an extraction process rarely used today, even though it results in excellent products. This process is described in Rakthaworn et al. [13]. It can be used on heat-sensitive materials, but it is too time-consuming, labor-intensive, and costly. A framed glass plate is spread with animal fat, and flower petals are applied. The essential oil migrates from the petals to the plate, and as the petals are depleted in a few days, new ones are applied. When the fat is totally saturated with aromatic oil, it is called pomade. If the pomade is dissolved in alcohol, it is possible to separate the aromatic material from the waxy components. The alcohol is driven off, and the final product is termed the absolute. The absolute, which is alcohol soluble, is the finest and most expensive of all aromatic products.

Expression or cold pressing is used on fruit to yield oils. Cold pressing was discussed in a paper by Cakaloglu [14]. It involves puncturing the skins of the whole fruit or the fruit peel and mechanically pressing the oil out. A small amount of juice is also extracted, which is separated from the oil with a centrifuge. Orange, grapefruit, lemon, lime, tangerine, mandarin, and bergamot can be extracted by this method. It is usually very cost-effective since the fruits used are high-volume flavor materials.

Orange oil may contain a high level of limonene, a terpene that may be undesirable. Limonene has little olfactory value and easily oxidizes into sensitizing materials. Distillation removes most of the limonene in a process called folding. Commonly, ten drums of orange oil are processed and nine drums of limonene removed, with the resultant product termed orange oil 10X. The folded oil has a stronger odor value and can reduce stability issues in many applications.

Steam distillation (Fig. 4) is a common extraction process for aroma compounds, working well for most heat-tolerant materials. Adding steam to the biomass creates an azeotrope that lowers the boiling points of the compounds being extracted. Steam vaporizes the essential oils present. A condenser cools the liquid, which is separated into oil and water phases.

For materials that are too sensitive for steam distillation, solvent extraction can be used. Plant matter is placed in a static extractor on perforated trays, and a benzene-free nonpolar solvent is passed through the material 3–15 times. Heat is applied to remove the solvent, and the material left behind is referred to as concrete.

The concrete can be gently heated with ethyl alcohol, and a rotative filter is used to remove most of the waxes and terpenes. The solution is chilled, and the remaining waxes are precipitated. The alcohol is evaporated under a vacuum, and the

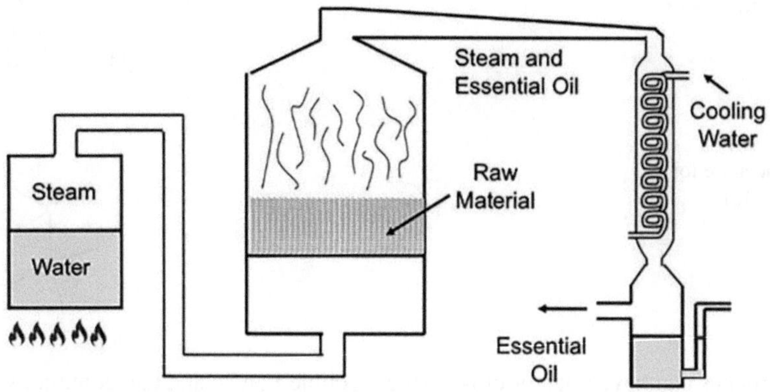

Fig. 4 Steam distillation

resulting product is termed the absolute. The absolutes of certain flowers, such as jasmine, rose, and tuberose, are among the most beautiful and expensive components of fine fragrances.

An even more delicate process uses CO_2 supercritical extraction, explained by Moreira [15]. CO_2 is easily compressed into a supercritical liquid phase at ambient temperatures and has lipophilic solvent properties. The solubility of supercritical liquids changes with temperature and pressure, which allows the fine-tuning of the extraction. A second solvent can also be added to further refine the process. Dissolved waxes can be present in the product, like what results from standard solvent extraction. Many unique olfactive products can be achieved by this process.

There are several innovative techniques being used for essential oil extraction, primarily on a laboratory scale. Examples are microwave extraction, turbo distillation, ultrasound-assisted extraction, and gravity-driven hydrodiffusion. See Bousbia [16] for details.

It is a common misconception among consumers that natural products are safer than synthetic chemicals, but this is not necessarily true. Natural products are complex mixtures that can include skin irritants, phototoxins, and other undesirable materials.

Isolates, useful individual materials, can be extracted from essential oils. It is possible to obtain Eugenol from clove leaf oil or eucalyptol from eucalyptus oil. Orange oil contains several olfactive components such as octanal, decanal, valencene, and sinensal. There can be difficulty in separating some of these products in high yield if the boiling points are too close for distillation to be effective.

Many plants biochemically synthesize a family of chemicals called terpenes to use as building blocks. Many aroma chemicals can be extracted from these materials at low cost, particularly if they are large-production food crops. Head-to-tail unions of isoprene yield a wide variety of terpenes (Fig. 5).

Common historical names have endured for terpenes: $n = 2$ is a monoterpene, $n = 3$ is a sesquiterpene, $n = 4$ is a diterpene, and $n = 6$ is a triterpene.

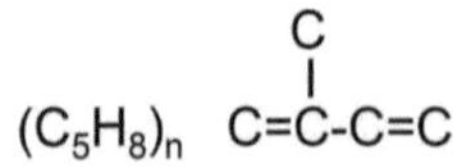

Fig. 5 Isoprene

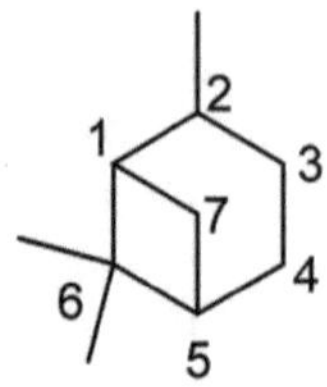

Fig. 6 IUPAC nomenclature for pinene [18]

Terpenes can have open or closed rings, double bonds, and can contain hydroxy or carboxy groups. Many transitions between different structures occur easily, so it is easy to synthesize numerous derivatives. For a fuller discussion of terpenes, see Silvestre and Gandini [17].

The stereochemistry of natural products can be complex and critical for their properties. The flat depiction of terpenes, particularly a puckered bicyclic terpene like pinene, does not do justice to the complexity of its three-dimensional structure. The pinene family of terpenes consists of two structural isomers, α- and β-pinene. Both exist as (+)- and (−)-enantiomers. The IUPAC nomenclature for pinene is shown in Fig. 6. The complexity of pinene chemistry can be explored in Nyamwihura and Victor [18].

The subject of essential oils is vast. The classic guide is Guenther [19] in six large volumes. It is old but has never been superseded.

Aroma Chemicals

Aroma raw materials are all organic compounds, so their chemistry is based on the properties of carbon. The size of aroma molecules ranges from C5 to C17. Helvetolide, with a molecular weight of 284, is the largest known fragrance ingredient. Nitrogen, oxygen, and sulfur are common heteroatoms. These elements are also known as osmophores, groups that, when present in a molecule, are responsible for its odor. Some minimal polarity is necessary to interact with the odor receptors in the nose.

Esters are the largest family of aroma chemicals. They are produced by the reaction of a carboxylic acid with an alcohol. At high pH and elevated temperature, the reverse of the esterification reaction occurs, which is termed hydrolysis. Hydrolysis results in a total change in odor quality. Hydrolysis also occurs, but at a reduced rate, at low pH. Low-molecular-weight esters often have a fruity odor, examples being the odors of apples (methyl butyrate) or bananas (3-methylbutyl acetate).

Alcohols contain a hydroxyl (OH) group and can be aliphatic or aromatic. If the hydroxyl group is directly attached to an aromatic ring, it is a phenol. They can be

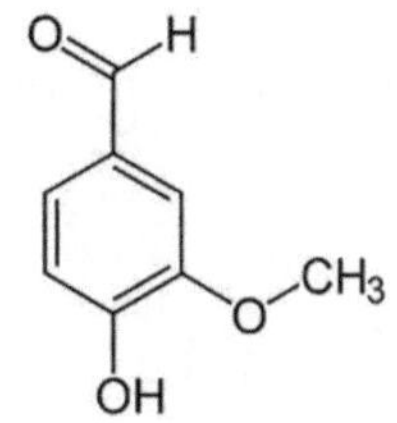

Fig. 7 Vanillin [20]

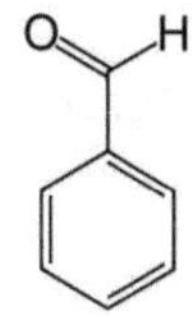

Fig. 8 Benzaldehyde [21]

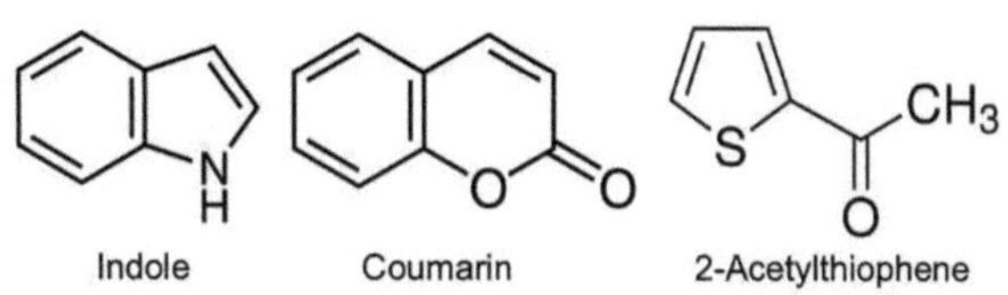

Fig. 9 Heterocyclics

very unstable and can create severe color changes, and are very sensitive to oxidation. An important example is vanillin (Fig. 7), a major component of vanilla.

Aldehydes and ketones are closely related. Aldehydes can be aromatic or aliphatic and are easy to oxidize or reduce. Benzaldehyde (Fig. 8), a cherry/almond scent, is particularly reactive because the functional group is attached directly to the aromatic ring and there are no other reactive sites on the molecule. Benzaldehyde readily exhibits autoxidation to benzoic acid, an odorless white powder.

Aldehydes can react with a primary amine to form a Schiff base. In perfumery, an important example is hydroxycitronellal and methyl anthranilate reacting to make Aurantiol. The reaction can take place slowly if both components are present separately in a fragrance, contributing to the natural ageing of the oil. Alternatively, the Schiff base can be created by a reflux reaction, and the resultant product used as an ingredient. Ketones are much less reactive than aldehydes but are easily reduced to secondary alcohols.

Heterocyclics are cyclic compounds that contain one or more heteroatoms: oxygen, nitrogen, or sulfur. They are identified by the prefixes Aza- for nitrogen, Oxa- for oxygen, and Thia- for sulfur. They are typically five- or six-member rings, and rings can be fused. Rings can be aromatic or nonaromatic. Some common heterocyclic structures are shown in Fig. 9.

Nitriles contain triple-bonded nitrogen, $C \equiv N$. Nitriles occur naturally in plant and animal sources, for example, fruit pits, especially almonds. A nitrile sometimes has an odor like its corresponding aldehyde, but nitriles are more stable than aldehydes to oxidation and extremes of pH.

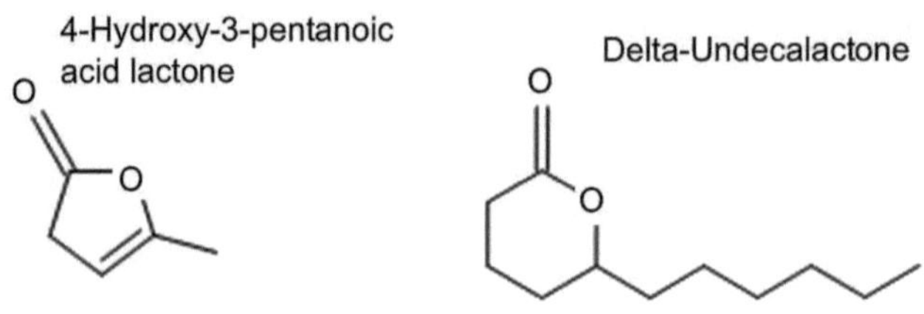

Fig. 10 Lactones

Fig. 11 Musks [22]

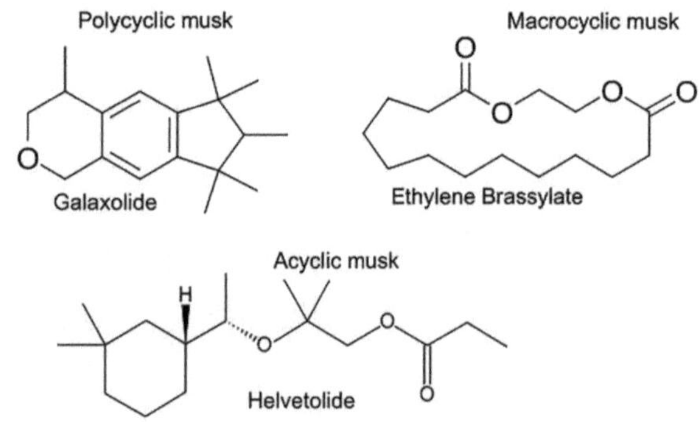

Lactones are cyclic esters. Common examples are γ or δ hydroxy acids forming 5- or 6-member rings by intramolecular esterification. Figure 10 shows some examples of lactones used in perfumery. Lactones react under the same conditions as noncyclic esters.

Originally, musk came from the male musk deer. Naturally derived musk is rare and expensive, and animal products are often banned. Many synthetic compounds are made to satisfy the demands of perfumery. The first successful replacements were nitro musks. Musk xylene, musk ketone, and musk ambrette were historically important but experienced regulatory problems. More recent molecules have been polycyclic musks, macrocyclic musks, and acyclic musks. Examples of each are shown in Fig. 11.

The study of aroma chemicals is vast. The interested reader is directed to Arctander [23] for the classic reference.

Synthesis, Green Chemistry, Fermentation, and Upcycling

Many reactions used in aroma chemical synthesis are easily found in organic chemistry textbooks, such as the Diels–Alder Reaction, Friedel-Crafts Reaction, Aldol Condensation, and Perkins Condensation. We will limit our discussion to one historic example, the Reimer–Tiemann Reaction and the successful synthesis of vanillin, and the use of BINAP to create stereospecific molecules. The specific example

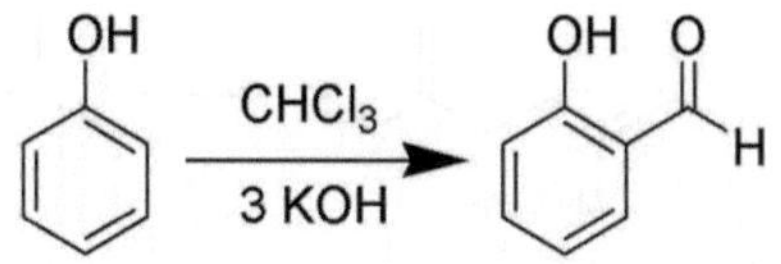

Fig. 12 Reimer–Tiemann reaction. (Public domain)

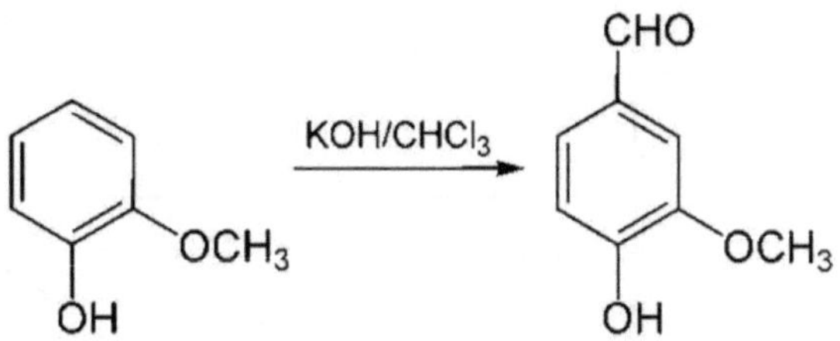

Fig. 13 Vanillin synthesis. (Public domain)

will be how Ryōji Noyori and his co-workers used rhodium complexes of BINAP for the synthesis of (−)-menthol.

One of the most important historic syntheses is the Reimer–Tiemann Reaction (Fig. 12). The Reimer–Tiemann Reaction occurs when phenol is treated with $CHCl_3$ in the presence of KOH or NaOH. An aldehyde group (−CHO) is introduced at the ortho position of the benzene ring, leading to the formation of o-hydroxybenzaldehyde.

In the late nineteenth century, natural vanillin was rare and expensive, making it an important target for synthesis. Haarmann's student friend Wilhelm Kubel had worked with tree sap and accidentally noticed a pleasant vanilla scent during his experiments. Building on this work, Haarmann and Tiemann successfully synthesized vanillin in 1874. With the help of Karl Reimer, Tiemann found a way to produce synthetic vanillin on an industrial scale using eugenol from clove oil. The synthesis of vanillin is shown in Fig. 13. Ferdinand Tiemann started a company in 1874 based on his synthesis of vanillin and later a sweet violet compound (Jonon). In 1882, he became a Professor at the University of Berlin. His company is now part of Symrise.

The stereospecific effects of menthol are significant, so the synthesis must be carefully devised to avoid a racemic mixture or the incorrect stereoisomer. One important method involves the use of BINAP (Fig. 14). It was developed by Professor Ryoji Noyori, who shared the Nobel Prize in Chemistry in 2001 [24]. Noyori used asymmetric hydrogenation using catalyst complexes of rhodium with the BINAP ligand. The synthesis is shown in Fig. 14, with the step creating the stereospecific product clearly indicated.

Using more benign and environmentally friendly processes is a major trend in the creation of aroma chemicals. Every manufacturing process must now feature the application of green chemistry in the synthesis of new and existing ingredients. Fermentation has also become an important source of aroma chemicals. Upcycling uses waste as the feedstock for fragrance ingredients, and entire commercial fragrances are being created solely from upcycled materials.

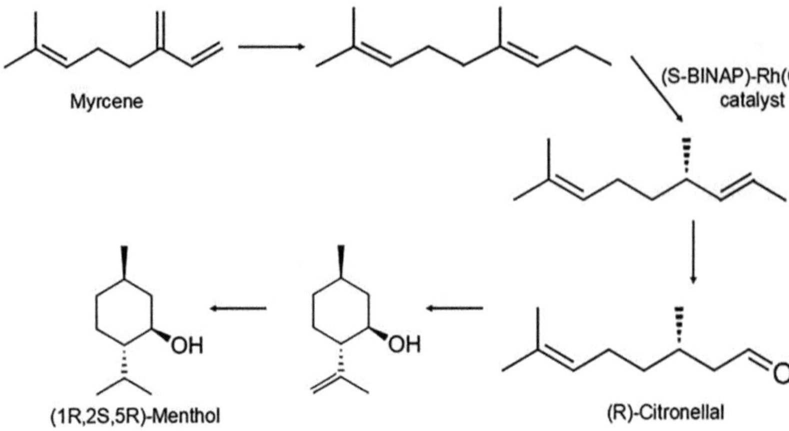

Fig. 14 Stereospecific synthesis of menthol

Green Chemistry began with a book authored by John Warner and Paul Anastas with 12 principles to reduce or eliminate hazardous substances in the manufacture of chemical products [25]. They are shown in Fig. 15.

Virtually every manufacturer of aroma chemicals applies these principles. The most common quantitative value is the E-Factor, the mass ratio of waste to finished product. A metric tool, Green Motion [26], can assess the effectiveness of the methods used on a numerical scale. Green Motion assesses the health, safety, and environmental impact of manufactured ingredients. It quantifies the following: raw material, solvents, hazard and toxicity of reagents, reaction, process, hazard and toxicity of finished product, and waste.

More recently, IFRA (International Fragrance Association) developed the IFRA Green Chemistry Compass [27]. It aligns the 12 Principles of Green Chemistry with the IFRA-IOFI Sustainability Charter: responsible sourcing, environmental footprint and climate change, well-being of employees, product safety, transparency, and partnerships. An Excel Workbook is available that calculates based on eight questions that cover ingredients, processes, waste, energy, catalysts, and solvents. This is summarized in Fig. 16.

Biologically based synthesis of aroma chemicals can rely on microbial processes (fermentation) or on bioconversions of natural precursors using microbial cells or enzymes (biocatalysis). Fermentation can provide products characterized as "natural," which is a valuable designation. There are many potential biomass substances to use as starting materials, such as corn or sugar cane. Natural biochemical pathways can be studied to provide genes and enzymes that make desirable molecules. To make an economically viable product, it is necessary to construct efficient metabolic pathways made of multiple genes that function in a cascade. To be fit for commercialization, the organism must make the molecule at a much higher concentration than what is found naturally in a plant.

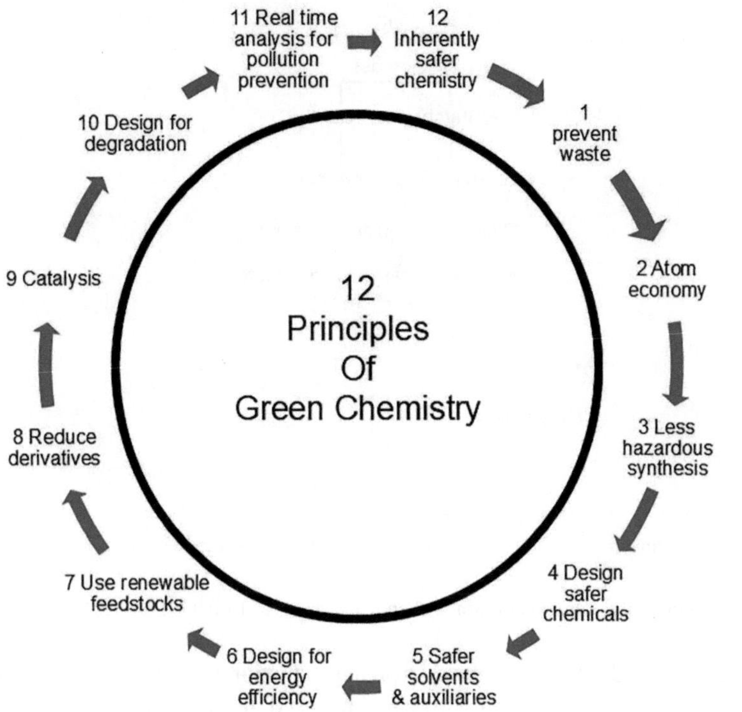

Fig. 15 Green chemistry

Principles of Green Chemistry

IFRA-IOFI Sustainability Charter Pillar	1. Waste Prevention	2. Atom Economy	3. Less Hazardous Chemical Synthesis	4.Designing Safer Chemicals	5. Safer Solvents & Auxiliaries	6. Design for Energy Efficiency	7. Use of Renewable Feedstocks	8. Reduce Derivatives	9. Catalysis	10. Design For Degradati on	11. Real Time Pollution Prevention	12. Safer Chemistry for Accident Prevention
Responsible Sourcing							Question 1					
Environmental Footprint and Climate Change	Question 4					Question 2			Question 3	Question 5		
Well-Being of Employees			Question 6		Question 7							Question 6
Product Safety				Question 8								
Transparency and Partnerships												

Fig. 16 Green chemistry applied to fragrance

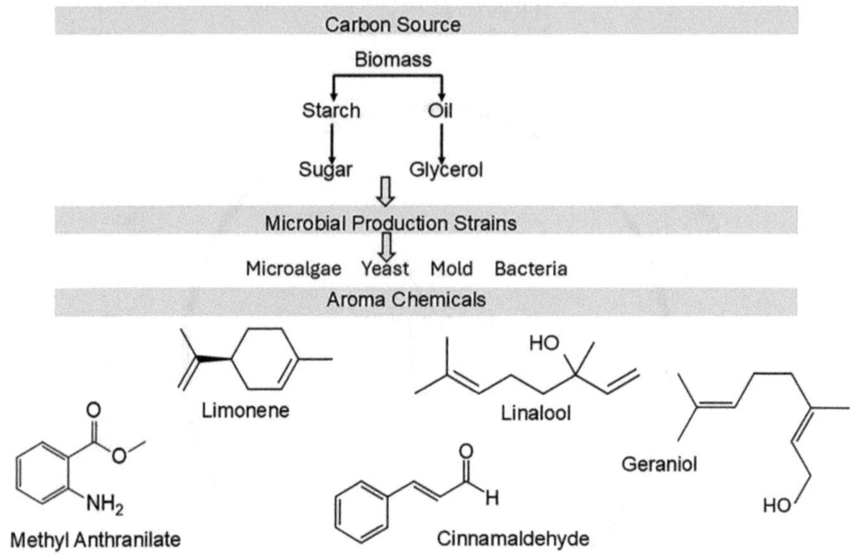

Fig. 17 Microbial production of aroma chemicals. (Derived from [29])

A generalized scheme for the microbial production of aroma chemicals is shown in Fig. 17. Examples of microorganisms used to make aroma chemicals are γ-Decalactone (a peach scent) from *Yarrowia lipolytica* (a yeast) and benzaldehyde (an almond scent) from *Ischnoderma benzoinum* (a fungus) [28].

Using raw materials in a circular process is environmentally friendly [30], which is an extremely important trend. It can be illustrated by turpentine. It begins with a pine tree plantation that provides raw materials for a paper mill. The intended product is paper, but an important by-product is crude sulfate turpentine. The crude sulfate turpentine can then be subjected to a desulfurization process and fractionization. The material can then be chemically manipulated into a variety of aroma chemicals such as linalool, citronellol, and myrcene. The plantation land is re-facilitated, new pine seedlings are planted, and the circle is completed.

Plants are an important source of aroma chemicals, but their active compounds are often present in minor quantities, thus making their isolation expensive. Additionally, the supply of natural products is often subject to change and can create severe shortages. Most fragrance suppliers have partnered with biotech firms to produce novel raw materials or existing materials manufactured in new ways.

Upcycled ingredients are derived from waste products or byproducts of other industries, reducing waste and promoting resource efficiency. Upcycling is another popular source of raw materials, and entire fragrance formulas have been made using materials sourced from waste. Examples of materials that have been upcycled include leftover apple juice pulp, turmeric leaf, and wastewater from cinnamon bark distillation. Other useful waste streams come from the overproduction of farm products such as spinach; the pulp and residue remaining after production, such as that of blueberries or cranberries, have been crushed for juice extraction. One of the first fragrances totally constructed of upcycled materials was I Am Trash Les Fleurs Du Dechet by Etat Libre d'Orange. Its composition included rose petals already subject to extraction, previously distilled sandalwood chips, and leftover apples from the food industry.

Applications: Hydroalcoholics, Emulsions, and Surfactant Systems

Hydroalcoholics are typically in the form of perfume, cologne, and aftershave. The basic ingredients are fragrance oil, alcohol, and water. Color, chelating agents, antioxidants, and ultraviolet (UV) absorbers are often added. The alcohol is composed of ethanol, water, and a denaturant. A composition of approximately 95% alcohol and 5% water is usually used, since it is the low-boiling azeotrope. The denaturant is a poison added to discourage drinking the alcohol. Denaturing exempts alcohol from the significant taxes imposed on the ingestible form. The denaturant is selected from materials that cannot readily be distilled out.

Aftershaves are a common and low-cost delivery system for male fragrances. The alcohol level is limited to 60–65% to avoid stinging, since minor cuts and abrasions often occur during shaving. Emollient oils such as glycerin and propylene glycol are added for sensory effects, and healing agents for cuts.

Fragrances are incorporated in a wide range of products, from eau de toilette to candles, laundry detergent, plastic toys, and toilet paper. In each instance, the effect of fragrance components on the application must be considered and can range from a color change to a viscosity drop among many other possibilities.

Since aroma perception is clearly important, changes in aroma must be addressed when a fragrance is incorporated in a base. Saunders studied how the composition of fragrance vapor is always distorted by unequal association with a substrate [31]. He explained a method where the distortion can be removed by a correction factor for each constituent. A fragrance mixture was placed in a base, and the composition of the headspace was measured. In Table 2, a = headspace of fragrance, b = headspace over fragranced base, c = a/b, d = original fragrance formula, and e = adjusted fragrance formula. Restated, if your fragrance formula is d and you want it to smell the same in another medium, your formula becomes e. Of course, the ratios change when different substrates are involved.

An emulsion has water, oil, liquid crystal structures, and interfaces where individual fragrance materials can partition based on chemical and physical properties. Polarity is a key property determining where an aroma molecule will partition in an emulsion. One metric is the solubility parameter, first proposed by Hildebrand in 1936, where the solubility parameter is equal to the square root of the heat of vaporization of the compound divided by its molar volume [32]:

Table 2 Adjusting fragrance for substrate (from [31])

	a	b	c	d	e
Limonene	31.9	15.9	2.0	2.0	4.0
Linalool	13.2	12.9	1.0	2.0	2.0
Linalyl acetate	12.3	7.7	1.6	2.0	3.3
PEA	6.2	22.6	0.3	3.5	1.1
Benzyl acetate	11.6	18.7	0.6	2.1	1.3

$$\delta = \left(\frac{\Delta H - RT}{Vm} \right)^{1/2}$$

A more recent measure of polarity is the Hanson Solubility Parameter (HSP) [33]. Hansen considers Dispersion, Polar, and Hydrogen bonding parameters in his calculations. These values for a molecule can be placed in a three-dimensional space, and the difference between their solubility properties is determined. The closer two molecules are in this "Hansen space," the more compatible they will be.

CLogP is the most common measure of aroma chemical polarity. P is the water/octanol partition component, log means the logarithm (which makes a more linear relationship), and c means it is calculated rather than obtained through wet chemistry. Most aroma chemical suppliers provide CLogP with their product specifications.

Another useful measure in determining the compatibility of molecules is the Effective Alkane Carbon Number (EACN), which is a measure of how one molecule views another [34]. A molecule having six carbons may have an EACN of 6, or it may not. For simple alkyl chains, it is easy: for hexane, heptane, octane, just use the number of carbon atoms in the oil, 6, 7, 8.

Cyclohexane has six carbons but behaves as if it has just three carbons. Benzene also has six carbon atoms but behaves like an oil with zero carbons. For a mixture of two oils, it is easy to form a new EACN value using the weighted average of the two values, and this can be extended to mixtures with more ingredients.

Now, we can see what happens to aroma molecules in a base. Figure 18 shows where some typical aroma molecules will be positioned at an interface based on polarity [35].

The concept of polarity easily explains the partitioning into different phases of aroma chemicals. In a typical water-in-oil emulsion (Fig. 19), aroma chemicals can migrate to the center of the micelles, into the water/oil interface, into the water phase for some especially polar materials, or be incorporated into the liquid crystal structures. This uneven distribution of the aroma molecules will necessarily affect the perceived odor in the air over the base.

In a surfactant system, the surfactant molecules form spherical (Fig. 20) or worm-like micelles (Fig. 21) [37].

Fragrance oils contain solvents and solubilizers as well as aroma chemicals, and they have an important role in how fragrances are incorporated in finished products.

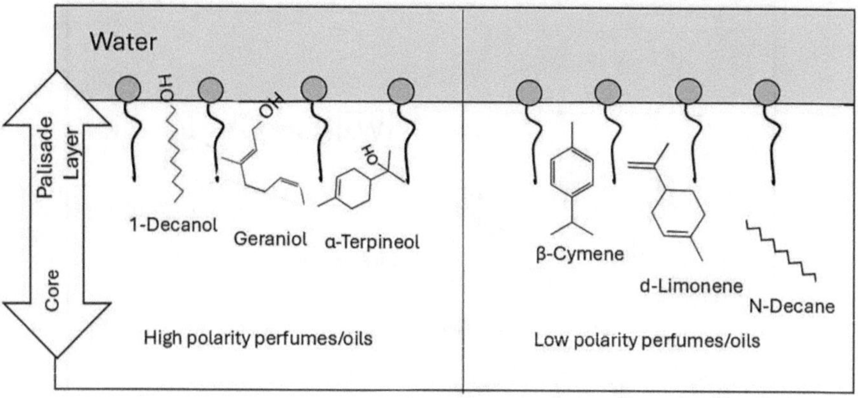

Fig. 18 Fragrances and interfaces. (Derived from [35])

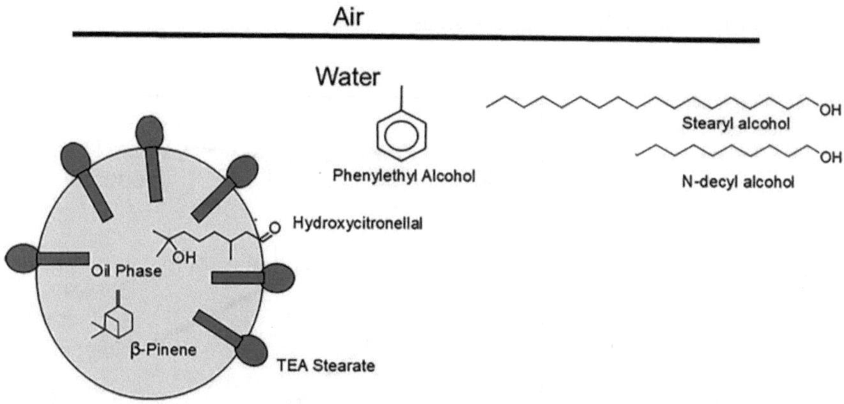

Fig. 19 Fragrance in emulsion system. (Based on [36])

Dipropylene glycol is polar and will help a fragrance dissolve clearly into a shampoo. Dioctyl adipate is a nonpolar and a possible choice for a fragrance in a candle. If a fragrance needs added water solubility, a nonionic surfactant such as polysorbate 20 may be necessary.

Stability of Fragrances in Diverse Applications

The stability of fragrance oils in various applications is a fundamental requirement. There are two aspects: the stability of the fragrance itself in the medium, and the effect of the fragrance on the finished product with properties such as color and viscosity. Chemical stability can involve fragrance components reacting with other

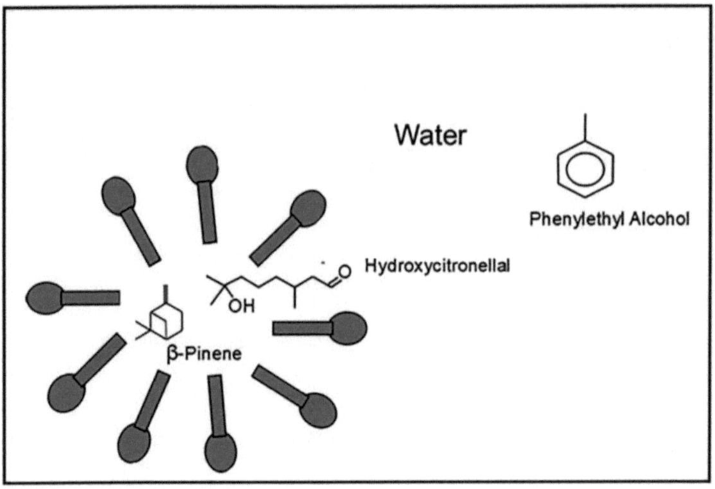

Fig. 20 Fragrance in micellar surfactant systems

Fig. 21 Fragrance in worm-like surfactant systems. (From [37])

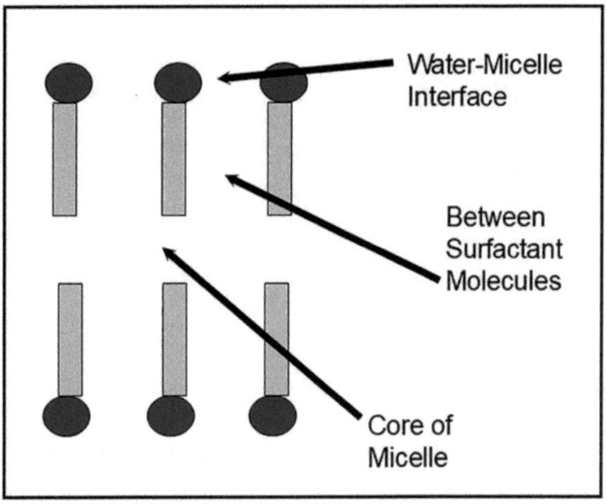

fragrance components, decomposition, and product base ingredients or impurities. Resultant issues include loss of olfactive strength, creation of undesirable off notes, coloration changes, and effects on physical stability.

Problematic reactions include Schiff base formation, acetal and ketal formation, ester hydrolysis, oxidation reactions, alcohol dehydration, and metal-catalyzed reactions. Aldehydes and ketones reacting with alcohols are important in the maceration of fine fragrance, which can result in odor loss by complete acetal formation (Fig. 22).

Turek [38] studied the stability of essential oils, and some key reactions are shown in Fig. 23.

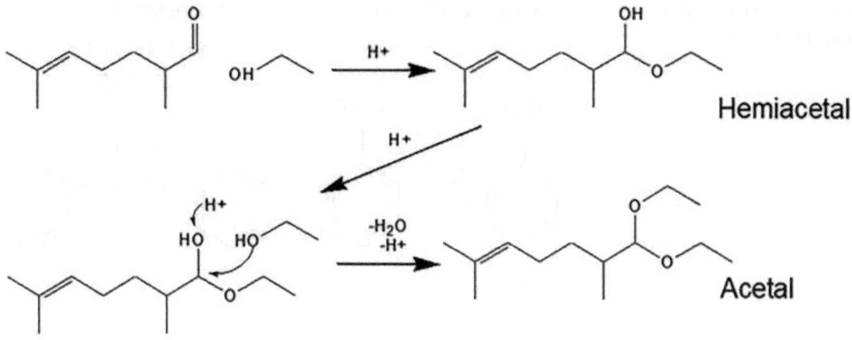

Fig. 22 Acetal formation in fine fragrance

Fig. 23 Conversion reactions in essential oils. (Derived from [38])

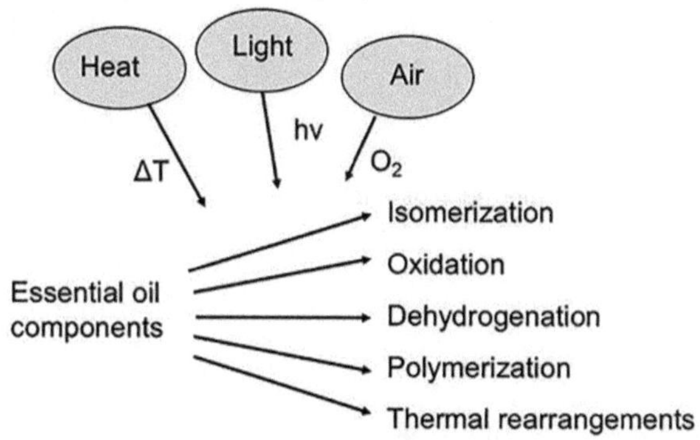

UV light can turn some aroma molecules that turn stable molecules into free radicals. Chromophores are configurations that can alter the energy of delocalized systems, putting them into the visible spectrum. These chromophores are often molecules with conjugated double bonds or azo groups. Figure 23 shows how a stable molecule, benzaldehyde, turns into a free radical and can then interact with a colored compound. It is natural to color a product red when the key ingredient smells like cherry, but it often creates an unstable combination (Fig. 24).

P-Cymene is a common product of terpenoid decomposition and can be used as a Marker to indicate essential oil or fragrance quality (Fig. 25).

There are many oxidation reactions of essential oils that can lead to the formation of polymers, peroxides, acids, aldehydes, alcohols, etc. [38]. Aldehydes are one of the least stable fragrance ingredients, and their conversion to carboxylic acids can result in the disappearance of the aldehyde note. Oxidation can also create terpenoid hydroperoxides, which are one of the main causes of allergenic reactions in fragrances.

Vanillin can undergo oxidative coupling that forms pink compounds (Fig. 26).

To prevent stability issues, use antioxidants such as BHT and vitamin C, chelants like EDTA and citric acid, and UV absorbers like benzophenones.

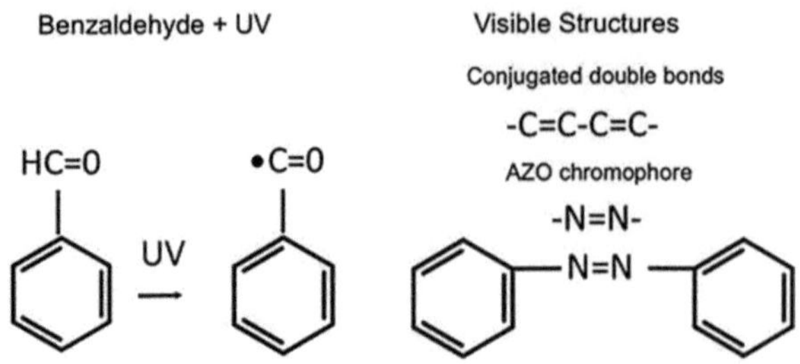

Fig. 24 Benzaldehyde unstable with UV

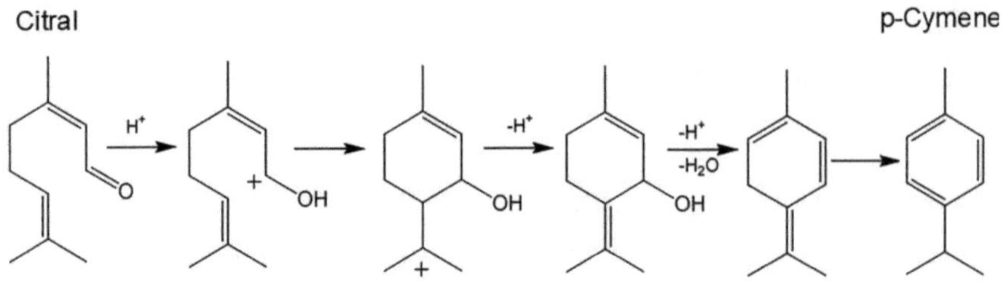

Fig. 25 Decomposition of citral

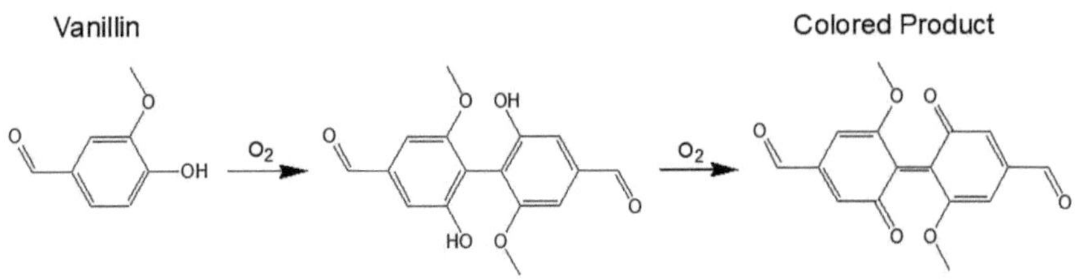

Fig. 26 Color creation in vanillin

Physical Chemistry of Fragrances

Many characteristics of fragrances are based on their cohesive properties. Evaporation rate, boiling point, vapor pressure, diffusivity, and molar mass are important examples. The thermodynamics of cohesion is well understood for simple systems, but harder to apply to complex mixtures such as fragrances. A fragrance is only detected when it is in the vapor phase, so any liquid fragrance must evaporate to have an effect. Evaporation properties can be understood by applying the laws of the physical chemistry of mixtures.

A fragrance material must be volatile, have an acceptable odor threshold, and consist primarily of C, H, and O atoms with occasional N and S. Volatility demands a range of C5 to C17. Raoult's law explains the composition of the vapor pressure over a complex solution. The vapor pressure of component A, P_A, is given by its

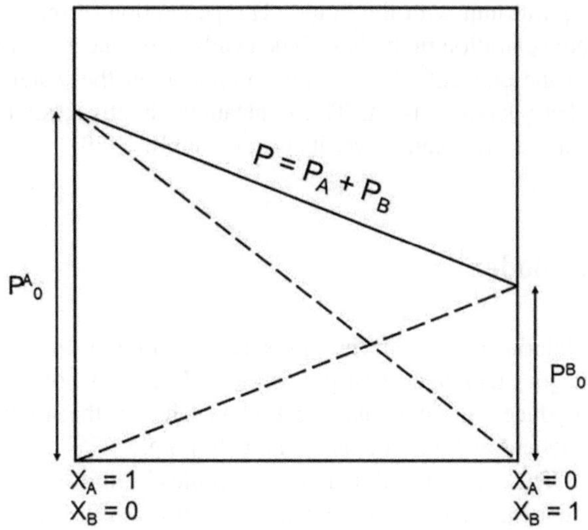

Fig. 27 Raoult's law

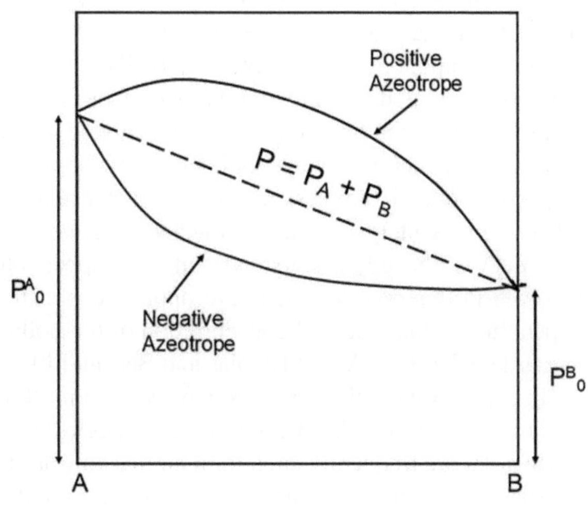

Fig. 28 Nonlinear behavior of Raoult's law

vapor pressure as a pure solution times its mole fraction in the solution, X_A. For a simple mixture of A and B, Ptotal = $P_AX_A + P_BX_B$. For complex perfume mixtures, the process is expanded to include each component. This is shown in Fig. 27.

A solution that obeys Raoult's law is described as an ideal solution. Nonideal behavior can result in higher or lower pressures, and this is one reason why fragrance mixtures sometimes produce unexpected properties (Fig. 28).

Alcohol and water make a negative azeotrope, which is why the most affordable commercial product is 95% alcohol and 5% water. Perfume fixatives function by their impact on the nonlinear behavior of azeotropes.

Henry's law states that at constant temperature, the amount of a gas that dissolves in a liquid is directly proportional to the partial pressure of that gas in equilibrium with the liquid. The perception of the fragrance is proportional to its concentration in the base. The headspace concentration of a volatile is proportional to the concentration of the compound in the water phase by the dimensionless Henry's law constant. This constant is directly related to the ratio of the vapor pressure of the volatile over its water solubility [39].

Malodor

Malodor may seem the opposite of a fragrance company concern but is in fact a major challenge often put to them. Malodors come from many sources. Humans produce oral, foot, and sweat odors. There is the smell of cat litter, cigarette smoke, kitchens and bathrooms, and much more.

The chemistry of malodor molecules is different from most fragrance molecules, frequently containing nitrogen or sulfur. Sulfur imparts the skunk excretion of its distinctive aroma, and nitrogen creates the smell of a rotten egg or cigarette smoke. Because they have unique chemistry, it is often possible to destroy a malodor without destroying a fragrance at the same time.

Most compounds are susceptible to oxidation, and this can lead to the decomposition of the malodor, removing the odor. Chlorine dioxide is a commercial example, which, while commercially useful, is not compatible with the use of fragrance.

Zwaardemaker pairs are conjugates of odorants that neutralize each other's respective odors. Examples are ethyl mercaptan and eucalyptol, chlorine and vanilla, tobacco and wintergreen oil, skatole and cedarwood. The reason for this neutralizing effect is not fully understood, but there appears to be no chemical interaction involved in this process. There are optimal concentrations of the "counter-smell" depending on the type and concentration of the pollutant. For a detailed review of Zwaardemaker's work, see Nicolas and Bensafi [40].

To some extent, all fragrances may assist in masking malodor, but some materials are much more effective than others. Aldehydes, orange oil, minty, fruity, and green notes are frequently cited for their masking ability. Fragrance companies have also designed non-aroma chemicals that target malodor. A classic example is Veilex 1 (1-cyclohexylethyl butyrate), patented in 1980 by Bush Boake Allen (now part of IFF). Veilex molecules may bind to the same receptors in the nose as malodor molecules, eliminating the perception of the malodor [41].

Other commercial malodor counteracting molecules are Meelium and Metazene. L.B. Kilgore developed and patented [42] an ester now known as lauryl methacrylate. The alkyl chain consists of about 60 to 65% lauryl (C12), 25–30% myristyl (C14), and 10% palmityl (C16). Its structure is shown in Fig. 29.

Two pervasive materials used in malodor products are zinc ricinoleate (Fig. 30) and cyclodextrin (Fig. 31). They both capture malodor molecules but use very different mechanisms.

Fig. 29 Metazene

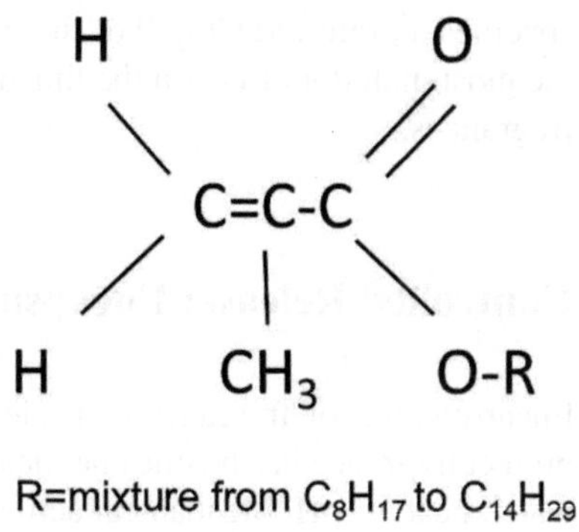

Fig. 30 Zinc ricinoleate

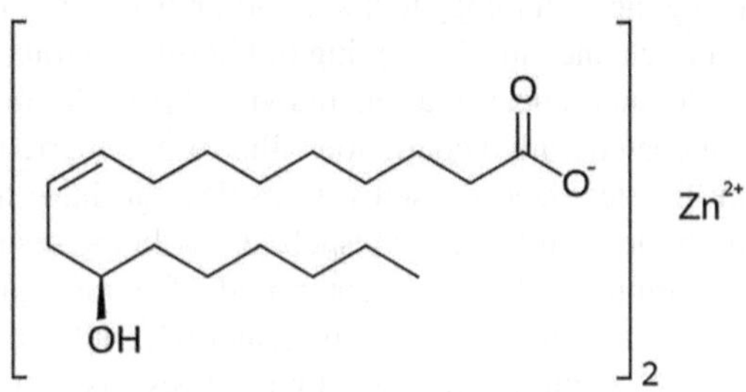

Fig. 31 Cyclodextrin

	A	B
α	0.57nm	1.37nm
β	0.78nm	1.53nm
γ	0.95mn	1.69nm

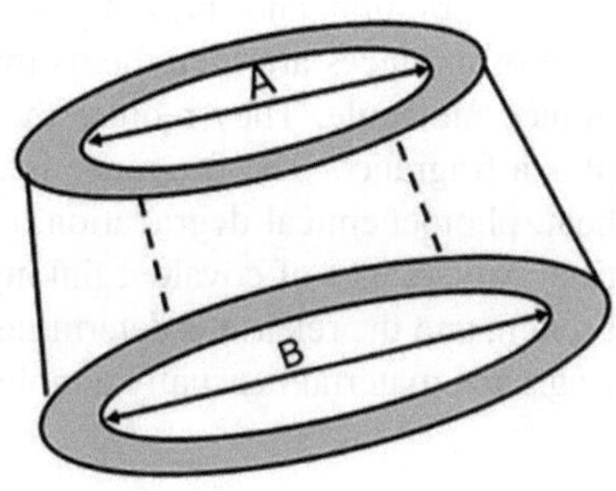

Zinc binds very strongly to nitrogen and sulfur, both characteristically found in malodor molecules. When a malodor molecule can bind to the zinc, the chains of zinc ricinoleate can fold to totally encapsulate it. Zinc ricinoleate can also bind to many fragrance molecules, which is one limit on its use. It is also very hard to solubilize in many applications. Since ricinoleic acid comes from castor oil, this material has been positioned as a naturally derived product.

Cyclodextrin comes in three different forms (alpha, beta, gamma) and is shaped like a truncated cone. Cyclodextrin is relatively polar on the outside and nonpolar on the inside, so it can capture aroma chemicals or malodor molecules in its interior and be compatible with water-based delivery systems. This material is used extensively in P&G's Febreze line and is covered by many patents. For example, see ref. [43]. In addition to cyclodextrin as an active material, this patent also specifies a preferred Clog P of three or less for fragrances. It is unclear which has the most malodor effect on the finished consumer products, the cyclodextrin or the fragrances.

Controlled Release: Encapsulation and Profragrances

Encapsulation of fragrance can release fragrance under specific triggers and can protect fragrance in a hostile chemical environment. Common triggers of release are water, heat, or pH. Mechanical action can also function as a release mechanism. In 1955, the National Cash Register Company received a patent for encapsulation, using the technology to make carbon paper. The revolutionary fragrance use was the "scratch and sniff" sampling of Giorgio fragrance in 1983.

In microencapsulation, individual particles or droplets of solid or liquid material are coated with a continuous film of polymeric material to produce micrometer- to millimeter-sized capsules. Capsules can have a single liquid area inside a shell, multiple liquid areas in a shell, or the liquid distributed through a matrix (Fig. 32).

Common shells are composed of gums (gum arabic, sodium alginate, carrageenan), celluloses (carboxymethylcellulose, methylcellulose), carbohydrates (starch, dextran, sucrose), lipids (beeswax, stearic acid, phospholipids), and proteins (gelatin, albumin). Microencapsulation methods include coacervation and solvent evaporation, and spray-drying.

Profragrances are ingredients that undergo a chemical change to liberate a fragrance molecule. They require the covalent linking of two molecules, an anchor plus a fragrance. The fragrance is released when the bond is broken. Triggers are heat, photochemical degradation, pH change, hydrolysis, and enzymatic degradation. Advantages of covalent linking are that there is slow release when the link is broken, and the release is determined by reaction rates, not fragrance volatility. The fragrance material is usually a molecule with a high odor impact, like aldehydes.

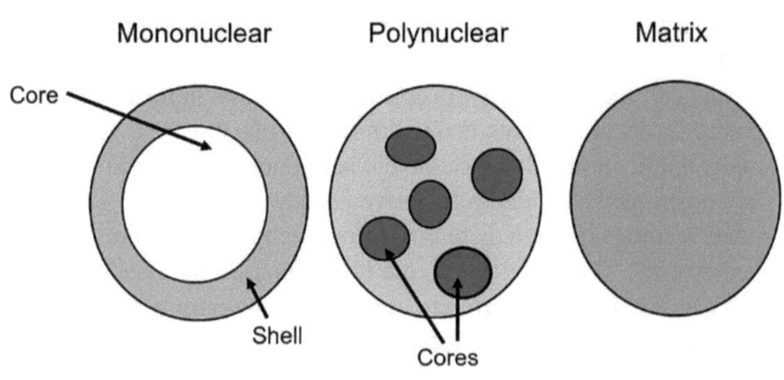

Fig. 32 Microencapsulation

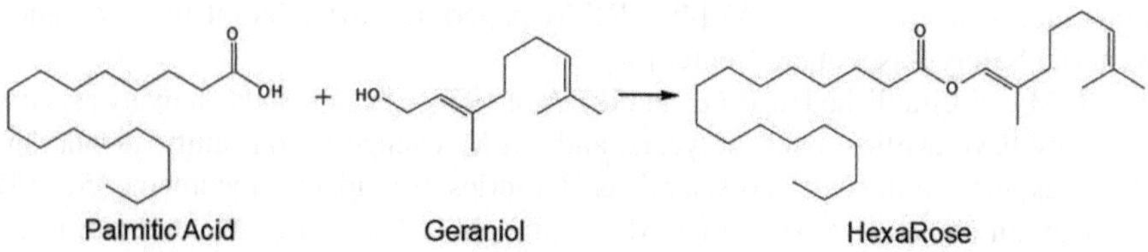

Fig. 33 HexaRose. (Derived from [44])

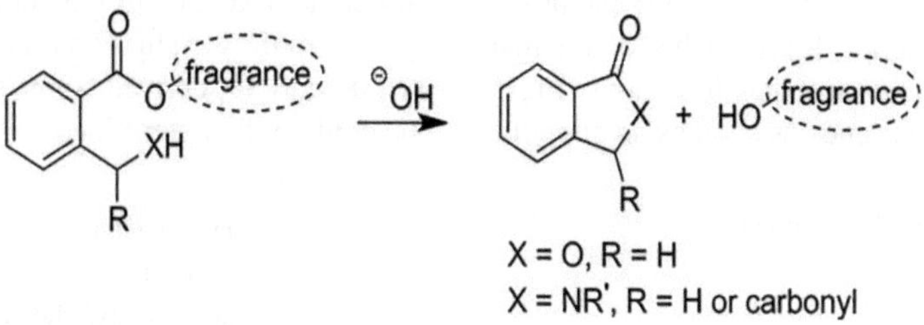

Fig. 34 Controlled release from benzoates. (Derived from [45])

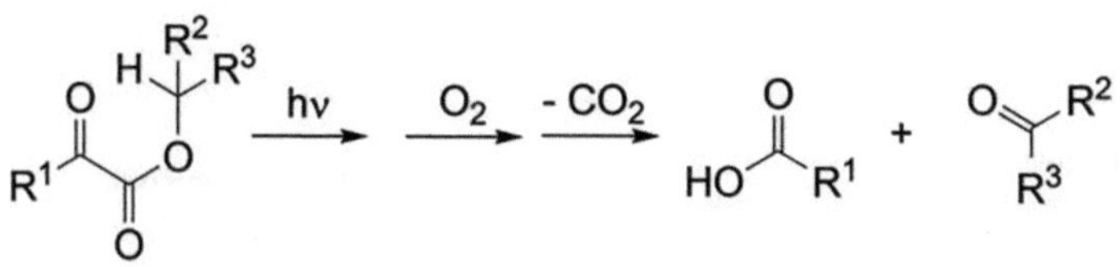

Fig. 35 Release of aldehydes and ketones. (Derived from [46])

The first commercialized profragrances were esters. HexaRose (Firmenich) is an ester composed of palmitic acid and geraniol [44], shown in Fig. 33.

Figure 34 shows the general principle of the controlled release of fragrance alcohols using ortho-substituted benzoates [45].

Figure 35 shows another profragrance example [46] using the release of aldehydes and ketones from 2-Oxo-(2-phenyl) acetates.

Fragrance Safety: The Role of RIFM (Criteria II Paper and 49th Amendment)

Guidance for safe fragrance use begins with the scientific studies performed by the Research Institute for Fragrance Materials (RIFM) [47]. RIFM's conclusions are vetted by the independent oversight of the Expert Panel for Fragrance Safety [48]. Specific guidelines for fragrance applications are issued by the International

Fragrance Association (IFRA) [49]. RIFM papers are available at the Fragrance Material Safety Assessment Center [50].

RIFM was established in 1966, and IFRA in 1973. The demands on many aspects of safety have evolved over the years, and RIFM science is constantly advancing. RIFM experts, in the early 2000s, released a series of guidance documents [51–53] that explain the science behind RIFM's approach. In October 2008, an entire issue of Regulatory Toxicology and Pharmacology [54] was devoted to Quantitative Risk Assessment (QRA), an approach specifically focused on skin safety.

RIFM tests for several end points (Table 3).

The Criteria II paper [55] established a unified approach to all the endpoints of RIFM science. It established a systematic way how a safety evaluation is conducted. The decision tree has four steps (Figs. 36, 37, 38, and 39, derived from [55]), and the process ends as soon as an endpoint can be confidently determined.

Table 3 RIFM endpoints

Environmental
Genotoxicity
Phototoxicity/ Photoallergy
Repeat Dose
Reproduction/ Developmental
Respiratory
Sensitization

Fig. 36 Evaluate available data

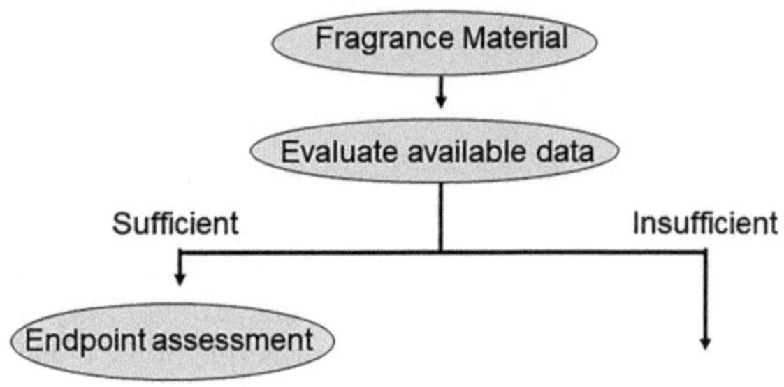

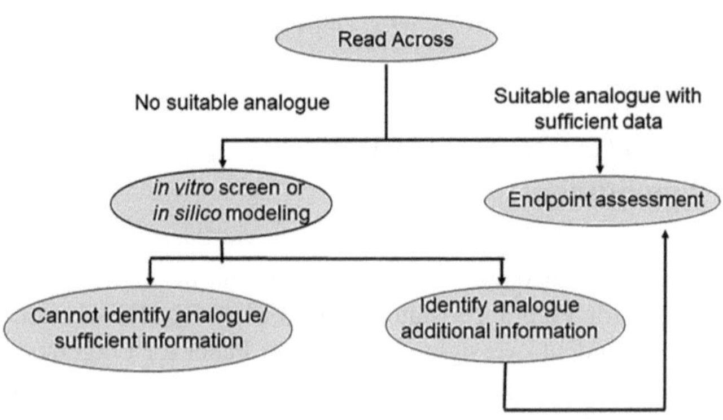

Fig. 37 Read-across, in vitro, or in silico methods

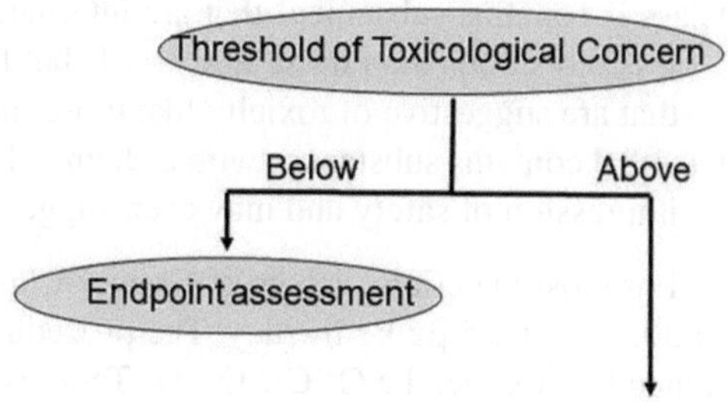

Fig. 38 Exposure threshold

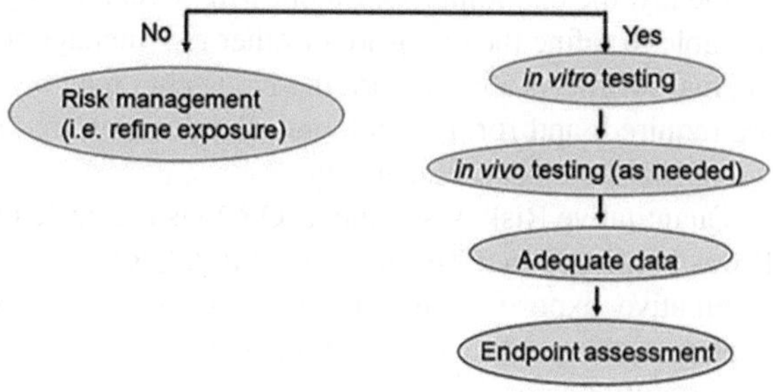

Fig. 39 Generate required data

Step 1 is to evaluate all available safety data on the material under consideration. If there is sufficient high-quality data, an endpoint assessment can be made. If insufficient data exists, proceed to Step 2.

Read-across uses information from an analogue to fill data gaps. The two materials must be similar in structure, properties, and metabolic by-products. The information must be of high quality and have an acceptable level of uncertainty for the purpose of filling in the data gap. RIFM uses clustering to assist in read-across, see Date et al. [56].

A wide range of in silico tools is available to assess the genotoxic potential of a material. For a review of the methods used in silico toxicology, see Richard [57]. There is also a Skin Absorption Model (SAM) described by Shen et al. [58]. If an appropriate analogue can be attained at this point or the in vitro tests are conclusive, the endpoint assessment can be made. If not, we must proceed to Step 3.

The Threshold of Toxicological Concern (TTC) in the context of fragrance safety was introduced in the Criteria II paper. It derives from work performed by the food industry, with the cooperation of the FDA, to address chemicals present at very low levels. It assumes that a chemical present at a low level with low expected toxicity will be safe. As with any toxicological concept, there is an additional level of consideration, which in this case is the Cramer Class of the material. The Cramer Class itself was derived from a decision tree [59].

The Cramer Classes are as follows:

Class I contains substances of simple chemical structure with known metabolic pathways and innocuous end products, which suggest a low order of oral toxicity.

Class II contains substances that are intermediate. They possess structures that are less innocuous than those in Class 1, but they do not contain structural features that are suggestive of toxicity like those in Class 3.

Class III contains substances with a chemical structure that permits no strong initial impression of safety and may even suggest a significant toxicity.

For most materials, values of Cramer Classes I, II, and III have exposure limits of 30, 9, and 1.5 μg/kg bw/day. The procedure for finding the Cramer Class determined by Toxtree, the OECD QSAR Toolbox, and expert judgment is explained by Bhatia [60].

Ideally, the endpoint assessment will be reached by Step 3. Otherwise, it may be possible to refine the exposure or other risk management calculations. In vitro testing may work. As a last resort, in vivo testing may be needed. However, many steps are required, and for any endpoint, the decision tree gives the result with the best science in the most efficient way.

Quantitative Risk Assessment (QRA) is the basic tool to establish dermal safety. It was published in 2008 and refined into QRA 2 with revised epidemiological and cumulative exposure information. The goal is to virtually eliminate fragrance allergy in the public. The starting point is Acceptable Exposure Level = NOEL/ Uncertainty Factor.

NOEL is the No Observable Effect Level, the highest dose or exposure level of a substance or material that produces no noticeable toxic effect. The uncertainty factor considers all the possible individual variables and includes safety factors. Toxicology is a highly conservative science, and the safety factors are large numbers that significantly reduce the recommended use levels for an ingredient.

The QRA 2 form of this relationship is.

$$\text{Acceptable Exposure Level}\left(\text{AEL}\right) = \text{WoE NESIL} / \text{SAF}$$

where WoE is the Weight of Evidence, NESIL is the No Expected Sensitization Induction Level, and SAF is the Sensitization Assessment Factor.

NESIL data are available from a variety of sources and may be highly variable or of questionable value. The Weight of Evidence of NESIL is the value based on a careful analysis of all available data as determined by the RIFM Expert Panel, an independent group of experts.

The SAFs for QRA 2 are shown in Fig. 40. The total SAF is the multiplicative product of the individual components.

The Dermal Sensitization Threshold (DST) applies the concept of the TTC to dermal sensitization [61]. The DST establishes a level below which there is no appreciable risk for the induction of sensitization. The levels derive from reactions with peptides: nonreactive substance 900 μg/cm2 and reactive substance 64 μg/cm2.

IFRA sets quantitative standards in a variety of product categories. The 49th amendment added subcategories addressing phototoxicity and systemic toxicity. Phototoxic restrictions differentiate leave-on and rinse-off products. It is known as systemic toxicity, when toxic effects occur at multiple sites.

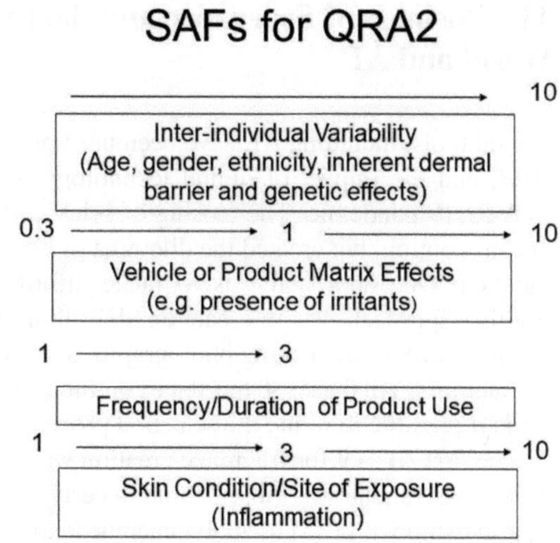

Fig. 40 SAFs for QRA2

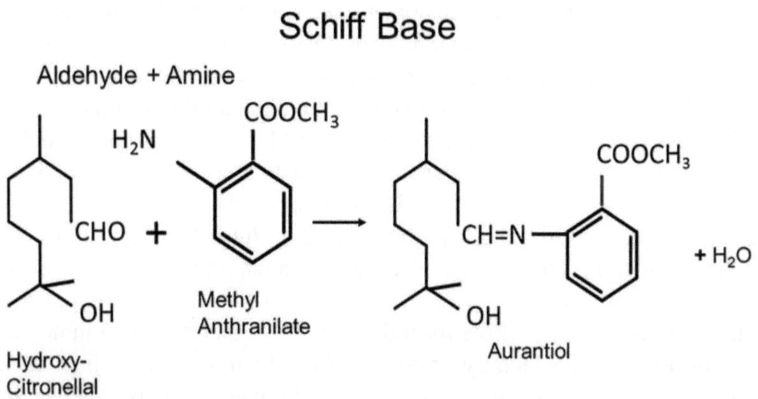

Fig. 41 Schiff base

A Schiff base is a reaction product of an aldehyde and an amine (Fig. 41). One requirement of the 49th Amendment is that Schiff bases must be considered completely disassociated. For example, Aurantiol must be considered 56.4% hydroxycitronellal unless proven otherwise.

Complexities enter when natural products are considered, what RIFM describes as Natural Complex Substances (NCSs). A paper by Api et al. [62] explains the RIFM approach to NCSs.

The result of applying RIFM's evaluations is a certificate of conformity for every fragrance placed in commerce, providing safe levels of use in different applications. The goal is to minimize or eliminate adverse reactions to fragrance in the general population.

The Business of Fragrance and the Impact of the Digital World and AI

Digital tools, including AI, have been used publicly in the fragrance industry since 2018, and the impact of digital technology was accelerated dramatically by the COVID-19 pandemic. The closing of brick-and-mortar stores made selling online the only option, but created the dilemma of how to sell fragrances if they cannot be sampled. One approach was to make affordable sample kits widely available. Another approach was to create an algorithm powered by questions the customer would answer or by picking photographs, and the fragrances were selected based on the answers. Analysis of the data combined with online comments and reviews made it possible to refine the selection process.

The first AI tool for fragrance creation was Symrise's Phylra [63], launched late in 2018. Givaudan was close behind in early 2019 with Carto [64]. Symrise created a partnership with IBM to apply machine learning to fragrance development. IBM's algorithm studies existing fragrance formulas and compares the ingredients to other datasets, such as geography and customer age, to target specific market segments. During 2019, Brazilian cosmetics company O Boticário worked with Symrise to create the first fragrance designed with the assistance of artificial intelligence.

Carto uses a wide touch screen where perfumers can develop their fragrances through interacting with Givaudan's Odor Value Map, which is not publicly available. Carto does not replace perfumers but is meant to assist them. In 2021, Givaudan acquired Myrissi, which had an AI tool capable of translating fragrances into color patterns and images. Givaudan also acquired DigiPulse™, which is an olfactive social listening tool. The findings are incorporated into the fragrance creation process.

Firmenich launched its AI platform d-lab in 2018, and its first commercial success was developing new laundry fragrances. In 2020, it introduced the first laundry fragrances that unite Artificial Intelligence with human creativity. Also, in 2020, it announced its first AI-created flavor with a grilled beef taste. In 2021, Firmenich introduced Scentmate™ to streamline the process by which customers request fragrances [65].

The first website to claim it used "Algorithmic Perfumery" for "the world's first AI-guided scent creation" was Every Human [5]. It has a fragrance library, but there is also an option to create an individual scent. Many online companies use some form of questionnaire or sampling to sell fragrance.

The art and science of fragrance is changing in an accelerated fashion as we enter the world of AI. Digital tools have changed fragrance creation and marketing in fundamental ways. Consumers have a wealth of online resources to explore everything from fragrance trends to chemical safety. Scientific advances keep innovation alive in the range of fragrance applications brought to the market. Fragrance continues to be, as it has for centuries, a key component of our lives.

Glossary

Accord A blend of fragrance notes that create a new scent profile.

Aldehyde An organic compound in which a carbon atom shares a double bond with an oxygen atom, a single bond with a hydrogen atom, and a single bond with another atom or group of atoms

Algorithm A sequence of instructions used to solve specific problems.

Azeotrope A mixture of two or more liquids whose proportions cannot be changed by simple distillation

BINAP Compound used in asymmetric synthesis

Chelant A molecule that can bond to a positively charged metal ion through a coordinate bond.

Chypre A fragrance having top notes of citrus, middle notes of florals, and base notes of oakmoss and patchouli

CLogP Calculated logarithm of P, the water/octanol partition coefficient. It is a measure of polarity.

EACN Effective Alkane Carbon Number, a measure of the hydrophobicity of an oil

Enantiomers A pair of molecules that are mirror images of one another and cannot be superimposed

Enfleurage A process that uses odorless fats to capture natural fragrant compounds

Fixative An ingredient used in perfumes to prolong its duration

Genotoxic Having the ability to damage an organism's DNA or genetic material

In silico Done by a computer program

Olfactory Relating to the sense of smell

Osmophore A functional group in a molecule that gives a compound its odor

Photochemical The branch of chemistry concerned with the chemical effects of light

Phototoxin A type of abnormal skin reaction induced by chemicals when exposed to ultraviolet radiation.

Stereoisomer Molecules that have the same molecular constitution but different three-dimensional spatial arrangement of the atoms.

Surfactant Surface active agent, a molecule that is part polar, part nonpolar

Hydroperoxides Compounds of the form ROOH where R is any organic group

Thermogravimetric analysis A technique that measures how a material's weight changes as it is heated or cooled

References

1. Balasubramanian, N. (2015). *Scented oils and perfumes in chemical technology in antiquity, rasmussen, ACS symposium series* (pp. 219–244). American Chemical Society. https://doi.org/10.1021/bk-2015-1211.ch008
2. By Jü – Own work, Public Domain, https://commons.wikimedia.org/w/index.php?curid=44527111
3. Kraft, P., Ledard C., & Goutell, P., (2007). From Rallet No 1 to Chanel No 5 vs Mademoiselle Chanel No 1, Cosmet. & Toil., 32, (10).

4. Mazzeo, T. J. (2010). *The secret of chanel no 5*. Harper Collins.
5. https://www.everyhuman.com/us/products
6. Carles J. A method of creation in perfumery, researches Part1 December 11, 1961, Part 2 December 12, 1962, Part 3 December 13, 1963.
7. Martínez-Guido, S. I., et al. (2014). *A multiobjective optimization approach for the development of a sustainable supply chain of a new fixative in the perfume industry*. American Chemical Society (ACS). https://doi.org/10.1021/sc500409g
8. Aggarwal, P., Dollimore, D., & Alexander, K. (1997). The use of thermogravimetry to follow the rate of evaporation of an ingredient used in perfumes. *Journal of Thermal Analysis, 49*, 595–599. https://doi.org/10.1007/BF01996741
9. Phang, P. The analysis of perfume fixatives by simultaneous TG–DTA studies. *Thermochimica Acta, 340–341*, 139–145. https://doi.org/10.1016/S0040-6031(99)00260-9
10. Lu, W., Lin, W., Huang, A., & Shu, C. (2022). Determination of the ambience duration of lavender essential oil with three perfume fixatives using the thermokinetics approach. *Journal of Thermal Analysis and Calorimetry, 147*(13), 7551–7561. https://doi.org/10.1007/s10973-021-11029-1
11. Lewczyk, D. C., et al. (2020). Kinetic treatment of evaporation via Thermogravimetric analysis: The case of D-limonene. *Industrial and Engineering Chemistry Research, 59*, 15069–15074. https://doi.org/10.1021/acs.iecr.0c02864
12. Mishra, A., & Parida, S. (2021). Extraction methods of essential oils from aromatic plants. In P. Bhadra & S. Parida (Eds.), *Aromatherapy and its benefits* (pp. 71–83). ISBN: 978-81-940943-7-1). Renu Publishers. https://doi.org/10.30954/NDP-AMTHY.2021.5
13. Rakthaworn, P., et al. (2009). Extraction methods for tuberose oil and their chemical components. *Kasetsart Journal (Natural Science), 43*, 204–211.
14. Cakaloglu, B., Ozyurt, V. H., & Otles, S. (2018). Cold press in oil extraction. A review. *Ukrainian Food Journal., 7*(4), 640–654.
15. Moriera, R. C., et al. (2023). Supercritical CO_2 as a valuable tool for aroma technology. *Journal of Agricultural and Food Chemistry, 71*, 9201–9212. https://doi.org/10.1021/acs.jafc.3c01023
16. Bousbia, N., et al. (2009). Comparison of two isolation methods for essential oil from rosemary leaves: Hydrodistillation and microwave hydrodiffusion and gravity. *Food Chemistry, 114*, 355–362. https://doi.org/10.1016/j.foodchem.2008.09.106
17. Silvestre, A. J. D., & Gandini, A. (2008). "Terpenes: Major sources, properties and applications", monomers, polymers and composites from renewable resources. In M. N. Belgacem & A. Gandini (Eds.), (pp. 17–38). Elsevier. https://doi.org/10.1016/B978-0-08-045316-3.X0001-4
18. Nyamwihura, J., & Victor, I. (2022). The pinene scaffold: Its occurrence, chemistry, synthetic utility, and pharmacological importance, (Review Article). *RSC Advances, 12*, 11346–11375.
19. Guenther, E. (1948–52). *The essential oils, Vol I–IV*. Van Nostrand.
20. By NEUROtiker – Own work, Public Domain. https://commons.wikimedia.org/w/index.php?curid=3223789
21. https://en.wikipedia.org/wiki/Benzaldehyde#:~:text=Benzaldehyde%20(C6H5,Ball%2Dand%2Dstick%20model
22. Wikimedia Commons. https://commons.wikimedia.org/wiki/File:Galaxolide.svg, https://commons.wikimedia.org/wiki/File:Ethylene_brassylate.svg, https://commons.wikimedia.org/wiki/File:D-Helvetolide.svg
23. Arctander, S. (1994). *Perfume and flavor chemicals*. Allured Publ.
24. Ryoji Noyori–Nobel Lecture. NobelPrize.org. Nobel Prize Outreach AB 2024. Thu. 31 Oct 2024. https://www.nobelprize.org/prizes/chemistry/2001/noyori/lecture/
25. Anastas, P. T., & Warner, J. C. (1998). *Green chemistry: Theory and practice*. Oxford University Press.
26. Tony Phan, T. V., Gallardo, C., & Mane, J. (2015). GREEN MOTION: A new and easy to use green chemistry metric from laboratories to industry. *Green Chemistry, 17*, 2846–2852.
27. https://ifrafragrance.org/priorities/ifra-green-chemistry-compass.

28. Gupta, S., et al. (2015). A biotechnological approach to microbial based perfumes and flavours. *Journal of Microbiology & Experimentation, 2*(1), 11–18.
29. Choi, K. R., & Lee, S. Y. (2023). Systems metabolic engineering of microorganisms for food and cosmetics production. *Nature Reviews Bioengineering, 1*(11), 832–857. https://doi.org/10.1038/s44222-023-00076-y
30. Ewing, T. A., Nouse, N., Van Lint, M., Van Haveren, J., Hugenholtz, J., & Van Es, D. S. (2022). Fermentation for the production of biobased chemicals in a circular economy: A perspective for the period 2022–2050. *Green Chemistry, 24*(17), 6373–6405. https://doi.org/10.1039/d1gc04758b
31. Saunders, H. C., (1973). An approach to fitting a perfume to the polarity of its substrate, Perf. & Flav., 88 No11.
32. https://en.wikipedia.org/wiki/Hildebrand_solubility_parameter
33. Abbott, S., & Hansen, C. M. (2008–13). *Hansen solubility parameters in practice.* Hansen-Solubility.com.
34. Bouton, F., Durand, M., Nardello-Rataj, V., Serry, M., & Aubry, J. M. (2009). Classification of terpene oils using the fish diagrams and the equivalent alkane carbon (EACN) scale. *Colloids and Surfaces A: Physicochemical and Engineering Aspects, 338*(1–3), 142–147. https://doi.org/10.1016/j.colsurfa.2008.05.027
35. Kamada, M., Shimizu, S., & Aramaki, K. (2014, September). Manipulation of the viscosity behavior of wormlike micellar gels by changing the molecular structure of added perfume. *Colloids and Surfaces A, 458*(20), 110–116. https://doi.org/10.1016/j.colsurfa.2014.01.003
36. Herman, S. J. (1994). Fragrancing emulsions. *C&T, 109*(8), 71–75.
37. Herman, S. J. (1994). Fragrance in emulsion and surfactant systems. *C&T, 121*(5), 59–68.
38. Turek, C., & Stintzing, F. C. (2013). Stability of essential oils: A review. *Comprehensive reviews in food science and food safety, 12*(1), 40–53. https://doi.org/10.1111/1541-4337.12006
39. Fieber, W., Frank, S., & Herrero, C. (2018, February). Competition between surfactants and apolar fragrances in micelle cores. *Colloids and Surfaces. A, 5*, 39. https://doi.org/10.1016/j.colsurfa.2017.12.031
40. Nicolas, S., & Bensafi, M. (2021). A historical review of olfactometry the invention of the olfactometer by Hendrik Zwaardemaker in 1888 and pioneering work in olfactometry at the turn of the twentieth century. *L'Année psychologique, 121*(3), 311–351. https://doi.org/10.3917/anpsy1.213.0311
41. Payne R., Boden R., Sierra M., & Tabert, M. (2013). Cyclohexylethyl carboxylic acid ester compositions and method for using the same for reducing malodors., US 8,506,943 B2.
42. Kilgore, L. B., (1951). Deodorant compositions. US 2,544,093.
43. Trinh, T et al. (1998). Composition for reducing malodor impression on inanimate surfaces, U 5,783,544.
44. Herrmann, A. (2010). *The chemistry and biology of volatiles* (p. p337). Wiley. https://doi.org/10.1002/9780470669532
45. Herrmann, A. (2017). Profragrance chemistry as an interdisciplinary research area and key technology for fragrance delivery. *Chimia, 71*(7/8), 414–419. https://doi.org/10.2533/chimia.2017.414
46. Levrand, B., & Herrmann, A. (2007). Controlled light-induced release of volatile aldehydes and ketones by Photofragmentation of 2-Oxo-(2-phenyl)acetates. *Chimia, 61*(10), 661–664. https://doi.org/10.2533/chimia.2007.661
47. https://www.rifm.org
48. http://fragrancesafetypanel.org
49. https://ifrafragrance.org
50. http://fragrancematerialsafetyresource.elsevier.com
51. Ford, R. A., Domeyer, B., Easterday, O., Maier, K., & Middleton, J. (2000). Criteria for development of a database for safety evaluation of fragrance ingredients. *Regulatory Toxicology and Pharmacology, 31*(2 Pt 1), 166–181. https://doi.org/10.1006/rtph.1999.1362

52. Cadby, P. A., Troy, W. R., & Vey, M. G. (2002). Consumer exposure to fragrance ingredients: Providing estimates for safety evaluation. *Regulatory Toxicology and Pharmacology, 36*(3), 246–252. https://doi.org/10.1006/rtph.2002.1581
53. Salvito, D. T., Senna, R. J., & Federle, T. W. (2002). A framework for prioritizing fragrance materials for aquatic risk assessment. *Environmental Toxicology and Chemistry, 21*(6), 1301–1308. https://doi.org/10.1002/etc.5620210627
54. Api, A. M., et al. (2008). Dermal sensitization quantitative risk assessment (QRA) for fragrance ingredients. *Regulatory Toxicology and Pharmacology, 52*(1), 3–23. https://doi.org/10.1016/j.yrtph.2007.10.008
55. Api, A. M., et al. (2015). Criteria for the research Institute for Fragrance Materials, Inc. (RIFM) safety evaluation process for fragrance ingredients. *Food and Chemical Toxicology, 82*(Suppl), S1–S19. https://doi.org/10.1016/j.fct.2014.11.014
56. Date, M. S., O'Brien, D., Botelho, D. J., et al. (2020). Clustering a chemical inventory for safety assessment of fragrance ingredients: Identifying read-across analogs to address data gaps. *Chemical Research in Toxicology, 33*(7), 1709–1718. https://doi.org/10.1021/acs.chemrestox.9b00518
57. Richard, A. M. (1998). Structure-based methods for predicting mutagenicity and carcinogenicity: Are we there yet? *Mutation Research, 400*(1–2), 493–507. https://doi.org/10.1016/s0027-5107(98)00068-2
58. Shen, J., Kromidas, L., Schultz, T., & Bhatia, S. (2014). An in silico skin absorption model for fragrance materials. *Food and Chemical Toxicology, 74*, 164–176. https://doi.org/10.1016/j.fct.2014.09.015
59. Cramer, G. M., Ford, R. A., & Hall, R. L. (1978). Estimation of toxic hazard–a decision tree approach. *Food and Cosmetics Toxicology, 16*(3), 255–276. https://doi.org/10.1016/s0015-6264(76)80522-6
60. Bhatia, S., Schultz, T., Roberts, D., Shen, J., Kromidas, L., & Marie, A. A. (2015). Comparison of Cramer classification between Toxtree, the OECD QSAR toolbox and expert judgment. *Regulatory Toxicology and Pharmacology, 71*(1), 52–62. https://doi.org/10.1016/j.yrtph.2014.11.005
61. Safford, R. J. (2008). The dermal sensitisation threshold- a TTC approach for allergic contact dermatitis. *Regulatory Toxicology and Pharmacology, 51*(2), 195–200. https://doi.org/10.1016/j.yrtph.2008.02.010
62. Api, A. M., et al. (2022 Jan 15). The RIFM approach to evaluating natural complex substances (NCS). *Food and Chemical Toxicology, 159*(Suppl 1), 112715. https://doi.org/10.1016/j.fct.2021.112715
63. https://www.symrise.com/scent-and-care/competence-platforms/philyra/
64. https://www.givaudan.com/file/156296/download
65. https://www.scentmate.com/

Nutraceuticals, Nutricosmetics, Pharmacocosmetics, and Cosmoceuticals

Kenneth Alexander

Abstract There is a huge demand for Cosmetics and Personal Care Products, including those containing natural ingredients, by people of all ages including men, women and children. The growth of the cosmetic industry has caused it to turn to more natural ingredients as well, especially in the personal care arena. Personal Care and Cosmetic products play a significant role in people's daily lives. This chapter deals with those ingredients for the maintenance of skin, nail and hair health known as nutraceuticals, nutricosmetics, pharmacocosmetics, and cosmoceuticals.

Keywords Phytochemicals · Supplements · Micronutrients · Herbals · Cosmoceutics · Botanicals

Introduction

Humans have long used herbs for many purposes, such as food, medication, and cosmetics. The word "Cosmetic" is a translation of the word "kosmo tikos" in Greek, which signifies having the ability to plan and decorate. Cosmetics are a group of health and beauty products utilized to improve and may alter someone's appearance, and additionally care for the body and facial parts. In almost every civilization on earth, cosmetics have a tradition that spans more than 7000 years. The beauty industry has powerfully shaped consumer perceptions of beauty. Human attitudes and desires for lasting beauty have driven entrepreneurs to develop natural products to fill the current demand.

K. Alexander (✉)
Pharmacy, The University of Toledo College of Pharmacy and Pharmaceutical Sciences - Professor Emeritus, Grand Rapids, OH, USA
e-mail: kenneth.alexander@utoledo.edu

K. Alexander, P. Romanowski (eds.), *Cosmetic Science*, AAPS Introductions in the Pharmaceutical Sciences 15, https://doi.org/10.1007/978-3-032-03596-7_10

In daily life, cosmetics are used by people of many different ethnicities and civilizations. The ability to express oneself creatively is of paramount importance. This importance of self-identity is said to be the main driver of cosmetics' current popularity. The primary purpose of makeup is to provide the user with a fresh, presentable appearance. Despite the enormous acceptance of the cosmetics business, many Western Nations still view cosmetics as little more than simple makeup. The US FDA plainly declared that cosmetic items that are primarily meant for use on the body of humans can change appearance, encouraging beauty, cleanliness, and either enhancing rather than endangering the building or functionality of the body" [1, 2].

Bark, leaf, flowers, roots, fruits, and seeds are only a few of the plant's many components that contain bioactive substances. As a result, plants play an important role in cosmetology and are directly used by the majority of civilizations to treat a variety of illnesses. Since they are less expensive, have fewer side effects, and are an excellent source for a safe future, plant-based goods are becoming more and more popular [3]. Plants have long been recognized as potential sources of a wide range of chemical molecules known as phytochemicals [4]. Herbs and spices are among the various useful foods that include a plethora of biofunctional molecules. They are typically used as preservatives and flavoring enhancers [5]. The region of the Himalayas is the epicenter of several herbal plant species [6]. Natural pigments are very important economically since they are more advantageous and have important characteristics, including being anticancer and antioxidant agents. Plants produce phytochemicals to protect themselves from microbial invasions. They can also be used therapeutically in pharmaceutical and nutraceutical products [7].

The desire to emulate certain ideals of beauty often leads individuals to seek physical perfection. However, on a more humorous side, one might be shown, whoever the "mentor is," they still aspire for perfection. Commercialization is the fundamental cause of this concern within us. Commercialization exploits people's dissatisfaction with their own appearance. Hence, it is evident that the beauty industry has done a magnificent job of brainwashing our thoughts on this subject.

Although many people view cosmetics as beneficial, cosmetics do carry some danger and toxicity. Since a large population uses cosmetics frequently, safety should be guaranteed and included in the products while they are being created. Moreover, excessive use of cosmetics can have hazardous and long-lasting negative effects, some of which may even be permanent. The following are a few safety issues that may be connected to personal care items and cosmetics:

- Scrapes and irritated skin around the eyes.
- Pollution of goods made outside of the United States.
- Intolerance to product ingredients or allergic reactions to such ingredients.
- Bacterial spread on the skin surface.
- Fire risks associated with aerosol products.

The effects of cosmetics cannot last longer than a certain amount of time. Several regulatory authorities, including the US FDA, suggest washing off the chemical before bedtime.

Nutricosmetics

Nutricosmetics is the term used to describe the category of supplements that aid in the upkeep and improvement of the nutrition of the skin. Various micronutrients and macronutrients are thought to contribute to these effects. The following instances illustrate the goal of nutricosmetics.

- A number of micronutrients, including beta carotene and omega 3-fatty acids not only support healthy skin but also to a lesser extent, healthy hair.
- Ascorbic acid aids in the reduction of free radicals, hence enhancing the antioxidant action. Even skincare products are divided into different markets based on the year, location, intended use, and type of therapy, as well as the degree of targeting. The market for nutricosmetics is expanding as the benefits and targeting capabilities increase.

  Growing demands for nutricosmetics include the following:

- Increased motivation to "age well and look well."
- Increased desire to use antiaging products and the emergence of the spa culture.
- Consistent scientific evidence for the main elements.
- Finding an appropriate method to preserve beauty among the aging population.
- Increase in demand for particularly useful products and substances.
- More knowledgeable about the usage of various products (especially those that promote inner beauty); pursuing a holistic regimen for aging as well as beauty.
- Demand for natural ingredients because they have fewer adverse effects [8].

Skin Aging Considerations

The skin is the largest organ of the human body (see chapter on Skin Physiology in this volume) and is perpetually aging by both internal and external factors [9]. The internal factors are a part of the natural aging process within cells. It can be accelerated by external factors such as ultraviolet (UV) radiation, tobacco, inadequate nutrition, and hormonal imbalances. These external factors cause the production of free radicals and inflammation that fragment and degrade collagen and elastic fibers [9]. Over time, both these internal and external influences lead to increased skin roughness, wrinkling, pigmentation change, telangiectasias, loss of elasticity, and decreased firmness, which gives skin its aged appearance [9, 10].

It accounts for approximately 15% of the body's weight. The outermost layer is called the epidermis. It acts as a physical barrier that protects us from biotic and abiotic aggressions (radiation, xenobiotics, microorganisms, etc.). The dermis is below. It is rich in water and the so-called extracellular matrix protein complex, where collagen fibers, elastic tissue, and other materials are found. This layer provides mechanical strength and elasticity. The innermost layer is the subcutis and is the key for insulation, mechanical protection, and contributes to thermoregulation.

Skin health is, therefore, crucial for preventing disorders of a different nature and gravity, ranging from infections to cancer. This organ is also important aesthetically as some attributes of its appearance, such as color, wrinkling, elasticity, and hydration, have clear social implications and can have an impact on mating possibilities and even on socioeconomic status [11–14].

Skin aging is continuously influenced by various internal and external factors such as the biologic progression of cells, ultraviolet (UV) radiation, tobacco, nutritional deficiencies, as well as hormonal imbalances that lead to the degradation of skin cells. Through the degradation of skin cells, free radicals and inflammation weaken the body's repair mechanisms and result in collagen and elastic fiber breakdown. The appearance of aging skin is highlighted by skin roughness, wrinkling, pigmentation change, telangiectasias, loss of elasticity, and decreased firmness. All of these processes are accelerated by these internal and external factors. Throughout the years, nutraceuticals have been studied to delay and fight against these internal and external factors, many of which are found in foods and byproducts consumed naturally. The state of science is improving day by day. The science of dermatology, trichology, and other fields is being used to make advancements in aesthetic technology.

Nutraceuticals

Nutraceuticals are oral dietary components naturally found in foods and believed to have a medical or health benefit [15]. Dr. Stephen Defelice, who combined the words "nutrition" and "pharmaceutical," coined the term in 1989 [16]. Nutraceuticals refer to substances derived from food sources that offer health benefits beyond basic nutrition [17]. These products, with claims that vary across regulatory jurisdictions, may purport to mitigate chronic illnesses, enhance health, decelerate the aging process, prolong life span, or bolster the body's structure or functionality [18–21]. Nutraceuticals are essentially biologically active compounds discovered in food, encompassing attributes of both nourishment and medication. These naturally occurring bioactive or chemical compounds function not only as nutritional supplements but also showcase properties that promote health, cure diseases, or possess preventive capabilities [22, 23]. They constitute essential components such as lipids, vitamins, carbohydrates, proteins, and minerals. Nutraceuticals have demonstrated potential efficacy in treating a range of diseases, including cardiovascular diseases, diabetes, atherosclerosis, cancer, and problems associated with neurological conditions [23–26]. Various food products such as prebiotics, probiotics, dietary fibers, fatty acids (polyunsaturated), antioxidants, spices, herbs, nutrients, and dietary supplements can be considered as nutraceuticals. Byproducts from food and agricultural sources are also valuable sources of nutraceuticals (e.g., leaves, peels, stems, seeds, and flowers). Nutraceuticals possess both nutritional and therapeutic value. Nutraceuticals may be considered helpful for the prevention of various diseases such as overweight, cardiovascular disease, cancer, osteoporosis, arthritis, increased sugar, and cholesterol levels.

The demand for nutricosmeceutics is growing due to the following reasons [27–31]:

- Increased motivation to "age well and look well."
- Increased desire to use antiaging products and the emergence of the spa culture.
- Consistent scientific evidence of the main elements.
- Finding an appropriate method to preserve beauty among the aging population.
- Increase in demand for particularly useful products and substances.
- Patients are more knowledgeable about the usage of various products (especially those that promote inner beauty); pursuing a holistic regimen for aging and beauty.
- There is greater demand for natural ingredients because they have fewer adverse effects.

Foods and Components Used for Their Nutricosmetic Effects

Foods

- Peptides derived from marine and terrestrial animal protein sources from collagen
- Glycosaminoglycans, marine fish, and bone matrix
- Polyphenols
- Fruits (apples, grapes, berries, citrus, etc.), vegetables (tomatoes, onions, cabbage, radishes, etc.), and cereals
- Soy, chocolate, coffee, and tea
- Polyunsaturated oils, nuts, and seeds
- Fatty acids (PUFAs) cold-water fish
- Carotenoids
- Fruits (mangoes, apricots, citrus, papayas, etc.), vegetables(green vegetables, tomatoes, carrots, pumpkins, peppers, etc.), salmon, and egg yolk
- Squalene
- Olive oil
- Coenzyme Q10
- Oils (soybean, corn, olive, and canola), nuts, and seeds
- Meat, poultry, and fish

Vitamins

- Vitamin A—green vegetables, sweet potatoes, carrots, apricots,mangoes, and palm oil (provitamin A carotenoids), liver, fish oils, milk, and eggs (preformed Vitamin A)

- Vitamin C—raw red and green peppers, oranges, acerola, grapefruits, kiwi, strawberries, broccoli, Brussels sprouts, etc.
- Vitamin E—nuts, seeds, vegetables (tomatoes, spinach, asparagus, Swiss chard, and broccoli), corn, soy, and avocados

Minerals

- Copper—nuts, seeds, grains and seafood, and meat
- Selenium—Brazil nuts, grains, seafood, organ meats, poultry, and dairy products
- Zinc—cereals, nuts, legumes, meat, seafood, and eggs

Medicinal Uses of Herbs

Historically, herbs were used for both medical and cosmetic purposes. They were utilized in both fresh and dried forms. They can be applied to the body directly after being mashed, either with or without the addition of other substances. In actuality, this was the only purpose for them in ancient times. But today, there are a few herbal product manufacturers in India who produce extracts, decoctions, infusions, tinctures, steam distillates, etc. The following are some ways that herbs can be used [32].

Infusions

These are essentially powerful herbal teas that can be made in stainless steel or China clay pots. The use of aluminum vessels is discouraged because they can contaminate infusions. By briefly macerating the raw medication using cold or boiling water, fresh infusions are created. The readily soluble components of crude medicines are present in these diluted solutions.

Decoctions

The herb and water are boiled to make these decoctions. In this procedure, the raw medication is boiled for a predetermined amount of time in a specific amount of water, cooled, and then filtered or refined. This method works well for extracting components that are heat- and water-stable. The "quath" or "kawath" Ayurvedic extracts are commonly prepared using this method. The initial ratio of crude medicine to water is fixed, e.g., 1:16 or 1:4, and throughout the extraction process, boiling reduces the volume to 1/4 of what it was initially. The purified extract is then either used as is or subjected to additional processing.

Leaf Extracts and Flavorings

Extracts are often made using highly alcoholic hydrocarbon solvents. Extraction, as the term is used in the pharmaceutical industry, entails the consistent use of certain solvents to separate the medicinally active components of animal or plant cells from their inactive or toxic parts extraction techniques. The somewhat impure liquids, semi-solids, or powders produced in this way by plants are only fit for external or oral consumption.

Decoctions, Infusions, Fluid Extracts, Tinctures, Pilular or Semi-Solid, and Powdered Extracts

These are some of the preparation classes that fall under this category. These remedies are sometimes referred to as "galenicals" after the Greek physician Galen, who lived in the second century. Standardized extraction techniques are used to obtain the therapeutically required amount of crude medicine and to remove the inert material using a selective solvent called a "menstruum." The extract obtained in this manner might be ready for immediate use.

Floral Waters

These beverages are prepared similarly to infusions. You can use the same ratios of herbs to water [18, 19, 33, 34].

There are many benefits of natural cosmetics over conventional cosmetics, which include the following [34]:

- They generally cause fewer side effects or allergic reactions.
- They blend easily with skin and hair.
- These cosmetics are far more efficient than other small-quantity cosmetics.
- Plant extracts can enhance texture and performance of cosmetic formulations.
- Easily accessible and present in a wide range of plants. They have more stability, purity, and efficacy with their herbal constituents.
- Herbal cosmetics are easy to handle, cost effective, and store for a longer period of time.

Foods for Skin Health and Beauty

The relationship between some dietary compounds and skin health as well as appearance is undeniable and has been long-known (e.g., vitamin C and vitamin A). Vitamin C deficiency can result in impaired wound healing and skin fragility, due to the role of the vitamin as an antioxidant and in collagen synthesis. It is well known that retinoic acid (a form of vitamin A) and its derivatives can modulate the expression of genes involved in cellular differentiation and proliferation at the skin level. Hence, their many therapeutic and cosmetic applications. In recent decades, additional compounds have been used for cosmetic purposes. These compounds aim to combat photoaging characterized by undesirable signs such as wrinkles, telangiectasia (spider veins), or hyperpigmentation. Another target is skin color, which is not only due to melanin but also due to carotenoids and hemoglobin derivatives. In this scenario, a new buzzword, "nutricosmetics," is becoming popularized in the commercial, technical, and scientific literature. Although there is no consensus definition on it, the term revolves around the oral consumption of products containing food components seeking a cosmetic benefit. Related terms are "beauty pills," "beauty from within," "beauty foods," "nutraceuticals for skincare," or "oral cosmetics" [20, 21, 36, 37].

Aesthetics and Public Health

An advantage of nutricosmetics is that their typical components can contribute to improved health either systemically or at least in other locations in addition to the skin. Due to a series of insightful and brilliant works, it is now becoming particularly timely in terms of making the most of foods, byproducts, or wastes. This approach can result in the production of diverse bioproducts for the agrofood, health, and cosmetic industries. This includes bioactive-rich extracts for formulation, bioplastics for packaging, biosurfactants, biogas, or soil amendments, contributing to a circular economy. Last but not least, Research, Development, and Implementation (R+D+I) on nutricosmetics will need to include life cycle assessments (LCA) to evaluate environmental impacts associated with all stages of the life cycle of the products.

It is apparent that aesthetic benefits (specifically color imparted by carotenoids) can be linked to the consumption of health-promoting foods, such as fruits and vegetables. In turn, this could be somehow harnessed in public health to promote healthy dietary habits in the future as envisioned by Whitehead et al. [17, 38] and others since the introduction of this concept. In this case, a clear association between the consumption of carotenoid-rich foods and skin color change can be established. Future studies establishing clear aesthetic associations with other health-promoting food components will certainly contribute to the foundations of this emerging new strategy.

Generally, the term "cosmetics" evokes a range of products usually resulting from the formulation and processing of several ingredients. More importantly, there are several food groups featured in recommended reference diets (cereals, fruits, vegetables, legumes, nuts, unsaturated oils, etc.) that are rich sources of compounds with putative cosmetic effects and could be considered natural dietary nutricosmetics. The current need to optimize resources and contribute to protecting the environment in alignment with the United Nation's Sustainable Development Goals [18, 39], the Paris Agreement [19, 40], or the European Union's Green Deal [20, 41], among other frames, should pave the way for the development of innovative products. Such products cater to the preferences of growing numbers of sustainability-concerned consumers, a group that is expected to grow in the near future.

Nutricosmeceutical

The word "nutricosmeceutical" was coined by merging the terms "nutraceuticals" and "cosmeceuticals." It is a term used by the industry to define supplements, functional foods, and drinks that include active substances that can improve human attractiveness and health. All this despite the fact that it is not recognized by the FDA or EU regulations, the nutricosmetics market is benefiting greatly from excellent economic conditions and the increase in the older population. From 2017 to 2025, the global market for nutricosmetics will expand at a 5.0% Compound Annual Growth Rate (CAGR), reaching USD 7.93 billion. The market for nutricosmetics is divided into three categories: ingredient type, application (skincare, haircare, and nailcare), and geography. As the average age in Europe rises, cosmeceuticals for the elderly are expected to grow in popularity. However, herbal supplements will become more popular in the United States, which has been recognized as the region's greatest impact, rendering the commercial driver. Skincare is expected to increase at a 4.5% Compound Annual Growth Rate (CAGR) to $194,961 million in 2024 [10]. The global market for snail beauty products (snail extracts, snail mucus, snail oil, snail serum, or snail filtrate) has grown due to an increasing demand for natural cosmetics. Snails are a popular meal among many people. Secondary metabolites are a good source of ingredients that can be employed. Escargot (snail meat) is high in protein, despite their poor fat content and mineral sources. However, according to HMSO, the n6/n3 and PUFA/SFA ratios were in a healthy range. The pharmacologic properties of snail mucin components have been extensively researched and documented. Snail mucus is used in cosmetics to promote the production of collagen, elastin, and dermal components, which help to heal photoaging symptoms and decrease free radical damage.

The importance of beautification to mankind has been known since prehistoric times, and the desire to look beautiful and healthy has been developing in today's society. In recent times, fitness, good health [41, 42], looks, and way of presenting oneself are all being counted as one of the qualities of personality. People today are often evaluated based on these factors. Therefore, there is a necessity and desirable requirement to give more attention to one's looks and beautification for social acceptance and professional success, which has also been overemphasized by the media.

Cosmeceuticals

Today, the term 'cosmetics' is widely recognized. It plays a part in maintaining the health of the skin in a natural and organic way. There is a clear distinction between cosmetics and cosmeceuticals. The word cosmeceutical is a combination of the words cosmetic and pharmaceutical and means a combination of cosmetics and pharmaceuticals [9]. The word "cosmeceuticals" was first introduced by Albert Kligman in 1984 [21] to refer to substances that persists for both their cosmetic and therapeutic effects. Cosmeceuticals or Nutricosmetics are cosmetic products that consist of biologically active ingredients and provide drug-like or medical benefits. It is a topical product with therapeutic benefits that protect against skin damage.In essence, a cosmeceutic is a skincare item that includes a physiologically active ingredient that is supposed to have medicinal skin benefits. This does not imply that all cosmetic products are made entirely of synthetic chemicals. It encompasses cosmetic actives with therapeutic, disease-fighting, or healing properties, thereby serving as a bridge between personal care products and pharmaceuticals [10, 15]. The global cosmetics industry is shifting away from synthetic products in favor of natural ones for hair and scalp care since they often have fewer adverse effects. In some spray-oriented cosmetics, natural plant extracts are employed as an anti-inflammatory and a fragrance. Hence, it is transformed into an actual cosmeceutical. Although the term is used in science, regulatory agencies such as the FDA do not recognize it as a separate category. This idea is also opposed by several government agencies in charge of drug regulation.

Cosmeceuticals are the fastest-growing segment in the beauty industry. Cosmeceuticals are topically applied, because they contain substances that are known to improve the biological activity and function of the skin. They improve appearance by providing nutrients that are necessary for the skin. They typically claim to improve skin tone, texture, and radiance, while reducing wrinkling [10]. The Food, Drug, and Cosmetic Act defines cosmetics according to their intended use, as "an article intended to be rubbed, poured, sprinkled, or sprayed on, or introduced into, or otherwise applied to the human body for cleansing, beautifying, promoting attractiveness, or altering the appearance. Products included in this definition include skin moisturizers, lipstick, finger nail polishes, eye and facial make-up preparations, permanent waves, hair colors, shampoos, deodorants, and toothpaste, as well as many other products which are intended to be used as cosmetics [33–35].

Cosmeceutically active ingredients are constantly being developed by big and small corporations engaged in pharmaceuticals, biotechnology, natural products, and cosmetics. Desirable features of cosmeceutical agents include efficacy, safety, formulation stability, novelty, patent protection, and they are metabolized within skin and are inexpensive to manufacture [35].

There are more than 400 cosmeceutical manufacturers, including companies that supply the cosmeceutical and/or manufacture cosmeceuticals to the US market. The biggest companies in the industry for finished products are Procter and Gamble,

Johnson and Johnson, L'Oréal, Estee Lauder, Avon, and Allergan, which nearly represent half of the US market [9]. The efficacy of a cosmetic product depends not only on the active ingredients but also on the delivery system to improve its efficiency. Skin is more than an assembly of several layers of cells as corneocytes, keratinocytes, and fibroblasts while complementing each other. Chemical exposure, pollution, stress, infrared and ultraviolet radiation and abrasion are involved in skin aging. The consequences can be visible or invisible: inflammation, burns, edema, long-term illness, actinic damage, and premature aging. Most cosmeceuticals claim to provide antiwrinkle and firming, moisturizing and lifting, skin toning and whitening activity, and the antiaging category of skincare products, which presently shows the highest commercial growth rates [36].

Characteristics of Cosmeceuticals

Cosmeceuticals can be characterized as follows:

1. The product has pharmaceutical activity and can be used on normal or near-normal skin.
2. The product should have a defined benefit for minor skin disorders (cosmetic indication).
3. As the skin disorder is mild, the product should have a low-risk profile [37].

Skin Cosmeceuticals

Skin Cosmeceuticals comprise most of the products that are part of our everyday grooming. Our skin is the largest organ of our body, which separates and protects the internal environment from the external environment, including air pollution, exposure to sunlight, etc. It is essential to protect and preserve the skin for good health. UV radiations coming from the sunlight penetrate the skin and accelerate damage due to free radicals, which consist of wrinkling, inflammation, and hyperpigmentation. The collagen and elastin fibers of the skin, due to prolonged exposure to UV radiation, are broken down by enzymes, collagenase, and elastase, and thus the texture of the skin deteriorates. Collagen and elastin are responsible for maintaining the elasticity and integrity of the skin.

Numerous plant extracts and antioxidants that are obtained from natural sources are able to prevent aging and also improve the appearance of the skin [38].

Botanicals Applicable as Cosmeceuticals for the Skin

A few of the most important botanicals pertaining to cosmeceutical use include teas, soy, pomegranate, date, grape seed, pycnogenol, horse chestnut, German chamomile, curcumin, comfrey, allantoin, and aloe. All are documented to treat dermatologic conditions. Only green and black tea, soy, pomegranate, and date have published clinical trials for the treatment of parameters of extrinsic aging [39].

Cosmeceuticals have either drug-like or some medicinal benefits. They are therefore able to affect the biological functioning of the skin by boosting collagen growth by combating the harmful effects of free radicals, and as such, they are able to maintain the keratin structure and make the skin healthier [40].

Types of Skin Cosmeceuticals

A. Based on their function, cosmeceuticals are classified as follows:

   - Skin-whitening and/or depigmenting cosmeceuticals
   - Moisturizing cosmeceuticals
   - Antiwrinkle cosmeceuticals
   - Sunscreen cosmeceuticals
   - Antiphotoaging cosmeceuticals

B. Based on various products presently on the market, cosmeceuticals can be divided into the following broad categories [41]:

   - Antioxidants
   - Growth factors
   - Peptides
   - Metals
   - Anti-inflammatories/botanicals
   - Polysaccharides
   - Pigment lightening agents

Cosmeceutical Ingredients with Their Sources

The cosmeceutical ingredients with their sources from which they are obtained are shown in Table 2.

Preservatives for Cosmeceuticals

- Formic acid, benzoic acid, and other organic acids.
- Alcohols: Isopropyl alcohol and ethanol.
- Aldehydes include cinnamic aldehyde and formaldehyde.
- Mercury: Thiomersol and nitromersol.
- Phenolics: Cresol and phenol.
- Esters: Methoxy p-hydroxybenzoate and ethyl p-hydroxybenzoate.
- Surface active agents: Benzalkonium chloride and cetylpyridinium chloride.
- Several substances, including vanillin and ethyl vanillin.

Hair Cosmeceuticals

Hair plays an important role in personal appearance and identity. Both men and women regard an abundance of hair as ideal, and providing them with positive self-image attributes such as beauty, strength, virility, youthfulness, and confidence. Conversely, the lack of hair is associated with negative attributes, and people have gone to extremes to conceal their baldness.

Historically, Egyptians used artificial braids to hide their baldness, and outlandish hair concoctions (e.g., snake oil) were prolific. In more recent history, minoxidil topical solution became the first clinically approved safe and effective hair-growth stimulant. A 2% concentration of minoxidil topical solution was approved in 1986 for men and in 1991 for women. A higher, more effective concentration (5%) is now available for use [42]. A hair cosmeceutical product includes conditioning agents, special care ingredients, and hair growth stimulants. Conditioning agents are intended to impart softness and gloss, to reduce flyaway, and to enhance disentangling facility. A number of ingredients may be used, mostly fatty ingredients, hydrolyzed proteins, quaternized cationic derivatives, cationic polymers, and silicones [43].

Shampooing is the most common form of cosmetic hair treatment. The diversity of qualities expected from a shampoo by today's consumer surpasses the primary function of cleansing. Current shampoo formulations are adapted to the variations associated with hair quality, haircare habits, and specific problems related to the superficial condition of the scalp. Hand in hand, test methods are developed to evaluate the efficacy of haircare products so that consumers are offered products that perform as claimed. Through the development of cosmetics with pharmaceutically active compounds, products are evolving that are becoming more similar to topical therapeutic agents (cosmeceuticals). The efficacy of the drive for a change in lifestyle includes the growing interest in herbs. This concept is based on the idea that plants have a huge potential for use as therapeutic agents. However, the continued use of herbs as a lifeblood of cosmetics is the quality of its materials. Thus, they must be used with caution. The best way to reduce skin issues is with cosmetic items.

The future is bright for the cosmetics industry. For beauty salons, yearly shopping positions range from $100 USD to $1500 USD, and a smaller percentage of annual purchase positions. Several of the ingredients used to make cosmetics have the potential to cause negative side effects when applied to the human body. Hence, utilizing cosmetics may not affect the body. The best cosmetic options for sensitive skin are mild, fragrance-free formulations. There is a huge demand for cosmetics and a lot of room for growth in the use of cosmetics in the coming years. The use of cosmetics has multiplied in the personal care system. While cosmetics are utilized by people of all sexes and play a significant role in daily life, their scope also plays a significant role. Cosmeceuticals that claim to act as hair growth stimulants should be measured by the standards set by the drugs minoxidil and, more recently, oral finasteride [44]. A liquid shampoo that contains biologically obtained protein detergent has been patented by Jack J. Mausner, which provides unusual mildness to the hair and scalp and also possesses hair conditioning properties [45].

A haircare cosmetic composition for the treatment of hair, in particular oily hair, has been obtained, which is based on an extract of Yarrow (Achillea Millefolium L). This composition includes an appropriate carrier and a nonallergenic dry extract of Yarrow, obtained by oxidation of the flower tops of Yarrow. The extract contains less than 0.5% by weight of polyphenolic derivatives [46]. A method for the treatment of androgenic alopecia is patented by Buck, where Liquor Carbonis Detergens (Coal Tar Solution) is topically administered [47].

Cosmeceuticals with Other Uses

The skin underneath the eye lacks subcutaneous fat and also oil glands. Therefore, products to be formulated for this particular area need to be extremely gentle. This delicate skin should be provided protection with plenty of moisture to replenish and repair. This will help reduce the early signs of premature aging. There are numerous cosmeceutical eye creams available in the market with natural emollients and beneficial nutrients. For instance, Zia Natural Skincare offers two different eye creams. The Ultimate Eye Crème contains antioxidants, vitamins A, C, and E, green tea, and tiare flower in a nongreasy base. Jensen said, "The Essential Creme is for allergic and puffy eyes, and is great for recovery after sad movies." The functional ingredients include butcher's broom, chamomile, and vitamin E [48]. The other functional ingredients include butcher's broom, chamomile, and vitamin E, antioxidants such as vitamins A, C, and E, green tea and tiare flower, Ginkgo biloba, and also cucumber, calendula, and a-bisabolol, an active constituent of chamomile, to calm irritated skin. A key ingredient in the eye-lifting moisture cream, which treats puffiness, irritation, and also protects against future skin damage, is yeast, which helps to plump up the wrinkles. The eye-wrinkle cream helps forestall the signs of aging and generally contains wheat germ, corn oil, squalene, and carrot extract. Eye firming fluid has aosain, an algae extract from seaweed that helps the skin to maintain elasticity [49].

Nanotechnology in Cosmeceuticals

Nanotechnology is regarded as the most imminent technology of the twenty-first century. It is contemplated to be a big boon for the cosmetic industry [50]. The term nanotechnology is derived from the combination of two words: the Greek word "Nano," which means billionth, and the word "technology." Nanotechnology or Nanoscaled technology generally refers to be at a size below 100 nm (10^{-9} m) [51]. Cosmeceuticals are known as the fastest-growing segment of the personal care industry. Although knowing the enormous benefits of nanoparticles, there is very little information on the long-term and short-term effects on the environment and the human organism itself. Safety concerns have been raised due to the reported toxicity and possible dangers posed by nanomaterials [52].

Major Classes of Nanocosmeceuticals

Moisturizers—When moisturizers are applied to the skin, a thin film of humectant material is formed, which retains moisture and gives a better appearance to the skin. Liposomes, nanoemulsions, and SLNs are widely used moisturizing formulations because of their prolonged effects. These are considered to be the most useful products for the management of various skin conditions (e.g., atopic dermatitis, psoriasis, and pruritus).

Sunscreens—Sunscreens are widely used to protect the skin from harmful radiation on exposure from the sun [53]. Sunscreens are used to provide protection against the adverse effects of ultraviolet UVB (290–320 nm) and UVA (320–400 nm) radiation. According to the United States Food and Drug Administration, the protection factor against UVA should be at least one-third of the overall sun protection factor. Titanium dioxide (TiO_2) and zinc oxide (ZnO) minerals are frequently employed in sunscreens as inorganic physical sun blockers. As TiO_2 is more effective in UVB and ZnO in the UVA range, the combination of these particles assures a broad-band UV protection [54].

Antiaging Products—Chemical products, pollution, stress, irradiation from infrared (IR) and ultraviolet (UV) sources, and abrasion are involved in skin aging. Collagen plays an important role in skin rejuvenation and wrinkle reversal effects. The quantity of collagen in the skin decreases along with increasing age. The aging of the skin manifests itself in many ways: drying out, loss of elasticity and texture, thinning, damaged barrier function, appearance of spots, modification of surface line isotropy, and finally, wrinkles. Most of the cosmeceuticals have been developed with claims of antiwrinkle, firming, moisturizing, lifting, skin toning, and whitening activity. Antiaging products are the main cosmeceuticals in the market currently being made using nanotechnology. L'Oréal has employed nanotechnology in products such as Revitalift antiwrinkle cream, which contains nanosomes of Pro-Retinol A, and claims that it instantly retautens the skin and reduces the appearance of wrinkles. Application of retinol can increase epidermal water content, epidermal hyperplasia, and cell renewal while enhancing collagen synthesis. Retinol also interferes with melanogenesis and inhibits matrix metalloproteinases, which are involved in collagen breakdown [55].

Haircare—Haircare is another promising field for nanotechnology. Companies are using nanotechnology in haircare products. Research is ongoing to discover the mechanisms by which nanoparticles can be used to prevent hair loss and to maintain shine, silkiness, and health of hair. Unlike ordinary hair straightening products, nanoemulsions in hair cosmetics do not destroy the outer structure of the hair fibers, called cuticles, to penetrate into the hair strands [56]. Sericin nanoparticles are used in many cosmeceuticals and are observed to be an agent that imparts gloss and softness and promotes maintenance of coloration in dyed hair [57].

Skin Cleanser—The art of cleansing has made great progress and has improved over the thousands of years by simply scraping the skin to an exercise in relaxation and improving skin's health and appearance. Soaps were the basic cleansing agents; however, they also have undergone a series of changes [58]. Skin cleansers may be an important adjunct to the regimen of those who use cosmetics, have sensitive or compromised skin, or utilize topical therapies. Cleansers emulsify dirt, oil, and microorganisms on the skin surface so that they can be easily removed [59]. Silver nanoparticles are used as a skin disinfectant and for decontamination. Nano Cyclic Inc. has produced Nano Cyclic cleanser pink soap, which is a scientifically balanced blend of nanosilver and natural ingredients that claims to kill harmful bacteria and fungi, fight acne, and diminish age spots and sun-damaged skin. It also consists of sericin and collagen [60].

Other Cosmeceuticals and Their Common Myths and Misconceptions

Cosmeceuticals are opened to and have been subjected to many misconceptions. Many consumers believe that cosmeceuticals are regulated and tested as drugs. They also believe that the ingredients and final products have been tested for safety and the claims made in advertisements are valid. Cosmeceuticals and pharmaceutical ingredients have never been closer together, and their regulatory environments are vastly different due to the distinct Congressional mandates given the Food and Drug Administration. Physicians are in a good position to help patients and potential cosmeceutical users understand the benefits and realistic limitations of these products [61].

Toxicity of Cosmeceuticals

"Natural" is the term which is most frequently used for cosmeceuticals and willingly or unwillingly connotes safety. Vitamin E has been shown to cause an increase in contact dermatitis, while the antioxidant p-hydroxyanisole increases skin pigmentation. Some peptide components have also been shown to be carcinogenic. Perhaps the deliberate adulteration and the incorporation of harmful products like steroids and retinoids are supposedly the greatest danger. Microbial contamination has been reported with unfavorable consequences [62].

Regulatory Aspects of Cosmeceuticals

Regulatory agencies have not yet formally recognized cosmeceuticals.

Although there is no legal class called cosmeceuticals, this term has gained application and recognition to designate the products at the borderline between cosmetics and pharmaceuticals. Cosmeceuticals are not subject to review by the FDA or any other equivalent agency. The regulation of cosmeceuticals has not been harmonized between the United States, European, Asian, and other countries. Hence, there is a need for implementing increased regulation of cosmeceuticals that would require proof of safety, which is not mandatory for cosmetics [63].

Cosmeceuticals

The term "cosmeceutical" was introduced by dermatologist Dr. Albert Kligman in 1984 and is derived from a combination of the words cosmetic and pharmaceutical.

What Are Cosmeceuticals?

Cosmeceuticals are products that have both cosmetic and therapeutic (medical or drug-like) effects, and are intended to have a beneficial effect on skin health and beauty. Like cosmetics, they are applied topically as creams or lotions but contain active ingredients that have an effect on skin cell function. In some cases, their action is limited to the skin surface (e.g., exfoliants), while others can penetrate to deeper levels, either enhancing or limiting normal skin functions.

Cosmeceuticals are available "over-the-counter" (without prescription) and are generally used as part of a regular skincare regimen to help improve skin tone and texture, pigmentation, and fine lines.

Most moisturizers restore barrier function and water content to the skin, improving the appearance of aged or dry skin. Cosmeceuticals should ideally deliver the active ingredient in a biologically effective form to the skin and reach the target site in sufficient quantity to have an effect.

What Are Some Common Ingredients in Cosmeceuticals?

- Sunscreens—These are probably the most important ingredient in cosmeceuticals because they protect against sun damage, photoaging, and skin cancers [64]. (A-Z of skin sun protection and sunscreens).
- Antioxidants—External factors such as ultraviolet (UV) radiation, pollution, and smoking, as well as internal factors including normal cellular metabolism, can generate molecules called free radicals, which are damaging to the skin. Antioxidants "mop up" these free radicals, thereby reducing inflammation and protecting the skin against sun damage and skin cancers. Some studies suggest that combinations of antioxidants can be more effective than single-ingredient formulations. Examples of antioxidants include:
  - Alpha-lipoic acid: It has anti-inflammatory and exfoliating effects. In 2020, the Australasian College of Dermatologists issued a statement that this was for personal use only as a disclaimer for this ingredient.
  - Vitamin C (L-ascorbic acid): It stimulates collagen repair and can improve fine lines, reduce inflammation, and pigmentation. Although it is found in a number of cosmeceutical products, many are not effective because the vitamin C is unstable when exposed to air, heat or light, is in too low a concentration, or is in a form that cannot be absorbed or metabolized by the skin.
  - Nicotinamide (vitamin B3): An antioxidant that improves skin barrier function. It can reduce fine lines, wrinkles and hyperpigmentation, and improve skin texture. It may also play a role in skin cancer prevention.
  - Vitamin E (alpha tocopherol): Another antioxidant that reduces UV damage and skin cancer. It also works synergistically with vitamin C in reducing collagen breakdown.
  - N-Acetyl-Glucosamine (NAG): It may help fade pigmentation and prevent sun damage (photo damage).
  - Ubiquinone (CoQ10): A naturally occurring antioxidant that reduces collagen breakdown due to sun exposure.
- Hydroxy acids: These can be classified according to their molecular structure into alpha-hydroxy acids (AHAs), polyhydroxy acids (PHAs), and beta-hydroxy acids (BHAs). Hydroxy acids improve skin texture and reduce the skin signs of aging by hydrating the skin and promoting the shedding of dead skin cells from the outer layer of the skin (epidermis).

- AHAs: They are often called "fruit acids" as many are derived from natural fruit sources. AHAs include glycolic acid, lactic acid, citric acid, mandelic acid, malic acid, tartaric acid, and lactobionic acid.
- PHAs: These include gluconolactone and lactobionic acid.
- The main BHA used is salicylic acid, which is particularly useful in people with oily or acne-prone skin because of its fat solubility and ability to penetrate pores.

- Retinoids (vitamin A)—These are natural or synthetic forms of vitamin A that can partially reverse skin changes induced by sun exposure. Common retinoids include tretinoin or retinoic acid, retinol, and retinaldehyde. They act as antioxidants, protecting cells from free radicals, as well as activating specific genes and proteins. Topical tretinoin has been shown to improve the appearance of photodamaged skin by reducing fine lines and wrinkles, skin looseness (laxity), and excess pigmentation, as well as improving skin texture. Tretinoin can cause side effects such as burning, stinging, redness, and flaking.
- Skin lightening agents—These help inhibit the production of melanin (the main skin pigment) to reduce skin discoloration and pigmentation. Examples include:

  - Hydroquinone. This has been the agent of choice for skin lightening for many years. Concerns regarding skin darkening, loss of pigmentation, and possible carcinogenicity have resulted in it being banned from over-the-counter products in some countries. However, these concerns have mainly arisen from animal studies using long-term, high doses and are probably not relevant to topical application in humans. The Australasian College of Dermatologists in 2020 recommended this ingredient for personal use only as a disclaimer for this ingredient.
  - Ascorbic acid (vitamin C)
  - Kojic acid
  - Azelaic acid
  - Licorice extract (glabridin).

- Botanicals—These include plant extracts from leaves, roots, fruits, berries, stems, bark, and flowers. Botanicals may have antioxidant, anti-inflammatory, and/or skin soothing properties; however, their effects remain largely untested or unproven. Examples include soy, curcumin, silymarin, pycnogenol, ginkgo biloba, green tea extract, grape seed extract, aloe vera, witch hazel, allantoin, and ferulic acid.
- Peptides and proteins—Peptides are short chains of amino acid sequences that are the building blocks of larger proteins. Cellular "messengers" formed from amino acids can imitate normal biological signals that either stimulate repair or inhibit processes that accelerate skin aging. Examples include the pentapeptide Pal-KTTKS.

- Growth factors—These proteins help control chemical signals between and within cells. They are important in wound healing and repair of damaged tissue, and may help to repair skin damage from sun exposure. Studies suggest that the use of multiple growth factors can stimulate collagen and elastin production and improve the appearance of photo-damaged skin.

Do cosmeceuticals really work? Ideally, cosmeceuticals should be clinically tested to ensure they have a proven benefit and can substantiate their claims; however, the cosmeceutical industry is largely unregulated. Unlike medicines, cosmeceuticals are not subject to review by the Food and Drug Administration (FDA) in the United States [66] or the Therapeutic Goods Administration (TGA) in Australia [67]. Although they are usually tested for safety, they do not have to undergo testing to ensure the claims they make regarding efficacy (effectiveness) are accurate. Unfortunately, many creams do not live up to their advertised hype. What is the future for cosmeceuticals? Research is continuing into new delivery systems (such as liposomes) and new active ingredients. The potential risks and side effects which can be seen with the use of Cosmeceuticals are given in Table 1. The various abnormalities that can occur in the absence of various minerals are given in Table 2. Ingredients that are deemed unsafe and their reason for discontinuance in Cosmetics are given in Table 3.

Table 1 Potential risks and side effects of cosmeceuticals

Potential risks	Side effects
Skin irritation	Cosmeceuticals containing active ingredients like retinoids or alpha-hydroxy acids may cause skin irritation, characterized by redness, itching, or burning sensation
Allergic reactions	Some individuals may experience allergic reactions to ingredients commonly found in cosmeceuticals, leading to symptoms such as rash, swelling, or hives
Photosensitivity	Certain cosmeceutical ingredients, like retinoids and certain essential oils, may increase skin sensitivity, resulting in sunburn or skin damage if adequate sun protection is not used
Overexfoliation	Excessive use of exfoliating cosmeceuticals, such as scrubs or chemical peels may lead to overexfoliation, causing skin dryness, sensitivity, or even inflammation
Skin barrier disruption	Active ingredients like strong acids or high concentrations of certain compounds may disrupt the skin barrier function, leading to increased susceptibility to infections or irritants
Pigmentation changes	Some cosmeceuticals, particularly those containing hydroquinone or certain acids, may cause changes in skin pigmentation, including hyperpigmentation or hypopigmentation

Table 2 Abnormalities that can occur in the absence of various minerals

Name of minerals	Major abnormalities occurred in the absence of the mineral
Zinc	Zinc deficiency has a bad impact on dermal, nervous system, immunity, gastrointestinal tract, skeletal, and reproductive systems [68]; in children, lack of zinc may reduce the number of lymphocytes (T-cells), in the blood and peripheral lymphoid tissues [69]; major populations remain at high risk of HIV [70] Acevedo-Murillo et al. (2019); viral infection cases enhance with zinc deficiencies
Selenium	Less selenium use has increased the risk of mortality, poor immune function [71]; in the absence of selenium, patients with AIDS, the health condition may worsen [72]; heart-damaging effects of the cytomegalovirus may occur [73]; the antioxidant activity of glutathione will not work properly to remove excess damaging radicals [74]
Copper	A big loss of immunity in the absence of copper [75]; patient may increase antiviral activities—Impaired intestinal absorption of nutrients [76]
Magnesium	Decreased immune function [76]
Iodine	Goiter and hypothyroidism [76]; delayed physical development and impairment of mental function in adults [77]; deficiency leads to brain damage [78]; thyroid hormone secretion is highly influenced by iodine [79]
Iron	Physical build-up, immune strength, and behavior worsened, apart from increased infection rate [80]
Manganese	For normal immunity function, blood sugar regulation, and development, along with the function of the brain [81]; skeletal defect, epilepsy, and abnormal absorption of carbohydrate and lipid [82]

Table 3 Cosmetic ingredients that are unsafe and the reasons for their discontinuance

Substance	Reason
Chloroacetamide (a preservative)	Because of sensitization (development of allergic reactions)
Ethoxyethanol and ethoxyethanol acetate (a solvent)	Because of reproductive and developmental toxicity
HC Blue No. 1 (a hair-coloring ingredient)	Because of possible carcinogenicity
p-Hydroxyanisole (an antioxidant)	Because of skin depigmentation
4-Methoxy-m-phenylenediamine, 4-methoxy-m-phenylenediamine HCl, and 4-methoxy-m-phenylenediamine sulfate (hair dye ingredients)	Because of the possibility of carcinogenicity
Pyrocatechol (used in hair dyes and skincare preparations)	Because of carcinogenic and co-carcinogenic potential (CIR describes this substance as unsafe for leave-on products and considers available data insufficient to assure safety for use in hair dyes)

Herbals and Skincare

Why Is Skincare Important?

Skin is often the first thing noticed and the last thing cared for, but your skin does more than meets the eye. Acting as our body's suit of armor, our skin shields us from the elements and battles unwanted invaders.

So, why focus on skincare? Because the health of your skin is a daily reflection of your overall well-being. It is about more than just looking good—it's about feeling good and embodying confidence in your own skin. What Are Some Signs Your Skin Health Is Lacking?

Ever looked in the mirror and wondered if your skin was trying to tell you something? Like any aspect of our health, our skin sends signals when it's not at its best. Understanding these signs is crucial because they are often the first indicators that something within needs a little more attention. Some of the most common signals that might suggest your skin health could use a boost include:

1. Dullness and Lack of Radiance: Healthy skin glows naturally. If your skin looks consistently dull and tired, it might be calling for more nutrients or hydration.
2. Dryness and Flakiness: This can be more than just a seasonal issue. Persistent dryness indicates that your skin barrier may be compromised, needing reinforcement from nourishing and hydrating ingredients.
3. Redness and Irritation: Occasional sensitivity can happen, but ongoing redness or irritation suggests underlying inflammation or sensitivity to current skincare products or environmental factors.
4. Increased Breakouts: Sudden or persistent breakouts can be a sign of stress, hormonal imbalance, or inadequate cleansing practices that leave pores blocked.
5. Fine Lines and Wrinkles: While natural aging is a beautiful sign of life's experiences, premature wrinkles may indicate oxidative stress or poor elasticity, pointing to a need for antioxidant-rich and collagen-supportive nutrients.
6. Uneven Skin Tone or Hyperpigmentation: Dark spots or uneven coloring can result from sun exposure, aging, or scarring, showing that your skin may require healing components to even out its appearance.

Practical Use of Herbs in Skincare

Herbs are nature's versatile powerhouses, used across the globe in both kitchens and medicine cabinets for their vast array of flavors, scents, and health benefits. What exactly are herbs? Simply put, they are plants valued for their aromatic, savory, or medicinal qualities. In the culinary world, herbs such as basil, parsley, and mint add depth and excitement to dishes. In the realm of medicine, herbs have been foundational to traditional remedies, trusted for thousands of years.

The use of herbal remedies in skincare traces back to ancient civilizations—from the Egyptians, who utilized aloe vera to soothe and moisturize the skin, to the

Greeks and Romans, who praised the healing powers of lavender and rosemary. These plants have long been revered not just for their aromatic qualities but for their ability to heal, protect, and nourish the skin.

The secret lies in the natural compounds found within herbs, which can have antioxidant, soothing, and hydrating properties. This rich history not only underscores their enduring relevance but also highlights their proven efficacy. It reminds us that sometimes the best solutions come straight from the earth.

You now understand the importance of skincare, and you have the scoop on herbs, but which ones are truly the best for maintaining and enhancing skin health? Some of the most beneficial herbs that have been celebrated for their ability to nourish and support skin will be presented here.

Like any skincare product, herbs can have their quirks, too. While side effects are rare when it comes to natural herbs, everyone's skin is a unique landscape. What soothes one person's skin might irritate another person's.

Skin sensitivity is an important factor to keep in mind. Even natural products can cause reactions, especially if you have sensitive skin. It's a good idea to do a patch test on yourself before introducing a new herb into your routine. Apply a small amount on your wrist or behind your ear to see how your skin feels about it after 24 h.

The purity of ingredients is extremely important. With herbs, the source matters. One should opt for high-quality, pure extracts that ensure you're getting the herb's benefits without unwanted additives. Lesser quality options might not give you the full effect and can sometimes harbor contaminants that don't play well with skin. This is why it is important to know about any interactions of herbs with other ingredients.

Herbs are friendly, but they like to know who they're being mixed with. Herbal extracts can interact with other products, affecting efficacy or causing irritation. Make sure to check compatibility, especially if you're using prescription skincare.

`Taking these considerations into account can help ensure your herbal skincare is as smooth and beneficial as possible. The goal is to enhance your skin's health, not complicate it.

When making or choosing an herbal skincare product, make sure to pick the ones containing the best oils and avoid chemical preservatives. Certified organic, cold-pressed oils such as olive, sesame, and almond are plant-based favorites for daily use, while some also enjoy beef tallow or sheep lanolin for rough, chapped skin. For salves, pick beeswax (or vegan carnauba wax) instead of petroleum-based paraffin.

Herbs and Their Healing Properties

The following herbs and their contributing activities should be considered or evaluated for their need in a product to be formulated for topical use:

Antibacterial: Cornflower, Echinacea, Rosemary, Lavender
Antioxidant: Cornflower, Rosemary, Rose

Anti-inflammatory: Chamomile, Chickweed, Rosemary, Comfrey, Echinacea, Birch, Licorice, Calendula, Lemon Balm, St. John's Wort.

Calendula, Chickweed, Chamomile blend really well with Thyme and Lavender, making a lovely soothing formula.

Astringent: Astringent herbs will help to tone and tighten the skin, temporarily closing pores and improving the texture. This result is only temporary; however, it is impossible to permanently close pores that are either stretched through the release of excess oil or hereditary. The following herbs make a great inclusion for toners: Horsetail, Sage, Birch, Ladies Mantle, Lemon Balm, Rosemary, Sage, Witch Hazel, Yarrow, and Basil are great examples of astringent herbs.

How to Use Herbs

Holistic Approach: Use herbs for a natural skincare routine, avoiding harsh chemicals and nourishing your skin from within.

Eat Herbs: Consume nettles, burdock, and sea vegetables to improve skin health. These superfoods are packed with essential nutrients that support a radiant complexion.

Drink Herbal Teas: Hydrate and detoxify with herbal teas like chickweed, red clover, and chamomile. They help regulate hormones and reduce stress, promoting healthier skin.

Herbal Steams and Washes: Cleanse and hydrate your skin using herbs such as lavender, peppermint, and rosemary. These herbs reduce bacteria and inflammation while providing aromatherapy benefits.

Choose Quality Ingredients: Opt for certified organic, cold-pressed oils like olive, sesame, and almond. Avoid chemical preservatives and choose natural waxes for salves.

Avoid Harmful Ingredients: Skip conventional cosmetics that use harmful chemicals. Herbal skincare supports your skin's natural beauty and health. Can Herbs Replace Your Skincare Routine?

Drink Your Herbs: Everyone knows that hydration is a key strategy for improving skin health. Why not include herbal tea (hot or iced) to enhance elimination, regulate hormones, and address stress?

Moisturize with Herbs: Nourish your skin with herbal oils, salves, and creams that penetrate the skin and provide for deep healing. Herbal oils and creams with Calendula, St. John's Wort, and aloe vera will both deeply nourish and heal your skin.

Wash with Herbs: Herbal steams and washes are a wonderful way to clean and hydrate your skin while relaxing your body and mind. Many fragrant herbs are rich in essential oils, which can cut down unwanted skin bacteria and support lower levels of inflammation. Plus, they provide aromatherapy benefits to reduce stress and promote mental clarity!

Eat Your Herbs: Nourish your skin from within by incorporating herbal superfoods into your diet. These nutrient-rich plants provide essential building blocks for healthy skin and nails. Herbalists emphasize the gut's crucial role in skin health, as the same type of tissue lines both the digestive system and the skin's surface. By optimizing gut health and overall nutrition with herbal foods, you can achieve a radiant complexion. These include:

1. *Nettles*—They are nature's original superfood. It is rich in protein (25% by dry weight), chlorophyll, iron, vitamins A and C, potassium, manganese, silica, and calcium. Nettles make a delicious spring vegetable that can be consumed dry as tea or hair wash year-round, supporting both internal and external health. It is protein-rich and provides essential amino acids for skin and nail health. It is nutrient-dense.

2. *Burdock*—This inulin-rich root vegetable is widely found in Japanese cuisine. It is a natural prebiotic that supports a healthy gut microbiome. Bitter compounds in burdock root also stimulate the production of digestive enzymes, which help break down fats and optimize nutrient absorption.

3. *Sea Vegetables* —"Seaweeds" (algae) such as kombu (brown kelp) and nori (red algae) are rich in minerals such as iodine, iron, and potassium, vitamins A and B-complex, as well as omega-3 fatty acids. These nutrients help the body build healthy skin and nails and can support healing from scars and blemishes. The fiber content of sea vegetables helps support bowel health, which naturally carries toxins out of the body.

4. *Chickweed*—This spring weed is a cooling herb that both nourishes with trace minerals and supports healthy elimination through the liver and kidneys, to clear toxins that lead to breakouts.

5. *Red Clover*—It is rich in phytoestrogens and isoflavones (a type of antioxidant), which protect the skin from oxidative stress. Red clover tea can help regulate hormonal imbalances that lead to blemishes and wrinkles, and promote collagen health.

6. *Chamomile*—It is one of the world's most famous herbs for supporting resting and digesting, which are crucial for overall skin wellness. It has a calming effect to reduce stress, which can improve skin conditions such as acne and eczema. Chamomile is also rich in anti-inflammatory essential oils (including chamazuline) that can assist in soothing irritated skin. Chamomile is not only great for a relaxing tea but also excellent for calming the skin. Known for its soothing properties, chamomile helps maintain the skin's appearance, reducing the look of redness and irritation. It's particularly beneficial for those with sensitive skin types, promoting a more even skin tone and a calmer complexion.

7. *Lavender*—Perhaps the most well-known herbal fragrance is the many varieties of lavender, which can be used fresh or dry, or used as a few drops of essential oil in a steam or bath. It is calming and relaxing. Lavender's sooth-

ing fragrance helps reduce stress and promote mental clarity. Its anti-inflammatory properties help calm irritated skin and reduce inflammation. The antibacterial effects of lavender reduce unwanted skin bacteria, promoting clearer skin. Lavender is loved for its soothing scent and its ability to support skin health. This coveted herb helps to maintain the skin's natural barrier, promoting hydration and reducing the appearance of stress and fatigue on the skin. Lavender is ideal for nightly skincare routines, as it helps to calm both the skin and senses, preparing you for a restful sleep.

8. *Peppermint*—Fresh peppermint, or peppermint essential oil, livens the mood and freshens and revitalizes the skin. By opening pores, peppermint supports the skin in releasing dirt and toxins and absorbing beneficial herbal compounds. It is invigorating and refreshing. Peppermint's fresh scent livens the mood and the skin.
9. *Rosemary*—It is good for more than just cooking, rosemary contains a plethora of bacteria-bashing compounds and inflammation-calming essential oils that reduce skin bacteria, helping to keep the skin clear as well as making it beneficial for soothing irritated skin. Famous for supporting the scalp, rosemary is also great for the rest of the body. In aromatherapy, it promotes mental clarity and reduces stress through its invigorating scent. Rosemary rejuvenates and revitalizes the skin. Rich in antioxidants, it naturally shields your skin from daily environmental challenges, enhancing your complexion's resilience and vibrancy. Its energizing aroma adds a refreshing touch to your skincare routine, making it as therapeutic as it is beneficial.
10. *Calendula*—This bright yellow or orange blossom is a top pick for skin health. From babies' bottoms to moms' nipples to gardeners' rough hands, calendula provides top-notch soothing and healing action for various skin issues. Reduces inflammation and promotes skin regeneration. It is gentle enough for all ages and skin types. Calendula is hailed for its soothing properties and is particularly beneficial for those with sensitive or irritated skin. It helps to maintain the skin's comfort, reducing the appearance of irritation and supporting the skin's healing processes. Calendula is a gentle yet effective herb that can help keep the skin looking healthy and calm.
11. *St. John's Wort*—It is well known for its mood-lifting properties when used internally. St. John's Wort has been used for millennia to calm painful, irritated skin and promote healing. It contains compounds that reduce inflammation and pain. The fresh flowers lend a bright red hue to infused oil and can soothe lip sores, sunburns, and scars. The infused oil's bright red hue indicates its potency and effectiveness.
12. *Aloe Vera*—Many moisturizing creams contain aloe vera gel, which has been widely researched for supporting burns, eczema, psoriasis, and many other skin conditions. Promotes healing and reduces the appearance of scars and blemishes. Aloe vera is a skincare superstar known for its hydrating

qualities. This herb is packed with a gel-like substance that's rich in vitamins and minerals, making it a wonderful natural moisturizer. It supports the skin by maintaining hydration and soothing dry or sensitive areas, helping keep the skin looking fresh and revitalized.

13. *Green Tea*—Green tea is loaded with antioxidants that help support the skin's resilience during exposure to environmental stressors. This herb aids in maintaining skin vitality, contributing to a fresher and more vibrant complexion. It is also known for its properties that support the natural filtering processes of the skin, helping to maintain clear and clean skin.
14. *Turmeric*—Turmeric is renowned not only in cooking but also in skincare for its outstanding natural properties that support skin health. It contains curcumin, a compound that helps maintain the skin's natural glow and can reduce the look of dullness. Turmeric supports the skin's vitality and is a great addition to any skincare routine for its enriching benefits.
15. *Sea Buckthorn*—Sea Buckthorn is packed with nutrients, including vitamins and omega fatty acids, which are essential for maintaining the health and vitality of the skin. This powerful herb supports skin hydration and can contribute to the maintenance of skin elasticity, helping to keep the skin looking youthful and radiant.
16. *Licorice Root*—Licorice root is a hidden gem in skincare, known for its ability to support skin health in several ways. This herb helps to maintain the skin's brightness and can reduce the appearance of uneven skin tone. Licorice root contains natural compounds that support the skin's balance, aid in maintaining a clear and even complexion.
17. *Neem*—Long revered in traditional remedies, neem excels in nurturing the skin's protective layer, ensuring it remains clean and clear. Especially effective for managing oily or blemish-prone skin, neem helps to maintain a balanced and healthy appearance, minimizing imperfections and promoting a smoother complexion.
18. *Ginseng*—Ginseng is a dynamo of vitality for your skin. Known for promoting skin elasticity and firmness, it effectively diminishes the signs of aging, such as fine lines and wrinkles. Its energizing properties invigorate the skin, imbuing your face with a lively and youthful glow that feels as good as it looks.
19. *Witch Hazel*—Witch hazel does wonders for refining skin texture. It's a natural toner that promotes a smoother, more toned complexion, optimizing pore clarity without overstripping. Ideal for incorporation into cleansers and toners, witch hazel helps keep your skin looking pristine and even-toned, providing a soothing and cleansing effect that feels utterly refreshing.
20. *Basil*—Basil wraps up our list with its notable skin-loving properties. This culinary favorite does more than enhance your pesto. It is rich in essential

oils and antioxidants, which are perfect for helping to maintain skin clarity and tone. Basil's natural oils can be especially helpful in managing the skin's oil production, making it a boon for those with oily skin aiming to achieve a more balanced complexion.

21. *Hibiscus*—Hibiscus is rich in antioxidants and has natural alpha-hydroxy acids (AHAs) that help exfoliate the skin, improve skin tone, and reduce hyperpigmentation.
22. *Gotu Kola*—This herb is known for its ability to improve skin elasticity and reduce scarring. It's often used in products aimed at healing wounds and reducing the appearance of stretch marks and wrinkles.
23. *Holy Basil (Tulsi)*—Tulsi has antimicrobial and anti-inflammatory properties. It helps in treating acne, reducing redness, and improving skin texture.
24. *Thyme*—Thyme has antibacterial properties, making it effective against acne-causing bacteria. It also has antioxidants that protect the skin.
25. *Comfrey*—Comfrey contains allantoin, which helps skin regeneration and wound healing. It is effective for treating scars, burns, and inflammation.
26. *Reishi*—Reishi mushrooms have powerful anti-inflammatory and antioxidant properties. They help reduce redness, hydrate the skin, and improve overall skin tone.
27. *Turkey Tail*—Turkey Tail mushrooms are rich in antioxidants and polysaccharides. They help in skin regeneration and protect against environmental stressors.
28. *Chaga*—Chaga mushrooms are rich in antioxidants, which help fight free radicals and prevent premature aging. They also have anti-inflammatory properties that can soothe the skin and reduce redness.
29. *Cordyceps*—Cordyceps mushrooms are known to boost collagen production and improve skin elasticity. They also help in hydrating the skin and reducing the signs of aging.
30. *Maitake*—Maitake mushrooms contain beta-glucans that help retain moisture in the skin, keeping it hydrated and plump. They also have anti-inflammatory properties that can benefit irritated skin.
31. *Shiitake—Shiitake mushrooms contain kojic acid, which is known for its skin-lighteni*ng properties. They also help with antiaging due to their antioxidant content and improve overall skin health by providing essential nutrients.
32. *Lion's Mane*—Lion's Mane mushrooms are known for their neuroprotective and anti-inflammatory properties. They can improve skin texture and reduce inflammation.

Here are some easy and effective ways to add herbs to your daily skincare routine that accomplish your goals:

Do It Yourself Herbal Masks—Whip up a simple herbal mask using ingredients like chamomile or turmeric. Mix these herbs with a bit of honey or yogurt, apply to your face, and let nature work its magic. It's a fun, wallet-friendly way to treat your skin.

Add Herbs to Your Cleanser—Boost your daily cleanser by adding a few drops of herbal extracts like lavender or rosemary. This small tweak can enhance the effectiveness of your cleansing routine, making it more soothing and refreshing.

Infuse Your Moisturizer—Mix a little aloe vera, sea buckthorn oil into your favorite moisturizer. These herbs can provide extra hydration and nourishment, giving your skin a healthy, radiant glow.

Sip Your Way to Glowing Skin—Don't just apply herbs—drink them. Herbal teas packed with skin-loving ingredients like chamomile and green tea can support your skin from the inside out. For a potent herbal boost, try a mixture of a delicious, silky peach supplement with 100% vegan ingredients like aloe vera, sea buckthorn, and centella asiatica. This blend helps keep your skin hydrated, collagen-rich, and glowing.

Herbal Steams—Treat your skin to an herbal steam. Add herbs like rosemary or lavender to hot water and let the steam open your pores, cleanse deeply, and leave your skin feeling refreshed and invigorated.

Conclusion

[67, 83],However, they claim to have a drug-like effect. The term cosmeceutical has no meaning under the law. Cosmeceuticals are not subject to FDA review or approval. However, cosmeceuticals are topical pharmaceutical-cosmetic hybrids that contain biological actives intended to enhance the beauty and health of the skin. The great demand for skin cosmeceuticals has led to the development of products to counteract the signs of aging skin, to decrease erythema, and to even out the tone and pigmentation of the skin. These cosmeceuticals can help to protect the skin from photo damage and, in some ways, repair it through stimulation of new collagen production.

It has also been noted from the literature that hair cosmeceuticals usually consist of actives which are hair growth stimulants, or used to improve hair aesthetic properties such as texture, shine, and manageability. Apart from skin and hair cosmeceuticals, actives which have also been reported to improve the condition of the applied area have been added in various cosmetics such as under-eyes, lips, and nail products. Nanotechnology can enable targeted ingredient delivery. Nanotechnology represents the main and important technologies of the twenty-first century, which offer excellent opportunities for both research and business. The rapid spread and commercialization of nanotechnology in cosmeceuticals has given rise to great technical and economic aspirations but also questions about the emerging risks to the health and safety of consumers. Cosmeceutical products based on nanotechnology should be designed and sold in a way that fully respects the health of consumers and the environment.

The drive for a change in lifestyle includes the growing interest in herbs. This concept is based on the idea that plants have a huge potential for use as therapeutic agents. The lifeblood for herbal use is cosmetic-quality materials; thus, they must be used with caution. The best way to reduce skin issues is with the use of appropriate cosmetic items. The future is really bright for the cosmetics industry. For beauty salons, yearly shopping positions range from middle class to the very rich. These products can range from 10 USD to 1500 USD per year, with an even smaller percentage of annual purchase positions. Several of the ingredients used to make cosmetics have the potential to cause negative side effects when applied to the human body. Hence, utilizing cosmetics may not affect the body. The best and current option for beauty goods for those who need to take care of their sensitive skin is cosmetics. There is a huge demand for cosmetics and a lot of room for growth in the use of cosmetics in the coming years. The use of cosmetics has multiplied in the personal care arena. As a result, it can be summed up by noting that while cosmetics are utilized by people of all sexes and play a significant role in daily life, their scope likewise plays a significant role.

Glossary

Acid-Alkaline This is the pH scale of the residue ash that forms when a particular food is oxidized.

Acupuncture The Chinese practice of inserting fine needles into the body at specific points to energize, unblock, or reroute subtle energies, known as chi.

Agribusiness Generic term referring to the business of agriculture and its associated politics.

Alkaline see Acid-Alkaline.

Allicin The active component of garlic.

Antioxidants These substances, found commonly in high-chlorophyll plants, especially those high in flavonoids, protect cells from free-radical damage.

Antispasmodic A herb that limits, corrects, or prevents excessive involuntary muscular contractions.

Aphrodisiac Any herb that increases sexual desire and potency. Some herbs are gender-specific, and some are shared.

Ayurveda The healing modality of ancient East India, based on the comprehension of constitutional typing into three basic doshas (or dispositions): pitta (fire), vata (air), and kapha (moisture).

Carminative A herb that expels gas from the gastrointestinal tract through excitation of internal peristalsis.

CFCs Abbreviation of chlorofluorocarbons, any of various usually gaseous substances, commonly used in refrigeration, thought to be harmful to the ozone layer of the upper atmosphere.

Detoxification This is the general term indicating a systematic cleansing of the body and mind in total or specific systems.

Endocrine System Glands that govern and regulate bodily functions, comprising the pituitary, thyroid, parathyroid, thymus, adrenals, liver, pancreas, testes, and ovaries.

Enzymes Substances secreted in the body, and present in live foods, that aid digestion.

Estrogen This is the female hormone whose production in the body is lessened during menopause.

Fecal Pertaining to the physical material of the stool.

Fibroids These are nonmalignant growths that form from muscle and connective tissue.

Flavonoids Flavonoids are classified as antioxidants, as they are generally capable of free-radical scavenging.

High-Density Lipoproteins These are the proteins containing a high ratio of protein to fat that carry cholesterol from tissue to theliver. Low-density lipoproteins are considered to be more harmful.

HIV Human immunodeficiency virus, the retrovirus considered to be the precursor to AIDS.

Immunity The capacity and function of the body to fend off foreign bodies (fungi, viruses, bacteria), or to disarm and eject them.

Lymphatic System This is the system of glands that secretes white blood cells.

Maceration A medicinal oil that has been made by infusing a base oil with a herb or a number of herbs over a period of time. A typical maceration would be of St. John's Wort in olive oil, used for cuts and bruises.

Microflora Tiny plant cells.

Mucilaginous Having the quality of mucus, that is, viscous, thick, and sticky. Internally used to soothe raw membranes.

Mucous The quality of thickness and stickiness in plant and flesh foods. Something described as mucous is said to be binding.

Organophosphates Chemical fertilizers.

Oxygenators Substances that bring oxygen into the system.

Phagocytosis A mechanism by which a phagocyte unleashes chemicals that engulf and digest bacteria and other pathogens (and dead cells).

Polysaccharide A large number of sugars linked together where energy is stored in living tissue.

Probiotics Substances that increase the presence of "friendly" bacteria in the intestines.

Progesterone The female hormone of the second phase of the menstrual cycle; it prepares for and supports pregnancy.

Rubefacient A category of herb defined by its capacity to draw blood to the surface of the body, causing redness by stimulating capillary action; for instance, cayenne pepper.

Steroids In the body, essential fatty acids are converted into hormone-like substances called prostaglandins; steroids block their production in order to relieve clotting and pain.

Stimulant Any herb that increases energy and function in the body.

T-Cells Also called T-lymphocytes, these are immune cells that play several roles in the body's defense. T-cells are so-called because they mature in the thymus.

Uterine Pertaining to the uterus.

Vibration In the esoteric sciences, all beings are considered to have an energetic quality, known as their vibration.

Virus Any of a group of minute infective and disease-producing agents consisting of a core of nucleic acid (DNA or RNA) enclosed in a protein shell. Viruses are acellular and able to function and reproduce only if they can invade a living cell to use the cell'ssystem to replicate themselves. In the process, they may disrupt or alter the host cell's own DNA.

Vitamin Any of a number of organic compounds, essential in small quantities for the normal functioning of the body.

References

1. Devi, N., Kumar, A., Garg, A., Hussain, A., & Khathuriya, R. (2018). A review on herbal cosmetics. *World Journal of Pharmaceutical Research, 7*(8), 298–310.
2. Surya, M., & Gunasekaran, S. (2021). A review on recent scenario of cosmetics. *International Journal of Pharmaceutical Sciences Review and Research, 68*(1), 190–197.
3. Sharma, B., & Alam, A. (2022). Phytochemical profiling, antioxid-ant potential and antimicrobial activities of Dalbergia sissoo Roxb. *Annals of Phytomedicine, 11*(1), 383–388.
4. Kumari, S., Sindhu, M., Singh, S., Goel, N., Rani, I., & Panghal, M. (2022). Traditional and modern usage of Nigella sativa L. (Black cumin). *Annals of Phytomedicine, 11*(1), 450–454.
5. Balyan, P., Akhter, J., Kumar, P., & Ali, A. (2022). Traditional and modern usage of Nigella sativa L. (Black cumin). *Annals of Phytomedicine, 11*(2), 255–265.
6. Gaurav, A., Gautam, S., Madhusoodan, A., Kharayat, N., Sharma, N., & Ramakrishanan, M. (2022). Qualitative and quantitative phytochemical screening and in vitro cytotoxicity study of Zanthoxylum armatum DC. and Pleurospermum angelicoides (DC.) Benth. ex C.B. Clarke: Important medicinal plants of the upper Himalayan region. *Annals of Phytomedicine, 11*(2), 411–420.
7. Pandhi, S., Kumar, A., & Rai, D. (2022). Efficacy evaluation of extraction technologies for guava (Psidium guajava L.) leaves extract. *Annals of Phytomedicine, 11*(1), 413–418.
8. Dini, I., & Laneri, S. (2019). Nutricosmetics: A brief overview. *Phytotherapy Research, 33*(12), 3054–3063.
9. Bellad, K. A., Nanjwade, B. K., Kamble, M. S., Srichana, T., & Idris, N. F. (2017). Development of cosmeceuticals. *World Journal of Pharmacy and Pharmaceutical Sciences, 6*(4), 643–691.
10. Kadam, V. S., Chintale, A. G., Deshmukh, K. P., & Nalwad, D. N. (2013). Cosmeceuticals an emerging concept: A comprehensive review. *International Journal of Research in Pharmacy and Chemistry, 3*(2), 308–316. ISSN 2231-2781.

11. Mohiuddin, A. K. (2019). Cosmetics in use: A pharmacological review. *Journal of Dermatology & Cosmetology, 3*(2), 50–67.
12. Nomakhosi, M., & Heidi, A. (2018). Natural options for management of melasma. *Journal of Cosmetic and Laser Therapy, 20*(7), 470–481.
13. Olson, K. (2009). Cosmetics in Roman antiquity: Substance, remedy, poison. *Classical World, 102*, 291–310.
14. Pande, S., Meshya, N., & Viral, D. (2010). Herbs play an important role in the field of cosmetics. *International Journal of PharmTech Research, 2*(1), 632–639.
15. Majid, M., & Prakash, L. (2006). *Cosmeceutical: A revolution in the making* (pp. 1–6). Sabinsa Corporation.
16. Dini, I., & Laneri, S. (2019). Nutricosmetics: A brief overview. *Phytotherapy Research, 33*(12), 3054–3063. Epstein, H. (2009). Cosmeceutical vehicles. *Clinics in Dermatology, 27*(5), 453–460.
17. Singh, J., & Sinha, S. (2012). Classification, regulatory acts and applica-tions of nutraceuticals for health: A review. *International Journal of Pharma and Bio Sciences, 2*(1), 177–187.
18. Ashwlayan, V., & Nimesh, S. (2018). Nutraceuticals in the management of diabetes mellitus. *Pharmacy & Pharmacology International Journal, 6*, 114–120.
19. Aquila, G., Marracino, L., Martino, V., Calabria, D., Campo, G., Caliceti, C., et al. (2019). The use of nutraceuticals to counteract atheroscler-osis: The role of the notch pathway. *Oxidative Medicine and Cellular Longevity, 2019*, 5470470.
20. McClements, D. J. (2019). Nutraceuticals: Super foods or superfads? In *Future foods* (pp. 167–201). Copernicus.
21. Sarris, J., Byrne, G. J., Stough, C., Bousman, C., Mischoulon, D., Murphy, J., et al. (2019). Nutraceuticals for major depressive disorder – more is not merrier: An 8-week double-blind, randomized, controlled trial. *Journal of Affective Disorders, 245*, 1007–1015.
22. Hussain, S. A., Panjagari, N. R., Singh, R. R., & Patil, G. R. (2015). Potential herbs and herbal nutraceuticals: Food applications and their interactions with food components. *Critical Reviews in Food Science and Nutrition, 55*(1), 94–122.
23. Dureja, H., Kaushik, D., & Kumar, V. (2003). Developments in nutraceuticals. *Indian Journal of Pharmacology, 35*(6), 363–372.
24. Jain, S., Purohit, A., Nema, P., Vishwakarma, H., & Jain, P. K. (2022). A brief review on nutraceuticals and its application. *Asian Journal of Dental and Health Sciences, 2*(1), 7–13.
25. Sharma, A., Kumar, P., Sharma, P., & Shrivastav, B. A. (2013). Comparative study of regulatory registration procedure of nutraceuticals in India, Canada and Australia. *International Journal of Pharmaceutical Quality Assurance, 4*(4), 61–66.
26. Smarta, R. B. (2017). Paradigm shift from pharmaceuticals to nutraceuticals *Nuffoods Spectrum.*
27. Maxwell, J., Smith, D., Brewster, M., & Eggleton, S. (2012). Food as pharmaceuticals as wellness products evolve, the distinction between food and medicine blurs. *R&C Worlds Express.*
28. Research and Markets. (2010). Biotech for wellness: Driving successful R&D and licensing in nutraceuticals through new business models and collaboration. Dublin: Research and Markets.
29. Kumar, P., Kumar, N., & Tushar, O. (2016). Nutraceuticals- A critical supplement for building a healthy India. *World Journal of Pharmacy and Pharmaceutical Sciences, 5*(3), 579–594.
30. Olaiya, C. O., Soetan, K. O., & Esan, A. (2016). The role of nutraceuticals, functional foods and value added food products in the prevention and treatment of chronic diseases. *African Journal of Food Science, 10*(10), 185–193.
31. Sosnowska, B., Penson, P., & Banach, M. (2017). The role of nutraceuticals in the prevention of cardiovascular disease. *Cardiovascular Diagnosis and Therapy, 7*(1), S21–S31.
32. Whitman, M. (2001). Understanding the perceived need for complementary and alternative nutraceuticals, lifestyle issues. *Clinical Journal of Oncology Nursing, 5*(5), 190–194.
33. UN Sustainable Development Goals 2020.
34. Paris Agreement http://bigpicture.unfccc.int/
35. European Green Deal http://www.eprs.ep.parl.union.eu/

36. Bijauliya, R. K., Alok, S., Kumar, M., Chanchal, D. K., & Yadav, S. (2017). A comprehensive review on herbal cosmetics. *International Journal of Pharmaceutical Sciences and Research, 8*(12), 4930–4949.
37. Shah, C. S., & Quadry, J. S. (1985). *Text book of pharmacognosy* (Vol. 5, pp. 2007–2089). B.S. Shah Prakashan.
38. Kokate, C. K., Purohit, A. P., & Goyal, S. B. (2004). *Pharmacognosy* (29th ed., pp. 167–250). Nirali Prakashan.
39. Daliu, P., Santini, A., & Novellino, E. (2019). From pharmaceuticals to nutraceuticals: Bridging disease prevention and management. *Expert Review of Clinical Pharmacology, 12*, 1–7.
40. Durazzo, A., Lucarini, M., & Santini, A. (2020). Nutraceuticals in human health. *Food, 9*(3), article 370.
41. Dureja, H., Kaushik, D., Gupta, M., Kumar, V., & Lather, V. (2005). Cosmeceuticals: An emerging concept. *Indian Journal of Pharmacology, 37*(3), 155–159.
42. Cosmetic DA. (2008). Is it a cosmetic, a drug or both (or is it soap?). *Plastic Surgical Nursing, 28*(4), 195–197. U.S. Food and Drug Administration, Office of cosmetics and colours; Centre for food safety & applied nutrition.
43. Rao, M. Y., & Shayeda. (2012). *Cosmeceuticals* (1st ed., pp. 1–20). BS Publications. Chapter-1.
44. https://www.pinterest.com/pin/472737292108910152/
45. Thornfeldt, C. (2005). *Cosmeceuticals containing herbs: Fact, fiction and future* (pp. 873–880). American Society for Dermatologic Surgery.
46. Verma, A., Gautam, S. P., Devi, R., Singh, N., & Singh, H. L. (2016). Cosmeceuticals: Acclaiming its most fascinating position in personal care industry. *Indian Research Journal of Pharmacy and Science, 3*, 506–518.
47. Anuradha, S. N., Vilashene, G., Lalithambigai, J., & Arunkumar, S. (2015). "Cosmeceuticals": An opinion in the direction of pharmaceuticals. *Asian Journal of Pharmaceutical and Clinical Research, 8*(2), 64–69.
48. Elsener, P., & Maibach, H. (2000). *Cosmeceuticals: Drugs vs cosmetics* (10th ed., p. 57). Marcel Dekker Inc.
49. Structure of Hair. http://designessentials.nl/hair-structure
50. Treub, R. M. (2001). The value of hair cosmetics and pharmaceuticals. *Dermatology, 202*(4), 275–282.
51. US Patent- US4140759A; Inventor-Jack J. Mausner; Protein Shampoo; 20th February, 1979.
52. US Patent-US4948583A; Inventor-Jean-Francoi Grollier, Georges Rosenbaum; A cosmetic composition for treatment of hair, in particular oily hair, based on extract of yarrow (Achillea Millefolum L); 14th August, 1990.
53. US Patent- US5609858A; Inventor-Carol J. Buck; Method for treatment of androgenic alopecia; 3rd November, 1997.
54. Grace, R. (2018). *Cosmeceuticals: Functional food for the skin*. New Hope Network.
55. Kaul, S., Gulati, N., Verma, D., Mukherjee, S., & Nagaich, U. (2018). Role of nanotechnology in cosmeceuticals: A review on recent advances. *Journal of Pharmaceutics, 2018*, 1–2.
56. Logothetidis, S. (2012). *Nanostructured materials and their applications* (p. 2). Springer Publication.
57. Lohani, A., Verma, A., Joshi, H., Yadav, N., & Karki, N. (2014). *Nanotechnology-based cosmeceuticals* (pp. 1–16). Hindavi Publishing Corporation.
58. Smijs, T. G., & Pavel, S. (2011). *Titanium dioxide and zinc oxide nanoparticles in sunscreens: Focus on their safety and effectiveness* (pp. 95–112). Dovepress Publication.
59. US Patent- US20120164196; Inventor- M. D Carmen, V. Pereda, A. Polezel et al.; Sericin cationic nanoparticles for application in products for hair and dyed hair; 28th June, 2012.
60. Mukhopadhyay, P. (2011). Cleansers and their role in various dermatological disorders. *Indian Journal of Dermatology, 56*(1), 2–6.
61. Kuehl, B. L., Fyfe, K. S., & Shear, N. J. (2003). Cutaneous cleanser. *Skin Therapy Letter, 6*(3), 1–4.
62. "NanocyclicCleanserPink"https://www.lunese.com/ProductDetails.asp?ProductCode=CY-15P

63. Newberg, A. E. (2009). Cosmeceuticals: Myths and misconceptions. *Clinics in Dermatology, 27*(5), 446–452.
64. Srikanth, P. V. S., Desu, P. K., Rajawardhan Reddy, M., Brahmaiah, B., & Nama, S. (2013). Mini review on cosmeceuticals. *Asian Journal of Pharmeceutical Science and Technology, 3*(2), 60.
65. Ayre, A. *Cosmeceuticals-The new face of health and personal care industry*. The Pharmatutor.
66. A-Z of Skin Sun protection and sunscreens. https://www.today.com/style/skin-care-glossary
67. US Food and Drug Administration. (1978). *"Hypoallergenic" cosmetics [Internet]*. USFDA. [Updated 2020; cited 22 Apr 2021]. Available from: https://www.fda.gov/cosmetics/cosmetics-labelling-claims/hypoallergenic-cosmetics
68. Therapeutic Goods Administration (TGA) in Australia. https://www.tga.gov.au/sites/default/files/tga-education-introduction _0_0.pdf
69. Verstraeten, E., Cluydts, R., Pevernagie, D., & Hoffmann, G. (2004). Executive function in sleep apnea: Controlling for attentional capacity in assessing executive attention sleep disordered breathing. *Sleep, 27*(4), 2004.
70. Prasad, A. S., Fitzgerald, J. T., Bao, B., Beck, F. W., & Chandrasekar, P. H. (2000). Duration of symptoms and plasma cytokine levels in patients with the common cold treated with zinc acetate. A randomized, double-blind, placebo-controlled trial. *Annals of Internal Medicine, 133*(4), 245–252. https://doi.org/10.7326/0003-4819-133-4-200008150-00006. PMID: 10929163.
71. Acevedo-Murillo, J. A., García León, M. L., Firo-Reyes, V., Santiago-Cordova, J. L., Gonzalez-Rodriguez, A. P., & Wong-Chew, R. M. (2019). Zinc supple-mentation promotes a Th1 response and improves clinical symptoms in fewer hours in children with pneumonia younger than 5 years old. A randomized controlled clinical trial. *Frontiers in Pediatrics, 7*, 431. https://doi.org/10.3389/fped.2019.00431. PMID: 31803694; PMCID: PMC6874056.
72. Rayman, M. P. (2012). Selenium and human health. *The Lancet, 379*(9822), 1256–1268.
73. Stone, A. A., Schwartz, J. E., Broderick, J. E., & Deaton, A. (2010). A snapshot of the age distribution of psychological well-being in the United States. *Proceedings of the National Academy of Sciences of the United States of America, 107*(22), 9985–9990. 2010/06/01. https://doi.org/10.1073/pnas.1003744107
74. Xiaoguang, D., Wang, X., & Geng, M. (2018). Alzheimer's disease hypothesis and related therapies. *Translational Neurodegeneration, 7*, 2. https://doi.org/10.1186/s40035-018-0107-y
75. Lobo, V., Patil, A., Phatak, A., & Chandra, N. (2010). Free radicals, antioxidants and functional foods: Impact on human health. *Pharmacognosy Reviews, 4*(8), 118–126. https://doi.org/10.4103/0973-7847.70902. PMID: 22228951; PMCID: PMC3249911.
76. Li, J., Mau, R. L., Dijkstra, P., Koch, B. J., Schwartz, E., Liu, X. J., Morrissey, E. M., Blazewicz, S. J., Pett-Ridge, J., Stone, B. W., & Hayer, M. (2019). Predictive genomic traits for bacterial growth in culture versus actual growth in soil. *The ISME Journal*. Received: 15 December 2018/Revised: 1 April 2019/Accepted: 3 April 2019; © International Society for Microbial Ecology 2019. https://doi.org/10.1038/s41396-019-0422-z
77. Institute of Medicine. (2001). *Dietary reference intakes for vitamin a, vitamin K, arsenic, boron, chromium, copper, iodine, iron, manganese, molybdenum, nickel, silicon, vanadium, and zinc*. National Academy Press, Food and Nutrition Board.
78. Zimmermann, M. B., & Crill, C. M. (2010). Iodine in enteral and parenteral nutrition. *Best Practice & Research. Clinical Endocrinology & Metabolism, 24*(1), 143–158. https://doi.org/10.1016/j.beem.2009.09.003
79. Laurberg, P., Cerqueira, C., Ovesen, L., Rasmussen, L. B., Perrild, H., Andersen, S., Pedersen, I. B., & Carlé, A. (2010). Iodine intake as a determinant of thyroid disorders in populations. *Best Practice & Research. Clinical Endocrinology & Metabolism, 24*(1), 13–27. https://doi.org/10.1016/j.beem.2009.08.013
80. Wolff, J. (1989). Excess iodide inhibits the thyroid by multiple mechanisms. *Advances in Experimental Medicine and Biology, 261*, 211–244.

81. Walker, A. R. P. (1998). The remedying of iron deficiency: What priority should it have? *The British Journal of Nutrition, 79*, 227–235.
82. Takeda, A. (2000). Movement of zinc and its functional significance in the brain. *Brain Research Reviews, 34*, 137–148. and Takeda, A. (2003). Manganese action in brain function. *Brain Research Reviews, 41*, 79–87.6.
83. Nkwenkeu, S.-F., Kennedy, G., Philippe, S., & Zayed, J. (2002). Oral manganese intake estimated with dietary records and with direct chemical analysis. *Science of the Total Environment, 287*, 147–153.
84. Handbook of Nutraceuticals and Natural Products. (1922). In P. Balakrishnan & S. Gopi (Eds.), *Biological, medicinal, and nutritional properties and applications* (Vol. 2). John Wiley & Sons.

Worldwide Ethnic Considerations for Cosmetic and Personal Care Products

Kenneth Alexander

Abstract This chapter deals with the various ethnic peoples worldwide. It limits itself to those having religious dietary restrictions and how it impacts their daily lives. This includes their choice of cosmetics and personal care products. This chapter concentrates on the Islamic, Jewish, Jains, Buddhist, and peoples of color other than Caucasians or white skinned individuals. It details what and why certain ingredients are forbidden. It also demonstrates the need for forensic proof that the ingredients are not contaminated or adulterated prior to incorporation into any cosmetic or personal care products. It also provides information as to why certain ingredients for cosmetics and personal care products must be excluded during certain religious holidays occurring throughout the year. It contains a skin-coloring system which was devised to aid in preparing or manufacturing cosmetic and personal care products for people of color. This technology has provided the industry a better means to serve the needs of people of color.

Keywords Ethnic diversity · Islam · Judaism · Jainism · Buddhism · Dietary laws related to cosmetic and personal product use restrictions · People of color

Introduction

Ethnic diversity is one of the social complexities found in most contemporary societies. Historically, it is the legacy of conquests that brought diverse peoples under the rule of a dominant group.

K. Alexander (✉)
Pharmacy, The University of Toledo College of Pharmacy and Pharmaceutical Sciences - Professor Emeritus, Grand Rapids, OH, USA
e-mail: kenneth.alexander@utoledo.edu

K. Alexander, P. Romanowski (eds.), *Cosmetic Science*, AAPS Introductions in the Pharmaceutical Sciences 15, https://doi.org/10.1007/978-3-032-03596-7_11

Ethnicity refers to the differentiation of groups of people who have shared cultural meanings, memories, and descent produced through social interaction. Ethnicity is considered to be shared characteristics such as culture, language, religion, and traditions, which contribute to a person or group's identity.

Definition of Ethnicity

- An ethnic group, or an ethnicity, is a category of people who identify with each other based on similarities such as common ancestry, language, society, culture, or nation.
- Ethnicity is usually an inherited status based on the society in which one lives.
- Membership of an ethnic group tends to be defined by a shared cultural heritage, ancestry, origin myth, history, homeland, language or dialect, symbolic systems such as religion, mythology and ritual, cuisine, dressing style, art, and physical appearance.
- An "ethnic group" has been defined as a group that regards itself or is regarded by others as a distinct community by virtue of certain characteristics that will help to distinguish the group from the surrounding community.

Terminology

- The term "ethnic" is derived from the Greek word "ethnos" which was borrowed from the Latin as "ethnicus."
- The inherited English language term for this concept is "folk," used alongside the latinate people since the late Middle English period.

This chapter will deal with various ethnic peoples worldwide dealing with either their religious dietary restrictions, how they originated and how it impacts their daily lives. This also includes their choice of cosmetics and personal care products. In this latter regard, it is the source of the individual ingredient(s) present in the product which either allows or prohibits the use of the cosmetic or personal care product itself. Since it is impossible to deal with all the various ethnic peoples, this chapter will concentrate on the following: Islam, Judaism, Jainism, Buddhism, and people with skin color other than Caucasian or white skinned individuals. Table 1 provides an overview of the various major religious groups of the world, which we call earth, and their dietary restrictions [1].

Table 1 Religious diet restrictions obtained from web site at interfaith@chapman.edu

RELIGIOUS DIETARY RESTRICTIONS

In order to be accomodating for diverse religions as a general rule, avoid any pork products (e.g. ham, bacon, pork, prosciutto) and avoid foods with alcohol as an ingredient. Alongside any alcoholic beverage served, ensure that a non-alcoholic beverage is also served. When planning events we encourage you to reach out to your guests for dietary restrictions as not every practitioner practices the same.

							MORE INFO
BAHA'I	✔	✔	✔	✔	✔	✘	Some are vegetarians, fast from March 2-20, and abstain from food and drink from sunrise to sunset.
BUDDHISM	Preferable to be vegetarian and refrain from meat				✔	✔	Many people will not eat meat or fish, and monks have additional restrictions.
PROTESTANT CHRISTIANITY	✔	✔	✔	✔	✔	✔	No apparent dietary restrictions, but moderation and some forms of fasting are observed.
EASTERN ORTHODOX CHRISTIANS	✔	✔	✔	✔	✔	✔	Meat and fish are at times restricted in observance of a fast.
HINDUISM	✘	✘	✘	✘	✔	✘	Vegetarian diet, while fasting is observed on certain days and certain foods are forbidden.
ISLAM	*HALAL	✘	*HALAL	✔	✔	✘	Anything with pork and lard is forbidden, and Halal foods are allowed.
JUDAISM	*KOSHER NOT WITH DAIRY	✘	*KOSHER NOT WITH DAIRY	NO SHELLFISH	NOT WITH MEAT	✔	Certain foods are restricted during Passover such as leavened products i.e. bread. Eating and drinking during fast days are prohibited.
MORMON	✔	✔	✔	✔	✔	✘	Obstain from alcoholic drinks, tobacco, coffee, and non-herbal teas.
ROMAN CATHOLICISM	✔	✔	✔	✔	✔	✔	Meat is restricted on Fridays of Lent, Ash Wednesday, and Good Friday (fish is permitted). Fasting is practiced.
SIKHISM	*Halal and *Kosher in some sects				✔	✘	Lacto-vegetarian diet in temples, while not forbidden from meats (individual choice).

✔ Food groups allowed

✘ Food groups not allowed

*HALAL refers to foods that are allowed to be eaten according to Islamic law, and how an animal is slaughtered

*KOSHER is a word used to describe food and drink that complies with Jewish religious dietary law, and refers to how an animal is slaughtered

Contact the Fish Interfaith Center at interfaith@chapman.edu for any questions.

Part 1: Islam and Halal Considerations

Besides Islam, many other religions also forbid certain foods or have specific requirements related to food, pharmaceuticals, cosmetics, and personal care products and include Jewish, Jain, Hindu, and Buddhist dietary laws. Food has always been the subject of taboos and obligations. Which foods we prefer and what we consider fit for (human) consumption differs depending on the place and time we live as well as the faith to which we adhere. Religious dietary laws are important to the observant Jewish, Muslim, Jain, and Buddhist populations [2]. However, not all the faithful comply with the religious dietary laws.

On this planet, which we call earth, there are many peoples who have certain ethnic and cultural practices which are dictated either by regional custom or religious belief. Cosmetics and personal care products have been around for a long time. These products are used daily by many people, and their consumption is on the rise every year. The use of these products is considered a necessity for personal hygiene, improved attractiveness, skin and hair protection from harmful ultraviolet light and pollutants, and slowing down of the aging process [3].

Due to advancements in technology, the cosmetic industry is constantly looking for new and effective products that are readily available, cheap, and safe. At the same time, information regarding the identity and the source of the ingredients used in cosmetics is not always readily available. Therefore, verification of the authenticity and acceptability of the ingredients may be needed [4].

Impact of Halal Cosmetics and Personal Care Sales

There are an estimated 2 billion Muslims in the world [5–8]. In the Middle East, the growth of halal cosmetics has increased by 12% annually and the value of cosmetic product sales in the Middle East was estimated at US$2.1 billion in 2007. In Saudi Arabia, the total sales of cosmetic products reached USD 1.3 billion in 2006 [9]. Halal products including cosmetic products have the potential of being offered not only to Muslims but to the world at large [10]. Today, there is an increased awareness concerning halal cosmetics. Consumers buy halal cosmetics if and when the products are available [11]. The global market for halal cosmetics is projected to expand at a CAGR of 9.6% from 2020 to 2025. This has been shown to result in a substantial demand for halal cosmetic products and developments [12]. This projection will be stifled if the halal-related issues and cosmetic health risks affecting the cosmetic supply chain are not effectively managed [13].

The halal cosmetic industry in Malaysia is worth approximately USD800 million, with an annual growth rate of 24 per cent [14]. In 2011, there were 114 local manufacturers in the industry certified as halal [14]. In 2013, the export of Malaysian halal cosmetic products was approximately at RM 464.45 million, with destinations being the major markets including the United States, Middle East, China, ASEAN, and Europe. Countries with large Muslim populations are expected to have a high demand for halal cosmetics as the worldwide socioeconomic conditions have improved significantly [15]. The growing market size of Muslim populations internationally is expected to create opportunities for the halal cosmetics and personal care industry [16]. The demand from the younger generation as a conscious consumer effort towards halal cosmetic products has increased the demand for halal products in general [17]. In addition, Muslims are becoming more aware and concerned over the presence of chemical and non-halal ingredients in their cosmetics [18]. Malaysia's first online beauty store to sell exclusively halal-certified beauty products named Pretty Suci, sells their halal brands and ships internationally. This is an admirable attempt to make halal beauty products more globally accessible [19].

Halal and Its Significance

The term "halal" refers to what is permitted by Islamic law [20]. It is a basic need for Muslims and encompasses all materials used in everyday life including cosmetics. Muslims want to be assured that the ingredients, handling, processing, distribution, transportation, and types of cosmetics used are halal-compliant. The halal aspects of cosmetics and personal care products cover: ingredients, all the processes involved in production right up to delivery to consumers, safety, and product efficacy evaluations [21, 22]. In order to verify halal compliance of cosmetic products, a method of detecting halal and non-halal ingredients is very important and critically needed. Halal cosmetic standards, halal certification, and the halal logo can be used as benchmarks for halal compliance. The importance of cosmetics and personal care products from the halal perspective should govern future opportunities for Muslims and other people who use these products [23].

There is currently a growing demand for halal cosmetics among the Muslim population. Due to their purchasing power, Muslim consumers today spend more on cosmetics and personal goods than ever before [24]. Cosmetics are used for beautifying purposes without affecting the body to make it look attractive, and covers a wide range of products including cleaning body parts, enhancing features, and changing skin tones and colors such as makeup, perfume, tooth paste, shampoo, and deodorant [24]. Cosmetics come in many ranges of products such as soap, face mask, powder, shampoo, lotions, toothpaste, and perfumes [25]. Muslim consumers now prefer to choose halal cosmetics and this trend is escalating [26, 27].

In many countries, manufacturers tend to use lard as a substitute for oil because it is cheaper and easily available. The use of pork and lard is a serious matter from the perspective of several religions, namely, Islam and Judaism [28]. Muslims require the products they use to be halal while the Jews require them to be kosher [2]. The concern is the same: they are concerned that the products might contain ingredients that are questionable with regard to their religious practices [29].

Concept of Halal Cosmetics

In Islam, halal means "allowed or permitted to be consumed by Islamic law" [30–32] as indicated in the Quran [33] and teachings of the Prophet Muhammad. In Surah Al-Maidah:4 as mentioned in the Quran:

They asked you what is lawful to them. Say: 'The good things are lawful to you, as well as that which you have taught the birds and beasts of pray to catch, teaching them of what Allah has taught you, eat what they catch for you, pronouncing upon it the name of Allah, and fear Allah, Allah is swift reckoning

The verse has clearly stated what is allowable in Islam, and what we must prevent from harm, and not prohibited in Islam. It tells us what everything in this world is halal unless something shows itself to be otherwise [34].

According to the Quran Surah:5 Al-Maaidah verses [82, 83], halal is a Quranic term meaning "permitted, allowed or lawful" [33]. In the same verses, the term halal and thoyyib ("good") are also included. According to the Qu' ran [20], the term halal means "permissible for consumption and used by Muslims" whereas 'haram' is anything that is unlawful or forbidden. Halal (lawful) and 'haram' (unlawful) are clearly shown in Islam to be serious matters. The word 'thoyyib' means the object it qualifies must meet the standards of quality, safety and wholesomeness [33] as allowed in Islam. It does not only cover the requirements of the religion but also sets a strict adherence to quality and hygiene compliance which is in line with good manufacturing practices (GMP) of the cosmetic industry [17, 36]. While the emphasis of GMP is on the standards of halal cosmetic and personal care products, it also extends to 'thoyyib', to produce clean and hygienic items [36]. Halal covers everything from raw material sourcing to distribution of products, right up to the delivery to consumers [21]. Halal is about trust, responsibility, respect, and strict compliance [22].

The Halal Malaysia Portal [37] has also defined halal as a product(s) or services which have been recognized by the authority, according to "sharia"' or Islamic law, in which the product or services has a halal logo. Therefore, the term "halal" is not limited to the ingredients alone, but also the manufacturing processes, storing, packaging, and delivery of the products according to sharia law. The realm of halal may extend to all consumables such as pharmaceuticals, cosmetics, personal care products, and toiletries. The concept of halal cosmetics also covers all aspects of the management system. Halal cosmetics are products that must not have any human parts or ingredients nor contain any animals or their parts forbidden to Muslims and must be slaughtered according to sharia law. There are no genetically modified organism(s) (GMO) which are declared as "najs"; no alcohol drinks (khamar); no contamination from "najs" during preparation, processing, manufacturing, and storage; and safe for consumers [30]. Additionally, there is not any use of animal fats detected by testing, dangerous chemicals, or other ingredients considered as unacceptable for Muslim consumers [38]. All halal beauty products are also recognized as clean, safe, and of high quality.

Since halal requirements extend to cosmetic and personal care products, it is then natural that Muslims would want to be certain that the cosmetic and personal care products they use are halal-compliant. At the same time, Muslims and non-Muslims involved in the production and supply of these products need to understand the meaning of halal cosmetic and personal care products and the requirements of halal laws.

Issues Regarding Halal Cosmetics

Despite the significant market demand of cosmetic and personal care products, the industry is very competitive and fast paced, requiring rapid product innovations [39]. Several factors are associated with halal cosmetics including: halal principles, product reliability, and product commitment, intention to choose halal cosmetics products, trust, environmental friendly, attitude, religious beliefs, and product innovations [40]. The potential growth for halal cosmetics and personal care products has been fueled by the growing demands for safe, natural products that fit consumers' religious and cultural beliefs [41]. The manufacturing of halal cosmetics is seen as a sort of innovation in the skincare and cosmetics industry. Manufacturers of halal cosmetics are now forced to introduce new technologies in the areas of research and development, formulation, and production.

The buying behaviors among Muslim consumers have changed drastically whereby today they buy products that emphasize hygiene, safety, and standards [42]. Due to this change in buying behavior, product lines must now include new halal products dedicated for different sexes, age groups, and ethnicity while catering to the various age and lifestyle demographics. Manufacturers should take into account the specific religious requirements that reflect on the various stages of product development. Marketing campaigns should promote the benefits of halal cosmetics to be used instead of campaigns that only portray the physical characteristics of the products. These marketing campaigns will increase awareness among both Muslim and non-Muslim consumers toward the benefits of Halal cosmetic products [42].

Halal Cosmetic Standards, Certification, and Regulation

In Malaysia, cosmetic products have to comply with local regulations and meet local quality control requirements [43, 44]. To get halal certification from the Department of Islamic Development Malaysia [45], products must fulfill the requirements of Malaysian Standard MS 2200:2008 [36] and Halal Certification Procedure Manual which require strict factory inspection and audit [46]. The certification ensures that halal cosmetics are of high quality and are ethical products (products that are compliant with Islamic law) which keep within the parameters designed for health and safety and benefits of users regardless of age, faith, or culture. Consumers should look for the halal logo/mark that certifies the product as halal-compliant [47]. The certification body responsible for granting the halal logo (Fig. 1) in Malaysia is JAKIM. The halal label issued by JAKIM is a registered trademark under the Trade Mark Act 1975. Besides JAKIM, there are other certification bodies worldwide that certify and award the halal logo. However, these halal certification bodies vary in their setup in implementing, inspecting, awarding, and monitoring halal certification [48].

Fig. 1 Malaysian halal logo

Impact of Halal Ingredients in Cosmetic Products

The global cosmetics industry is dominated, and to a certain degree monopolized, by non-Muslim companies. This poses serious challenges to the issue of halal ingredients in cosmetic products manufactured by the company's products competing with producers of kosher goods as well as other socially conscious products which have an established brand in certain countries [49, 50]. Due to anti-Islamic sentiment fueled by the 9/11/2001 incident as well as ongoing conflicts in the Middle East, some Western consumers have negative associations with firms having Islamic ties in which the name of Islam could trigger an aggressive response. However, the demand for halal cosmetics remains strong due to increased awareness and knowledge among Muslim consumers [51]. For the manufacturers of cosmetic products, innovation in halal cosmetics and the benefits of halal cosmetics must be emphasized.

Detection of Non-Halal Ingredients

The detection of non-halal ingredients in determining the halal status of any cosmetic and personal care product is important to safeguard the integrity of the halal products and confidence of the consumers. In the last few years, detection methods for non-halal ingredients have been developed quite extensively to assist religious authorities in verifying halal compliance and to detect the presence of non-halal ingredients. Several detection techniques such as Fourier transform infrared (FTIR) spectroscopy, comprehensive two-dimensional gas chromatography hyphenated with time-of-flight mass spectrometry (GCxGC-TOF-MS), and gas chromatography mass spectrometry (GCMS) have been developed to detect gelatin, alcohol, fats and oils in cosmetic products [44, 52–56]. The methods developed could be extended to traditional cosmetic and personal care products that require halal certification.

Halal Cosmetic and Personal Care Products

In the past, many Muslims used cosmetic products without thinking of the need to meet the halal requirement. It seems that Muslims and non-Muslims do not fully understand the meaning and requirements of halal. They may think halal is only about the manner in which animals are slaughtered for consumption by Muslims [10]. The "halalness" of these products is very important as it might affect the worship and prayers of Muslims. There are many interpretations of halal. Generally, halal from the perspective of the cosmetic industry means the product does not contain porcine by-products and derivatives and alcohol. However, halal as a term that has a far wider meaning than that in scope and application. The manufacture and sale of cosmetic and personal care products are regulated in certain countries. The United States Food and Drug Administration (USFDA) in 2004 [57], the European Union (EU) Cosmetic Directive in 1976 [58], and the ASEAN Cosmetic Directive in 2008 [59] have laid down certain rules for the manufacture, labeling, and sale of cosmetic products. The basic requirements of these regulations concern the safety of consumers who use cosmetic products [60]. The requirement for safety also applies to halal cosmetics to ensure that the products are indeed not harmful to the user [36]. The Malaysian Standard MS 2200: Part 1: 2008 prescribes practical guidelines for halal cosmetics and the personal-care industry on the preparation and handling of halal cosmetic products [36]. The standard definition for cosmetic and personal care products is given as "any products or preparation intended to be placed in contact with various external parts of the human body (epidermis, hair system, nails, lips and external genital organs) or with the oral cavity to clean, perfume, change their appearance and/or correct body odors and/or protect them or keep them in good condition" [36, 43]. The products are not for treating or preventing diseases in human beings. Cosmetic and personal care products are halal if they comply with Islamic law.

Besides the products, accessories accompanying them, such as brushes, bottles, containers, packaging, and mirrors, must also comply with Islamic law. Islamic law states that these products must not contain human parts or ingredients derived from human parts or contain animal by-products that are forbidden to Muslims such as from pigs and dogs. Only by-products from animals permitted by and slaughtered according to Islamic requirements are permissible in cosmetic products, which includes chickens, cows, buffaloes, turkeys, sheep, and goats. Therefore, the source of the raw materials is very important in the formulation of halal cosmetic products. If genetically modified organisms (GMO) are used in the products, the GMO must not contain components forbidden by Islam. The products must be prepared, processed, manufactured, or stored and transported in a clean and hygienic condition. The product must not be contaminated with "najs" in any circumstances and condition. Najs are religiously prohibited dirt [22] or ritually unclean material. The product must be clean and not harmful to consumers. Halal products must be recognized as symbols of cleanliness, safety, and high quality [61]. This includes substantiation of its claims and performance. It should be recognized as a credible stamp of hygiene and standards [17]. Therefore, halal cosmetics can be used and accepted not only by Muslims but by non-Muslims as well. For non-Muslims, halal can become a mark of unquestioned conformance and quality in trade dealings with Muslims.

Since halal cosmetic production activities cover all aspects of the process and production of cosmetic products, production must be carried out under strict hygienic conditions in accordance with good manufacturing practices (GMP) [62] and public health legislation [43]. If the cosmetic and personal care products are not prepared or processed according to halal requirements, they are forbidden from being used by Muslims. All halal cosmetic production establishments are confined to halal processing only. They are required not to operate non-halal cosmetic processing to avoid mandatory ritual cleansing [36]. Segregation at every stage is required for halal cosmetic products including storing, displaying, and selling. All these activities must be labeled with a sign clearly carrying the word "halal" to prevent them from being mixed with or contaminated by things deemed non-halal.

Fast Moving Consumer Goods

Cosmetic products are considered Fast Moving Consumer Goods (FMCG) [62] due to their daily nature of consumption [63–65]. This FMCG motivated numerous cosmetic companies worldwide to compete for market dominance in the cosmetics industry [66–68]. Cosmetic companies are ambitious to present a diverse range of high-quality products to compete with established products [69]. To leverage product value and competitiveness, some of them have moved towards getting halal certification [70]. Getting halal certification requires the implementation of halal assurance or internal halal control requirements throughout the cosmetic supply chain [71]. This will guarantee product safety [72] and maintain the halal status of cosmetic products [73] (Husain, 2016). Implementing the halal requirements is not limited to securing halal certification and stops once the applicants obtain the product's halal certification. The halal requirements must be built-in and continuously complied with into each cosmetic product's life cycle. This halal built-in concept was first introduced in the Malaysian Standard 2424 in 2010 as the Halal Pharmaceutical General Requirement. It was again raised in the discussion of halal pharmaceuticals in 2016 [74, 75]. This concept was then followed in other halal standards, including halal cosmetics. The built-in concept emphasizes compliance with halal requirements in the entire management and control system. Ignoring the halal requirements along the supply chain may result in health risks and gradually undermine consumer trust in halal-certified products [76–79]. These halal requirements in the cosmetic product supply chain according to the halal built-in concept to reduce the safety risk and sustain the halal status of cosmetic production were regulated for halal cosmetic products in Malaysia.

Ingredients Used in Halal Cosmetic and Personal Care Products

Recently, halal aspects in the beauty industry have received great attention due to the revelation of the inclusion of halal and "haram" ingredients in cosmetics and personal care products. All ingredients, if they are used for halal cosmetics must be checked and must conform to halal requirements. This is to ascertain the purity, safety, quality, and source of the ingredients. According to the standard [36], "the sources of ingredients of halal cosmetic products can include halal animals (land and aquatic), plants, microorganism, alcohol, chemicals, soil, and water as long as they are not hazardous and najs." The presence of alcohol, specifically ethanol, in cosmetics is of very great concern among Muslim consumers. According to Malaysian Standard [36], industrial alcohol is permitted. However, sources from alcoholic drinks are prohibited. All land animals are halal, except those that are clearly forbidden such as pigs and animals not slaughtered according to Islamic law. Thus, collagen and placenta from these animals are not permitted. Since anything from human origin is not allowed in halal practice, thus, human placenta and cysteine from human hair is not permitted in halal cosmetics. All aquatic animals are halal as long as they are not poisonous, intoxicating or hazardous to health. Aquatic animals are those that live in water such as fish but animals that live both on land and water such as crocodiles, turtles, and frogs are forbidden in Islam [60]. It becomes an issue only if the plants are processed in an unhygienic manner or processed together with unlawful (haram) ingredients or if it contains najs [36]. The ingredients in cosmetic products must be stated on labels on the package for the information of consumers.

Safety Issues

Cosmetics and personal care products placed in the market must not cause any damage to human health when applied under normal or reasonably foreseeable conditions of use [36, 57, 58]. Safety of cosmetics and personal care products is regulated and required by law all over the world. The products are not allowed to be placed in the market unless their safety has been scientifically proven. Cosmetic products consist of various chemical-based and natural-based ingredients. Many of these compounds are for general use and their properties, use, and action are well documented. New ingredients must have proper documentation to prove their safety based on scientific evaluation of the ingredients [76] (Wenninger, 1995). Safety is one of the requirements of halal products. The safety requirements also fulfills the halal and "thoyyib" requirement under Islamic law [77]. A cosmetic product is halal if it is deemed safe [36]. Necessary safety assessments are required to ensure cosmetic products are safe for use by consumers and service providers such as hairdressers and beauticians. Care should be taken to avoid skin irritation and sensitization. If the products are applied around the scalp, face, and eyes, eye tolerance needs to be addressed as a major component of the safety assessment of cosmetic products. In ensuring the safety of the finished products, chemical, microbiological, and toxicity tests are the basic requirements that need to be carried out. For most cosmetic products, a product deemed safe must have a pH value in the range of 5.0–6.5 [77, 78]. Skin and hair have a natural pH; therefore, choosing products that are too high or too low in pH will affect the skin and hair by either nourishing or irritating it. Skin has an optimal pH of 5.5 which is mildly acidic in nature. Therefore, maintaining the pH of skin in this range is important because it would help skin stay healthy, fight blemishes, prevent infections and irritation, as well as slow down skin aging [79]. Skin care products in this pH range include cleansers, moisturizers, and color cosmetics. The pH range of hair is 4.5–5.5; therefore, the healthy range of shampoos should be 6.5. Hair care products with a pH higher than 7.5 would cause the hair to be dry and brittle. Cosmetic products must also not cause any allergic reaction to the skin and eyes. The pH level of 5.5–6.5 is the safest level and will not cause skin and eye irritation. The pH levels below 3.5 or higher than 9 would cause much irritation to the skin and eyes and should be avoided. As such, in vitro skin and eye irritancy tests conducted on these products should be non-irritant to mild-irritant in nature [81, 84]. The tested results as mildly irritant can be accepted as an indicator of a preliminary irritant testing and used as a screening tool only. Further evaluation and reconfirmation need to be carried out on human subjects (in vivo). If the product tests as irritant, it is not released for sale as it is not safe for users [43]. According to the Guidelines for Control of Cosmetic Products in Malaysia 2009, cosmetic products are safe if they comply with the toxic metals and total microbial allowable limits [43]. Hence, the toxic metals in cosmetic products such as lead, arsenic, and mercury shall be below 20, 1, and 5 ppm, respectively; and the total microbial count is less than 1000 cfu ml^{-1} or cfu g^{-1} [43]. Careful selection of ingredients is important to make sure that the finished products are safe at a given concentration. Stability of the products might affect their safety. Shelf-life studies and challenge testing for preservative effectiveness must be tests that are frequently conducted.

Non-Halal Ingredient Considerations

Detection of the ingredient authenticity in halal cosmetic products is important to determine that the products, especially the oils, fats, and proteins, are halal-compliant. Muslim consumers are concerned about the mixing of animal fats, especially any form of lard in food, cosmetics, and pharmaceutical products. This has been a cause of great concern not only to Muslims but followers of several other religions, for instance, Judaism [85–88]. Halal cosmetic products must not contain or be contaminated with porcine-derived products. Therefore, the detection of adulteration due to one or more of the ingredients is a very important criterion in halal verification.

Cosmetic products are complex and contain several components of highly processed products which are manufactured from ingredients of animal or plant origin which undergo various and usually multiple treatments either chemically or physically, then commercially processed.

They are then used as ingredients in the cosmetics, food, and other industries [89]. Examples of such products are amino acids, collagen, sorbitol, albumin, fatty acids, and enzymes. As such, cosmetic analysis has become quite difficult due to the usually high complexity of ingredient usage. Detection of halal/haram sources from raw materials and intermediate ingredients in these cosmetic products may be determined using several latest techniques of various state-of-the-art instruments. It is a critical aspect of ensuring ingredients as well as the final products are halal.

Future Opportunities

The most promising halal markets are the fast-growing economies of the Asia, Middle East, Europe, and America. With a growing consumer base, and increasing growth in many parts of the world, the industry is set to become a competitive force in world international trade. The halal industry has now expanded well beyond the food sector, further widening the economic potentials for halal cosmetics [90]. The Asia Pacific region is becoming a dominant region for halal cosmetics due to its considerable Muslim population and improving socioeconomic conditions. It is a good opportunity for companies to gain access to the global scene through e-commerce. The challenge is to find the right partners as once a new market is open, the competition is even greater as established beauty businesses will also enter the online business. Urbanization has significantly impacted consumers' lifestyles and preferences and sophisticated consumers are increasingly demanding convenience and functional products in terms of packaging, application, and delivery format. With increasingly affluent consumers able to pay for premium brands or more expensive products, manufacturers of halal products must introduce new products which offer convenience and functional benefits.

In the current market, personal care products no longer apply solely to women. Men, particularly those in urban areas, now also generate strong demand for personal care products that they believe will enhance their appearances. This has opened up new opportunities for manufacturers and they have responded by introducing a number of new halal cosmetic products and brands specifically aimed for men into the market.

Conclusion

Halal products are now considered a basic need for all Muslims. Halal cosmetics and personal care products cover all aspects of the production process beginning with the raw materials (ingredients) and on from there to encompass handling, processing, storage, distribution, transportation, and delivery to the consumer. It also includes the safety of product and product efficacy. In order to verify the halal compliance of products, the development of a method of detecting and authenticating non-halal ingredients is very important and critically needed [91–94]. The introduction of the new halal standard MS 2200:2008 for cosmetic and personal care products is an aspiration to set world standards to manufacture quality halal cosmetic products for consumers. Together with the halal certification and the halal logo/mark, halal standards can serve as a benchmark for halal compliance [95].

Part 2: Kosher Considerations

Comparison of the Kosher and Halal Laws

There is a major religious significance of the dietary laws for both Jews and Muslims. The kosher (kashrus) dietary laws determine which foods are "fit or proper" for consumption by Jewish consumers who observe these laws [96–98] (American-Israeli Cooperative Enterprise (2013), (Chabad.org) (Rich, Tracey (2011)). The laws are Biblical in origin, coming mainly from the original five books of the Holy Scriptures, the Torah also known as the Tenach, which has remained unchanged to the present day. At the same time that Moses received the Ten Commandments on Mount Sinai, Jewish tradition teaches that he also received the oral law, which was eventually written down many years later in the Talmud. This oral law is as much a part of Biblical law as the written text. Over the years, the meaning of Biblical "kosher" laws has been interpreted and extended by the rabbis to protect the Jewish people from violating any of the fundamental laws, and to address new issues and technologies. The system of Jewish law is referred to as "halacha" [99].

Halal dietary laws have been previously discussed for Muslims. They can be found in the Quran and 2003 Institute of Food Technologists the Sunna©, the practice of the Prophet Muhammad as recorded in the books of Hadith, the Traditions. Islamic law is referred to as Shariah and has been interpreted by Muslim scholars over the years. The basic principles of the Islamic laws remain definite and unaltered [100, 101]. However, their interpretation and application may change according to the time, place, and circumstances as with Jewish Law. Besides the two basic sources of Islamic law, the Quran and Sunna, as well as other sources of jurisprudence are used for determining the impressibility of food. This is especially true when a contemporary situation is not explicitly covered by the first two basic sources [101]. The first issue is "Ijma," meaning a consensus of legal opinion. The second is "Qiyas," meaning reasoning by analogy. In the latter case, the process of

"Ijtihad," or exerting oneself fully to derive and answer to the problem, is used. Current issues of genetically modified organisms (GMO), animal feed, hormones, and so on, are discussed in the light of these two concepts and several other lesser sources of Islamic jurisprudence. Unconventional sources of ingredients, synthetic materials, and innovations in animal slaughter and meat processing are some of the issues Muslim scholars are currently dealing with in helping consumers make informed choices.

In addition to Islam, many other religions also forbid certain foods or have specific requirements related to food. There are Jewish, Islamic, Hindu, and Buddha dietary laws. Food has always been the subject of taboos and obligations. Which food we prefer and what we consider fit for (human) consumption differs depending on the place and time we live and the faith to which we adhere. Religious dietary laws are important to observant Jewish and Muslim populations, although not all the faithful comply with the religious dietary laws.

Islamic dietary laws determine which foods are permitted for Muslims. Halal means permitted, whereas haram means prohibited. Several foods are considered harmful for humans to consume and are forbidden. This is expressed by the prohibition of the consumption of pork, blood, alcohol, carrion, and meat that has not been slaughtered according to Islamic prescriptions. Meat is the most strictly regulated food. The animal (of a permitted species) must be slaughtered by a sane adult Muslim by cutting the throat quickly with a sharp knife. The name of Allah must be invoked while cutting [102].

The question whether stunning is allowed remains an issue of debate, both within and beyond the Muslim and Jewish communities. The rules for foods that are not explicitly prohibited by the Quran may be interpreted differently by various scholars.

Jewish dietary laws (kashrut) determine which foods are fit for consumption by observant Jews (kosher). It is a complex and extensive system with many detailed prescriptions concerning the production, preparation, and consumption of food. The prescriptions are laid down in Jewish biblical and rabbinical sources. Kosher laws deal predominantly with three issues: prohibited foods (e.g. pork, shellfish, and rabbit), prescriptions for religious slaughter (shechita), and the prohibition on preparing and consuming dairy products and meat together. In addition, there are numerous prescriptions dealing with special issues such as wine and grape juice, cooking equipment, and Passover. Ruminants and fowl must be slaughtered by a specially trained religious slaughterer (shochet) using a special knife [103, 104]. Prior to the slaughter, the shochet makes a blessing. The animal is not stunned. Slaughtered animals are inspected for visible defects by rabbinical-trained inspectors, particularly the lungs. Red meat and poultry have to be soaked and salted to remove all the blood. Any ingredients derived from animal sources are generally prohibited because of the difficulty of obtaining them from kosher animals. The prohibition of mixing milk and meat requires that the processing and handling of all materials and products fall into one of three categories: meat, dairy, or neutral (pareve). To assure the complete separation of milk and meat, all equipment must belong to a specific category. After eating meat, one has to wait 3–6 h before eating dairy. There is some disagreement over what constitutes kosher between the Orthodox, Conservative, and Reform Jewish schools of thought.

It cannot be visibly determined whether food is halal or kosher (as with other credence quality attributes relating to organic food and fair-trade products). So, how does a consumer know which food is halal or kosher? There are basically three options:

- Buying from someone of known reputation (e.g., an Islamic butcher).
- Asking a religious leader which foods are permitted.
- Buying foods with a halal or kosher label.

In traditional societies, a combination of the first two options is often applied. People living in a religious community that runs all political, economic, and communal matters internally rely on religious leaders and food suppliers of known reputation. In previous times, this applied to many European Jewish communities, for whom the chief rabbi was often the final authority in kashrut supervision. In the 1930s in the Netherlands, only a few food manufacturers were under rabbinical supervision as most foods were prepared in the home.

Due to the increase in industrially manufactured foods and the growing geographical distance between production and consumption, which is the driver for the internationalization of the food market, the reliance on local suppliers and religious leaders is often no longer sufficient. Traditional local arrangements are also disrupted by migration. Today, consumers who seek kosher or halal foods are dependent on a label or trademark that identifies a product as being kosher or halal. The consumer has to trust the source and message of the communication.

These developments have resulted in a large number of kosher-certified products in US supermarkets and a growing number of halal-labeled or -certified products in Western European supermarkets.

The growth of halal certification in the Netherlands [105] fits into a general pattern of growing third-party certification and other regulatory arrangements involving a mix of private and public actors. Food safety regulation currently involves a large number of public and private organizations with complementary, overlapping, or competing roles. [106–111] The relations between public and private actors in food regulation are varied and complex and form an interesting field of study. In some cases, private regulation is largely independent of public regulation which includes the Marine Stewardship Council label for sustainable fish. In other cases, private regulation is encouraged or enforced by governmental actors including many industrial hygiene codes.

Following the Kosher Laws

Why do Jews follow the kosher dietary laws? Many explanations have been given. The explanation by Rabbi I. Grunfeld [99] summarizes the most widely held ideas about the subject. It is important to note that, unlike kosher laws, the health aspects of eating are an important consideration with halal laws. These laws are viewed by the Jewish community as being given without a need for explanation. Only in modern times have some people felt a need to try to justify them as health laws. For a discussion of why kosher laws are not health laws, one should see [112].

And ye shall be men of a holy calling unto Me, and ye shall not eat any meat that is torn in the field (Exodus XXII:30).

Holiness or self-sanctification is a moral term; it is identical with moral freedom or moral autonomy. Its aim is the complete self-mastery of man. [113]

To the superficial observer, it seems that men who do not obey the law are freer than law-abiding men, because they can follow their own inclinations. In reality, though, such men are subject to the most cruel bondage; they are slaves of their own instincts, impulses, and desires. The first step towards emancipation from the tyranny of animal inclinations in man is, therefore, a voluntary submission to the moral law. The constraint of law is the beginning of human freedom ....

Thus, the fundamental idea of Jewish ethics, holiness, is inseparably connected with the idea of law. The dietary laws occupy a central position in that system of moral discipline and is the basis of all Jewish laws. Unconventional sources of ingredients, synthetic materials, and innovations in slaughter and processing are some of the same issues Muslim scholars are now dealing with.

The three strongest natural instincts in man are the impulses of food, sex, and acquisition. Judaism does not aim at the destruction of these impulses, but at their control and indeed their sanctification. It is the law which spiritualizes these instincts and trans-figures them into legitimate joys of life.

Why do Muslims follow the halal dietary laws? The main reason for the observance of the Islamic faith is to follow the Divine Orders.

O ye who believe! Eat of the good things wherewith WE have provided you, and render thanks to ALLAH if it is He whom ye worship. (Quran II:172)

God reminds the believers time and again in the Holy Scripture to eat what is "Halalan Tayyiban," meaning "permitted and good or wholesome." "O, Mankind! Eat of that which is Lawful and Wholesome in the Earth ...." (Quran II:168) "Eat of the good things. We have provided for your sustenance, but commit no excess therein." (Quran XX:81) Again in Sura 6 of the Quran, entitled "Cattle," Muslims are instructed to eat the meat of animals upon which Allah's name has been invoked. This is generally interpreted as meaning that an invocation has to be made at the time of slaughtering an animal. *"Eat of that over which the name of Allah hath been mentioned, if ye are believers in His revelations."* (Quran VI:119).

While Muslims eat what is permitted specifically or by implication, albeit without comment, they avoid eating what is specifically disallowed, such as: *"And eat not of that whereupon Allah's name hath not been mentioned, for lo, It is an abomination. Lo! The devils do inspire their minions to dispute with you. But if ye obey them, ye will be in truth idolators."* (Quran VI:121) The majority of Islamic scholars are of the opinion that this verse deals with proper slaughtering of the allowed animals. Since Muslim dietary laws relate to Divine permissions and prohibitions, if anyone observes these laws, he or she is rewarded in the hereafter, but if anyone violates these laws, he or she may receive punishment accordingly. The rules for those foods that are not specifically prohibited may be interpreted differently by various scholars. The things that are specifically prohibited are just a few in number, and are summarized in the following verses:

> *Forbidden unto you are: carrion and blood and swine flesh, and that which hath been dedicated unto any other than Allah, and the strangled, and the dead through beating, and the dead through falling from a height, and that which hath been killed by the goring of horns, and the devoured of wild beasts save that which ye make lawful, and that which hath been immolated to idols. And that ye swear by the divining arrows. This is abomination.* (Quran V:3)

Although these permissions and prohibitions as a divine injunction are enough for a Muslim to observe the laws, it is believed that the dietary laws are based on health reasons that suggest impurity or harmfulness of prohibited foods.

The Kosher and Halal Market

Why are we concerned about kosher and halal in the secular world? Because both kosher and halal are important components of the food business [2, 111, 114, 115]. Most people, even in the food industry, are not aware of the breadth of foods that are under religious supervision. This section provides background on the economic aspects that make it important for the food industry to have a better understanding of kosher and halal.

The kosher market, according to Integrated Marketing, an advertising agency specializing in the kosher food industry comprises almost 75,000 products in the United States. In 2001, about 165 billion dollars worth of products were estimated to have a kosher marking on them. The deliberate consumers of kosher food; that is, those who specifically look for the kosher mark, are estimated to be more than ten million Americans and they are purchasing almost 7 billion dollars worth of kosher products. Annually, almost 10,000 companies produce kosher products and the average US supermarket has 13,000 kosher products. Fewer than 1/3, and possibly as low as 20%, of kosher consumers are Jewish which translate to almost 900,000 year-round consumers. Other consumers who, at times, find kosher products helpful in meeting their dietary needs include Muslims, Seventh Day Adventists, vegetarians, vegans, as well as people with various types of allergies, more particular to dairy, grains, and legumes. In general, these are the consumers who value the quality of kosher products, even though there is rarely a one-to-one correlation between kosher and these consumers' needs. Hebrew National's slogan, *"We report to a higher authority"* and *"You don't have to be Jewish to love Levy's Rye Bread,"* are two of the more famous campaigns used to advertise kosher products to non-kosher consumers. Ad-Week magazine in the early 1990s called kosher *"the Good Housekeeping Seal for the '90s."* By undertaking the kosher certification, companies can incrementally expand their market by opening up new markets. It should be noted that although many supermarkets define the kosher consumer as someone who only purchases products with kosher supervision symbols on the package, but there are products that do not always need to have a supervision mark, as we will describe. This also includes most of the 6 to 8 million Muslims in North America who observe halal laws. However, the food industry has, for the most part, ignored the specific needs of these and other ethnic consumer groups. This information might assist kosher supervision agencies in addressing these issues.

The Muslim population in the United States is developing a stronger marketplace presence each year. Over the past 30 years, many halal markets and ethnic stores have sprung up, mainly in the major metropolitan areas. Most of the 6 to eight million Muslims in North America observe halal laws, particularly the avoidance of pork, but the food industry has for the most part ignored their concerns. There are excellent opportunities to be realized in the North American halal market. An even more compelling opportunity exist on a worldwide basis as the food industry moves to a more global business model. The number of Muslims in the world is more than 1.3 billion people, and trade in halal products is about 150 billion dollars [116]. Many countries of South Asia, Southeast Asia, the Middle East, and Northern Africa have predominantly Muslim populations. Although only about 15% of India's population is Muslim, it is the second largest Muslim country in the world, after Indonesia. In many countries, halal certification has become necessary for products to be imported.

Although many Muslims purchase kosher food in the United States, these foods, as seen in the section on Halal, do not always meet the needs of the Muslim consumer. The most common areas of concern for the Muslim consumer, in meeting their dietary needs when considering purchasing kosher products, are the use of various questionable gelatins [117] in products produced by more lenient kosher supervisions and the use of alcohol as a carrier for flavors as well as a food ingredient.

Halachos for Using Cosmetics and Personal Care Items for Passover

Pesach, known to non-Jews as Passover, prohibits certain foods (chometz) to be consumed or used during this eight-day holiday. These foods include products made with wheat, oats rye, and other grains as well as other products derived from these basic ingredients.

In addition to pharmaceutical companies, certain rabbis contact many cosmetic companies and bases the following chometz-free list on their research:

L'Halacha, which includes all non-food items not fit for canine consumption, known in Hebrew as "nifsal mei'achilas kelev" or something that one would not feed his dog, may be used on Pesach. This includes all cosmetics, soaps, ointments, and creams. Nonetheless, people have acted stringently with regard to these items. Below are several reasons why people are strict:

1. Many products, including shaving lotion and perfume, contain denatured alcohol which can be restored to regular alcohol [118] [65] (see Igros Moshe O.C. 3:62).

According to some opinions, one should not use such products, if chometz-based. The Sefer Kovetz Halachos Pesach has a brief discussion regarding the different opinions) [119]. The list notes products which do not use chometz-based alcohols.
2. The Biur Halacha [120] writes in the name of the Gra that one should be strict and not use non-kosher soap all year (sicha k'shtiya). Although we are not accustomed to this stringency, many individuals have adopted this chumra during Pesach and do not permit the use of chometz items even if they are used externally.
3. Some are of the opinion that we do not say "nifsal" (food is unfit for canine consumption) applies to a chometz item that has a good fragrance even if it is a non-food item, since the purpose of having a good fragrance is still being fulfilled. According to this, deodorants that contain a chometz derivative could not be used on Pesach, even if it is inedible [121].
4. Lipstick is often inadvertently ingested when eating food. If it contains chometz, it is halachically permissible to apply to the lips since the lipstick is unfit for canine consumption. Nevertheless, most women prefer not to consume even a trace of anything prohibited. The Pesach list provides chometz-free lipsticks.
5. Mouthwash and toothpaste contain sorbitol and other ingredients which may be derived from chometz. Although, l'Halacha, these items are permissible to use since they are nifsal mei'achilas kelev, many prefer not to use them since they are taken orally [122]. The Pesach list provides chometz-free brands of such products.
6. Historically, it has been the custom to follow stringent opinions regarding Hilchos Pesach. In addition, it seems to have been a prevalent custom to restrict the use of items which may contain chometz, even when they are clearly nifsal mei'achilas kelev. The Pesach list provides accurate information for those who wish to continue to follow the more strict opinion and prevailing custom when using such products.

Keeping Kosher/Kashrut

Although the details of kashrut are extensive, the laws all derive from a few fairly simple, straightforward rules:

1. Certain animals may not be eaten at all. This restriction includes the flesh, organs, eggs and milk of the forbidden animals.
2. Of the animals that may be eaten, the birds and mammals must be killed in accordance with Jewish law.
3. All blood must be drained from meat and poultry or broiled out of it before it is eaten.
4. Certain parts of permitted animals may not be eaten.
5. Fruits and vegetables are permitted, but must be inspected for insects (which cannot be eaten).

6. Meat (the flesh of birds and mammals) cannot be eaten with dairy. Fish, eggs, fruits, vegetables and grains can be eaten with either meat or dairy. (According to some views, fish may not be eaten with meat).
7. Utensils which also includes pots and pans and other cooking surfaces) that have come into contact with meat may not be used with dairy, and vice versa. Utensils that have come into contact with non-kosher food may not be used with kosher food. This applies only where the contact occurred while the food was hot.
8. Grape products made by non-Jews may not be eaten.

The first thing to know about buying kosher food applies whether you are hosting or being hosted: look for the kosher symbol (hechsher) (Fig. 2).

Fig. 2 Kosher certification symbols used on products in the United States

All kosher-prepared foods have a symbol somewhere on the label from one of a number of kosher-certifying agencies. This symbol means that the item is prepared according to kosher dietary guidelines. The letter "K" by itself is not a kashrut symbol. A majority of the kosher products you will see on store shelves will bear one of these marks and will be a lovely gift for your hosts.

The first rule of cooking for someone who keeps kosher is to involve them in the process. Always ask what you can do to make them comfortable eating in your home. Depending upon the level of observance of the person you are hoping to host, the answers may fall along the spectrum ranging from:

1. Anything that does not have any meat or meat by products. This can include dairy products, pasta, kosher fish are those with fins and scales and anything else you would serve to a lacto-ova vegetarian or piscetarian.
2. Non-cooked cold items (vegetable salads, canned tuna salad, and other items that have not been cooked).
3. Non-cooked cold items as along as all ingredients have a hechsher or kosher symbol.
4. Non-cooked cold items as along as all ingredients have a hechsher or kosher symbol and the dishes and utensils used in preparation and for serving are all either new or unused disposables.
5. Hot food if it is bought from a kosher establishment and served on paper plates with plastic silverware.
6. Hot food if it is bought from a kosher establishment, triple wrapped by the establishment before heating/reheating in the oven with the seal intact until serving and served on paper plates with plastic silverware.

The Basic Laws and Key Concepts of Kashrut

Kashrut, which is the Hebrew word for kosher, refers to the Jewish laws that deal with what foods can and cannot be eaten and how those foods must be prepared. The word comes from the Hebrew root that means proper or correct. Foods are not blessed by rabbis to make them kosher. Rather, someone (a "mashgeach") supervises the production of processed foods, meat, and poultry to make sure that all the laws of kashrut have been followed. If all the ingredients and procedures meet the rules of kashrut, then the food is certified kosher.

In Jewish law there is no such thing as "kosher-style." Kosher is not a style of cooking but rather a set of standards that need to be followed in order for food to be certified kosher. Jews observe the laws of kashrut because there are commandments in the Torah which is also known as the Old Testament or Tanach. Rabbis have tried to make these laws meaningful and have developed interesting explanations about why they exist, although the Torah does not specify reasons for their observance.

Some interesting explanations that have been proposed in rabbinic literature include some of the following:

- The simple act of eating is elevated into a religious ritual.
- The dinner table is compared to the altar in the temple.
- Ritualizing the slaughter of animals reflects the reverence for animal life.
- Rules about what you can eat and not eat are a reminder to distinguish between sacred and profane, between right and wrong, or between good and evil.
- There are potential health benefits from some of the Kashrut laws.

General Rules of Kashrut

- Certain animals are forbidden from being eaten at all. The most common ones include any food product derived from a pig and all shellfish.
- Of the animals that may be eaten, fowl and meat must only be from animals that were slaughtered according to Jewish law.
- All fruits and vegetables are permitted but need to be inspected for bugs. (This is mostly of concern for certain vegetables such as broccoli or berries (e.g. strawberries).
- Dairy and meat cannot be eaten together. Poultry is considered to be meat.
- Grains, eggs, fruits, vegetables, and fish are called PAREVE which means that they are neutral, neither meat nor dairy, and can be eaten with either food groups.
- Fish cannot be eaten on the same plate as meat but can be eaten either before or after.
- There is a waiting time between eating meat and then eating dairy food. The Orthodox standard is 6 h between eating meat and dairy. Therefore, if one has a meat sandwich at 1 PM, you could not eat a diary dinner or have a dairy snack until 7 PM. This law is more custom-based and depends on family of origin. Some Modern Orthodox individuals wait 3 h. Individuals of Dutch ancestry only wait 1 h.
- There is only a short waiting time of 1/2 h between eating meat after eating a dairy food. A note of mention is that some people will wait 6 h after eating certain hard cheeses.
- Utensils used for meat and dairy always need to be separated. This means that if a meat spoon was used to stir a creamed soup, that spoon is no longer considered to be a kosher utensil. There are methods to remedy this and make the utensil kosher once again.
- Similarly, utensils that have come into contact with non-kosher food may not be used with kosher food. Therefore all utensils, pots, pans, toasters, microwave oven, etc., that will be used for kosher food needs to remain separate and cannot be used for non-kosher foods as well.

Kosher Certification

Some of the acceptable kosher symbols are given in Fig. 2. The most common and most widely accepted symbol is the OU, which is the certification from the Orthodox Union of Rabbis based in New York. The other certifications are from rabbinical supervisions in other cities. Most Orthodox individuals will accept most of these certifications. There will always be some individuals who will reject any particular certification.

Conclusion

Kosher and halal are so close to their original religious roots that the only difference is in their interpretation of their method of slaughter of animals as well as their definition of what sea food they can eat. Minor differences also exist with mixing dairy and meat and fish in a given meal. Both have strict rules for ingredient inclusions in cosmetics and personal care products that make production, sale, and use adhere to specific certification procedures.

Part 3: JAINISM and BUDDHISM

A vegan and alcohol/drug free lifestyle stems from a compelling ethical and moral value system. This is where one makes a conscious effort not to harm any animals and not to harm one's own body, mind, and soul. As a result, one avoids the use of all animal products such as meat, fish, chicken, eggs, milk, cheese, ice-cream, butter, ghee, and all other dairy products as well as silk, fur, pearls, and leather. One also refrains from all types of addictive substances such as alcohol and drugs.

Introduction

The subcontinent of India, a cradle of civilization, is also the Transparency birthplace of the three great religious traditions of the world: Hinduism (Vedic religion), Buddhism, and Jainism. The Jain religion and its philosophy being one of the oldest, which predates recorded history as referenced in the Hindu scriptures, has become one of the essential spiritual traditions of the South Asian religious fabric. It is an original system, quite distinct and independent from other systems and philosophical traditions of India.

Jainism is a way of life based on ahimsa (non-violence), compassion, and simple living [124–140]. Jain philosophy is ennobling and optimistic. It is where the soul of man and woman evolves to Godhood when the soul purifies itself by removing its impurities such as anger, ego, deceit, and greed, arising out of attachment and aversion. This process is explained in the philosophy of karma in Jainism. It embraces ancient techniques and philosophies handed down by the Jinas, who are also known as Arihantas or Tirthankaras. These are the main prophets in the Jain faith which number 24. The lineage can be traced back to prehistoric times. Lord Mahävir was the last Tirthankar (599 BC to 527 BC), and Lord Pärshvanäth was his predecessor (850 BC). Whereas Lord Ädinäth was the first Tirthankar and is considered prehistoric time.

Jains believe in the philosophy of karma, reincarnation of the worldly soul, heaven and hell as a reward and punishment for one's deeds. It is the liberation (Nirvana or Moksha) of the self or soul from life's misery of birth and death in a way similar to the Hindu and Buddhist beliefs. Though there are multiple similarities in these South Asian religions, as well as some fundamental differences still do exist.

The cardinal principle of Jainism is Ahimsa. Jains believe that every living being is a master of his/her own destiny. They rely a great deal on self-effort and self-initiative for both their worldly requirements and their salvation or liberation. The Jain scriptures also state that the universe and all its entities such as soul and matter are eternal (there is no beginning or end). No one has created them and no one can destroy them. All entities continuously change their form. Jains believe that the soul is ignorant of its true nature and is bounded by karma from eternity. It is due to karma that the soul migrates from one life cycle to another and passes through many pleasant and painful situations. The ignorant soul continues to attract and bind new karma due to its action of attachment and aversion. [140, 141].

Jainism: A Religion of Compassion and Ecology

To overcome the suffering, Jainism lays out the spiritual path that consists of integrated trinity: Right Conviction or Faith (Samyak Darshan), Right Knowledge (Samyak Jnän), and Right Conduct (Samyak Chäritra). They must coexist in a person if one is to make any spiritual progress on the path of liberation. Initially one needs to acquire the proper knowledge of the true nature of soul and other reality such as matter, karma, and other non-living substances. When one is totally convinced of one's knowledge, at that moment this knowledge will remove the ignorance about one's own nature. This stage of spirituality is called realization of truth or attainment of Right Conviction and Right Knowledge. Popularly it is known as the self-realization stage or Samyaktva.

The realization of truth leads to Right Conduct. There are various spiritual stages that exist in practicing the Right Conduct called Gunasthänas. Right Conduct includes:

1. Compassion and non-violence (Ahimsa) towards all living beings.
2. Pluralism of views (Anekäntaväda or Syädväda) towards all faiths and ideas.
3. Non-possession (Aparigraha) or limitation of possessions and non-possessiveness.
4. Self-purification, self-control, austerity, penance, asceticism, and meditation.

With regard to truth, Jain philosophy firmly states that the whole truth cannot be observed from a single viewpoint by the common people. To understand the true nature of reality, it is essential to acknowledge the multiple perspectives of each situation or idea. We must strive to be open-minded and embrace the positive thoughts and vantage points of other human beings, religions, and philosophies. This concept is called Anekäntaväda (non-absolutism). In summary, according to Jainism, the ultimate goal of the soul is to achieve liberation through understanding and realization. This is accomplished through the supreme ideals of non-violence, equal kindness, reverence for all forms of life, non-possessiveness, conquering all passions such as attachments and aversions, and through the philosophy of non-absolutism (Anekäntaväda). Above all, these ideals translate into a religion of universal love and compassion not only towards human beings but also towards all forms of life and environment.

Jains believe in leading a simple life. They are against rituals and animal sacrifice. They believe that every living being has a soul. They follow the principle of non-killing, non-violence, and non-injury to any living being. They treat even unconscious killing of an ant as sin, and discourage it. According to Jain literature, Jains are vegetarians or more precisely vegans. Their diet is governed by certain principles. The Jain diet practices are different than other religions. Important features of the Jain diet include the fact that they do not eat meat. They eat vegetables, squash, beans, peas, tomatoes, and fruits. They avoid eating root vegetables, such as onions, garlic, and beets.. They do not eat chicken, eggs, turkeys, or any poultry items nor do they eat fish, lobster, oysters, or any kind of seafood. According to custom, they do not drink beer, wine, liquor, whiskey, or any kind of alcohol. They prefer to drink boiled water; do not eat honey; and do not eat food stored overnight, etc.

There has been considerable change in the diet and dietary practices of the modern Jain community. The old Jain strict dietary practices are not followed in current days. They have adopted a "balanced diet" i.e. a diet that contains adequate amounts of all the necessary nutrients required for healthy growth and activity, for a person depending on the age, height, build of the body, profession, exertion and illness, etc. This may, on occasion, violate the principle of non-killing, non-violence, and non-injury to any living being.

Jain Cosmetics and Personal Care Products

Since Jains are pure vegans, their needs for cosmetic and personal care products are very restrictive in the ingredients that are used in their formulation. No animal products should be used in their production. No insect products such as beeswax, honey, and carmine can be used as well. Thus, the cosmetic and personal care formulator is extremely limited in their ingredient selection. However, more modern Jains have realized this problem and have accepted the concept that in order to live in this modern age, they accept the products that mostly adhere to their beliefs.

Buddhist Religious Principles

Like many religions, Buddhism has dietary restrictions and food traditions. Buddhists—those who practice Buddhism—follow the teachings of the Buddha or "awakened one" and adhere to specific diet laws [142]. Siddhartha Gautama, or the "Buddha," founded Buddhism in the fifth to fourth century B.C. in the eastern part of India. Today, it is practiced worldwide. Several forms of Buddhism exist globally, including Mahayana, Theravada, and Vajrayana. Each type has slightly different interpretations of Buddha's teaching, particularly when it comes to dietary practices. This includes Vegetarianism; however, not all Buddhists adhere to this concept. Five ethical teachings govern how Buddhists live. One of the teachings prohibits taking the life of any person or animal. Many Buddhists interpret this to mean that you should not consume animals, as doing so would require killing. Buddhists with this interpretation usually follow a lacto-vegetarian diet. This means they consume dairy products but excludes eggs, poultry, fish, and meat from their diet. On the other hand, other Buddhists consume meat and other animal products, as long as the animals are not slaughtered specifically for them. Nonetheless, most dishes considered Buddhist are vegetarian, despite not all traditions requiring lay followers of Buddhism to follow this diet [145].

Another ethical teaching of Buddhism prohibits intoxication from alcohol given that it clouds the mind and can lead you to break other religious rules. Still, lay followers of the religion often disregard this teaching, as some traditional ceremonies incorporate alcohol. Aside from alcohol, some Buddhists avoid consuming strong-smelling plants, specifically garlic, onion, chives, leeks, and shallots, as these vegetables are thought to increase sexual desire when eaten cooked and anger when eaten raw [146].

Buddhist also fast. Fasting refers to abstaining from all or certain types of foods or drinks. The practice, specifically intermittent fasting, is becoming increasingly popular for weight loss, but it is also often done for religious purposes. Buddhists are expected to abstain from food from noon until the dawn of the following day as a way to practice self-control [147, 148]. However, as with the exclusion of meat and alcohol, not all Buddhists or lay followers of the religion fast.

Like other religions, Buddhism has specific dietary practices that followers may or may not practice. Some Buddhists may fast or refrain from consuming animals, alcohol, and certain vegetables. Every diet, including the Buddhist diet, has pros and cons to consider. A Buddhist diet follows a primarily plant-based approach. A plant-based diet is rich in fruits, vegetables, nuts, seeds, whole grains, legumes, and beans, but it may also include some animal products.

This diet provides important compounds, such as antioxidants, phytochemicals, vitamins, minerals, and fiber, which have been associated with a decreased risk of heart disease, type 2 diabetes, and certain types of cancer [149–152]. Aside from these health benefits, following a plant-based or vegetarian diet may also benefit your waistline. One study demonstrated that Buddhists who followed a vegetarian diet for 11–34 years had less body fat than those who followed the diet for 5–10 years—and even less body fat than those who followed it for 3–4 years [144].

The downsides to this diet is that vegetarian diets which restrict the intake of meat can be deficient in certain nutrients if they are not planned appropriately, even if they permit eggs and dairy. Studies have found that Buddhist lacto-vegetarians had calorie intakes similar to those of non-vegetarian Catholics. However, they had higher intakes of folate, fiber, and vitamin A and consumed less protein and iron [144, 151]. Consequently, they had lower levels of iron and vitamin B12. Low levels of these nutrients can cause anemia, a condition characterized by a lack of oxygen-carrying red blood cells [144, 151, 152]. Aside from iron and vitamin B12, other nutrients that vegetarians may be lacking include vitamin D, omega-3 fatty acids, and zinc [153]. Still, it is possible to consume a nutritionally adequate vegetarian diet by planning properly and taking supplements to fill any nutritional gaps.

The pros and cons of fasting can be considered. Fasting is an important practice in Buddhism. Buddhists generally fast from noon to dawn of the following day. Depending on preferences and schedule, fasting for approximately 18 h every day to either be a pro or con of the Buddhist diet.

Consuming the entire daily calorie intake before noon can not only be physically difficult but also interfere with the individuals social and professional life. On the other hand, fasting may be convenient and helpful for weight loss. In a 4-day study in 11 overweight adults, those fasting for 18 h had better blood sugar control and increased expression of genes involved in autophagy—a process that replaces damaged cells with healthy ones—compared with those fasting for 12 h [154, 155]. While these results are promising, longer studies are necessary to make definitive conclusions about whether the practice is superior to a standard reduced-calorie diet for weight loss and other health benefits [155–158].

While not all Buddhists are vegetarians, many choose to follow a vegetarian or lacto-vegetarian diet. Examples of foods to eat and avoid on a lacto-vegetarian diet are as follows:

Foods to eat

Dairy: yogurt, cottage cheese, and milk.
Grains: bread, oatmeal, quinoa, and rice.
Fruits: apples, bananas, berries, grapes, oranges, and peaches.

Vegetables: broccoli, tomatoes, green beans, cucumber, zucchini, asparagus, and peppers.
Starchy vegetables: potatoes, corn, peas, and cassava.
Legumes: chickpeas, kidney beans, pinto beans, black beans, and lentils.
Nuts: almonds, walnuts, pecans, and pistachios.
Oils: olive oil, flaxseed oil, and canola oil.

Foods to avoid

Meats: beef, veal, pork, and lamb.
Fish: salmon, herring, cod, tilapia, trout, and tuna.
Eggs and poultry: eggs, chicken, turkey, duck, quail, and pheasant.
Pungent vegetables and spices: onions, garlic, scallions, chives, and leeks.
Alcohol: beer, wine, and spirits.

Given that the Buddhist diet consists primarily of plants, it may lack certain vitamins and minerals, particularly iron and vitamin B12. Fasting, while an important component of Buddhism, may not be for everyone. While it is not a requirement of practicing Buddhism, many follow a vegetarian or lacto-vegetarian diet that also excludes alcohol and pungent vegetables and spices.

Part 4: Other Ethnic Cosmetic and Personal Care Considerations

Introduction

Consumers around the world spend over $330 billion a year on fragrances, cosmetics, and toiletries. The global beauty business pervades our lives. It influences how we perceive ourselves and what it means to be beautiful [159]. However, the global industry has been criticized for its lack of diversity and inclusion. In order to bring awareness to this movement, some businesses have brought change to center stage. Naomi Campbell became the first black woman on the cover of Vogue in 1989. Today, Fenty Beauty pushed the boundary of the makeup industry toward more inclusivity and diversity regardless of complexion and gender [160]. Even MAC Cosmetics launched its M-A-C VIVA GLAM program to support healthy futures and equal rights for women, girls, and the LGBTQ communities. Diversity goes beyond the centerstage or a marketing campaign. Consumers seek out products that are personalized for them in addition to pushing inclusivity and transparency from the brand. The cosmetic and personal care industry has different spices of color, texture, shape, and size which create a unique blend of cultures and backgrounds that we can all learn from [161]. The challenge of diversity is not only ensuring representation of historically underrepresented groups, it is also creating an environment where "we" is everyone [162].

Cosmetic use, hygienic approaches, and many basic dermatological principles differ for ethnic individuals [163]. Still, comparatively few publications highlight these variations or discuss appropriate management with these products. Among many ethnic patients, issues related to skin hydration, restoration of even pigmentation, hair removal, and acne care remain problematic yet not fully addressed [164]. There are some dermatological conditions that may be rare in Caucasian skin but are much more common in the ethnic patient [165].

Diversity and Inclusion

Diversity in beauty means recognizing that all ethnicities, races, cultures, skin tones, body types, and backgrounds are beautiful [166]. It also means giving the same space and opportunities [167] in all different settings such as product development, scientists, marketing, and supply chain. Beauty companies must enhance organizational success by valuing their employees and consumers. Ulta Beauty and Sephora have used their platform to show their inclusivity by pledging 15% of shelf space to Black-owned businesses. But is this enough to address the lack of diversity? Thanks to #PullUpOrShutUp challenges, beauty companies have noted the lack of organizational inclusion efforts. Brands such as L'Oréal USA, Ouai, and Morphe posted their employee statistics on social media along with their call for action [168]..

Inclusion creates an environment that acknowledges, welcomes, and accepts different approaches, styles, perspectives, and experiences. It also requires individuals to become culturally competent. One must work on becoming aware of their of inequities and develop skills and abilities to recognize, accept, and adapt to cultural differences and similarities [169, 170].

The beauty industry has been criticized for neglecting people of color and glamorizing a certain ideal representation. The predominance of white western concepts of beauty among multiple ethnic groups in the United States has raised a great concern [171]. In order to bring awareness of the problem, beauty companies have researched ways to include everyone. This does not mean highlighting a person of color to show consumers a brand constitutes diversity and inclusion. It means a company must hire diverse groups of people internally for positions such as chemists to product development to marketing [172].

Colorism and racism have had a huge impact on the black community. The majority of cosmetic products sold are marketed for white people and light-complexioned minorities. Eurocentric beauty demeans the physical features of black consumers [173]. This results in drastic changes in black regimens including skin lightening, chemical hair relaxers, and eye color surgeries. Insensitivity to one's culture has also been an issue within the industry. The Asian-Pacific Beauty industry was valued at $163.92 billion in 2020 due to the "skincare at home" dynamic of the COVID-19 pandemic [174]. Even though the innovation was effective, respect is still lacking toward our peers. Tarte Cosmetics made an insensitive joke against Asian people on social media. The meme used the terms "ching chong" as a racial slur to describe how a person's brain functions in the morning. This is an example of racial insensitivity against another ethnic group. Thanks to social media, Tarte Cosmetics deleted the meme, issued an apology, and recognized the need for racial language education.

The United States has become progressively more multicultural in which the non-Caucasian ethnic population is growing at record rates. The US Census Bureau projects that by the year 2056, greater than 50% of the US population will be of non-Caucasian descent. [175] For centuries, the esthetic ideal was based on artist and anatomist drawings. We now have defined mathematical anthropometric measurements of the ideal facial structure; however, most of these measurements have been based on the Caucasian face. In recent years, ethnocentric variability in these ideals have been described and are constantly changing with the increasing presence of ethnic persons in the media, changing the standards of beauty. [176].

The term ethnic skin has been used in the medical literature to describe the skin of color (Table 2). Traditionally, the Fitzpatrick skin-type model for types III-VI (Table 3) was introduced [177]. Similar terms used include dark skin, brown skin, and pigmented skin. These terms do not define any particular race, ethnicity, or culture, nor do they adequately describe all skin types and pigmentation. Among African Americans, there is wide range of skin pigmentation. There are also differences in skin pigmentation among Asians in Japan and China as compared with slightly darker skin types in Koreans and people of South-East Asia, such as Thailand, Singapore, and Cambodia. Skin-type definition initially proposed by Fitzpatrick does not adequately define the multicultural variations we see every day. Until a more inclusive term is defined, we will use the term "ethnic skin" to refer to patients with pigmentation darker than that of the Caucasian race (Table 3).

Table 2 Ethnic groups that comprise persons defined as having "ethnic skin"

Skin type	Persons defined within the group
"Black"	African, black persons of African descent, including African-American, Caribbean-American, and Latin-American persons
Latino or Hispanic	Persons of Spanish and indigenous central/south American descent, including central Americans, south Americans, and Caribbean-American persons of Spanish descent, including Cuban, Puerto Rican, and Dominican
East Asian	Chinese, Japanese, Korean
Southeast Asian and Pacific islander	Filipino, Vietnamese, Cambodian, Thai, Malaysian, Laotian, Burmese, Hmong descent, Polynesian, Micronesian
Australoid	Australian aborigine, Melanesian descent (new the Republic of Guinea, Papua, Solomon Islands)
Native Americans	More than 560 recognized tribes, including Inuit.
East Indian	Indian, Pakistani, Bangladesh, Sri Lanka
Middle eastern	Iranian, Iraqi, persons from Saudi Arabia and the Arabian peninsula (including Kuwait, Bahrain, Oman, Qatar, the United Arab Emirates, Yemen), Lebanese, Afghani, Jordanian, Syrian, Israeli, Turkish, north African (Egypt, Morocco, Algeria, Libya)

Table 3 Fitzpatrick skin phototype classification system

Type	Ethnic groups that comprise persons defined as having "ethnic skin"[a]	Skin color
I	Burn and no tan	White
II	Burn and minimal tan	White
III	Burn then tan well	White to light olive
IV	Tan, minimal to no burn	Light brown
V	Tan, no burn	Brown
VI	Tan, no burn	Dark brown

Modified from. Fitzpatrick TB: Skin phototypes 2002, 20th World Congress of Dermatology, Paris, July 1–52002.
[a]Thirty minutes unprotected sun exposure in peak season (spring or summer) depending on the latitude

Table 4 Lancer Ethnicity Scale (LES)

Geography	Fitzpatrick skin phototype	LES skin type
Persons of African background Central, east, west African	V	5
Eritrean and Ethiopian	V	5
North African, Middle East Arabic	V	5
Sephardic Jewish	III	4
Persons of Asian background Chinese, Korean, Japanese, Thai, Vietnamese	IV	4
Filipino, Polynesian	IV	4
Persons of European background	II	3
Celtic	I	1
Central, E. European	III	2
Nordic	I–II	1
N. European (general)	I	1–2
S. European, Mediterranean	III	3–4
Persons of Latin/central/south American background central/ south American Indian	IV	4
Persons of north American background native American (including Inuit)	II	3

Modified from Wolbarsht ML, Urbach F: The Lancer Ethnicity Scale.
Lasers Surg Med 25:105–106, 1999.

Traditionally, there are nine geographic races, each with particular genetic similarities. These geographic races include Europeans (which include Middle Eastern and Mediterranean persons), Eastern Indians, Asians, American Indians, Africans, Melanesians, Micronesians, Polynesians, and Australian Aborigines. We modify these schema into categories in which, ethnic persons share similar anatomic characteristics as shown in Table 4 [179].

Ethnic patients have natural features that are unique and thus have different cosmetic concerns. The cosmetic concerns of ethnic patients differ as the result of differences in skin pathophysiology [180], mechanisms of aging, and unique facial structure. For example, photoaging in the African-American population is not as pronounced as that seen in Caucasian patients, yet dyschromia is a significant problem in this subset of patients. The cosmetic industry is beginning to understand these distinctive features and is developing a large market of products for the ethnic patient. There is no longer a single standard of beauty. We must now adapt to the more diverse population and understand how to accommodate the diversity of beauty of persons of different ethnicities in the United States. Ethnic patients do not necessarily want a Westernized look because what constitutes beauty is determined by racial, cultural, and environmental influences [171, 181, 182]. Leaders in skin care must understand these differences and adapt their practices accordingly.

Future for Ethnic Skin and Personal Care Products

One of the leading segments for the beauty industry is the expanding market for multicultural or ethnic products. In recent years, demand for multicultural beauty products has grown rapidly. The market for this sector is expected to become even more diverse. Combining this fact with the growth of interracial couples, there are now new possibilities for cosmetic products, such as foundation makeup.

The number of Americans of mixed race grew by 32% from 2000 through 2010. Multicultural beauty products are outpacing the growth of the overall market for cosmetics and toiletries. The millennial generation, those born between 1980 and 2005, are characterized by their ethnic and cultural heterogeneity. A high number of millennials are immigrants or children of immigrants. There have been predictions that minority birth rates will surpass those of Caucasians, who will no longer be the majority in the United States by 2043. Forecasts say that in 2050, Hispanics will account for 28% of all the US population, African Americans for 17%, and Asians for 8%. Around the globe, countries are more homogeneous and less heterogeneous.

Therefore, products are needed to address consumers' individuality. For example, a cosmetic company created a foundation makeup with 22 shades to match a tonal palette that would represent all women globally. This was done with nine years of research, measuring skin shades from over 7000 women all over the world [183]. They computerized 700 virtual shades that gave the widest cross-section and simplified the range to 22 final choices that represented the global palette.

Developing global products requires studying and researching a range of hair and skin types with different characteristics [184]. Variation in pigmentation requires the collection and analysis of consumer data [185]. It requires an understanding of the heritage of different races. In addition to physiological needs, one benefit of multicultural beauty is that multicultural companies are more inclined to also meet the cultural, religious, or philosophical needs of the market they are attempting to attract, such as Halal, Kosher, or Hindu. The common thread running through this multicultural segment is that the ingredients in the formulation and packaging do not contain animal ingredients and are not tested on animals.

Ethnic products may require greater investment by a company's consumer and market research teams, as well as additional resources.

Understanding the needs of women of different heritage and color is essential. The variety of makeup routines reflects ethnic origin and cultural heritage. This determines whether a woman feels positive towards her natural skin color. For many women, skin color is a major factor in their cultural identity [187]. Foundation makeup must be formulated to meet these specific needs.

It was just 20 years ago that major companies may have had a line of perhaps 12–15 shades to encompass all types of skin tones. Today, these same companies have a separate ethnic line that may cover 12–15 shades, in addition to one for Caucasian skin. They have finally realized that diversity and complexity exists within the ethnic segment of the marketplace and are responding to their consumer needs.

Glossary for Halal, Kosher, Jainism, and Other Ethnic Considerations

Part 1 Halal

Ijma Meaning a consensus of legal opinion.
Qiyas Meaning reasoning by analogy.
Quran Referring to the Islamic Bible or Koran.
Ijtihad Or exerting oneself fully to derive and answer to the problem.
Sunna The practice of the Prophet Muhammad.
Hadith The Islamic book of the Traditions.
Halal It is a Quranic term meaning allowed or permitted to be consumed by Islamic (Muslim) religious dietary laws.
Haram Is anything that is unlawful or forbidden to be consumed by Islamic (Muslim) religious dietary laws.
Thoyyib Means the object it qualifies must meet the standards of quality, safety and wholesomeness as allowed in Islam.
Halalan Tayyiban Meaning permitted and good or wholesome.
Fast Moving Consumer Goods (FMCG) Consumer products which are considered as such due to their daily nature of consumption.

Part 2 Kosher

Pesach Known to non-Jews as the holiday of Passover.
chometz Certain foods prohibits to be consumed or used during the 8 days of Passover.

nifsal mei'achilas kelev Something that one would not feed his dog may be used on Pesach.

L'Halacha Includes all non-food items not fit for canine consumption which can be used during Passover.

Shochet A specially trained Jewish religious slaughterer who uses a special knife for Ruminants and fowl which must be slaughtered.

Sicha k'shtiya Hebrew term indicating one should be strict and not use non-kosher soap all year.

Chumra Hebrew meaning advice.

Hilchos Pesach Passover dietary and use laws.

Hechsher The kosher symbol on all products that are deemed kosher.

Kosher-style In Jewish law, there is no such thing. It is either kosher or not.

Torah The scrolls on which the Old Testament is written. It is also referred in printed for as the Tanach or as the Hebrew Bible.

OU An abbreviation which indicates kosher certification from the Orthodox Union of Rabbis based in New York.

Cholov Isroel This is the term for "kosher milk" though the Orthodox Union (OU) considers milk produced in the United States to be kosher without certification. This standard is usually held only by ultra-orthodox individuals in certain communities.

"D" Alongside a kosher label means that it is a product that may contain dairy ingredients.

"Pareve" Alongside a kosher label means that the food is parve and can be eaten with dairy or meat.

"P" Means that the food is kosher for Passover and can be eaten during Passover and as well as all year around.

Halachah Hebrew word for Jewish law.

Kosher Means fit for ritual use or proper for eating.

Kashrus Hebrew word for kosher.

Treif Yiddish word for not kosher.

Mashgiach Is the person who supervises the production of kosher food or the slaughtering of animals according to Jewish law.

Brachah Means a blessing. Blessings are said both before and after meals. Some blessings are short and recited before eating a particular food such as an apple. There are different blessing for different food groups.

Ritual Hand Washing Is a practice of washing hands and reciting a blessing before a meal in which bread is eaten.

Birkat Hamazon A blessing for Grace after Meals. This is a somewhat lengthy blessing which is recited after a meal in which bread has been eaten. If bread was not eaten during the meal, a shorter blessing can be recited. Sometimes, individuals avoid eating bread at a meal so that they do not have to wash their hands before eating and recite the Birkat Hamazon afterwards. This practice may interact with a desire to avoid carbohydrates in eating disordered individuals.

Pareve Refers to neither dairy nor meat foods such as fruits, vegetables, grains, and fish.

Dairy Refers to milk products or any food items that has an ingredient derived from a dairy product.

Milchig Is the Yiddish (Jewish) word for dairy.

Meat Includes beef, poultry, veal, or any food that has an ingredient that is derived from a meat product. Foods that are pareve but cooked in a meat pot are also considered to be meat.

Fleishig Is the Yiddish (Jewish) word for meat.

Part 3: Jainism

Jainism A vegan and alcohol/drug free lifestyle stems from a compelling ethical and moral value system.

Ahimsa a belief that every living being is a master of his/her own destiny without violence.

Jinas An ancient people who are also known as Arihantas or Tirthankaras and whom embraced ancient techniques and philosophies.

Karma A belief in the philosophy of reincarnation of the worldly soul.

Nirvana or Moksha Liberation of the self or soul from life's misery of birth and death.

Samyak Darshan The spiritual path that consists of the Right Path of Conviction or Faith and is part of the trinity.

Samyak Jnän The spiritual path that consists of the Right Path to Knowledge and is part of the trinity.

Samyak Chäritra The spiritual path that consists of the Right Path of Conduct and is part of the trinity.

Samyaktva Known as the stage of self-realization.

Gunasthänas The various spiritual stages that exist in practicing the Right Conduct.

Aparigraha The limitation of possessions and non-possessiveness.

Anekäntaväda A non-absolutism in which one strives to be open-minded and embrace the positive thoughts and vantage points of other human beings, religions, and philosophies.

Vegan Is a shortened version of vegetarian**.** *Vegan: The most restrictive form of the vegetarian diet, vegans do not consume any animal products of any kind. This includes meat, seafood, dairy like cheese and milk, and eggs.*

Vegetarianism Various forms include the lacto-ovo-vegetarian diet, ovo-vegetarian diet, and vegan diet.

Lacto-Vegetarian The lacto-vegetarian diet is a variation of vegetarianism that excludes meat, poultry, seafood, and eggs. Unlike some other vegetarian diets, it includes certain dairy products, such as yogurt, cheese, and milk.

Lacto-Ovo-Vegetarian A diet that prohibits meat, poultry, and seafood products, this is the most common type of vegetarianism. Dairy and eggs are allowed. If you follow a lacto-vegetarian diet, then you also refrain from eating eggs in addition to meat products. Ovo-vegetarians consume eggs but not dairy.

Pescetarian A pescetarian does not eat meat or poultry but does eat fish, lobsters, oysters, or any kind of seafood. They may also include dairy and eggs in their diet if they wish.

Flexitarian If you cannot quite commit to cutting out all meat, a flexitarian diet allows for occasional meat or seafood. A flexitarian follows a lacto-ovo-vegetarian or even vegan lifestyle most of the time, only consuming meat or seafood every once and a while.

Buddhist diet Usually follow a lacto-vegetarian diet. This means they consume dairy products but exclude eggs, poultry, fish, and meat from their diet. However, others may include meat and alcohol since they are dependent on what others provide for them.

Jain diet Individuals who eat vegetables and fruits; avoid eating root vegetables; do not eat meat; do not eat chicken, eggs, turkeys, or any poultry items; do not eat fish, lobsters, oysters, or any kind of seafood; do not drink beer, wine, liquor, whiskey, or any kind of alcohol; prefer to drink boiled water; do not eat honey; and do not eat food stored overnight, etc.

Part 4: Other Ethnic Considerations

Diversity Recognizing that all ethnicities, races, cultures, skin tones, body types, and backgrounds as one.

Inclusion An environment that acknowledges, welcomes, and accepts different approaches, styles, perspectives, and experiences.

Transparency The ability to see more clearly whether it be data, procedures or information concerning a given subject.

Dyschromia The absence of color as it refers to the white race.

Ethnic Skin Refers to patients with pigmentation darker than that of the Caucasian race.

Eurocentric The emphasis placed on catering to or marketing to white people and light-complexioned minorities.

Colorism Preferential treatment of one colored race over all others.

Racism Preferential treatment of one race over all others.

Millennials Those born between 1980 and 2005 who are characterized by their ethnic and cultural heterogeneity.

Fitzpatrick Scale Skin-type scale and delineation initially proposed by Fitzpatrick.

Lancer Ethnicity Scale (LES) A modification of the Fitzpatrick skin typing scale

References

1. Religious Diet Restrictions obtained from web site at interfaith@chapman.edu
2. Regenstein, J. M., Chaudry, M. M., & Regenstein, C. E. (2003). The Kosher and Halal Food Laws. *Comprehensive Reviews in Food Science and Food Safety, 2*, 111–127.

3. Mitsui, T. (1996). *New cosmetic science*. Elsevier.
4. Lockley, A. K., & Bardsley, R. G. (2000). DNA-based methods for food authentication. *Trends in Food Science and Technology, 11*, 67–77.
5. MITI. (2006). *Development of the halal industry. IMP3 Third Industrial Master Plan 2006–2020* (pp. 593–613). Ministry of Inter-national Trade and Industry, Kuala Lumpur.
6. HDC. (2009, April). *Halal and the current economic Malaise*. Halal Industry Development Corporation. VIBE.
7. Pew Research Center. (2011, January 27). The future of the global Muslim population. Retrieved September 9, 2015, from http://www.pewforum.org/2011/01/27/the-future-of-the-global-muslim-population/.
8. Pew Research. (2013). *The world's muslims: Religion, politics and society*. Pew Research Centre. Retrieved from http://www.pewforum.org/2011/01/27/the-future-of-the-global-muslim-population/
9. Kamaruzaman, K. A. (2008). Halal cosmetics between real concern and plain ignorance. *The Halal Journal, 3–4*, 26–28.
10. Muhammad, R. (2007). Re-branding halal. *The Halal Journal, 5–6*.
11. Jabatan Kemajuan Islam Malaysia. Kuala Lumpur, Malaysia, Perniagaan, R., & Kamaruzaman, K. A. (2008). Halal cosmetics between real concern and plain ignorance. *The Halal Journal, 3*(4), 26–22.
12. Statista. (2018). Halal cosmetics: Market value worldwide.
13. Research and Markets. (2021). Halal cosmetics market: Global industry trends, share, size, growth, opportunity and forecast 2021–2026. https://www.researchandmarkets.com/reports/5311861/ halal-cosmetics-market-global-industry. Accessed on 11th Jan 2025.
14. Halal Industry Development Corporation (2014). Opportunities in halal economy. Halal Industry Development Corporation. Retrieved from http://www.kccci.org.my/attachments/article/1921/opportunies_in_Halal_Economy.pdf.
15. Boo, S.L. (2013). 'With halal makeup, piety in a jar'. The Malaymail online. Retrieved from http://m.themalaymailonline.com/malaysia/article/with-halal-makeup-piety-in-a-jar.
16. Ariyanchira, S. (2017). Halal and stylish: The emerging markets of halal cosmetics. Acubiz Consulting Sdn Bhd. Retrieved on December 2017 from https://acubiz.com.my/emerging-markets-halal-cosmetics/
17. Amat, S. H. (2006, November 15). Halal – new market opportunities. Paper presented in the ninth Efficient Consumer Response (ECR) Conference, KLCC, Kuala Lumpur Malaysia. Retrieved on May, 2009 from http://www.islam.gov.my/portal/lihat.php?jakim=2140.
18. Houlis, A. (2017). Why the halal cosmetic industry has staying power. . Retrieved from https://fashionista.com/2017/05/halal-cosmetics-makeup-brands.
19. Ng, Y. (2017). Malaysia's first halal beauty online store is here. Female Retrieved on December 2017 http://femalemag.com.my/beauty/malaysias-first-halal-beauty-online-store/.
20. Al-Qardawi, Y. (1995). *'The lawful and the prohibited in Islam* (pp. 1–78). Islamic Book Trust.
21. Che Man, Y. B., & Sazili, A. Q. (2010). Food production from the halal perspective. In I. Guerrero-Legaretta & Y. H. Hui (Eds.), *Handbook of poultry science and technology* (Vol. 1.: primary processing). Wiley.
22. Che Man, Y. B., Bojei, J., Abdullah, A. N., & Latif, M. A. (2007). Halal food. In F. Arshad, N. M. R. Abdullah, B. Kaur, & A. M. Abdullah (Eds.), *50 Years of Malaysian Agriculture: Transformational issues challenges & direction* (pp. 195–266). Penerbit Universiti Putra Malaysia.
23. Ali, Q., Salman, A., Yaacob, H., Parveen, S., & Financial Cost Fahmi S. (2017). Halal labeling effect on muslim consumers attitude and behavior and social influence: Factors affecting the adoption of halal cosmetics in Malaysia. *Academy of Entrepreneurship Journal, 25*(2), 1–17.
24. Swidi, A., Wie, C., Hassan, M. G., Hosam, A. A., & Kassim, A. W. K. (2010). The Mainstream Cosmetics Industry in Malaysia and the emergence, growth, and prospects of

halal Cosmetics. In *Proceedings of the Third International Conference on International Studies* (pp. 1–20).
25. Kumar, S., Massie, C., & Dumonceaux, M. D. (2006). Comparative innovative business strategies of major players in cosmetic industry. *Industrial Management & Data Systems, 106*(3), 285–306. https://doi.org/10.1108/02635570610653461
26. Winter, R. (2005). *A consumer's dictionary of cosmetic ingredients* (6th ed.). Three Rivers Press.
27. Daud, N. M., Aziz, H. A., Baharudin, N. H., & Shamsudin, S. F. (2012). Identifying the determinant attributes of halal cosmetics product that influences it. *Journal of Applied Science Sciences Research, 8*(1), 301–313.
28. Bonne, K., & Verbeke, W. (2008). Religious values informing halal meat production and the control and delivery of halal credence quality. *Agriculture and Human Values, 25*(1), 35–47.
29. Khattak, H. (2009). Halal certified cosmetics and personal care products – where purity comes first. *Halal Digest, 1*, 1–3. Retrieved on February, 2011 from http://www.infanca.org/newsletter/2009.01.htm
30. Kaur, K., Osman, S., & Maziha, S. (2014). Predicting working women purchasing behavior of malaysian halal cosmetic products by using theory of planned behavior predicting working women purchasing behavior of Malaysian Halal Cosmetics. *International Academic Research Journal of Business and Management.*
31. Khan, M. M., Asad, H., & Mehboob, I. (2017). Investigating the consumer behavior for Halal endorsed products: Case of an emerging Muslim market. *Journal of Islamic Marketing., 8*, 00–00. https://doi.org/10.1108/JIMA-09-2015-0068
32. Naseri, R., Rahmiati, F., & Esa, M. M. (2021). Consumer attitude and online purchase intention: a segmentation analysis in Malaysian Halal Cosmetic Industry. *International Journal of Academic Research in Business and Social Sciences, 11*(3). https://doi.org/10.6007/IJARBSS/v11-i3/8977
33. Din al-Hafiz, A. H. (2008). Surah Al-Maidah verse 87–88. In *Al-Quran dan Terjemahanya* (p. 122). Dar El-Fajr Publisher.
34. Alam, S. S., & Sayuti, N. (2011). Applying the theory of planned behavior (TPB) in halal food purchasing. *International Journal of Commerce and Management, 21*(1), 8–20.
35. Marina, A. M., Che Man, Y. B., Nazimah, S. A. H., et al. (2009). Chemical properties of virgin coconut oil. *The Journal of the American Oil Chemists' Society, 86*, 301–307. https://doi.org/10.1007/s11746-009-1351-1
36. DSM. (2008). *Malaysian standard MS 2200: Part 1:2008 Islamic consumer goods – Part 1: Cosmetic and personal care – General guidelines* (pp. 1–6). Department of Standards Malaysia. Printing Department, SIRIM Berhad.
37. The Halal Malaysia Portal. (2017). Definition of halal. Halal Malaysia Official Portal. Retrieved from http://www.halal.gov.my/v4/index.php?data=bW9kdWxlcy9uZXdzOzs7Ow==&utama=panduan&ids=gp1.
38. Aoun, I., & Tournois, L. (2015). Building holistic brands: An exploratory study of Halal cosmetics. *Journal of Islamic Marketing, 6*(1), 109–132.
39. Kumar, Ranjit Research Methodology. (2005). *a step-by-step guide for beginners* (2nd ed.). Sage.
40. Majid, M. B., Sabir, I., & Ashraf, T. (2015). *Consumer purchase intention towards halal cosmetics & personal care products in Pakistan.* Global of Research in Business & Management.
41. The Halal Journal, Issue #23 July–August and Issue #25 September-October (y).
42. Yeo, B. L., Mohamed, R. H., & Muda, M. (2016). A study of Malaysian customers purchase motivation of halal cosmetics retail products: Examining theory of consumption value and customer satisfaction. *Procedia Economics and Finance, 37*, 176–182.
43. NPCB. (2009a). *Guidelines for control of cosmetic products in Malaysia (Revision 02).* National Pharmaceutical Control Bureau, Ministry of Health.

44. Hashim, P., Shahab, N., Masilamani, T., Baharom, R., & Ibrahim, R. (2009). A cosmetic analysis in compliance with the legislative requirements, halal and quality control. *Malaysian Journal of Chemistry, 11*, 1081–1087.
45. JAKIM. (2020). *Malaysian manual procedure for halal certification (Domestic) 2020 (MMPHC 2020)*. Department of Islamic Development Malaysia (JAKIM).
46. JAKIM. (1993). *Garis Panduan Makanan* (2nd ed.). Minuman dan Bahan Gunaan Orang Islam.
47. Alias, S. S. M. A. A. S. (2019). Malaysia model: Challenges in halal certification. *Journal of Chemical Information and Modeling, 53*(9), 1689–1699.
48. Khan, M., & Haleem, A. (2016). Understanding "Halal" and "Halal" certification and accreditation system: A brief review. *Saudi Journal of Business and Management Studies, 1*(1), 32–42.
49. Izberk-Bilgin, E., & Nakata, C. C. (2016, May–June). A new look at faith-based marketing: The global halal market. *Business Horizons, 59*(Issue #3), 285–292.
50. Izberk-Bilgin, E. (2012). Infidel brands: Unveiling alternative meanings of global brands at the nexus of globalization, consumer culture, and Islamism. *Journal of Consumer Research, 39*(4), 663–687.
51. Fahmi, S. (2017). *Halal labeling effect on muslim consumers attitude and behavior*. https://doi.org/10.2991/icoi-17.2017.26
52. Rohman, A., & Che Man, Y. B. (2009). Analysis of cod-liver oil adulteration using Fourier transform infrared (FTIR) spectroscopy. *Journal of American Oil Chemistry Society, 86*, 1149–1153.
53. Rohman, A., Che Man, Y. B., & Sismindari. (2009). Quantitative analysis of virgin coconut oil in cream cosmetics preparations using Fourier Transform Infrared (FTIR) spectroscopy. *Pakistan Journal of Pharmaceutical Science, 22*, 415–420.
54. Rohman, A., Che Man, Y. B., Ismail, A., & Hashim, P. (2010). Application of FTIR spectroscopy for the determination of virgin coconut oil in binary mixtures with olive oil and palm oil. *Journal of American Oil Chemistry Society, 87*, 601–606.
55. Rohman, A., & Che Man, Y. B. (2010). FTIR spectroscopy combined with chemometrics for analysis of lard in the mixtures with body fats of lamb, cow, and chicken. *International Food Research Journal, 17*, 519–526.
56. Norakasha, R., Hashim, D. M., Che Man, Y. B., Shuhaimi, M., & Noorfaizan, A. (2009). Potential use of amino acids analysis for distinguishing bovine and porcine gelatins. In P. Hashim, A. Ismail, A. Q. Sazili, N. N. Mohd, S. A. Zaki, & N. K. Mokhtar (Eds.), *Proceeding of the third IMT-GT International Symposium on Halal Science and Management 2009* (pp. 14–18). Universiti Putra Malaysia.
57. USFDA. Chapter VI: Cosmetics. In The Federal Food, Drug and Cosmetic Act (FD&C Act). U.S. Food and Drug Administration, USA.
58. EU Cosmetic Directive. (1976). 76/768/EEC. The European Union.
59. *The Association of Southeast Asian Nations (ASEAN) ASEAN Documents Series 2008*. ASEAN Secretariat, July 2009 General information for ASEAN appears on-line at the ASEAN. Website: www.asean.org
60. Antignac, E., Nohynek, G. J., Re, T., Clouzeau, J., & Toutain, H. Safety of botanical ingredients in personal care products/cosmetics. *Food and Chemical Toxicology, 49*(2), 324–341. https://doi.org/10.1016/j.fct.2010.11.022
61. Rajikin, M. H., Omar, B., & Sulaiman, S. (1997). Pemakanan dan Kesihatan. Kuala Lumpur, Malaysia: Dewan Bahasa dan Pustaka Publisher; NPCB. (2009b). Guidelines on good manufacturing practice for cosmetic, Annex 1, Part 9. Guidelines for control of cosmetic products in Malaysia (Revision 02). National Pharmaceutical Control Bureau, Ministry of Health
62. Okdinawati, L., Simatupang, T. M., Imran, A., & Lestari, Y. D. (2021). Value co-creation model of halal logistics services. *Jurnal Ilmiah Teknik Industri, 20*(1), 45–60. https://doi.org/10.23917/jiti.v20i1.13028

63. Susanti, A. T. Y. (2020). Literature review: Legal aspect of consumer protection for cosmetic users. *Advances in Economics, Business and Management Research, 140*, 572–575. https://doi.org/10.2991/aebmr.k.200513.110
64. Bocken, N. M. P., Harsch, A., & Weissbrod, I. (2022). Circular business models for the fastmoving consumer goods industry: Desirability, feasibility, and viability. *Sustainable Production and Consumption, 30*, 799–814. https://doi.org/10.1016/j.spc.2022.01.012
65. Malthora, S. (2014). A study on FMCG (fast moving consumer goods) marketing. *International Journal of Innovative Research & Development, 3*(1), 366–369.
66. Hashim, A. J. C. M., & Musa, R. (2013). *Modeling the effects on the attitude of young adult urban muslim women towards halal cosmetic products: New insights for 6162*.
67. Hashim, C. M., Jihan, A., & Musa, R. (2014). Factors influencing attitude towards halal cosmetic among young adult urban muslim women: A focus group analysis. *Procedia – Social and Behavioral Sciences., 130*, 129–134. https://doi.org/10.1016/j.sbspro.2014.04.016
68. Kumar, S., Massie, C., & Dumonceaux, M. D. (2006). *Comparative innovative business strategies of major players in cosmetic industry*. Industrial Management and Data.
69. Noorham, N., Syed Anuar, S. N., & Guliling, H. (2020). The practices of local cosmetic product on the customer loyalty among Malaysian youth. *ESTEEM Journal of Social Sciences and Humanities, 5*, 1–10.
70. Radzi, N. F. N. M. W. J. W. M. (2021). Religion and cosmetics: Guidelines for preparing products aimed at the muslim world based on the interpretation of halal cosmetics in Malaysia. *Journal of Cosmetic Science, 72*(2), 139–154.
71. Sin, M. A. M., Razalli, M. R., & Ismail, R. M. (2019). The search of best practices of Malaysian halal cosmetics industry to improve firm performance. *International Journal of Islamic Marketing and Branding, 4*(3/4), 212. https://doi.org/10.1504/ijimb.2019.107285
72. Sumpin, N. A., Kassim, N. F., Zaki, M. I. M., Piah, Z. H. M., & Majid, M. A. A. (2019). Will the Real Halal Logo Please Stand Up? *Inter-national Journal of Academic Research in Business and Social Sciences, 9*, 1226–1234. https://doi.org/10.6007/ijarbss/v9-i9/6416
73. Ayob, A., Hussain, W., & Hirwani, W. M. (2016). Buying local or imported goods? profiling non-income consumers in developing countries. *International Review of Management and Marketing., 6*, 688–695.
74. Abd Razak, M. A., & Ramli, M. A. (2022). Halal-related food crimes: Concept and mitigation strategies. *Al-Qanatir, International Journal of Islamic Studies, 27*(2).
75. Biotech. (2016). Building the Bioeconomy 2016: Examining the national the Biotechnology industry development Strategies globally The report was commissioned by the Biotechnology Industry Organization. Pugatch Consilium 2016.
76. Tieman, M. (2009a, March–April). Halal compliant terminal: Facilitating imports and exports. *The Halal Journal*.
77. Tieman, M. (2009b, January–February). Halal Transportation: The building blocks of a halal transportation system. *The Halal Journal*.
78. Tieman, M. (2017). Halal risk management: Combining robustness and resilience. *Journal of Islamic Marketing, 8*(3), 461–475. https://doi.org/10.1108/JIMA-06-2015-0041
79. Tieman, M. (2011). The application of Halal in supply chain management: In-depth interviews. *Journal of Islamic Marketing, 2*(2), 186–195.
80. Wenninger, J. A. (1995). *International cosmetic ingredient dictionary* (6th ed.). Cosmetic, Toiletry and Fragrance Association.
81. Hashim, P., Che Man, Y. B., & Kassim, N. (2009a). Analysis of alcohol in cosmetic products by headspace gas chromatography—mass spectrometry. In Puziah Hashim, Amin Ismail, Awis Q. Sazili, Nabiha N. Mohd Zaki, Syariena Arshad, & Nurfadhilah K. Mokhtar (Eds.), Proceeding of the third IMT-GT international symposium on halal science and management 2009 (p. 110–114). Universiti Putra Malaysia.
82. Hashim, P., Shahab, N., Masilamani, T., Baharom, R., & Ibrahim, R. (2009b). A cosmetic analysis in compliance with the legislative requirements, halal and quality control. *Malaysian Journal of Chemistry, 11*, 1081–1108.

83. Anonymous. (2012). What you need to know about pH levels in your beauty products. Retrieved from http://seecosmetics.com/beautyfashion/what-you-need-to-know-ph-levels-in-your-beauty-products.
84. OECD. (2002). OECD Guidelines for testing of chemicals, No. 404 acute dermal irritation, corrosion revised test guidelines (p. 1–13). Retrieved from http://www.mattek.com/pages/pdf/OECD-404-Acute-Dermal-Irritation-Corrosion.pdf.htm.
85. Marikkar, J. M. N., Lai, O. M., Ghazali, H. M., & Che Man, Y. B. (2001). Detection of lard and randomized lard as adulterants in refined, bleached, deodorized palm oil by differential scanning calorimetry. *Journal of the American Oil Chemists' Society, 78*(11), 1113–1119. https://doi.org/10.1007/s11746-001-0398-5
86. Marikkar, J. M. N., Lai, O. M., Ghazali, H. M., & Che Man, Y. B. (2001). Detection of lard and randomized lard as adulterants in refined-bleached-deodorized palm oil by differential scanning calorimetry. *Journal of American Oil Chemistry Society, 78*, 1113–1119.
87. Marikkar, J. M. N., Lai, O. M., Ghazali, H. M., & Che Man, Y. B. (2002). Compositional and thermal analysis of RBD palm oil adulterated with lipase-catalyzed interesterified lard. *Food Chemistry, 76*, 249–258.
88. Marikkar, J. M. N., Ghazali, H. M., Man, Y. C., Peiris, T. S. G., & Lai, O. M. (2005). Distinguishing lard from other animal fats in admixtures of some vegetable oils using liquid chromatographic data coupled with multivariate data analysis. *Food Chemistry, 91*(1), 5–14.
89. CFIA. (2008). Highly processed products. Canadian Food Inspection Agency. AHPD-IE-2001-8-3 (p. 1–10).
90. Dalir, S., Olya, H.G.T., Al-Ansi, A., Rahim, A.A., & Lee, H.Y. (2020).
91. Elasrag, H. (2016). Halal industry: Key challenges and opportunities. SSRN Electronic Journal, 69,631. https://doi.org/10.2139/ssrn.2735417.
92. Gudowska, B. (2017). The power of quality in product management of cosmetics. The case of halal'beauty products. *Globalization, the State and the Individual, 3*(March), 29–39. http://wydawnictwo.wsei.eu/index.php/gsi/article/view/544
93. Jusoh, A., Kamarulzaman, L., & Zakaria, Z. (2017). Contemporary issues and development in the global halal industry. https://doi.org/10.1007/978-981-10-1452-9
94. Laluddin, H., Sikandar, S., Haneef, S., & Khalid, H. (2019). The scope, opportunities and challenges of halal industry: Some reflections. *International Journal of Economics, Management and Accounting, 27*(2), 397–421.
95. Othman, N. N., Zakaria, Z., & Aziz, A. (2020). Legal control for the safety of cosmetic products application use in Malaysia. *Journal of Halal Industry & Services, 3*(1–6). https://doi.org/10.36877/jhis.a0000080

KOSHER References:

96. American-Israeli Cooperative Enterprise. (2013). *American-Israeli Cooperative Enterprise.* Jewish Dietary Laws. From http://www.jewishvirtuallibrary.org/jsource/Judaism/kashrut.html
97. Chabad.org. (n.d.). Kosher fish list., from http://www.chabad.org/library/article_cdo/id/82675/jewish/ Kosher-Fish-List.htm
98. Rich, T. (2011). Kashrut: Jewish Dietary Laws. Retrieved February 8, 2013, from http://www.jewfaq.org/kashrut.htm.
99. Grunfeld, I. (1972). *The Jewish dietary laws* (pp. 11–12). The Soncino Press.
100. Chaudry, M. M. (1992). Islamic food laws: Philosophical basis and practical implications. *Food Technology, 46*(10), 92.
101. Chaudry, M. M., & Regenstein, J. M. (1994). Implications of biotechnology and genetic engineering for kosher and halal foods. *Trends in Food Science and Technology, 5*, 165–168.

102. Khan, G. M. (1991). *Al-Dhabah, slaying animals for food the Islamic way* (pp. 19–20). Abul Qasim Bookstore.
103. Hodkin, M. (2005). When ritual slaughter isn't Kosher: An examination of shechita and the humane methods of slaughter act. *Journal of Animal Law, 1*(1), 129–150.
104. Milne, E. L. (2007). Protecting islam's garden from the wilderness: Halal fraud statutes and the first amendment. *Journal of Food Law and Policy, 2*, 61–83. for an overview of Islamic dietary laws and their interpretations.
105. (<http://www.kosherholland.nl>)
106. Fuchs, D., Kalfagianni, A., & Havinga, T. (2009). Actors in private food governance: The legitimacy of retail standards and multistake-holder initiatives with civil society participation. *Agriculture and Human Values*. https://doi.org/10.1007/s10460-009-9236-3
107. Havinga, T. (2006). Private regulation of food safety by supermarkets. *Law and Policy, 28*(4), 515–533.
108. Levi-Faur, D., (2010). 'Regulation and Regulatory Governance', Jerusalem Papers on Regulation and Governance No. 1, Available at: <http://regulation.huji.ac.il>.
109. Marsden, T., Flynn, A. & Harrison, M., (2010). The new regulation and governance of food: Beyond the food crisis?
110. Van Waarden, F. (2006). Taste, tradition, transactions, and trust: The public and private regulation of food. In C. Ansell & D. Vogel (Eds.), *What's the Beef? The contested governance of European food safety* (pp. 35–59).
111. Meidinger, E. (2009). Private import safety regulation and transnation-al new governance. In C. Coglianese et al. (Eds.), *Import safety: Regulatory governance in the global economy* (pp. 233–253) at p. 234.
112. Regenstein, J. M. (1994). Health aspects of kosher foods. *Activities report and minutes of work groups and sub-work groups of the R & D Associates, 46*(1), 77–83.
113. Comprehensive Reviews in Food Science and Food Safety, Vol. 2, Issue s2,pp. 3–109 April 2003 (111CRFSFS).
114. Rosenthal, S. F. (1997). Food for thought: Kosher fraud laws and the religious clauses of the first amendment. *George Washington Law Review, 65*, 951–1013.
115. Epstein, G. S., & Gang, I. N., (2002). The Political Economy of Kosher Wars, Departmental Working Papers 200,227, Rutgers University, Department of Economics. Available at: <http://ideas.repec.org/e/pep1.html> (last visited 27 July 2010)
116. Egan, M. (2002). Overview of halal from Agri-Canada perspective. In *Presented at the Fourth Intl Halal Food Conference; April 21–23*. Sheraton Gateway Hotel.
117. Osman, N, & Abo-Dief, H. (2023). The origins, usage, and production methods of Halal and Kosher Gelatin. https://doi.org/10.1007/978-3-031-41459-6_10.
118. Uddin, S., Hossain, M. A., Suresh, S., Al Amin, M., & Johan, M. (2021). Halal and Kosher gelatin: Applications as well as detection approaches with challenges and prospects. *Food Bioscience., 44*(101), 422. https://doi.org/10.1016/j.fbio.2021.101422
119. gros Moshe O.C. 3:62.
120. Sefer Kovetz Halachos Pesach Chapter 12, footnote 7.
121. The Biur Halacha 326:10 B'shaar.
122. Maharam Shick, Orach Chaim 242.
123. Sefer Kovetz Halachos Pesach 12:11.

References for Jainism

124. Tattvärtha Sutra By Ächärya Umäsväti, (K. K. Dixit, Trans.). L. D. Institute of Indology.
125. Sutras, J., (1895). Part I & II, translated by Hermann Jacobi .
126. Ätmasiddhi-shästra by Shrimad Räjchandra, (Shree Brahmachäriji, Trans).

127. Ramappa, K. (1987). *A Handbook of Jainology, by Ächärya Shri Bhuvan-bhanuji Surishvarji.*, English Trans. Shri Vishva Kalyän Prakashan Trust.
128. Ramappa, K. (1987). *Ganadhar-väd by Ächärya Shri Bhuvan-bhanuji Surishvarji.*, English Trans. Shri Vishva Kalyän Prakashan Trust.
129. Muni Shi Nyayavijayaji Jain Darshan (Jain philosophy and religion), (Shri Nagin Shah, Trans.). Motilal Banärasidäs(1997).
130. Sancheti, A. L. *First steps to jainism.* M. Sujan Mal Ugam Kanwar Sancheti Trust.
131. Jaini, P. S. (1997). *The jain path of purification.* Motilal Banärasidäs Publications.
132. Mehta, S. T. U. *The Path of Arhat: A Religious Democracy.* Pujya Sohanalal Smäraka Pärshvanäth, Sodhapitha.
133. Rosenfield, C., & Segall, L.. Ahimsa is not a religion... It is a way of life. Guided by Shree Gurudev Chitrabhanuji.
134. Sangave, V. A. (1990). *Aspects of Jaina Religion.* Bhartiya Jnänpith Publications.
135. Sangave, V. A. (1991). *The Jaina Path of Ahimsa.* Padmashri Sumatibai Vidyäpith Trust.
136. Shah, H. (1998). *Primary principles of jainism compilation.* Jain Society of San Francisco.
137. Jain Theism by Hemant Shah. Academy of Philosophy, (1997).
138. Doshi, S. M. (1994). *Essence of Jainism.* Indira Memorial Trust.
139. Doshi, S. M. (1997). *Spiritual codes and restraints.* Indira Memorial Trust.
140. Jain, C. R. (2009). *What is Jainism Printed by K. Mittra.* London Press Ltd.
141. Jain Education International For Private & Personal Use Only www.jainelibrary.org).

References for Buddhism

142. Kalra S, Priya G, Grewal E, Aye TT, Waraich BK, SweLatt T, Khun T, Phanvarine M, Sutta S, Kaush U, Manilka, Ruder S, Kalra B. Diabetes management and the buddhist philosophy: Toward holistic care. The Indian Journal of Endocrinology and Metabolism. 2018 November-December;22(6):806–811. https://doi.org/10.4103/ijem.IJEM_285_17. Erratum in: The Indian Journal of Endocrinology and Metabolism. 2019 Nov-Dec;23(6):654. doi: 10.4103/2230-8210.276225. PMID: 30766823; PMCID: PMC6330864.
143. McCaffree, J. (2002, July 01). Dietary restrictions of other religions. *Journal of the American Dietetic Association, 102*(3), 912. https://doi.org/10.1016/S0002-8223(02)90212-9
144. Lee, Y., & Krawinkel, M. (2009). Body composition and nutrient intake of Buddhist vegetarians. *Asia Pacific Journal of Clinical Nutrition, 18*(2), 265–271.
145. Persynaki, A., Karras, S., & Pichard, C. (2017). Unraveling the metabolic health benefits of fasting related to religious beliefs: A narrative review. *Nutrition., 35*, 14–20. https://doi.org/10.1016/j.nut.2016.10.005. Epub 2016 Oct 14.
146. Kim, H., Caulfield, L. E., Garcia-Larsen, V., Steffen, L. M., Coresh, J., & Rebholz, C. M. (2019). Plant-based diets are associated with a lower risk of incident cardiovascular disease, cardiovascular disease mortality, and all-cause mortality in a general population of middle-aged adults. *Journal of the American Heart Association, 8*(16), e012865. https://doi.org/10.1161/JAHA.119.012865. Epub 2019 Aug 7. PMID: 31387433; PMCID: PMC6759882.
147. Satija, A., Bhupathiraju, S. N., Rimm, E. B., Spiegelman, D., Chiuve, S. E., Borgi, L., Willett, W. C., Manson, J. E., Sun, Q., & Hu, F. B. (2016). Plant-based dietary patterns and incidence of Type 2 diabetes in US men and women: Results from three prospective cohort studies. *PLoS Med., 13*(6), e1002039. https://doi.org/10.1371/journal.pmed.1002039. PMID: 27299701; PMCID: PMC4907448.
148. Qian, F., Liu, G., Hu, F. B., Bhupathiraju, S. N., & Sun, Q. (2019). Association between plant-based dietary patterns and risk of Type 2 diabetes: A systematic review and meta-

analysis. *JAMA Internal Medicine, 179*(10), 1335–1344. https://doi.org/10.1001/jamainternmed.2019.2195. PMID: 31329220; PMCID: PMC6646993.
149. Kane-Diallo, A., Srour, B., Sellem, L., Deschasaux, M., Latino-Martel, P., Hercberg, S., Galan, P., Fassier, P., Guéraud, F., Pierre, F. H., Kesse-Guyot, E., Allès, B., & Touvier, M. (2018). Association between a pro plant-based dietary score and cancer risk in the prospective NutriNet-santé cohort. *The International Journal of Cancer, 143*(9), 2168–2176. https://doi.org/10.1002/ijc.31593. Epub 2018 Aug 7.
150. Lee, Y., & Krawinkel, M. (2011). The nutritional status of iron, folate, and vitamin B-12 of Buddhist vegetarians. *Asia Pacific Journal of Clinical Nutrition, 20*(1), 42–49.
151. Turner J, Parsi M, & Badireddy M. Anemia. (Updated 2023 August 8). In: StatPearls [Internet]. StatPearls Publishing; 2025 Jan-. Available from: https://www.ncbi.nlm.nih.gov/books/NBK499994/
152. Craig, W. J. (2010, December). Nutrition concerns and health effects of vegetarian diets. *Nutrition in Clinical Practice, 25*(6), 613–620. https://doi.org/10.1177/0884533610385707
153. Jamshed, H., Beyl, R. A., Della Manna, D. L., Yang, E. S., Ravussin, E., & Peterson, C. M. (2019). Early Time-Restricted Feeding Improves 24-Hour Glucose Levels and Affects Markers of the Circadian Clock, Aging, and Autophagy in Humans. *Nutrients., 11*(6), 1234. https://doi.org/10.3390/nu11061234. PMID: 31151228; PMCID: PMC6627766.
154. Levine, B., & Kroemer, G. (2019). Biological functions of autophagy genes: A disease perspective. *Cell., 176*(1–2), 11–42. https://doi.org/10.1016/j.cell.2018.09.048. PMID: 30633901; PMCID: PMC6347410.
155. Headland, M., Clifton, P. M., Carter, S., & Keogh, J. B. (2016). Weight-loss outcomes: A systematic review and meta-analysis of intermittent energy restriction trials lasting a minimum of 6 Months. *Nutrients., 8*(6), 354. https://doi.org/10.3390/nu8060354. PMID: 27338458; PMCID: PMC4924195.
156. Cioffi, I., Evangelista, A., Ponzo, V., Ciccone, G., Soldati, L., Santarpia, L., Contaldo, F., Pasanisi, F., Ghigo, E., & Bo, S. (2018). Intermittent versus continuous energy restriction on weight loss and cardiometabolic outcomes: A systematic review and meta-analysis of randomized controlled trials. *Journal of Translational Medicine, 16*(1), 371. https://doi.org/10.1186/s12967-018-1748-4. PMID: 30583725; PMCID: PMC6304782.
157. Harris, L., Hamilton, S., Azevedo, L. B., Olajide, J., De Brún, C., Waller, G., Whittaker, V., Sharp, T., Lean, M., Hankey, C., & Ells, L. (2018, February). Intermittent fasting interventions for treatment of overweight and obesity in adults: A systematic review and meta-analysis. *The JBI Database of Systematic Reviews and Implementation Reports, 16*(2), 507–547. https://doi.org/10.11124/JBISRIR-2016-003248
158. Cho, Y., Hong, N., Kim, K. W., Cho, S. J., Lee, M., Lee, Y. H., Lee, Y. H., Kang, E. S., Cha, B. S., & Lee, B. W. (2019). The effectiveness of intermittent fasting to reduce body mass index and glucose metabolism: A systematic review and meta-analysis. *Journal of Clinical Medicine, 8*(10), 1645. https://doi.org/10.3390/jcm8101645. PMID: 31601019; PMCID: PMC6832593.

References for Other Ethnic Considerations

159. Jones, G. (2010). *Beauty imagined: A history of the global beauty industry hardcover.* Oxford University Press.
160. Beveridge, I. (2020). *Intercultural marketing: Theory and practice* (1st ed. [E-book]). Routledge.
161. Diversity at work: The practice of inclusion (2013–11-18). (2021). [E-Book]. Jossey-Bass
162. DeFrancisco, V. P., & Palczewski, C. H. (2007). *Communicating gender diversity: A critical approach* (1st ed.). Sage.

163. Johnson, L. C., & Corah, N. L. (1963). Racial differences in skin resistance. *Science, 139*, 766–767.
164. Porter, J. P., & Lee, J. I. (2002). Facial analysis: Maintaining ethnic balance. *Facial Plastic Surgery Clinics of North America, 10*, 343–349.
165. Reed, J. T., Ghadially, R., & Elias, P. M. (1995). Skin type, but neither race nor gender, influence epidermal permeability barrier function. *Archives of Dermatology, 131*, 1134–1138.
166. Igarashi, T., Nishino, K., & Nayar, S. K. (2005, June). *The appearance of human skin*. Columbia University.
167. Mohiuddin, Oru Brands Find Ethnic Beauty, Euromonitor International Resources, (2013, October).
168. Guichard, S., & Roulier, V. L.'. O. (2010). *Cosmetic dermatology: Products and procedures, chapter 22: Facial foundations* (p. 167). Blackwell Publishing Ltd..
169. McKenzie, K., & Crowcroft, N. S. (1996). Describing race, ethnicity, and culture in medical research. *BMJ, 312*, 1054.
170. Kaplan, J. B., & Bennett, T. (2003). Use of race and ethnicity in biomedical publication. *JAMA, 289*, 2709–2716.
171. Carter, E. L. (2003). Race vs ethnicity in dermatology. *Archives of Dermatology, 139*, 539–540.
172. Andersen, K. E., & Maibach, H. I. (1979). Black and white human skin differences. *The Journal of the American Academy of Dermatology, 1*, 276–282.
173. CB Insights. (2021). *How the inclusive beauty movement is redefining the industry*. CB Insights Research. https://www.cbinsights.com/research/what-is-inclusive-beauty/
174. Asia Pacific beauty and personal care Products Market: 2021–26: Industry Share, size, growth – Mordor intelligence. Asia Pacific Beauty and Personal Care Products Market | 2021–26 | Industry Share, Size, Growth – Mordor Intelligence. (n.d.). https://www.mordorintelligence.com/industry-reports/asia-pacific-beauty-and-personal-care-products-market-industry.
175. US Census Bureau. (2002). Population projections of the US by age, sex, race and Hispanic origin: 1995–2050. Current population report: 25–1130. US Government Press.
176. American Society of Aesthetic Plastic Surgery. (2005). *Cosmetic Surgery National Data Bank* (pp. 1–20). American Society for Aesthetic Plastic Surgery.
177. Berardesca, E., & Maibach, H. I. (1991). Sensitive and ethnic skin: A need for special skin care agents. *Dermatologic Clinics, 9*, 89–96.
178. Fitzpatrick, T. B. (1988). The validity and practicality of skin reactive skin type I through VI. *Archives of Dermatology, 124*, 869–871.
179. Lancer, H. A. (1998). Lancer Ethnicity Scale (LES). *Lasers in Surgery and Medicine, 22*, 9.
180. Berardesca, E., & Maibach, H. (1996). Racial differences in skin pathophysi ology. *The Journal of the American Academy of Dermatology, 34*, 667–672.
181. 24, A. W. R. A., & Reade, A. W. (2020, August 25). "More than one TOKEN Latina": VIVE cosmetics is all about REAL Latinx Representation. Beauty Independent. https://www.beautyindependent.com/vive-cosmetics-real-representation-latinx-community/.
182. Alexis, A. F., Sergay, A. B., & Taylor, S. C. (2007). Common dermatologic disorders in skin of color: A comparative practice survey. *Cutis, 80*, 387–394.
183. Barillaro, D. (2016, April). Multicultural beauty: Many new shades for Millennials' types of skin, Kosmetica World.
184. Weigand, D. A., Haygood, C., & Gaylor, J. R. (1974). Cell layers and density of Negro and Caucasian stratum corneum. *The Journal of Investigative Dermatology, 62*, 563–568.
185. Bamfield, P., & Hutching, M. G. (2010). *Chromic phenomena: Techno-logical applications of colour chemistry* (p. 192). Royal Society of Chemistry.
186. David Schlossman, Edward Bartholomey, Carl Orr, US 20100061947 A1, UV protective cosmetic product incorporating titanium dioxide and transparent iron oxides, March 11, 201.
187. Montagna, W., Prota, G., & Kenney, J. A. (1993). *Black skin structure and function* (pp. 21–54). Academic.

New and Future Technologies for the Creation of Cosmetics

Perry Romanowski

Abstract This chapter explores how new scientific and technological approaches are being applied in the cosmetic industry to improve product function, safety, and sustainability. It reviews a range of emerging methods, including advanced ingredient delivery systems, bio-based production through fermentation and genetic modification, artificial intelligence in product development, and new formats such as 3D printing. These innovations are presented with attention to both their scientific basis and the practical hurdles that limit their current use. Topics such as ingredient stability, production scalability, environmental impact, and regulatory oversight are addressed to give a balanced view of what these technologies may offer. This chapter also reflects on how shifts in consumer expectations are influencing which ideas move forward. Taken together, these developments signal a possible shift in how cosmetics are conceived, formulated, and evaluated.

Keywords Advanced delivery systems · Liposomes · Niosomes · Nanocapsules · Cosmetic biotechnology · Fermentation-based ingredients · Microbiome and postbiotics · Artificial intelligence in formulation · Personalized skincare · Green chemistry · 3D printing in cosmetics

Cosmetic consumers are constantly seeking new products to solve the problems related to beauty and personal care. This has motivated the industry to be constantly creating and introducing new products to the market each year. One estimate by the market research firm Euromonitor found that over 25% of all new brands launched in 2023 were in the beauty and personal care sector [1]. While the demand for new products continues, the development of innovative new technologies has not been able to keep up. The primary technologies behind the shampoos and skin moisturizers people were using in the 1980s and 1990s is not significantly different than the ones used in products today. From a product performance standpoint, there have not

P. Romanowski (✉)
Vice President, Element 44 Inc, Chicago, IL, USA

K. Alexander, P. Romanowski (eds.), *Cosmetic Science*, AAPS Introductions in the Pharmaceutical Sciences 15, https://doi.org/10.1007/978-3-032-03596-7_12

been many major innovations in the industry in decades. Currently, most of what passes for innovation in the cosmetic industry comes from packaging improvements and marketing stories. But there are some promising technologies that could help break the industry out of the innovation rut it seems to be in. This chapter will review some of those technologies.

Types of Innovation

Innovation can be defined as the process of applying scientific knowledge and methodologies to solve practical problems, leading to significant advancements in industry and technology [2]. Improving product performance is the primary type of innovation formulating chemists focus on, but it is only one type. In the cosmetic industry, innovation extends beyond product performance to encompass development driven by sustainability, regulatory compliance, and shifting consumer demands. For example, consumer demand for "eco-friendly" products has inspired innovation to help the industry move away from the use of petrochemicals to more sustainable products. Additionally, regulatory demands and consumer fears have created a need for alternative technologies from ones that have been traditionally used [3]. While these types of changes may not lead to consumer perceptible differences in product performance, they do represent real change and innovation in the industry.

Before getting further, it would be worthwhile to define what is meant by emerging technology. Emerging technology refers to innovations that are in the early stages of development and have the potential to significantly impact the industry and the products produced [4]. These technologies are characterized by rapid growth, the promise of solving existing problems, and the ability to disrupt established norms or create entirely new industries. What makes a technology "emerging" is its novelty combined with the uncertainty of its long-term influence. It is typically on the cutting edge of research and development, capturing interest because of the potential to transform processes, improve efficiencies, or introduce entirely new capabilities.

Emerging technologies are also defined by a combination of innovation and scalability—they offer potential new solutions that can be expanded for widespread use, but may not yet be fully understood or integrated into mainstream markets. Their impact is speculative, but if successful, they can significantly disrupt the current market. Additionally, the label "emerging" often suggests that the technology is in a dynamic, evolving phase, with many unanswered questions regarding technical feasibility, regulation, ethics, and societal impact. The inherent unpredictability makes these technologies both interesting to pursue and challenging to adopt.

In this chapter, we will look at a variety of emerging technologies that have the potential to advance performance improvements and innovation of cosmetics. Each section will cover a description of the technology, how it works, and what it can potentially do when applied to cosmetic products. Along the way, we will also discuss some of the challenges that inhibit innovation and pose ideas as to what the industry could do to help spur more significant innovations in product development. The technologies will be presented roughly in order of how close they are to coming to fruition and having a real impact on the industry. In truth, this is just a guess as future discoveries or developments may move one of these technologies to the forefront of the cosmetic industry.

Advanced Delivery Technologies

Advanced delivery systems in cosmetics refer to technologies designed to enhance the delivery, stability, and efficacy of active ingredients in formulations. These systems have evolved from simple emulsions to sophisticated encapsulation and controlled release techniques. Their development has been advanced by discoveries in materials science, nanotechnology, and dermatological research. They trace back to the mid-twentieth century, with liposomes being one of the first technologies introduced in the 1960s as carriers for drug delivery, later adapted for cosmetic applications [5].

Modern delivery systems have been used in cosmetic formulations to address challenges that have limited the efficacy and appeal of products. Traditional formulations often struggle with stability problems, degradation of active ingredients, inadequate penetration into the skin, and incompatibility between components. Advanced delivery systems, such as liposomes, nanoemulsions, and particulate systems, can stabilize sensitive actives like antioxidants and biomolecules, and protect them from oxidation or environmental degradation. These delivery systems can also enhance skin absorption by overcoming the barrier posed by the stratum corneum, increasing the chances that ingredients reach their target sites effectively [6].

Additionally, these systems enable the potential for controlled or sustained release of actives, reducing irritation and improving tolerability for sensitive users. Beyond functional improvements, modern delivery technologies also enhance the sensory experience of products, offering lightweight, fast-absorbing, and aesthetically pleasing textures that align with consumer expectations. By solving these issues, advanced delivery systems can potentially maximize the performance, safety, and user satisfaction of cosmetic formulations [6].

Encapsulation Technologies

Encapsulation technologies provide a unique way to deliver active ingredients in cosmetics by providing a protective barrier that enhances stability, bioavailability, and targeted delivery. The main types of encapsulates include liposomes, niosomes, polymeric nanocapsules, microsponges, and porous microspheres.

Liposomes and Niosomes

Liposomes, among the earliest encapsulation methods first described in 1961 by Bangham et al. [5], are spherical vesicles made of phospholipid bilayers. Their structure mimics natural cell membranes, allowing them to encapsulate both hydrophilic and lipophilic actives effectively.

When applied to the skin, liposomes merge with the lipid barrier, theoretically facilitating the release of their contents into deeper layers. This would make them ideal delivery systems for active ingredients or cosmeceuticals that are meant to interact with the production of biochemicals like Collagen or Elastin at the lower layers of the skin. A complete review of the current status of this technology can be found in Dragićević and Maibach [7].

Niosomes, like liposomes, are vesicular systems used for delivering active ingredients in cosmetic formulations, but they are composed of non-ionic surfactants instead of phospholipids. This structural difference makes niosomes more stable, cost-effective, and resistant to oxidative degradation, providing a longer shelf life compared to liposomes. Similar to liposomes, niosomes can encapsulate both hydrophilic and lipophilic substances, enhancing the penetration of active ingredients into the skin. Their bilayer structure allows for controlled and targeted delivery, improving the efficacy of the actives. Niosomes are particularly attractive in cosmetics due to their flexibility in formulation, as they can be tailored for specific skin types and conditions. While they share functional similarities with liposomes, their lower production cost and greater stability make them a valuable alternative for formulating advanced skincare products.

Nanocapsules

Nanocapsules are a specialized form of core-shell micro/nanocapsules, where an active ingredient or payload is encapsulated within a nanoscale shell made of polymers or other materials. These capsules typically range in size from 1 to 100 nanometers. Their core can contain solid, liquid, or gaseous substances either hydrophilic or lipophilic, while the surrounding shell serves as a protective barrier, ensuring stability and controlled release of the encapsulated materials [8].

The materials used to construct nanocapsules are diverse, including both natural and synthetic polymers. Common examples include poly(methacrylate), poly(ethylene glycol), cellulose derivatives, poly(lactide), poly(styrene), gelatin, and poly(lactide-co-glycolide). The choice of material depends on the desired characteristics of the nanocapsules, such as biocompatibility, stability, and the specific application they are intended for.

Developed to address limitations in traditional delivery systems, nanocapsules provide enhanced protection for active ingredients against environmental degradation, such as oxidation and photodegradation. Their small size allows them to penetrate deeper into the skin, delivering actives more effectively to targeted layers. This makes nanocapsules potentially useful in delivering sensitive ingredients like vitamins, antioxidants, and peptides, ensuring their stability and efficacy.

Microsponges

Microsponges are a delivery system consisting of porous, polymeric particles capable of entrapping active ingredients within their sponge-like structure. These particles, typically 10–25 microns in size, can theoretically allow for the controlled release of actives over extended periods. As the skin absorbs the product, microsponges gradually release the actives, providing prolonged efficacy. Microsponges can carry both hydrophilic and lipophilic ingredients, making them useful for a variety of cosmetic formulations. Their design also helps minimize irritation by ensuring gradual delivery, which is particularly beneficial for potent actives like benzoyl peroxide or retinoids. Additionally, microsponges enhance the sensory experience of formulations by reducing greasiness and improving the spreadability of creams and lotions [9].

Porous Microspheres

Porous microspheres are spherical delivery systems composed of polymeric materials with a network of interconnected pores that can entrap active ingredients. Various polymers can be used such as polymethyl methacrylate, polyethylene, nylon, and more [10]. These microspheres, typically ranging from 1 to 100 microns in size, are designed to provide controlled release and improve the stability of encapsulated actives. Their porous structure allows for the gradual release of ingredients, minimizing the risk of irritation while maintaining consistent efficacy. Porous microspheres are versatile and can carry both hydrophilic and lipophilic substances, making them suitable for a range of cosmetic applications, including sunscreens, anti-aging products, and oil-control formulations. Finally, like microsponges, their ability to improve the tactile properties of formulations, such as reducing greasiness or adding a smooth finish, enhances user satisfaction.

Limitations of Encapsulate Technologies

Unfortunately, despite their potential effectiveness in enhancing skin penetration and stabilizing active ingredients, encapsulation technologies face various limitations that have hindered their use in cosmetic formulations [11, 12]. These include skin barrier resistance, stability problems, performance difficulties, production challenges, and regulatory concerns.

Skin Barrier Resistance

The stratum corneum has evolved to provide significant resistance to penetration of most compounds. While these technologies are meant to overcome that barrier, they often remain confined to the superficial layers rather than reaching deeper levels where certain actives may be most effective [13]. The relatively large size of microsponges compared to other delivery systems may restrict penetration into deeper skin layers, making them more suitable for surface-level applications.

Stability Concerns

In theory, the design of all of these encapsulation technologies should lead to stable structures. However, in practice, the structural integrity of things like liposomes, niosomes, and nanocapsules can be compromised over time and during cosmetic product production. Issues such as leakage of encapsulated ingredients, lipid oxidation, and aggregation are common, impacting their effectiveness and product longevity [14].

Production Problems

The production process of making encapsulated materials is another significant hurdle [15]. Nanocapsules often require specialized techniques, such as nanoprecipitation or interfacial polymerization, which can be costly and difficult to scale. Microsphere production uses emulsion polymerization or spray drying, which can be complex and resource-intensive, leading to higher costs. Microsponge production also involves polymerization or solvent techniques but requires careful control to avoid issues like incomplete encapsulation or uneven release profiles.

Performance

While the use of liposomes and other encapsulation techniques has been tried by numerous formulators, there is little evidence that they substantially increase the performance of active ingredients. There is actually an ongoing debate about whether liposomes penetrate the skin intact or release their contents upon disintegration [14]. This lack of clarity and performance has inhibited the adoption of the technology.

Ingredient Compatibility

In a cosmetic formulation, compounds like ethanol or surfactants are often required to improve the penetration of encapsulates. However, these enhancers can also alter the structure of things like liposomes, niosomes, or microspheres. They can also increase the likelihood of skin irritation for some users. This has limited their effectiveness and prompted formulators to consider other options. Compatibility with actives is another concern, as some ingredients may interact with the microsphere material, potentially reducing stability or efficacy [16].

Size and Charge Challenges

Smaller liposomes may penetrate the skin more effectively, but can struggle to encapsulate certain ingredients [17]. Conversely, larger liposomes are less effective in penetrating the skin barrier. Charge differences also influence their interaction with the skin. Stability during storage is another concern, as aggregation or premature release of actives can occur. The size and rigidity of porous microspheres also limit their penetration into deeper skin layers, confining their effectiveness to surface-level applications.

Regulatory Challenges

Perhaps the most significant challenge affecting encapsulates is the regulatory uncertainty surrounding the use of nanotechnology and novel delivery systems in cosmetics. Governments have implemented strict safety evaluations [16]. Additionally, concerns regarding the environmental impact of non-biodegradable polymeric materials, such as polyethylene or polystyrene, have prompted regulatory scrutiny and consumer demand for more sustainable alternatives. In fact, "microbeads" were banned in the US for use in rinse-off personal care products [18]. Consumers also have been made to fear nanotechnology. This has complicated their commercial application despite the promising benefits.

Future of Encapsulation Technologies

Improvements in vesicle design for liposomes, niosomes, nanocapsules, and microsponges can greatly enhance their utility in cosmetic formulations. For liposomes and niosomes, optimizing particle size, surface charge, and bilayer composition can improve stability, skin penetration, and encapsulation efficiency. The incorporation of advanced surfactants and biocompatible polymers could enhance their compatibility with sensitive skin and broaden their application.

Nanocapsules can benefit from more durable shell designs and stimuli-responsive materials that enable precise, controlled release of active ingredients under specific conditions like pH or temperature changes. For microsponges, refining pore size and distribution can increase drug loading capacity and improve sustained release.

Using biodegradable and renewable materials in these systems would address environmental concerns while maintaining or improving performance. Additionally, innovations in hybrid systems, where vesicles combine multiple technologies, could also enable multifunctional formulations that target deeper skin layers, provide extended release, and reduce irritation, making these systems even more effective in meeting diverse cosmetic needs.

Biotechnology in Cosmetics

Biotechnology represents an area in which cosmetic formulation technology could find some potential, significant innovations. Initially rooted in pharmaceutical and biomedical research, biotechnology's integration into cosmetics can lead to the creation of bioengineered ingredients, personalized solutions, and environmentally friendly alternatives. Applications ranging from the synthesis of lab-produced alternatives to animal-derived ingredients to the use of microbial fermentation for active ingredient production are all possible. Biotechnology is opening new possibilities for addressing consumer's skin and hair problems. This section examines how biotechnology is being used in cosmetics, highlighting its benefits, practical applications, and the potential it holds for the future. By examining these developments, we can better understand the role biotechnology could play in shaping the next generation of beauty and skincare products.

What Is Biotechnology?

Biotechnology is the application of biological systems, organisms, or processes to develop beneficial products and solutions [19]. In cosmetics, it plays a key role in creating innovative ingredients and formulations to improve skin, hair, and overall appearance.

The roots of biotechnology trace back to ancient practices, such as fermentation for producing bread, wine, and cheese. These early techniques laid the groundwork for modern advancements, including genetic engineering and microbial synthesis. Over time, biotechnology has evolved to include methods like recombinant DNA technology, enabling the production of complex compounds such as growth factors, peptides, and enzymes [19]. This progression has significantly increased the possibilities for formulating products that are more targeted and potentially more effective for skincare and personal care.

Fermentation

Fermentation is a foundational process in biotechnology that uses microorganisms such as bacteria, fungi, and yeast to produce ingredients for cosmetics. These microorganisms are cultivated in controlled environments where they convert simple substrates like sugars into complex bioactive compounds [20]. The ingredients can be molecules currently used in cosmetics such as hyaluronic acid, squalane, and collagen, or they can be new materials like peptides as growth factors that may provide new performance benefits. The process can be efficient, scalable, and allows for the sustainable production of high-quality ingredients.

At the industrial level, fermentation is carried out in bioreactors—specialized vessels were designed to provide optimal conditions for microbial growth and product synthesis. These reactors maintain controlled temperature, pH, and aeration, ensuring consistent output. Bioreactors are useful because they can enable the production of a wide range of ingredients, from antioxidants to moisturizing agents. Their ability to scale up production while minimizing contamination and waste makes them invaluable in the cosmetic industry.

Genetic Engineering

Genetic engineering takes biotechnology a step further by modifying the DNA of organisms to produce specific, high-value compounds [21]. This method enables the creation of ingredients that may not naturally occur in significant quantities or that are otherwise difficult to extract. By inserting targeted genes into host organisms, such as bacteria or yeast, scientists can engineer these organisms to synthesize beneficial molecules with precision.

One of the first applications of using genetic engineering was the production of human insulin, achieved by inserting the human insulin gene into *Escherichia coli.* This principle has been adapted for cosmetics, allowing for the production of growth factors, enzymes, and other bioactive ingredients [22]. This technique not only expands the range of possible cosmetic ingredients, but also may reduce the environmental impact by providing a controlled and sustainable method of production. Following this approach, companies may be able to discover new compounds for innovation in skincare and personal care formulations.

Lab-Produced Ingredients

The production of cosmetic ingredients currently faces several challenges that biotechnology has the potential to address. These challenges include sustainability concerns, resource scarcity, variability in ingredient quality, ethical considerations, and environmental impact.

One of the most pressing challenges in the production of cosmetic ingredients is the sustainability of sourcing materials from natural resources. Over harvesting of plants, destruction of habitats, and depletion of finite resources are common consequences of relying on traditional methods of using plant materials to create ingredients. Biotechnology offers a promising solution by enabling the production of these ingredients in controlled laboratory environments through microbial fermentation, metabolic engineering, and synthetic biology. For example, yeast fermentation can be used to produce squalene, a popular emollient, providing an ethical and sustainable alternative to sourcing it from shark liver [23].

Biotechnology has also enabled the production of bioactive compounds that were previously difficult or impossible to obtain in large quantities. Some examples include lab-produced hyaluronic acid [24], collagen [25], and peptides [26] that offer benefits for skincare and haircare. These ingredients, produced through fermentation or other biotechnological methods, are not only suitable replacements for standard versions but also are consistent in quality, which is crucial for maintaining product reliability. Additionally, biotechnology has opened the door to creating new molecules designed specifically for cosmetic applications, enhancing the industry's ability to address diverse skin and hair needs. Table 1 shows some of the most common ingredients produced by biotechnology in the cosmetic industry.

Enhanced Product Performance

One of the reasons products developed using biotechnological methods are novel is that they can offer targeted and efficient solutions tailored to specific consumer needs. For example, bioengineered actives may be designed to penetrate the skin more effectively, ensuring that key ingredients reach the layers where they are most

Table 1 Common ingredients produced by biotechnology in the cosmetic industry

Ingredient	Type of ingredient	Use in cosmetics	Production using biotechnology
Squalane	Emollient	Moisturizes and softens skin	Fermentation of sugarcane using engineered yeast
Hyaluronic acid	Moisturizer	Hydrates and improves skin elasticity	Microbial fermentation of bacteria like Bacillus subtilis
Collagen	Protein	Enhances skin firmness and reduces aging signs	Microbial fermentation using genetically modified Pichia pastoris
Peptides	Signaling molecule	Stimulates collagen production & skin repair	Enzymatic processes or bioengineering for peptide extraction
Probiotics	Microorganism	Strengthens skin microbiome and barrier	Fermentation of kefir grains to produce beneficial microbes
Kojic acid	Depigmenting agent	Lightens skin and reduces hyperpigmentation	Fermentation of fungi like Aspergillus oryzae
Enzymes	Catalyst	Repairs skin damage and promotes exfoliation	Microbial fermentation or recombinant DNA technology
Growth factors	Protein	Promotes cell repair and skin regeneration	Recombinant DNA technology to produce proteins in bioreactors
Fragrances	Fragrance	Provides sustainable fragrance ingredients	Fermentation of sugarcane to produce molecules like Ambrofix
Hemisqualane	Emollient	Lightweight emollient for hair and skin care	Fermentation of sugarcane for a lightweight emollient
Bioethanol	Solvent	Used in disinfectants, fragrances, and sprays	Fermentation of renewable feedstocks such as sugarcane or corn

beneficial [26]. Similarly, biotechnological processes allow for precise control over ingredient properties, such as stability and bioavailability, enhancing the overall efficacy of formulations. This focus on performance ensures that products deliver measurable benefits, which is increasingly important in a competitive and informed marketplace.

Microbiome Technology

Microbial biotechnology involves the use of bacteria and other microorganisms to develop products that support skin health [27]. Probiotics, live beneficial bacteria, and postbiotics, the byproducts of microbial fermentation, are increasingly used in cosmetics with the goal of improving the skin microbiome [28]. These ingredients may help balance the skin's natural flora, reducing issues such as dryness, redness, and sensitivity.

Despite its potential, the current state of the science in this area remains limited, and many claims about probiotics and postbiotics in cosmetics lack rigorous clinical validation. The skin microbiome is highly complex and influenced by numerous factors, including genetics, environment, and lifestyle. While some studies suggest benefits from topical probiotics and postbiotics, clear evidence of their long-term efficacy and mechanisms of action is still lacking [27]. Additionally, maintaining the viability of live bacteria in cosmetic formulations and ensuring they deliver measurable results on the skin pose significant challenges, both technical and regulatory. A product that is currently leading the way in the use of this technology contains live Ammonia-Oxidizing Bacteria (AOB), specifically *Nitrosomonas eutropha* [29].

If this technology is going to lead to substantial breakthroughs, more robust clinical research is needed to establish consistent outcomes across different skin types and conditions. Improved formulation techniques will also be essential to stabilize live bacteria in traditional types of cosmetic formulas. Additionally, work will have to be done to optimize postbiotic activity so the materials remain effective when applied to the skin.

Genomic Insights

Genomics and proteomics are developing biological fields that could have an impact on the development of cosmetic products by providing a deeper understanding of individual skin biology. Genomics focuses on the study of genes and their functions, while proteomics examines the proteins expressed by these genes. Together, they offer insights into the genetic and molecular mechanisms underlying skin aging, hyperpigmentation, elasticity, and other conditions [30]. It is believed that by analyzing an individual's genetic profile, cosmetic formulations may someday be tailored to effectively treat specific concerns, such as sagging skin, collagen degradation, inflammation, or oxidative stress. This approach could potentially deliver more effective and personalized skincare solutions.

Proteomics complements genomics by providing a snapshot of the active biological processes within the skin. Proteins are the workhorses of cells, and their expression levels, modifications, and interactions can offer valuable insights into skin conditions like aging, hydration, and barrier function. Proteomics can help identify biomarkers for skin health and aid in the discovery of novel active ingredients or therapeutic targets for cosmetic formulations. For example, proteomic analysis may reveal changes in keratin or keratin-associated proteins associated with hair aging or damage, guiding the development of products that strengthen or repair the hair shaft and improve overall hair health [31]. While genomic insights hold promise for personalized skincare, the technology is still in its early stages, and claims of DNA-based formulations delivering highly targeted solutions warrant skepticism. Current genetic testing for skincare often relies on broad correlations rather than definitive, actionable links between genes and skin conditions [30]. As with the microbiome, skin health is complex and influenced by numerous factors such as the environment, consumer lifestyle, and epigenetics. It is possible that genetics alone may not provide a complete picture. Additionally, the cost and accessibility of such testing pose barriers to widespread adoption. Finally, even after identifying the important factors involved with an individual's DNA, making the translation to a cosmetic formulation is not clear. It is not yet known how to determine the exact dosage of an active ingredient based on a person's DNA sequence.

Proteomics also has its limitations. Protein expression is highly dynamic and influenced by external factors such as UV exposure, pollution, and lifestyle, making it challenging to draw consistent conclusions. Additionally, proteomic technologies are complex, costly, and require advanced bioinformatics tools for analysis, which limits their practical application in routine cosmetic product development [31]. As with many technologies, the product marketers are often ahead of what science can deliver.

To achieve meaningful breakthroughs, more research is needed to establish clear, scientifically validated connections between specific genetic markers and skin responses. Larger, diverse datasets are also required to refine predictive models, ensuring accuracy and relevance across different populations. Advances in technology, such as improved genomic sequencing and deeper integration with AI-driven analysis, should help improve the technology [32]. With these developments, genomic-based skincare may be able to move beyond marketing claims to deliver truly effective and individualized solutions.

Biodegradable Packaging

Biotechnology is affecting the cosmetic industry beyond formulations. It also has had an impact on sustainable packaging solutions. Innovations in bioplastics, derived from renewable sources like corn or sugarcane, are helping reduce the environmental impact of cosmetic packaging. Additionally, enzymes and other biological processes are being explored to create packaging materials that decompose naturally without leaving harmful residues [33]. These developments align with broader sustainability goals and offer brands a way to meet consumer expectations for eco-friendly practices.

The applications of biotechnology in cosmetics are potentially impressive and continue to grow. From lab-produced ingredients to genomic insights and sustainable packaging, this technology has the potential to change how beauty products are created and perform.

Challenges and Limitations

While biotechnology offers opportunities for innovation in cosmetics, several challenges and limitations remain that will have to be addressed. These obstacles range from economic and technical barriers to consumer and regulatory considerations [19].

Cost and Scalability

Developing biotechnological solutions for cosmetic products requires a significant investment. While techniques to produce ingredients in the lab and microbial technologies show promise, the high costs associated with their production can make them difficult for start-up companies or mass-market applications. Scaling these processes to meet global demand adds another layer of complexity, as facilities capable of producing biotechnological ingredients at a large scale are limited and often costly to establish. Addressing these challenges will be necessary to make biotechnology more widely adopted in the industry.

Consumer Perception

Despite the advantages of biotechnology, some consumers remain hesitant about lab-produced ingredients and products derived through biotechnological methods. Concerns about the "unnatural" nature of genetically modified organisms (GMOs) in beauty products or misconceptions surrounding their safety can hinder their acceptance in the marketplace [34]. For these technologies to have a significant impact, this challenge will need to be overcome.

Regulatory Hurdles

The introduction of biotechnological products brings regulatory challenges. Cosmetic ingredients and formulations already must meet stringent safety and efficacy standards, which can be particularly complex for lab-produced or bioengineered compounds. Regulatory frameworks often lag behind technological advancements, creating uncertainty for companies seeking to launch biotechnological products.

Biotechnology allows companies to make sophisticated claims about their products, such as enhanced collagen production or DNA repair, which often resemble medicinal claims. These claims, while appealing to consumers, bring increased scrutiny from regulatory bodies. For example, in Europe, products must comply with the Cosmetics Regulation, which primarily ensures safety but does not require the extensive clinical trials that pharmaceuticals undergo. However, if a product's claims suggest therapeutic effects, it may fall under medicinal regulations, requiring costly and lengthy approval processes. Similarly, in the United States, the FDA differentiates between cosmetics and drugs based on intended use, as established by labeling, advertising, or consumer perception. This dual categorization means a product can simultaneously fall under both regulatory frameworks, complicating compliance for companies and leading to case-by-case evaluations [35].

The banning of animal testing for cosmetics in most places around the world has made it particularly challenging to demonstrate safety of a truly new and novel ingredient. Navigating these requirements can be time-consuming and costly, especially for smaller brands with limited resources. Establishing clearer guidelines for biotechnological innovations in cosmetics will be essential if it is going to make a significant impact on the cosmetic industry.

Technical Constraints

Producing complex compounds through biotechnology is a technically demanding process. Challenges include ensuring the consistency of ingredients, maintaining stability during storage and use, and integrating these compounds into existing formulations without compromising performance. For example, lab-produced peptides may require specific environmental conditions to remain effective, adding complexity to both production and product design [36]. These technical hurdles highlight the need for continued research and development to improve the reliability and scalability of these technologies.

While the challenges of cost, scalability, consumer perception, regulation, and technical complexity are significant, they are not insurmountable. If these issues are properly addressed, biotechnology promises to help produce products that perform significantly better than the current offering and are more environmentally friendly.

Green Chemistry and Sustainable Formulations

Another area in which technology could impact cosmetic formulations and lead to new innovation is Green chemistry. The concept of Green Chemistry was introduced by *Anastas* and *Warner* in the early 1990s and is a scientific approach focused on designing chemical processes and products that minimize environmental and human health impacts. It emphasizes the use of renewable raw materials, energy efficiency, and the reduction or elimination of hazardous substances throughout a product's lifecycle [37].

Central principles of Green Chemistry include avoiding toxic byproducts, using biodegradable materials, and creating closed-loop systems that recycle waste into usable inputs. In cosmetics, green chemistry involves synthesizing safe, eco-friendly ingredients, reducing water and energy use, and prioritizing processes that reduce pollution and carbon emissions. This methodology couples innovation with sustainability, encouraging environmentally responsible practices across industries.

Impact on Formulating

Green chemistry emphasizes efficiency and optimization at every stage of a product's lifecycle. This focus often leads to formulations that are highly concentrated and made with bio-compatible ingredients. Some example of green chemistry applied to cosmetics would be as follows [38].

Biodegradable Ingredients in Cosmetic Formulations

Biodegradable ingredients have emerged as a key component in the creation of sustainable cosmetic formulations. They offer an alternative to the non-degradable traditional compounds that pose environmental challenges [39]. Biodegradable ingredients break down naturally into harmless substances such as water, carbon dioxide, and biomass, reducing their persistence in ecosystems. Derived from renewable sources like plant oils, sugars, and algae, biodegradable ingredients are designed to maintain efficacy while minimizing long-term environmental impact. Their applications range from emulsifiers and surfactants to exfoliants and packaging materials, reflecting their versatility in modern cosmetic formulations.

Examples of biodegradable ingredients include plant-based exfoliants like ground walnut shells or jojoba beads, which replace plastic microbeads banned in many regions due to their contribution to marine pollution. Other examples are coffee grounds or fruit peels which are upcycled ingredients from other industries. Similarly, sugar-derived surfactants, such as decyl glucoside, offer effective cleansing properties without the perceived ecological harm of synthetic detergents. Biodegradable polymers, such as polylactic acid (PLA), are increasingly used as thickeners, film-formers, or encapsulation materials [33]. These alternatives ensure that even rinse-off products, which typically enter wastewater systems, decompose without leaving harmful residues.

Waterless Formulations

Waterless formulations are concentrated formats that require fewer preservatives and packaging materials. A classic example is bar soap. These products eliminate or significantly reduce water content, focusing instead on concentrated, solid, or powdered forms that offer equivalent or enhanced efficacy. By minimizing water consumption and reducing the need for large, heavy packaging, waterless formulations contribute to resource conservation and a lower carbon footprint throughout a product's lifecycle.

More modern examples of waterless cosmetics include solid shampoos and conditioners, which are gaining popularity for their portability and reduced packaging waste [40]. Powdered cleansers and serums activate with minimal water, offering highly concentrated formulas with extended shelf lives due to the absence of water, which often requires preservatives. Stick-based products, such as sunscreens and balms, also exemplify waterless innovation, delivering targeted application with compact, travel-friendly designs. Brands are increasingly employing these formats to appeal to eco-conscious consumers seeking effective, sustainable alternatives to traditional liquid formulations.

While waterless formulations offer clear environmental benefits, they also present unique challenges for formulators. Achieving adequate performance, stability, and user experience without the emulsifying and textural benefits of water requires careful selection of alternative solvents, stabilizers, and active delivery mechanisms. Furthermore, consumer education is necessary to ensure proper usage, as these products often require different application methods. However, the potential of waterless formulations to change the beauty industry is promising as it could have a big impact on product waste and environmental sustainability.

Hurdles to Overcome

Although green chemistry holds promise to innovate cosmetics in terms of performance and reduced environmental impact, several challenges remain that hinder its widespread adoption. One major hurdle is cost. Sustainable ingredients and production methods often require higher upfront investment, which can be a barrier for brands and those targeting price-sensitive markets. This financial constraint can make it difficult for some companies to compete, especially when conventional alternatives are often cheaper and more readily available. Additionally, the lack of standardized definitions and regulations for terms like "natural" and "green" creates confusion in the marketplace. This regulatory inconsistency complicates product development and can lead to greenwashing, where brands exaggerate their sustainability claims without substantial evidence [38].

Another significant challenge lies in the technical aspects of incorporating green chemistry into formulations. Natural and biodegradable ingredients can be less stable which can pose a challenge for product longevity and consumer satisfaction. Additionally, natural ingredients often do not perform as well as their synthetic counterparts [41]. While consumer demand for sustainable products is growing, misconceptions about the performance of eco-friendly cosmetics persist. For green chemistry applications to have a significant impact on cosmetic product innovation, cosmetic companies will likely have to invest in educating customers and building trust to overcome these barriers to use.

AI and Formulating

Perhaps one of the biggest recent developments in society that may impact the future of cosmetic products is the emergence of artificial intelligence (AI) tools [42]. While AI is still in its early stages, it has the potential to make a significant impact on the cosmetic formulation process. By analyzing large datasets, AI offers insights into ingredient interactions, consumer preferences, and market trends. This technology may enable the creation of more effective, innovative, and personalized formulations, addressing diverse needs while reducing development time and costs. As a tool for enhancing efficiency and creativity, AI could be a powerful addition to the toolkit of every cosmetic formulator. This section explores its application in formulation and the ways it could influence product development in meaningful ways.

The Integration of AI in Cosmetic Formulation

To make AI useful for product development, there have been a number of different tools and strategies developed including machine learning, deep learning, and large language models.

Machine Learning and Its Role in Formulation

Machine learning is a branch of artificial intelligence that uses algorithms to identify patterns in data, allowing computers to make predictions or decisions without explicit programming. It works by feeding vast datasets into a model, which then "learns" from the data to recognize trends, relationships, or outcomes [43].

A cosmetic chemist can use machine learning to streamline and enhance the formulation process in several ways. First, the chemist inputs data, such as ingredient properties, concentrations, and performance metrics, into a machine learning model [44]. The model analyzes these datasets to predict how different ingredients interact, identifying combinations that optimize desired effects like hydration, anti-aging, or oil control. Machine learning can also evaluate large datasets of skin types, environmental conditions, and consumer preferences, allowing chemists to tailor formulations for specific target audiences. For instance, a chemist might use a model to predict the stability of a formula [45] or how effectively it penetrates the skin [46]. These insights can reduce development time by minimizing the need for repeated trial-and-error experiments, ultimately helping chemists create innovative and effective products faster and with greater precision.

Deep Learning for Ingredient Analysis

Deep learning is a subset of machine learning that uses artificial neural networks to analyze complex data and make predictions or decisions. In cosmetics, deep learning can be used to analyze molecular structures, ingredient interactions, and even images of skin conditions. This information could be used to suggest ingredient combinations that may achieve specific effects, such as reducing wrinkles or treating acne [42]. For cosmetic chemists, this technology provides a helpful way to uncover insights that were previously difficult or impossible to obtain, such as predicting the efficacy of new compounds or customizing formulations for individual needs.

Large Language Models in Product Development

Large language models (LLM) are advanced AI systems trained on massive amounts of text data to understand and generate human-like language. They work by analyzing patterns in the text, such as word relationships and sentence structures, to predict and create coherent, contextually relevant responses [47]. For cosmetic formulators, these models can impact product development in various ways. They provide instant access to ingredient databases, scientific literature, and market trends, saving hours of research time. A large language model can also generate ideas for new products or supporting tests, summarize the latest research and even suggest ways to solve stability or other formulation problems.

These tools also simplify communication with marketing teams or consumers, creating tailored product descriptions or personalized advice based on skin concerns. By integrating this technology, chemists can increase their productivity and creativity which could lead to more innovative products.

Automating Ingredient Discovery

Traditionally, ingredient discovery in cosmetics has relied on a combination of scientific research, trial and error, and natural world exploration. Researchers would begin by identifying specific needs or problems, such as reducing wrinkles or enhancing hydration. They then study natural sources like plants, minerals, or marine organisms, often guided by traditional medicine or ethnobotanical knowledge, to find substances with potential benefits. Next, in the laboratory, chemists conduct experiments to isolate, extract, and test new substances. Once a promising compound is identified, it undergoes various stages of optimization to improve performance in formulations. This process is time-consuming, requiring significant resources and often years of experimentation to identify and validate a single new ingredient.

By using AI, ingredient discovery can be achieved much faster. It is done by rapidly analyzing large datasets to identify novel compounds with desirable properties. For instance, Shiru's Protein Discovery.ai platform [48] employs AI to sift through millions of protein sequences, pinpointing those that can be harnessed for personal care products. This accelerates the development of innovative ingredients that are both effective and sustainable. Similarly, AI tools can analyze skin-specific conditions more precisely, offering insights that traditional methods might miss [49]. This not only reduces costs and minimizes reliance on time-intensive trial-and-error approaches but may also enable cosmetic chemists to create innovative, personalized products that align more closely with consumer needs and expectations.

AI-Driven Product Personalization

Personalization has become a popular trend in the marketing of modern cosmetics [50]. Using AI cosmetic brands can potentially meet the unique needs of individuals in more targeted ways than previously available. By combining various data points—such as skin type, lifestyle, environmental factors, and even genetic predispositions—AI can be used to better personalize products in ways that traditional methods may not be able to. This data-driven approach potentially allows for the development of customized products that could boost user satisfaction, meet regulatory standards, and strengthen brand loyalty.

Tailoring Formulations to Individual Needs

Creating formulas that are customized to consumers is fairly straightforward. The process begins by collecting relevant consumer information through digital surveys, skin diagnostic tools, wearable devices, or apps that analyze factors such as skin type, tone, hydration levels, sensitivity, and lifestyle habits [51]. These insights allow platforms to recommend targeted products or routines that address specific concerns.

Additionally, genetic testing and environmental data, such as local climate and pollution levels, can be integrated to gain a deeper understanding of specific needs. This information can be processed using AI algorithms to identify patterns and create profiles that form the basis for tailored formulations.

Once the consumer data are analyzed, an AI tool can help determine how to adjust formulas to address specific requirements. For example, a person with dry, sensitive skin living in a humid climate may benefit from a formula incorporating more glycerin or hyaluronic acid. AI systems may 1 day be able to evaluate the precise concentrations of these ingredients to optimize hydration without disrupting the skin barrier. Additionally, the AI tool can assess the compatibility of hyaluronic acid with specific preservatives like phenoxyethanol to ensure stability and safety. By using a library of ingredient data, including efficacy, safety profiles, and synergies, the system can recommend adjustments to enhance performance, such as slightly increasing the concentration of an ingredient like allantoin if signs of irritation are detected or reducing hyaluronic acid levels to prevent excessive stickiness in a humid environment. This personalized approach ensures each ingredient serves its intended function effectively while tailoring formulations to the individual's unique needs [52].

After generating potential formulas, the next step involves creating and testing prototypes. AI could be used to help accelerate this process by simulating how different combinations of ingredients perform under various conditions, reducing the need for repetitive physical testing. This step could also incorporate virtual modeling to predict the texture, absorption rate, and sensory feel of the product. Once these virtual tests are completed, the most promising formulations can be physically produced in small batches for lab-based performance and stability testing, ensuring the product meets performance and safety standards. Automated robotic laboratories for product manufacturing could be used more commonly than it is now [53].

The final stage of product customization is consumer testing, where prototypes are provided to a select group of users from the original dataset. Feedback from these tests is collected and analyzed to refine the formula further. AI plays a role here by identifying trends and common concerns from user feedback, ensuring the end product matches consumer expectations. This iterative process ensures the creation of a customized cosmetic product that should satisfy the individual needs but also adheres to regulatory standards and performs well.

AI in Sustainability and Ethical Innovation

As mentioned previously, sustainability and a concern for the impact of cosmetics on the environment will be a driver of innovation in the future. The integration of AI into cosmetics may not just improve product development but may also advance sustainability and ethical practices within the industry. By optimizing processes and reducing waste, AI can help brands address consumer demands for environmentally and socially responsible solutions.

Reducing Environmental Impact

One area where AI can play a role is in minimizing waste during the production of cosmetics [54]. By analyzing data about raw materials, cosmetic chemists can optimize formulations to minimize excess ingredient usage while still delivering effective products. This approach not only conserves resources but also helps lower the environmental impact of manufacturing.

Sustainable packaging is another area where AI can make a difference [55]. Advanced algorithms can assess materials for properties like recyclability, durability, and production efficiency. Brands can use this information to create packaging that meets sustainability targets, such as reducing single-use plastics or adopting biodegradable alternatives. These changes align with broader efforts to cut the industry's contribution to environmental pollution.

Supporting Circular Economy Initiatives

AI can also help cosmetic companies employ circular economy practices, where resources are reused and recycled to reduce waste [56]. For instance, lifecycle assessments powered by AI can evaluate a product's environmental impact from creation to disposal, highlighting areas for improvement. These insights can guide companies in selecting recyclable materials, cutting emissions during production, or designing refillable systems.

Additionally, AI can track how consumers use and dispose of products, offering data that supports better design for reuse or recycling [57]. This information enables brands to refine their sustainability efforts and ensure their products align with eco-friendly goals, ultimately reducing their impact on the planet. By integrating AI into sustainability initiatives, the cosmetics industry could find practical ways to tackle global environmental challenges.

Replacing Animal Testing

Animal testing has long been a contentious issue in cosmetics, and AI has the potential to provide an alternative. Predictive models powered by AI can assess the safety and efficacy of ingredients by simulating how they interact with human skin cells and tissues. These simulations rely on data from prior studies, molecular structures, and toxicological profiles to produce accurate results without the need for animal experiments. By analyzing massive amounts of data, predicting these models can theoretically be more efficient than humans. For example, when testing a chemical for toxicity, AI algorithms can study patterns in existing data to predict how a new substance might behave [58]. This process, called in silico testing, can model biological processes, simulate chemical reactions, and even predict long-term effects without the need for animals.

One of the most promising aspects of AI in this field is its ability to personalize predictions. For example, while animal tests provide generalized results, AI models can simulate how a chemical might affect different populations based on genetic, age, or environmental factors. This precision could lead to safer products for everyone.

The Challenges of AI in Toxicology

Although AI has enormous potential, there are many challenges to its full adoption [59]. The biggest hurdle lies in the quality of the data used to train AI models. Using incomplete or inaccurate data will produce unreliable results. Many existing toxicology datasets are inconsistent or poorly labeled, which hampers the ability of AI to make precise predictions.

Another challenge is gaining trust in AI predictions. Regulators and companies may hesitate to rely solely on AI, fearing errors or unforeseen risks. After all, replacing traditional methods with AI represents a paradigm shift, and such transitions take time and effort to implement.

The key to overcoming these challenges lies in curating high-quality datasets. Researchers must carefully clean, organize, and validate data to ensure they are accurate and reliable. This process is known as data curation, and it is essential for AI to make meaningful predictions.

In addition to better data, rigorous testing and validation of AI models are critical. AI predictions need to be transparent and reproducible, so regulators and companies can trust their outcomes. As these models grow more sophisticated, they are expected to become a widely adopted solution for safety assessments in the industry.

Challenges and Limitations

While artificial intelligence has the potential to lead to novel innovation in the cosmetics industry, its adoption is not without challenges. Issues related to privacy, technical accuracy, and cost can pose significant hurdles, particularly as the technology becomes more widely implemented. Addressing these challenges will be essential for adapting AI tools to the cosmetic industry.

Data Privacy and Ethical Concerns

AI's effectiveness relies heavily on data collection, often involving sensitive consumer information such as skin metrics, purchasing habits, and even personal images for facial analysis. While these insights are useful for personalization, they also raise concerns about privacy and data security [60]. Consumers and regulators will likely want assurances that information is handled responsibly and that companies are transparent about how data is collected, stored, and used.

Building consumer trust requires clear communication and compliance with privacy regulations, such as the General Data Protection Regulation (GDPR) or the California Consumer Privacy Act (CCPA) [61]. Brands must implement robust safeguards to protect user data and provide options for consumers to control how their information is used. Ethical considerations, such as avoiding bias in AI algorithms, are equally critical to ensure that solutions are fair and inclusive for all users.

Technical Challenges

Although AI has made significant advancements, limitations still exist in its ability to deliver consistent accuracy in certain areas of cosmetic formulation. For instance, predicting the behavior of complex ingredient combinations can be difficult, particularly for products designed to meet diverse skin needs or environmental conditions [62].

Moreover, AI models depend on the quality of the data they are trained on. If the input data is incomplete or biased, the results may be unreliable or fail to account for variations across demographics. Ensuring that AI systems are trained on diverse and comprehensive datasets is essential to overcome these limitations, but achieving this can be both time-intensive and resource-heavy.

Cost and Accessibility

The cost of implementing AI tools can be prohibitive, especially for small- and medium-sized brands with limited budgets. Developing or licensing sophisticated AI systems, hiring skilled personnel, and integrating these technologies into existing operations often require substantial investment.

Despite these challenges, the continued advancement of AI in cosmetics seems inevitable. By addressing privacy concerns, improving technical capabilities, and finding ways to make AI more accessible, the industry can ensure that these tools are used responsibly and effectively to benefit both brands and consumers.

3D Printing of Cosmetics

A newer area of innovation for cosmetic product creation is 3D printing [63]. This is a technology that builds objects layer by layer from digital designs and it is already having an impact on healthcare and pharmaceutical industries. By enabling the creation of complex structures with precision and efficiency, it has opened new possibilities for innovation and production across various sectors.

In cosmetics specifically, 3D printing has the potential to be a useful tool for creating customized products both for manufacturers and individual consumers. From custom formulations to intricate packaging designs, 3D printing offers opportunities to innovate on traditional approaches in beauty.

What Is 3D Printing?

There are several varieties of 3D printing technologies that can be used in cosmetic production. Each of them offer different advantages that can be used for specific product applications. Some work better for packaging and others can be used for directly printing a finished formula. The primary versions that have been used for making cosmetic products or related items include Micro 3D Printing, Selective Laser Sintering, Fused Deposition Modeling, Sterolithography, and Bioprinting [64].

Micro 3D Printing: Micro 3D printing is a manufacturing process that uses techniques like photopolymerization or laser-induced methods to build intricate, microscopic structures with resolutions as fine as microns. This technology works by layering materials in precise patterns to create detailed three-dimensional designs that would be impossible with traditional methods.

Selective Laser Sintering (SLS): This technology is used for creating complex designs without the need for molds. It uses a high-powered laser to selectively fuse powder materials to create a three-dimensional object. The process begins with a thin layer of powder spread across a build platform. A laser beam then traces the cross-section of the object being printed, heating the powder to just below its melting point. This causes the particles to fuse together through a process called sintering, creating a solid mass that forms part of the final object. Chanel employed SLS to produce their innovative 3D-printed mascara brush, Le Volume Révolution [65].

Fused Deposition Modeling (FDM): FDM works by extruding thermoplastic material through a heated nozzle, depositing it layer by layer to create three-dimensional objects. The process begins with a 3D model that is sliced into thin horizontal layers. The printer then follows these layer instructions, building the object from the bottom up by precisely depositing molten material that cools and solidifies [66]. This common 3D printing method is most suitable for prototyping packaging and applicators due to its cost-effectiveness and speed.

Stereolithography (SLA): Stereolithography (SLA) is a sophisticated 3D printing technology that uses photopolymerization to create three-dimensional objects from liquid resin by selectively curing liquid photopolymer resin with a UV laser. The process begins with a vat of liquid resin, and a build platform that is lowered into it. A UV laser beam traces each cross-section of the object being printed, causing the resin to cure and solidify precisely where the laser makes contact. After each layer is complete, the build platform lowers slightly, allowing fresh resin to flow over the previous layer, and the process repeats [66]. While these polymers are not suitable for skin, they can be used to create packaging prototypes with unique designs.

Bioprinting: This technique is used to create 3D-printed human skin models for product testing and development. It uses biological materials, such as cells, biomolecules, and biocompatible hydrogels, to create structures that mimic natural tissues. The process involves layering these materials in precise patterns, often guided by computer models, to replicate the complexity of biological systems. In cosmetic product development, bioprinting is used to create realistic skin models that closely mimic human skin's structure and behavior. These models are useful for testing product efficacy, evaluating safety, and studying skin interactions without relying on animal testing. Additionally, bioprinting allows for the development of advanced delivery systems, such as bioengineered patches, that enhance the effectiveness of active ingredients. Companies like Chanel [67] and JALA Group [68] have collaborated with biotech firms to produce skin models for studying imperfections and creating more effective skincare products.

Benefits of 3D Printing in Cosmetics

The integration of 3D printing into cosmetics has many advantages, making it a valuable tool for creating innovative cosmetics of the future. This technology has the potential to enhance creativity, efficiency, and sustainability in the beauty industry.

Customization and Personalization

3D printing allows for the creation of products that reflect individual preferences. Foundations can be precisely matched to an individual's skin tone, while face masks can be printed to fit unique facial contours. Similarly, serums and other skincare items can be customized with ingredients and concentrations tailored to address specific concerns, such as hydration, brightening, or anti-aging. Using micro 3D-printing techniques, companies are able to create customized tools, such as applicators or brushes. They can also use it to enhance ingredient delivery through tiny carriers or microneedles. Neutrogena has used this technology in their MaskiD product to create personalized face masks. It allows for precise placement of active ingredients in specific areas of the mask based on individual skin needs [69]. These personalized solutions allow consumers to receive products that feel uniquely theirs.

The ability to manufacture products on demand streamlines production, reducing both time and resource requirements. Prototyping can be done faster and more accurate with 3D printing, allowing brands to bring new products to market more quickly. Waste is also minimized, as materials are used precisely, avoiding the excess typically associated with traditional manufacturing methods.

Innovation in Design and Formulation

3D printing opens up possibilities for designs and textures that were previously unattainable. Intricate shapes, layered formulations, and multifunctional product structures are now achievable in manufacturing and can offer consumers products that are both visually appealing and functional. For instance, makeup palettes with detailed patterns or skincare products with active ingredients embedded in specific layers are examples of how this technology can redefine cosmetic design [70].

Environmental Impact

Sustainability is a growing concern in cosmetics, and 3D printing addresses this by optimizing resource usage. By using only the necessary amount of material for each product, waste can be significantly reduced during production. Additionally, 3D printing supports the development of eco-friendly packaging, such as refillable containers or designs that use biodegradable materials. These advancements may contribute to the broader effort of creating a more sustainable beauty industry.

Challenges and Limitations

While 3D printing offers an interesting opportunity for the cosmetics industry, several challenges and limitations remain. These range from cost and material concerns to regulatory and technical hurdles that can affect how widely the technology is adopted [66].

Cost and Accessibility

The initial investment required for 3D printing can be significant, particularly for advanced printers capable of producing high-quality cosmetic products. Small businesses and independent brands may find it challenging to afford the equipment, materials, and maintenance costs associated with this technology. Additionally, the expertise required to operate 3D printers effectively can increase operational expenses, creating further barriers for companies seeking to compete in the market.

Material Limitations

Ensuring that 3D printing materials are safe and compatible with skin is a critical consideration for the cosmetics industry. Not all materials used in 3D printing meet the safety requirements for products intended for direct application to the skin [63]. Developing biocompatible and non-toxic materials that perform well during printing is essential, but progress in this area remains slow. Furthermore, achieving consistency in texture, efficacy, and sensory appeal can be difficult when transitioning traditional formulations to a 3D-printed format.

Regulatory Concerns

Regulations surrounding cosmetics are designed to protect consumer safety, and 3D-printed products will have to be created following the same standards as traditionally manufactured items. Brands will have to ensure that formulations meet ingredient safety guidelines and that packaging provides clear and accurate labeling. The relatively new application of 3D printing in cosmetics adds complexity to compliance, as regulatory bodies may not yet have specific frameworks for evaluating 3D-printed products.

Technical Expertise

Successfully implementing 3D printing requires technical knowledge and specialized training. From designing digital models to troubleshooting printer issues, the process demands skills that may not currently exist within many cosmetic companies. Despite these challenges, the potential benefits of 3D printing in cosmetics make it an area worth exploring. By addressing cost, material, regulatory, and technical concerns, the industry can unlock the full potential of this transformative technology and ensure its long-term success.

The Future of 3D Printing in Cosmetics

The future of 3D printing in cosmetics is full of potential and may have an impact on the way cosmetics are produced in the future. Advancements in materials, integration with AI, reduced cost and innovations in retail experiences could all lead to an expanding role of 3D printing, making it a key driver of innovation and customization in the beauty industry.

Advancements in Materials and Technology

Emerging materials designed for 3D printing are expected to address some of the current limitations in safety and functionality. Biocompatible materials, specifically formulated for use in cosmetics, can ensure that printed products are safe for skin while delivering the desired performance. In addition, research into sustainable materials could result in options that are biodegradable or recyclable, aligning with consumer demand for eco-friendly products. Advances in printing technology itself will improve precision, speed, and scalability, potentially making 3D printing more efficient and practical for large-scale production.

Integration with AI and Data Analytics

The combination of 3D printing and AI has the potential to refine personalization and product optimization. AI can analyze detailed consumer data, such as skin conditions, lifestyle factors, and environmental exposures, to recommend product formulations that can be directly printed. This integration could result in a seamless process where consumers input their preferences or needs, and a 3D printer produces a product tailored specifically to them. Additionally, data analytics can help brands identify emerging trends and consumer demands, ensuring that 3D printing efforts align with market needs.

Potential for Retail Innovation

In-store 3D printing kiosks have the potential to transform how consumers shop for cosmetics. These kiosks could allow shoppers to create custom products on demand, from personalized lipsticks to skincare formulations designed for their specific needs. The ability to experiment with colors, finishes, and formulations in real time would create a unique and interactive shopping experience. Retailers could also use this model to reduce inventory costs, as products would be printed as needed rather than stocked in bulk.

The continued development of 3D printing technology promises to change the cosmetics industry. By advancing materials, integrating AI, expanding accessibility, and reimagining retail, 3D printing may move beyond novelty to become a core component of how beauty products are created and consumed.

Final Thoughts

The emerging technologies covered in this chapter highlight a changing and promising future for innovation in the cosmetic industry. These advancements, from sophisticated delivery systems to sustainable raw material alternatives, may be the key to addressing both consumer demands and environmental concerns. While historical innovations have typically focused on marketing or packaging, the integration of technologies such as nanotechnology, bioengineering, 3D printing, and AI-driven customization is setting the stage for significant changes in the products offered by industry. These technologies promise not only to enhance product performance, but also help move the industry toward more eco-conscious and ethically produced cosmetics.

Glossary

Active ingredients The extent to which an active ingredient is absorbed and used by the skin.

Antioxidants The way a product's attributes, such as texture or scent, are evaluated by users.

Bioavailability Gel-like materials used for their hydrating and soothing properties in cosmetics.

Bioengineered ingredients Referring to substances that easily mix with or absorb water.

Biotechnology Components in cosmetics that provide specific benefits, such as moisturizing, anti-aging, or protection.

Collagen Compounds that prevent or slow the oxidation of molecules, protecting skin from damage.

Consumer perception Components produced using biotechnology, offering sustainable and high-quality alternatives.

Controlled release Application of biological systems or organisms to create innovative cosmetic ingredients and solutions.

Cosmeceuticals A structural protein in the skin targeted by many cosmetics to improve firmness and elasticity.

Delivery systems Mechanism that delivers active ingredients over time to enhance efficacy and reduce irritation.

Encapsulation Cosmetic products with biologically active ingredients that provide medical or drug-like benefits.

Fermentation Technologies designed to transport active ingredients to their intended site of action in the skin.

Green chemistry A technique to protect active ingredients by enclosing them within a stable structure for targeted delivery.

Hydrogels The use of microorganisms to produce cosmetic ingredients like hyaluronic acid and peptides.

Hydrophilic The design of products and processes that minimize environmental impact and promote sustainability.

Lipophilic Referring to substances that mix well with oils or fats, used for delivering oil-soluble ingredients.

Liposomes Spherical vesicles made of phospholipids used for delivering both hydrophilic and lipophilic substances.

Micronization The process of reducing particle size to enhance the dispersion and absorption of ingredients.

Microsponges Nano-sized polymer shells designed to encapsulate actives, improving penetration and stability.

Moisturizers Porous polymer structures that control the release of actives and improve sensory experience in products.

Nanocapsules The use of nanoscale materials to improve the delivery, stability, and performance of cosmetics.

Nanotechnology Ingredients that hydrate the skin and prevent water loss, maintaining a healthy skin barrier.

Niosomes The process by which active ingredients move through the skin's layers to reach their target site.

Oxidative stress Engineering organisms at the genetic level to create or modify cosmetic ingredients.

Peptides Non-ionic surfactant-based vesicles offering cost-effective and stable delivery of active ingredients.

Photostability The outermost layer of the skin, often targeted in cosmetic formulations to improve ingredient absorption.

Polymeric systems Damage caused to the skin by free radicals, often targeted by antioxidant-based formulations.

Probiotics Short chains of amino acids used in cosmetics to promote skin repair and anti-aging effects.

Sensory properties The ability of a cosmetic ingredient or formulation to remain stable when exposed to light.

Skin penetration Delivery systems using polymers to stabilize, protect, and control the release of active ingredients.

Stratum corneum Practices aimed at minimizing environmental impact, such as using renewable or lab-produced materials.

Sustainability Live microorganisms used in cosmetics to promote skin health and balance the microbiome.

Synthetic biology Characteristics like texture and absorption that impact consumer experience with a cosmetic product.

References

1. Euromonitor. https://www.euromonitor.com/press/press-releases/august-2024/more-than-25-of-all-new-brands-launched-are-in-the-beauty-and-personal-care-industry-euromonitor-international
2. Rosenberg, N. (2010). Chapter 9: Innovation and the chain-linked model. In *Studies on science and the innovation process: Selected works by Nathan Rosenberg* (pp. 173–204). World Scientific Publishing Co. Pte. Ltd.
3. Bom, S., Jorge, J., Ribeiro, H. M., & Marto, J. (2019). A step forward on sustainability in the cosmetics industry: A review. *Journal of Cleaner Production, 225*, 270–290. https://doi.org/10.1016/j.jclepro.2019.03.255
4. Rotolo, D., Hicks, D., & Martin, B. R. (2015). What is an emerging technology? *Research Policy, 44*(10), 1827–1843. https://doi.org/10.1016/j.respol.2015.06.006
5. Bangham, A. D., & Horne, R. W. (1964). Negative staining of phospholipids and their structural modification by surface-active agents as observed in the electron microscope. *Journal of Molecular Biology, 8*, 660–668.
6. Tadros, T. F. (1992). Future developments in cosmetic formulations. *International Journal of Cosmetic Science, 14*(3), 93–111. https://doi.org/10.1111/j.1467-2494.1992.tb00045.x
7. Dragićević, N., & Maibach, H. I. (2024). Lipid-based vesicles (liposomes) and their combination with physical methods for dermal and transdermal drug delivery. In *Percutaneous absorption: Drugs, cosmetics, mechanisms, methods* (6th ed., pp. 519–540). Springer.
8. Elkalla, E., Khizar, S., Tarhini, M., Lebaz, N., Zine, N., Jaffrezic-Renault, N., et al. (2023). Core-shell micro/nanocapsules: From encapsulation to applications. *Journal of Microencapsulation, 40*(3), 125–156. https://doi.org/10.1080/02652048.2023.2178538
9. Mahant, S., Kumar, S., Nanda, S., & Rao, R. (2020). Microsponges for dermatological applications: Perspectives and challenges. *Asian Journal of Pharmaceutical Sciences, 15*(3), 273–291.
10. Alagusundaram, M., Chetty, C. M. S., Umashankari, K., Badarinath, A. V., Lavanya, C., & Ramkanth, S. (2009). Microspheres as a novel drug delivery system—A review. *International Journal of ChemTech Research, 1*(3), 526–534.
11. Paudel, K. S., Milewski, M., Swadley, C. L., Brogden, N. K., Ghosh, P., & Stinchcomb, A. L. (2010). Challenges and opportunities in dermal/transdermal delivery. *Therapeutic Delivery, 1*(1), 109–131.
12. Akl, M. A., Eldeen, M. A., & Kassem, A. M. (2024). Beyond skin deep: Phospholipid-based nanovesicles as game-changers in transdermal drug delivery. *AAPS PharmSciTech, 25*, 184. https://doi.org/10.1208/s12249-024-02896-6
13. Peralta, M. F., Guzmán, M. L., Pérez, A. P., et al. (2018). Liposomes can both enhance or reduce drugs penetration through the skin. *Scientific Reports, 8*(1), 13253. https://doi.org/10.1038/s41598-018-31693-y
14. Lombardo, D., & Kiselev, M. A. (2022). Methods of liposomes preparation: Formation and control factors of versatile nanocarriers for biomedical and nanomedicine application. *Pharmaceutics, 14*(3), 543. https://doi.org/10.3390/pharmaceutics14030543
15. MMDPC. (n.d.). *Liposome in cosmetics market: Key challenges that drive the industry*. Retrieved December 30, 2024, from https://www.linkedin.com/pulse/liposome-cosmetics-market-key-challenges-drive-industry-mmdpc/

16. Schlich, M., Musazzi, U. M., Campani, V., Biondi, M., Franzé, S., Lai, F., De Rosa, G., Sinico, C., & Cilurzo, F. (2022). Design and development of topical liposomal formulations in a regulatory perspective. *Drug Delivery and Translational Research, 12*(8), 1811–1828. https://doi.org/10.1007/s13346-021-01089-z
17. Carugo, D., Bottaro, E., Owen, J., et al. (2016). Liposome production by microfluidics: Potential and limiting factors. *Scientific Reports, 6*, 25876. https://doi.org/10.1038/srep25876
18. U.S. Congress. (2015). *Microbead-Free Waters Act of 2015, H.R. 1321, 114th Congress*. Retrieved December 30, 2024, from https://www.congress.gov/bill/114th-congress/house-bill/1321/text
19. Goyal, N., & Jerold, F. (2023). Biocosmetics: Technological advances and future outlook. *Environmental Science and Pollution Research, 30*, 25148–25169. https://doi.org/10.1007/s11356-021-17567-3
20. Pérez-Rivero, C., & López-Gómez, J. P. (2023). Unlocking the potential of fermentation in cosmetics: A review. *Fermentation, 9*(5), 463. https://doi.org/10.3390/fermentation9050463
21. Kurnaz, I. A. (2015). *Techniques in genetic engineering* (pp. 1–2). CRC Press. ISBN: 978-1-4822-6090-8.
22. Gomes, C., Silva, A. C., Marques, A. C., Sousa Lobo, J., & Amaral, M. H. (2020). Biotechnology applied to cosmetics and aesthetic medicines. *Cosmetics, 7*(2), 33. https://doi.org/10.3390/cosmetics7020033
23. Shalu, S., Karthikanath, P. K. R., Vaidyanathan, V. K., Blank, L. M., Germer, A., & Balakumaran, P. A. (2024). Microbial squalene: A sustainable alternative for the cosmetics and pharmaceutical industry – A review. *Engineering in Life Sciences, 24*. https://doi.org/10.1002/elsc.202400003
24. de Oliveira, J. D., Carvalho, L. S., Gomes, A. M. V., et al. (2016). Genetic basis for hyper production of hyaluronic acid in natural and engineered microorganisms. *Microbial Cell Factories, 15*, 119. https://doi.org/10.1186/s12934-016-0517-4
25. Chen, C.-X., Zhang, Y.-Y., Yang, J., Yan, M.-H., Jia, Y., & Jiang, S. (2024). An overview of progress in the application of recombinant collagen in cosmetics. *Journal of Dermatologic Science and Cosmetic Technology, 1*(4), Article 100059. https://doi.org/10.1016/j.jdsct.2024.100059
26. Aguilar-Toalá, J. E., Hernández-Mendoza, A., González-Córdova, A. F., Vallejo-Cordoba, B., & Liceaga, A. M. (2019). Potential role of natural bioactive peptides for development of cosmeceutical skin products. *Peptides, 122*, Article 170170. https://doi.org/10.1016/j.peptides.2019.170170
27. Nath, Y. A. (2021). Microbial biotechnology for bio-prospecting of microbial bioactive compounds and secondary metabolites. *Journal of Applied Biology and Biotechnology, 9*, 1–6. https://doi.org/10.7324/JABB.2021.92ed
28. Han, J. H., & Kim, H. S. (2024). Skin deep: The potential of microbiome cosmetics. *Journal of Microbiology, 62*, 181–199. https://doi.org/10.1007/s12275-024-00128-x
29. Habeebuddin, M., Karnati, R. K., Shiroorkar, P. N., Nagaraja, S., Asdaq, S. M. B., Khalid Anwer, M., & Fattepur, S. (2022). Topical probiotics: More than a skin deep. *Pharmaceutics, 14*(3), 557. https://doi.org/10.3390/pharmaceutics14030557
30. Bagatin, E., Martin, R. P., Varela, P., Gomes, C. P., Marins, M. M., Filippeli-Silva, R., Yarak, S., Soares, J. L. M., & Pesquero, J. B. (2022). Could in vivo histological and gene expression analysis of aged skin be useful for pointing new paths for cosmetics development? *Journal of Biomedical Research, 3*(1), 18–23.
31. Tinoco, A., et al. (2022). Biotechnology of functional proteins and peptides for hair cosmetic formulations. *Trends in Biotechnology, 40*(5), 591–605. https://doi.org/10.1016/j.tibtech.2022.02.007
32. Kalicińska, J., Wiśniowska, B., Polak, S., & Spiewak, R. (2023). Artificial intelligence that predicts sensitizing potential of cosmetic ingredients with accuracy comparable to animal and in vitro tests—How does the Infotechnomics compare to other "omics" in the cosmetics safety assessment? *International Journal of Molecular Sciences, 24*(7), 6801. https://doi.org/10.3390/ijms24076801

33. Mifsud, S., Refalo, P., & Rochman, A. (2023). Biodegradable polymers for cosmetic packaging: A technical and life cycle perspective. *Materials Research Proceedings, 28*, 2015–2024. https://doi.org/10.21741/9781644902479-217
34. Cervellon, M.-C., & Carey, L. (2011). Consumers' perceptions of 'green': Why and how consumers use eco-fashion and green beauty products. *Critical Studies in Fashion & Beauty, 2*(1–2), 117–138. https://doi.org/10.1386/csfb.2.1-2.117_1
35. Rinaldi, A. (2008). Healing beauty? *EMBO Reports, 9*(11), 1073–1077. https://doi.org/10.1038/embor.2008.200
36. Lintner, K. (2007). Peptides, amino acids and proteins in skin care. *Cosmetics & Toiletries, 122*(10), 26.
37. Anastas, P. T., & Warner, J. C. (1998). *Green chemistry: Theory and practice*. Oxford University Press. ISBN: 9780198502340.
38. Franca, C. C. V., & Ueno, H. M. (2020). Green cosmetics: Perspectives and challenges in the context of green chemistry. *Desenvolvimento e Meio Ambiente, 53*, 133–150. https://doi.org/10.5380/dma.v53i0.62322
39. Zacarias, C. H., Paludetti, M. F., de Oliveira, A. Á. S., & Federico, L. B. (2024). An integrated approach to address the biodegradability of cosmetic formulations as part of a corporate sustainability strategy. *International Journal of Cosmetic Science., 47*. https://doi.org/10.1111/ics.13033
40. Yu, C., & Anigbogu, C. (2020). How waterless beauty is changing consumer behavior and addressing sustainability. *Cosmetics & Toiletries, 135*(9), 51–52.
41. Liu, J. K. (2022). Natural products in cosmetics. *Natural Products and Bioprospecting, 12*(1), 40.
42. Xin, H., Virk, A. S., Virk, S. S., Akin-Ige, F., & Amin, S. (2024). Applications of artificial intelligence and machine learning on critical materials used in cosmetics and personal care formulation design. *Current Opinion in Colloid & Interface Science, 73*, Article 101847. https://doi.org/10.1016/j.cocis.2024.101847
43. Vatiwutipong, P., Vachmanus, S., Noraset, T., & Tuarob, S. (2023). Artificial intelligence in cosmetic dermatology: A systematic literature review. *IEEE Access, 11*, 71407–71425. https://doi.org/10.1109/ACCESS.2023.3295001
44. Chitre, A., Querimit, R. C. M., Rihm, S. D., et al. (2024). Accelerating formulation design via machine learning: Generating a high-throughput shampoo formulations dataset. *Scientific Data, 11*, 728. https://doi.org/10.1038/s41597-024-03573-w
45. COPTIS. (n.d.). *SCC77 exhibitor abstract: Innovations in cosmetic product development*. Retrieved December 30, 2024, from https://www.scconline.org/Portals/100/SCC%20Events%20Page/SCC77/Exhibition/Exhibit%20Abstracts/207%20COPTIS%20SCC77%20Exhibitor%20Abstract.pdf?ver=SEXd40R4oP9a355UqyQoOA%3D%3D
46. Değím, T., Hadgraft, J., Ilbasmiş, S., & Ozkan, Y. (2003). Prediction of skin penetration using artificial neural network (ANN) modeling. *Journal of Pharmaceutical Sciences, 92*(3), 656–664. https://doi.org/10.1002/jps.10312
47. Han, S., Wang, M., Zhang, J., Li, D., & Duan, J. (2024). A review of large language models: Fundamental architectures, key technological evolutions, interdisciplinary technologies integration, optimization and compression techniques, applications, and challenges. *Electronics, 13*(24), 5040. https://doi.org/10.3390/electronics13245040
48. Shiru. (2024). *Shiru launches ProteinDiscovery.ai marketplace touted as the "Amazon of proteins"*. Vegconomist. Retrieved December 30, 2024, from https://vegconomist.com/protein/shiru-proteindiscovery-ai-marketplace-amazon-proteins/
49. Chen, M., Zhou, P., Wu, D., Hu, L., Hassan, M. M., & Alamri, A. (2020). AI-skin: Skin disease recognition based on self-learning and wide data collection through a closed-loop framework. *Information Fusion, 54*, 1–9. https://doi.org/10.1016/j.inffus.2019.06.005
50. Krishnan, J., Ranjith, R., & Vijayakumar, K. (2023). Cosmetic products customisation and customer segmentation. In *2023 International conference on advances in computing, com-*

munication and applied informatics (ACCAI) (pp. 1–11). IEEE. https://doi.org/10.1109/ACCAI58221.2023.10200286
51. Cosmetics & Toiletries. (n.d.). *5 AI skin care technologies pushing the industry*. Retrieved December 30, 2024, from https://www.cosmeticsandtoiletries.com/formulas-products/tech-equipment-production/news/22868294/5-ai-skin-care-technologies-pushing-the-industry
52. Bhosale, U., Gaddam, S., Koutam, H., Sapre, P., Nallan, C., & Syed, S. M. (2023). Empowering consumers: The evolving paradigm of customization in personal care products. *Current cosmetic. Science, 2*(1), Article E201023222482. https://doi.org/10.2174/0126667797265268231011103859
53. Sunkle, S., Jain, D., Saxena, K., Patil, A., Singh, T., Rai, B., & Kulkarni, V. (2021). Integrated "generate, make, and test" for formulated products using knowledge graphs. *Data Intelligence, 3*(3), 340–375. https://doi.org/10.1162/dint_a_00096
54. Hasan, M. R., Shawon, R. E. R., Rahman, A., Al Mukaddim, A., Khan, M. A., Hider, M. A., & Zeeshan, M. A. F. (2024). Optimizing sustainable supply chains: Integrating environmental concerns and carbon footprint reduction through AI-enhanced decision-making in the USA. *Journal of Economics, Finance and Accounting Studies, 6*(4), 57–71.
55. Dube, M., & Dube, S. (2023). Towards sustainable color cosmetics packaging. *Cosmetics, 10*(139). https://doi.org/10.3390/cosmetics10060139
56. Morea, D., Fortunati, S., & Martiniello, L. (2021). Circular economy and corporate social responsibility: Towards an integrated strategic approach in the multinational cosmetics industry. *Journal of Cleaner Production, 315*, Article 128232. https://doi.org/10.1016/j.jclepro.2021.128232
57. Seyyedi, S. R., Kowsari, E., Ramakrishna, S., Gheibi, M., & Chinnappan, A. (2023). Marine plastics, circular economy, and artificial intelligence: A comprehensive review of challenges, solutions, and policies. *Journal of Environmental Management, 345*, Article 118591. https://doi.org/10.1016/j.jenvman.2023.118591
58. Santín, E. P., Solana, R. R., García, M. G., Suárez, M. M. G., Díaz, G. D. B., Cabal, M. D. C., Rojas, J. M. M., & Sánchez, J. I. L. (2021). Toxicity prediction based on artificial intelligence: A multidisciplinary overview. *Wiley Interdisciplinary Reviews: Computational Molecular Science, 11*(2), Article e1516. https://doi.org/10.1002/wcms.1516
59. Alves, V. M., Auerbach, S. S., Kleinstreuer, N., Rooney, J. P., Muratov, E. N., Rusyn, I., Tropsha, A., & Schmitt, C. (2021). Curated data in—trustworthy in silico models out: The impact of data quality on the reliability of artificial intelligence models as alternatives to animal testing. *Alternatives to Laboratory Animals, 49*(3), 73–82. https://doi.org/10.1177/02611929211029635
60. Stahl, B. C., & Wright, D. (2018). Ethics and privacy in AI and big data: Implementing responsible research and innovation. *IEEE Security & Privacy, 16*(3), 26–33. https://doi.org/10.1109/MSP.2018.2701164
61. Goldsteen, A., Ezov, G., Shmelkin, R., et al. (2022). Data minimization for GDPR compliance in machine learning models. *AI and Ethics, 2*, 477–491. https://doi.org/10.1007/s43681-021-00095-8
62. Winarto, L., & Wisesa, A. (2024). Analyzing the impact of artificial intelligence and sustainability on gen Z consumer purchase intentions: A case study of L'Oréal cosmetics Indonesia. *European Journal of Business and Management Research, 9*(5), 16–30. https://doi.org/10.24018/ejbmr.2024.9.5.461
63. Jiao, Y., Stevic, M., Buanz, A., Uddin, M. J., & Tamburic, S. (2022). Current and prospective applications of 3D printing in cosmetics: A literature review. *Cosmetics, 9*(6), 115. https://doi.org/10.3390/cosmetics9060115
64. Shahrubudin, N., Lee, T. C., & Ramlan, R. (2019). An overview on 3D printing technology: Technological, materials, and applications. *Procedia Manufacturing, 35*, 1286–1296. https://doi.org/10.1016/j.promfg.2019.06.089

65. 3D Printing Industry. (2018). *Chanel announces plan to mass-produce 3D printed mascara brush*. Retrieved December 30, 2024, from https://3dprintingindustry.com/news/chanel-announces-plan-mass-produce-3d-printed-mascara-brush-130715/
66. Liu, Z., Wang, Y., Wu, B., et al. (2019). A critical review of fused deposition modeling 3D printing technology in manufacturing polylactic acid parts. *International Journal of Advanced Manufacturing Technology, 102*, 2877–2889. https://doi.org/10.1007/s00170-019-03332-x
67. GLCNC Machining. (n.d.). *Chanel partners with LabSkin Creations to launch 3D bioprinted skin*. Retrieved December 30, 2024, from https://glcncmachining.com/chanel-partners-with-labskin-creations-to-launch-3d-bioprinted-skin/
68. PR Newswire. (2017). *JALA Group announces successful printing of Asian skin using 3D bioprinting technology*. Retrieved December 30, 2024, from https://www.prnewswire.com/news-releases/jala-group-announces-successful-printing-of-asian-skin-using-3d-bioprinting-technology-300407828.html
69. 3D Printing Industry. (2019). *Neutrogena to launch personalised 3D-printed face masks*. Retrieved December 30, 2024, from https://3dprintingindustry.com/news/neutrogena-to-launch-personalised-3d-printed-face-masks-146307/
70. Beskan, U., & Algin Yapar, E. (2019). Usage of 3D printer technology in medical and pharmaceutical fields: A review. *Universal Journal of Pharmaceutical Research, 4*(3), 37–40. https://doi.org/10.22270/ujpr.v4i3.277

Cosmetic Packaging

Allison Turquoise Kent-Gunn Garibay

Abstract This chapter presents a comprehensive examination of cosmetic packaging through the lenses of functionality, consumer behavior, sustainability, and innovation. Beginning with a historical overview of materials and design, this chapter traces the evolution of packaging technologies and their cultural significance. Core functional considerations, including product protection, usability, regulatory compliance, and quality assurance, are explored alongside aesthetic strategies that shape consumer perception and brand identity. A central focus is placed on sustainability, with in-depth analysis of life cycle assessment (LCA), material impacts, and circular strategies such as refill systems, lightweighting, and recycled content. Drawing from current case studies and emerging industry data, this chapter bridges academic rigor with practical applications, equipping readers with a critical framework to assess packaging choices in a rapidly evolving, sustainability-driven beauty industry.

Keywords Cosmetic packaging · Sustainability · Life cycle assessment (LCA) · Consumer behavior · Packaging materials · Quality assurance · Refillable systems

Introduction

Definition of Cosmetic Packaging

Cosmetic packaging serves a multifaceted role, providing not just containment but also presentation, protection, identification, information, convenience, and compliance throughout the product's lifecycle. It must safeguard the product from environmental factors and hazards—biological, physical, and chemical—while being

A. T. K-G. Garibay (✉)
3E Beauty Consulting, Los Angeles, CA, USA

K. Alexander, P. Romanowski (eds.), *Cosmetic Science*, AAPS Introductions in the Pharmaceutical Sciences 15, https://doi.org/10.1007/978-3-032-03596-7_13

cost-effective and aesthetically on-brand. It must be designed and tested to ensure the product's stability from initial production through user consumption, preserving its quality and effectiveness throughout its shelf life [6].

Importance of Cosmetic Packaging in the Beauty Industry

Cosmetic packaging often appears to be a secondary consideration, primarily viewed as a protective vessel for the product. However, its role extends far beyond mere functionality. Packaging serves as the primary form of communication between the product and prospective users. It is the first touchpoint a consumer has with a product, being the first aspect of a product they see and touch. Packaging design is crucial in shaping first impressions and conveying brand identity, often functioning as a powerful, although frequently underestimated, marketing tool.

From the choice of font and color to the packaging's silhouette and finish, every element is scrutinized by consumers, often on a subconscious level. These visual and tactile details contribute to the overall consumer experience and impact the product's perceived value. For instance, the weight and texture of a glass jar compared to a plastic container can influence perceptions of quality and prestige, even if these differences are not overtly acknowledged by the consumer and only occur subliminally.

Assessing the Physical Aspects of Packaging

To comprehensively evaluate and articulate the qualities of a package, it is useful to consider the VTMA framework for assessing cosmetic packaging:

1. Visual: This aspect encompasses the visual characteristics of the packaging, such as color, design, and shape. Visual perception plays a significant role in how a package is initially perceived and judged by consumers, and is typically the first touchpoint a consumer will have with a product.
2. Tactile: Tactile characteristics include the texture, weight, temperature, and rigidity of the packaging. These elements affect how the consumer interacts with the product, impacting their overall experience and satisfaction. Tactile perception is typically the second touchpoint a consumer has with a product.
3. Material: The material of the packaging can affect the visual, tactile, and sometimes auditory experience of a package. For example, a glass jar may convey a sense of luxury and quality due to its heavy weight in hand, while feeling cool to the touch. Additionally, it can produce a distinctive sound when placed on a surface. The material's inherent qualities contribute to consumer perceptions of the product's value and price point. Most recently, consumers have begun to rely on

packaging material to provide cues as to the environmental impact of a given package. Different materials, such as glass versus plastic, evoke varying cognitive responses and associations regarding prestige and sustainability.

4. Audio: Often overlooked, audio cues from packaging can subtly influence user experience. For instance, the "click" sound of a cap sealing a bottle provides reassurance that the packaging is securely closed, preventing spills. Audio cues can also be critical for accessibility, assisting consumers with visual impairments or dexterity challenges by providing additional sensory feedback. While most consumers are not conscious of the use of audio cues in cosmetic packaging, the cues have a subconscious effect.

Understanding these four dimensions—Visual, Tactile, Material, and Audio (VTMA)—enables a systematic evaluation of packaging, highlighting how each aspect contributes to consumer perception and overall product experience.

Historical Overview

Evolution of Cosmetic Packaging Materials

The evolution of cosmetic packaging materials reflects both technological advancements and changing consumer preferences throughout history.

Ancient Packaging The use of cosmetic packaging dates back to 2500 BC with kohl, one of the earliest known cosmetics. Initially, kohl was stored in shells or simple containers made from stone, ceramic, wood, or reed. These early containers were often paired with applicators, marking the beginning of functional and aesthetic considerations in packaging [46].

Ceramic Pots The introduction of ceramic pots allowed for more elaborate designs and diverse colors, signifying a shift from purely functional vessels to integral components of the cosmetic experience. This era marked a significant advancement in cosmetic packaging, as containers began to reflect artistic and cultural values, the impact of which continues to be seen in cosmetic packaging today [46].

Blown Glass During the Roman Empire, blown glass became a popular material for perfumes and ointments. This innovation not only enhanced the aesthetic appeal of packaging but also contributed to the preservation of contents [46].

Industrial Revolution The mass production of packaging materials did not gain momentum until the Industrial Revolution, which introduced more efficient glass manufacturing techniques and expanded the availability of packaged goods [46].

Milestones in Cosmetic Packaging Design

Plastic in Cosmetic Packaging

The widespread introduction of plastic in the 1950s revolutionized cosmetic packaging, offering unparalleled versatility, durability, and cost-efficiency[46].

Below are some of the key milestones and the role of various plastics in the beauty industry:

Transformative Advancement The widespread adoption of plastic was one of the most transformative moments in cosmetic packaging history. Its lightweight nature, low cost, and ability to be molded into virtually any shape allowed for more innovative packaging designs and mass production at scale.

Versatility Unlike glass or metal, plastic can be manufactured in different forms—from rigid to flexible—making it suitable for a wide variety of beauty products, including creams, lotions, sprays, and powders. Its adaptability paved the way for everything from squeezable tubes to airless pumps.

Durability and Portability Plastic packaging became popular not just for its cost-effectiveness but also for its durability. Unlike glass, plastic does not shatter, making it ideal for products that are frequently carried or exposed to the risk of impact.

Sustainability Concerns Despite its benefits, plastic has faced criticism due to its environmental impact. However, the perception that plastic is inherently unsustainable is not entirely accurate. In some cases, plastic can have a lower carbon footprint compared to glass and metal, especially when it comes to transportation emissions due to its lightweight nature. Advances in recycling technologies and the development of biodegradable plastics are also helping to mitigate its environmental impact. We will explore the environmental impact of plastic in greater detail later in this chapter.

However, it is important to note that plastic is a category, not a specific material, and many types of plastic are utilized in cosmetic packaging, all with varying physical qualities, limitations, and potential for recyclability.

Common Plastics in Cosmetic Packaging

PET (Polyethylene Terephthalate) Commonly used for bottles and containers, PET is lightweight, transparent, and easily recyclable. It is a go-to material for liquid products and is commonly used for skincare, haircare, and bodycare products. It is the most commonly recycled plastic in the USA, making it a popular choice for brands prioritizing recyclability [31, 43].

PP (Polypropylene) Known for its versatility, PP is used in a variety of cosmetic containers, including jars, caps, and airless pumps. PP innately has a gray milkiness to it and does not offer the same level of transparency as PET. PP is considered recyclable, although it has a lower recycling rate in the USA than PET [35].

HDPE (High-Density Polyethylene) HDPE is highly durable and resistant to chemicals, making it suitable for an array of liquid products. It offers squeezability, which makes it ideal for bottles meant to be used without pumps, such as shampoo and conditioner bottles. HDPE is the second most commonly recycled plastic in the USA [31].

LDPE (Low-Density Polyethylene) Often used in flexible packaging, like tubes and sachets, LDPE is valued for its pliability and durability [35].

MDPE (Medium-Density Polyethylene) A balance between HDPE and LDPE, MDPE offers greater flexibility while maintaining strength, often used in squeeze tubes [31].

Acrylic (PMMA) Acrylic plastic is used in luxury cosmetic packaging due to its glass-like clarity and smooth finish. It provides an upscale look while being less fragile than glass, making it a popular choice for high-end product lines. However, unlike the aforementioned plastic, acrylic is not commonly recycled in the USA [43].

ABS (Acrylonitrile Butadiene Styrene) ABS is a hard, rigid plastic used in packaging components like compact cases, lipstick tubes, and caps. Its toughness and ability to be molded into precise shapes make it ideal for intricate packaging designs. Like acrylic, ABS is not commonly recycled in the USA [31, 35].

Plastics have undoubtedly played a major role in shaping modern cosmetic packaging, offering not only cost-effective solutions but also enabling design innovation across product categories. Today, the focus on creating sustainable plastics and improving recycling systems continues to evolve, making plastic packaging both a challenge and an opportunity for the beauty industry.

Plastic The widespread adoption of plastic represents one of the most transformative advancements in cosmetic packaging. Plastics, known for their durability, versatility, and cost-effectiveness, have quickly become a staple in the beauty industry. However, plastic packaging has faced criticism for its environmental impact. It is crucial to note that plastic is not inherently unsustainable; in some cases, it can have a lower environmental footprint, especially in regard to carbon emissions, compared to alternatives like glass and metal.

Glass and Metal in Cosmetic Packaging

Glass

Glass has been a cornerstone of cosmetic packaging for centuries and represents one of the earliest materials used to contain and preserve beauty products [11]. Its introduction marked a milestone in the cosmetic industry due to several key factors:

Milestones in Preservation Glass is nonporous and impermeable, making it an ideal material for preserving product formulas, particularly those prone to oxidation or sensitive to light and air. This made glass an early go-to for perfumes, serums, and skincare products, where product integrity is crucial.

Aesthetic Appeal Glass conveys luxury and quality due to its weight, clarity, and smooth surface. It became the material of choice for premium brands, offering a high-end look that resonates with consumers seeking elegance in their beauty products.

Sustainability Glass is fully recyclable without degrading in quality, which has resulted in it becoming an increasingly popular choice for brands concerned with the environmental impact of their packaging. However, glass is generally more energy-intensive to produce than plastic, resulting in higher carbon emissions during production. Additionally, due to its weight, it also tends to generate more carbon emissions during transportation than plastic, which is lighter weight.

Metal (Aluminum, Steel, Tin)

Metals like aluminum, steel, and tin also played pivotal roles in the evolution of cosmetic packaging, contributing to both product protection and design innovation.

Aluminum Aluminum became a milestone material due to its lightweight, corrosion resistance, and ability to protect products from light and oxygen [37]. It is also highly recyclable, aligning with the industry's current sustainability goals. However, like glass, aluminum is generally more energy-intensive to produce and ship, resulting in higher carbon emissions.

Steel and Tin These materials were widely used in vintage cosmetic packaging for products like lip balms, solid perfumes, and ointments. Their durability and ability to be molded into intricate shapes allowed for innovative, decorative packaging designs, adding a new dimension to consumer appeal [46].

Both glass and metal have significantly shaped cosmetic packaging history, providing product protection long before the introduction of plastics. Due to concerns over plastic's environmental impact, the industry is seeing a return to these materials, which consumers perceive as being more elevated, luxurious materials, as well as more sustainable due to their higher recyclability rates in the USA.

New Material Developments Recent advancements in packaging materials include bioresins, plant-based polymers, and biodegradable or compostable materials. These innovations aim to address sustainability concerns and reduce the environmental impact of cosmetic packaging.

We will explore the environmental impact of each of these materials in more detail later in this chapter.

Impact of Historical Trends on Modern Packaging Practices

Modern cosmetic packaging is increasingly influenced by historical practices and materials. There is a growing trend toward reusing packaging and incorporating naturally derived packaging materials, such as plant-based polymers and paper. This shift reflects a broader movement toward sustainability and environmental responsibility.

However, adapting ancient packaging practices to contemporary consumer needs presents challenges. Modern consumers interact with beauty products differently from those in preindustrial societies, necessitating adjustments to historical designs to meet current product regulations, retail expectations, and consumer preferences.

The history of cosmetic packaging also underscores the impact of industrialization on the beauty industry. The nineteenth and early twentieth centuries saw significant growth in product volume and accessibility, driven by advances in manufacturing, transportation, and rising consumer incomes. The expansion of global markets and the emergence of iconic brands, such as Helena Rubinstein and L'Oréal, further shaped the industry's trajectory [22].

Despite this historical context, the beauty industry is diverse, with varying standards and practices across different societies and eras. As we revisit ancient packaging concepts, it is essential to balance historical inspiration with modern demands and sustainability goals.

Future Trends and Innovations

Emerging Technologies in Packaging Design and Manufacturing

Plastic Recycling: Mechanical Versus Chemical

The most common recycling process used for cosmetic packaging is mechanical recycling, which involves breaking down plastics into small granules, then remelting and reforming them into new products. However, mechanical recycling degrades the material's integrity over time, eventually rendering it unsuitable for reuse [4]. Generally, plastic can be recycled mechanically three to seven times before degrading, and no longer maintaining its material properties. The number of recycling cycles plastic can undergo is impacted by various factors, including the level of polymer degradation and the type of plastic. For example, PET and HDPE plastic tend to retain their integrity for more recycling cycles, compared to lower-density plastics, like LDPE [45].

In contrast, chemical recycling is a relatively new process that breaks down plastics at the molecular level, allowing them to be reassembled into various end products. Unlike mechanical recycling, chemical recycling does not compromise the integrity of the plastic, enabling continuous reuse. However, this process is expensive, energy-intensive, and raises environmental concerns, including high energy consumption and potentially hazardous waste [4].

Bioresins

Bioresins, or plant-based plastics, are derived from renewable plant sources such as sugarcane bagasse, corn husk, and soybeans. Just as plastic is a material category, and not a specific material, bioresin is likewise a material category which includes different kinds of plant-based plastics such as PE, PEF, PHA, and PLA. While bioresins provide an alternative to petroleum-derived plastics, they still come with environmental concerns. For instance, the cultivation of bioresin crops can impact land use, water consumption, and food supply, particularly in developing countries where many of these bioresins are harvested [14].

It is important to clarify that while bioresins are marketed as "more sustainable" due to their renewable origins, they largely retain the same attributes as their petroleum-derived counterparts. Additionally, like with most traditional resins, bioplastics are not generally biodegradable or compostable. The key distinction in their environmental impact lies in the early stages of their life cycles, particularly in raw material extraction and processing. Since bioresins retain the same physical properties as traditional plastics, they do not inherently improve recyclability, which remains a significant focus for brands and consumers [36]. However, there are advancements being made in the bioresin category, particularly in regard to compostability, which would provide end-of-life sustainability beyond mere recyclability [4].

Plant-Based Materials (Nonplastic)

Forecasted Shifts in Consumer Preferences and Market Demands

The rise of social media has dramatically influenced consumer perceptions of cosmetic packaging. Previously, packaging was seen primarily on store shelves, in print advertisements, or in home bathrooms. Today, packaging exists prominently in social media feeds and on brand websites, creating more visual touchpoints with consumers. This visibility offers brands a greater opportunity to make a lasting first impression, as packaging now serves as a brand's silent ambassador. Packaging is often the first element a consumer interacts with, even before testing the product itself, underscoring its critical importance.

Sustainability and Environmental Considerations in Future Packaging

Sustainability in Cosmetic Packaging

Sustainability is a dominant theme in the packaging industry, with many brands making sustainability a key component of their branding—albeit with varying degrees of success.

Before delving into the complexities of sustainable packaging, it is crucial to understand why it matters. The global personal care and beauty industry produces over 120 billion units of packaging annually, much of which is single-use and ends up in landfills [34].

Sustainable packaging is not a one-size-fits-all solution. A material or component may be sustainable in one regard (e.g., glass is highly recyclable) but unsustainable in another (e.g., glass production generates more greenhouse gases than plastic). Therefore, it is essential to consider the entire life cycle of packaging, from manufacturing to disposal, rather than focusing solely on material recyclability. Additionally, while sustainability is often viewed exclusively in regard to environmental impact, both economic sustainability and social sustainability must also be considered alongside environmental sustainability. The pursuit of environmental, economic, and social sustainability is often referred to as the Triple Bottom Line [28, 30].

A comprehensive approach to evaluating the sustainability of packaging is through a Life Cycle Analysis (LCA). An LCA systematically assesses the environmental impact of a product or material across its entire life cycle, encompassing the following key stages:

1. Raw Material Extraction: The extraction and processing of materials like plastic, glass, and aluminum. This phase is often the most environmentally taxing due to its high-energy requirements, accounting for the majority of a package's carbon emissions. This is largely due to the mining of bauxite, which is used to produce aluminum, as the mining process reduces biodiversity and threatens water sources in developing regions [26].
2. Manufacturing: Transforming raw materials into the final product. Waste generated during this phase, such as factory scrap, adds to the overall environmental burden.
3. Distribution: Transporting the packaging to various destinations, including manufacturing plants and retailers. In addition to the carbon emissions generated during transportation, this phase also creates packaging waste, as products must

be protected during transit with tertiary packaging, especially if housed in fragile packaging materials, like glass.

4. Consumer Usage: Consumer behavior and the associated waste as a result of using a product is an often overlooked phase within a product's life cycle. This phase is challenging to quantify due to the variability of consumer habits; however, one significant environmental factor is water usage, particularly for rinse-off products. The amount of water required to effectively rinse off a facial cleanser, for example, is yet another factor in a product's overall environmental impact.
5. Disposal: Also referred to as "end of life," this phase tends to get the most attention from brands and consumers alike. Ideally, the disposal of a product entails recycling, composting or reusing the product package; however, poor design or improper package disposal can result in a package ending up in a landfill. While recycling remains a primary sustainability strategy for the disposal phase, in practice, its effectiveness remains limited, with no guarantee that properly discarded and sorted materials will actually be recycled. For example, a general recycling guideline states if a package is under 2 inches by 2 inches, it is unlikely to be properly sorted once it enters a recycling facility, and therefore will likely end up in a landfill, regardless of the recyclability of its material [15].

Each phase contributes to the overall environmental impact, with raw material extraction typically being the most significant contributor due to its energy-intensive nature. This is why conducting a full LCA is essential—it can reveal the hidden environmental costs of materials that may initially seem sustainable, such as aluminum, which, despite its higher recycling rate, requires a significant amount of energy to convert mined bauxite to alumina and then finally into aluminum [18].

An LCA offers valuable insights, but it is important to note these analyses are not absolute. The findings of an LCA can vary based on the assumptions and data used, and while they are not an exact science, advancements in technology are improving the accuracy and efficiency of LCAs.

Common Packaging Sustainability Strategies

Recyclability

One of the most prevalent mistakes brands make is over-relying on recyclability as their primary sustainability strategy. While it may seem like an ideal solution, the reality is far from simple. In the USA, less than 8.7% of plastic waste was successfully recycled. However, it is essential to highlight that this recycling rate combines all plastics, some of which are not commonly recycled, thus skewing the recycling rate. Unfortunately, many brands and consumers solely look at the combined recycling rate of plastic, and not the individual recycling rate of commonly recycled plastic. For example, two of the most commonly recycled plastics in the USA possess recycling rates within a similar range to glass and aluminum, with PET bottles possessing a 29.1% recycling rate and natural HDPE bottles, a 29.3% recycling rate. In comparison, glass has a recycling rate of 31% and aluminum has a recycling rate of 34.9% [40–42]. However, it is important to note that glass and aluminum can withstand more recycling cycles than PET or HDPE plastic, which degrades with each cycle.

The success of recycling is influenced by a host of factors: material type, size, color, decoration, and, importantly, the demand for that recycled material. Even with the right conditions, the business model of recycling facilities—aiming for profit rather than altruism—means that if it is not cost-efficient to recycle certain materials, they will often end up in landfills, regardless of their recyclability. Thus, while many brands place heavy emphasis on designing recyclable packaging, focusing on impactful sustainability strategies that begin earlier in the product's life cycle can often have a greater impact.

One such strategy is lightweighting. Sustainability in packaging is not limited to the materials used but also extends to design choices. One underutilized strategy in sustainable packaging design is lightweighting, which involves reducing the amount of material and parts required for production. Over time, and at scale, this approach can significantly reduce waste and carbon emissions [32].

Example: Cocokind Retinol Body Cream Tube (Fig. 1)

This tube features a cap that is significantly smaller in size than industry standard tube caps. Additionally, the tube cap is fused directly to the tube sleeve, as opposed to being produced as a separate component part, increasing manufacturing efficiencies. Due to its lightweight design, the tube reduces the amount of material used by 55% (when compared to industry standard tubes).

Fig. 1 Cocokind's Retinol Body Cream Lightweight Tube, 2024

Recycled Content Incorporating recycled content into packaging has emerged as another favored sustainability strategy. This approach involves using previously recycled materials—such as postconsumer resin (PCR) plastics—instead of virgin materials. While recycled content can help close the loop on packaging waste, it also comes with challenges. For instance, recycled plastics tend to degrade in quality with each reuse cycle, which can limit their application in certain types of cosmetic packaging where aesthetics and performance are key [1, 17].

Plant-Based Materials

Bioresins

Plant-based materials like bioresins—derived from renewable sources such as corn, sugarcane, or soy—are increasingly being explored as alternatives to petroleum-based plastics. These materials can offer a renewable source for plastics and potentially reduce carbon emissions during production. However, plant-based plastics can raise other sustainability concerns, particularly around land use and water consumption in agriculture. Moreover, they may not necessarily be biodegradable or recyclable, depending on their chemical structure [29].

Brands opting for plant-based materials must take into account these complexities, ensuring that the use of such materials does not inadvertently result in environmental harm elsewhere in the supply chain.

Plant-Based Material (Nonresins)

Beyond plant-based plastics are plant-based packaging materials that are sourced from renewable sources, but do not possess the same molecular structure or qualities as plastic. These often come with an extensive list of limitations, including compatibility with liquid formulas, reduced barrier properties, and an increased propensity for damage and breakage during consumer use. However, the industry is investing significant resources to improve the quality, availability, and scalability of these materials [16].

Example: Half Magic PaperFoam (Fig. 2)

Refill and Reuse

In an effort to minimize packaging waste, many brands are embracing refillable packaging systems. This strategy encourages consumers to reuse the original packaging and simply purchase a refill, which typically features a lighter-weight component.

Fig. 2 Half Magic's Cheek Fluff PaperFoam Compact, 2024

While refill systems show promise, they also face certain barriers, such as convenience and consumer adoption. For refill systems, particularly "bottle in a bottle" systems, to be effective, consumers must purchase the same product multiple times for there to be a reduction in material use and waste, as most refill systems initially use more material and generate more waste, as the system features an inner bottle (the refill) and an outer bottle (the keepsake component), as opposed to a singular bottle, as is the case with most single-use packaging systems. As a result, in order for bottle-in-a-bottle refill systems to be effective, they need to align with consumers' loyalty to a given product, while also offering a financial incentive to ensure they're inclined to purchase the refill, instead of the entire refill system again.

However, within the refill category also exist pouches, which offer an immediate reduction in material use due to their lightweight design. Refill pouches require less consumer loyalty, and in some cases, the refill pouch design may allow consumers to use the product straight from the pouch itself, eliminating the need to ever purchase the rigid, refill component.

Example: Gntl Skin Wash (Fig. 3)

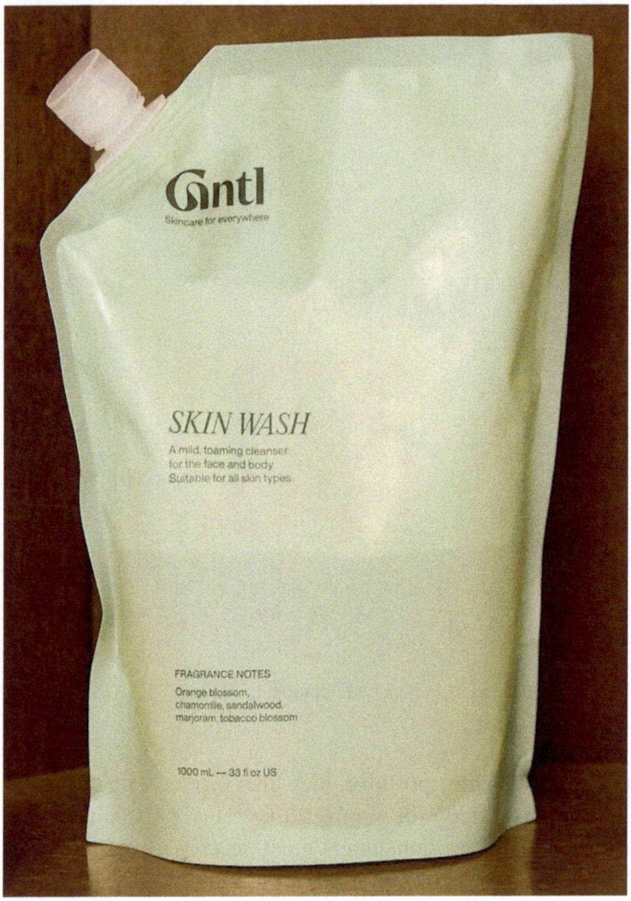

Fig. 3 Gntl's Skin Wash Pouch, 2024

Life Cycle Assessment (LCA) of Cosmetic Packaging Materials

In a recent LCA conducted by Babo Botanicals, the environmental impact of various packaging materials—plastic, glass, aluminum, and paper—was compared. While aluminum had the highest recycling rate, it also had the largest overall environmental burden due to its carbon emissions during the raw material extraction phase, which relies heavily on mining bauxite to produce aluminum. As a result, even with 20% PCR (postconsumer recycled) content, the aluminum package generated 608% more carbon emissions than the paper carton option, which proved to be the option with the lowest overall environmental burden (Fig. 4, 2024). This exemplifies the importance of looking at the holistic impact of a material or packaging design, instead of focusing solely on a given phase within the product life cycle.

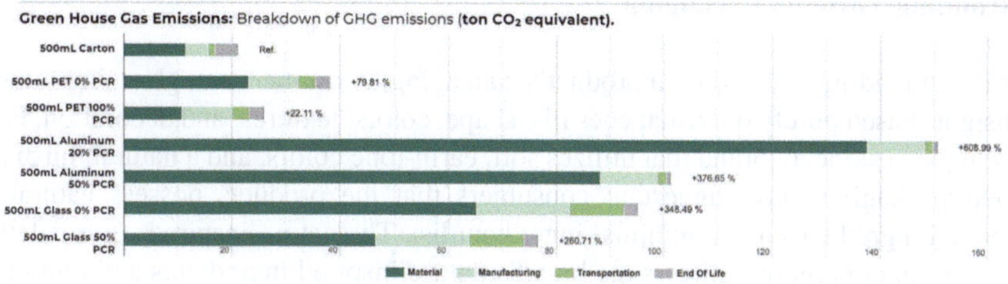

Fig. 4 Babo Botanicals life cycle analysis on greenhouse gas emissions, 2024

Case Studies Demonstrating Sustainable Packaging Solutions

Babo Botanicals

Babo Botanicals stands out for its use of packaging solutions that balance sustainability and brand identity. Leveraging findings from LCAs on various package options, the brand's emphasis on reducing environmental impact has led them to explore lightweight packaging, reduced material use, and even refill pouch systems. Their approach demonstrates how a brand can align its sustainability initiatives with its product ethos without sacrificing consumer appeal.

Cocokind

Cocokind offers another strong case study in sustainable packaging. The brand has adopted a multifaceted strategy that includes the use of postconsumer recycled (PCR) plastics and lightweight designs. In addition, Cocokind integrates sustainability education into their secondary packaging and throughout their online presence, encouraging consumers to engage with their sustainability journey by providing transparency about their packaging materials and end-of-life options.

Aesthetics and Design Principles

Role of Packaging Design in Brand Identity and Marketing

Packaging is not merely a vessel; it is a vital part of a brand's market positioning and communication to prospective customers. Many beauty consumers will not visit a brand's website or read through an "About" page, so packaging often serves as the primary communication tool between the brand and the consumer. As such, it plays a pivotal role in conveying brand values, product positioning, and market differentiation.

Branding Through Packaging

Without reading any text on a product's packaging, consumers can glean important insights based purely on visual cues like shape, colors, textures, and decoration. For example, a skincare brand that utilizes soft, earth-tone colors, and a natural finish in their packaging may suggest to consumers that the products possess naturally derived ingredients and a minimal ingredient list. The nature-inspired color palette and minimal branding mirrors the brand's nature-inspired ingredients and minimal ingredient list, creating subliminal associations in consumers' minds before they ever read the product name or pick up the package off l retailer shelves. The physical attributes of a package often create subconscious associations, which are essential in differentiating a product in a crowded marketplace and carving out a space, albeit a subconscious one, in the minds of prospective customers.

Trends in Packaging Aesthetics and Visual Communication

Minimalism

Minimalist design has become a dominant trend in packaging aesthetics, driven by consumer preferences, sustainability concerns, and cost. Today's consumers, often overstimulated by an onslaught of digital stimuli and an increasingly connected world, find comfort in the simplicity and calm that minimalist packaging provides. Less ornate, more subtle designs tend to appeal to a wider audience, aligning with values of decluttering and reducing consumption. Minimalism can also suggest a "no-fluff" approach, positioning a product and brand as focusing their efforts on efficacy over flashy graphics and attention-grabbing packaging.

Gender-Neutral

Gender-neutral packaging has emerged as a trend in response to evolving consumer preferences. Brands are opting for more neutral tones and shapes to create packaging that appeals across demographics, making products more accessible to a broader audience. This trend also aligns with the rising demand for inclusivity and gender-fluid products in the beauty industry, allowing consumers to express their identity visually with products that do not attempt to pigeonhole them into traditional roles or aesthetics.

Return to Playfulness

While minimalism dominates, some brands are reviving playful, fun elements in their designs. This can manifest through unique packaging silhouettes or bold colors and typeface. In a market saturated with subdued color palettes and minimal typeface, this approach can be an effective way to stand out, especially to Gen-Z consumers who tend to gravitate to highly individualistic branding.

Protection and Preservation of Cosmetic Products

The primary purpose of cosmetic packaging is to preserve the integrity of the product formula by safeguarding it from external elements such as air, light, moisture, and contamination. Packaging also protects the product during shipping and from improper handling by the consumer. The right packaging materials and designs can prevent oxidation, spoilage, and degradation, ensuring product efficacy and shelf life [45].

Convenience and User-Friendly Design Features

Consumers are more likely to repurchase products with convenient and intuitive packaging. Easy-to-use designs like pumps, twist-off caps, or airless packaging can enhance the user experience. For instance, airless packaging not only improves convenience but also protects the formula from air exposure, extending the product's life without needing additional preservatives.

Testing and Quality Assurance

Overview of Packaging Testing Methods

Packaging testing is a vital component of the product development process, ensuring both the safety and effectiveness of the final product. While the specific tests required before a product launch depend on the formula, packaging material, and format, several key assessments are commonly conducted [23].

Testing and Quality Assurance

While there is an array of packaging tests available, the most popular for cosmetic packaging are outlined below:

1. Stability Testing
   Purpose: Ensures that the product maintains its integrity over time under various storage and usage conditions.
   Conditions:

   - Accelerated testing: 40 °C ± 2 °C and 75% ± 5% RH (relative humidity) for 3–6 months.
   - Room temperature testing: 20–25 °C for 12 months.
   - Low-temperature testing: 4–8 °C for 12 months.
   - Freeze-thaw testing: −20 °C to 25 °C, cycling for 3–5 cycles.
   - Cause for concern: Formula separation, color change, texture degradation, or loss of efficacy [19].

2. Compatibility Testing
   Purpose: Verifies that the formula does not react with the packaging material, maintaining formula quality and stability.
   Conditions:

   - Storage in different environments: Testing at 40 °C and 75% RH, room temperature, and 4 °C for 1–3 months.
   - Contact time: Continuous contact between the formula and packaging material during the test period.
   - Packaging components: Cap liners, bottle walls, pumps, etc., are all tested for chemical interactions.

   Cause for concern: Chemical reactions such as leaching, discoloration, cracking, or formula changes [19].
3. Barrier Testing
   Purpose: Assesses the packaging's ability to protect the product from external factors such as oxygen, moisture, and light.
   Conditions:

   - Moisture vapor transmission rate (MVTR) testing: Packaging exposed to 50% RH at 25 °C for 24–72 h [39].
   - Oxygen transmission rate (OTR) testing: Packaging exposed to an oxygen-rich environment and tested at 23 °C and 50% RH for permeability over 24–72 h [2].
   - Light exposure testing: Formula exposed to UV light for 1–3 months, simulating light exposure during storage [7].

   Cause for concern: Formula degradation or loss of efficacy over time.

4. Leakage and Seal Integrity Testing
   Purpose: Ensures that the packaging maintains an airtight and leak-proof seal.
   Conditions:

   - Vacuum testing: Packaging is placed under a vacuum to check for leakage [25].
   - Pressure tests: The packaging is subjected to internal pressure to assess durability and sealing effectiveness [12].

   Cause for concern: Product leakage, contamination, or loss of integrity, as all of these can be caused by weak seals.
5. Transit and Drop Testing
   Purpose: Evaluates how well the packaging and product withstand shipping, handling, and consumer use.
   Conditions:

   - Vibration testing: Simulates transportation by vibrating at various frequencies and directions over a period of hours.
   - Drop tests: Packaging is dropped multiple times from specific heights (typically 1–2 m) onto different surfaces.
   - Compression testing: Simulates stacking during shipping by applying weight to the packaging.

   Cause for concern: Visible signs of damage or product leakage, which can occur during transportation if a product is not properly packed or protected [10].
6. Fill Weight and Volume Testing
   Purpose: Verifies that the packaging contains the correct amount of product as labeled.
   Conditions:

   - Fill level tests: Packaging is filled and weighed under normal conditions (room temperature, controlled filling environment).
   - Net content tests: Ensures the actual fill volume matches the label claims after sealing and during shelf life testing.

   Cause for concern: Inconsistent fill levels outside of industry standard variances, as this can result from misaligned filling systems or environmental factors like temperature and pressure [21].
7. Material Safety Testing
   Purpose: Ensures that packaging materials are nontoxic and do not pose a risk to consumers.
   Conditions:

   - Extractables and leachable testing: Packaging material is exposed to the formula at 40 °C or 60 °C for 1–3 months, then analyzed for chemical leaching.

- Migration testing: Packaging is subjected to food-grade safety tests, often at elevated temperatures (e.g., 60 °C) for 10 days, to assess potential chemical migration into the product.

Cause for concern: Chemical migration from packaging materials, as this can pose safety risks to consumers [27].

8. Weight Loss Testing
   Purpose: Determine the product weight loss due to evaporation or leakage over time.
   Process: Packaging is placed in controlled environmental conditions (e.g., 40 °C and 75% relative humidity) for 1–3 months to measure any loss in product weight due to evaporation or leakage. In order for the package to be considered acceptable, product weight loss should not exceed 1% per month [24].
   Cause for concern: Changes in product quality or shelf life, which can be caused by packaging with poor barrier properties, which may allow moisture or volatile ingredients to escape.
9. Dispensing Testing
   Purpose: Determine the compatibility and proper dispensing of a product formula with a package's dispensing system.
   Process: Conducted by dispensing the product repeatedly under various conditions, such as different angles, pressures, and product viscosities, to ensure consistent product release.
   Cause for Concern: If the packaging fails to dispense the product smoothly or evenly, the dispensing system may not be suited for the formula texture or viscosity.
10. Restitution Testing
    Purpose: Determines the amount of product that can be dispensed compared to the product residue left inside the package.
    Process: Typically determined by weighing the remaining product after use.
    Cause for concern: Low restitution can result in product wastage and negatively impact consumer experience, especially for expensive or premium products.

Common Challenges When Pairing Formulas and Packaging

1. Ingredient-Packaging Interactions
   Challenge: Ingredients such as acids, essential oils, or preservatives can react with packaging materials (plastics or metals in particular), causing instability or physical changes in the formula.
   Solution: Use inert materials like glass, certain plastics with barrier coatings, or lined metal to avoid chemical reactions.
2. Volatile Formulas
   Challenge: Volatile components (e.g., alcohols or fragrances) can evaporate or permeate through packaging materials, affecting the formula's stability.

Solution: Use airtight or barrier packaging that limits evaporation or absorption by the material.

3. Oxidation of Formulas
   Challenge: Light or oxygen-sensitive ingredients (e.g., L-ascorbic acid or retinol) can oxidize, leading to loss of efficacy and formula color changes.
   Solution: Use airless pumps, opaque or tinted packaging, or nitrogen-flushed containers to limit exposure to oxygen and light.
4. Packaging Sustainability Versus Formula Integrity
   Challenge: Sustainable materials may lack the same protective properties as traditional plastics or metals, compromising product safety or longevity.
   Solution: When considering packaging materials that do not possess weakened properties, ensure there is ample time for testing, both in the lab and in application, to ensure the sustainability of the packaging materials, and do not compromise the integrity or efficacy of the product formula.
5. Texture and Viscosity Challenges
   Challenge: Thick or viscous formulas may be difficult to dispense from traditional packaging like dip-tube pumps or tubes.
   Solution: Select dispensing systems that pair well with a formula's texture and viscosity, such as bottles with the appropriate size orifice, or modify the formula's viscosity to improve user experience if changing the packaging is not an option.
6. Preservation and Contamination Risk
   Challenge: Formulas with minimal preservatives are more prone to microbial growth if the packaging does not provide adequate protection, such as jars that require consumers to dip their fingers, or concealer applications that introduce bacteria from a consumer's face into the concealer formula.
   Solution: Reduce the amount of environmental and consumer exposure the formula requires, and consider selecting packaging that minimizes the introduction of microbial growth, such as airless pump systems.
7. Aesthetic and Branding Constraints
   Challenge: Brand-driven packaging preferences (e.g., clear glass) may not provide adequate protection for sensitive formulas.
   Solution: Combine aesthetic appeal with functional design, such as incorporating protective layers or UV coatings into packaging. While a brand may be driven by their aesthetic vision, if you fear their packaging and decoration choice will ultimately hinder a product's performance, safety, or shelf life, speak up.
8. Temperature Sensitivity
   Challenge: Products that are sensitive to heat (e.g., certain organic formulas) can degrade if exposed to high temperatures during shipping or storage, especially if a product is intended for international distribution.
   Solution: Use insulated packaging materials or secondary packaging to minimize exposure to extreme temperatures, and ensure the proper tertiary packaging and packout is used, especially when the product travels long distances by land, sea, or air [9].

Importance of Quality Control in Packaging Manufacturing

Rigorous quality control ensures that packaging meets the necessary functional and aesthetic standards. Brands need to ensure that packaging is not only visually consistent but also that it performs effectively, sealing in the product and protecting it from contamination or loss of integrity. Any defects, both functional and aesthetic, can lead to consumer dissatisfaction.

Packaging Material Challenges

Metal compatibility Not compatible with reactive ingredients, which limits its application. However, this can be mitigated by the inclusion of a thin plastic barrier on the inside of the metal component, reducing the likelihood of the formula and metal reacting.

Glass compatibility Compatible with a wide array of formulas and ingredients; however, glass packaging has limitations around its design capabilities. For example, while glass is generally compatible with high alcohol formulas, glass bottles generally cannot be paired with airless pump systems due to vacuum seal limitations. Hence why both packaging material and design must be considered [44].

Plastic compatibility Not compatible with high alcohol formulas, like fine fragrance, as both the high alcohol content and the fragrance oils can slowly dissolve certain types of plastic. Additionally, high alcohol content may cause the stabilizers or additives in the plastic to start leaching [20].

One of the most frequent considerations in packaging selection is whether to pair a formula with an airless or dip-tube pump. While dip tubes are more affordable, if a formula includes ingredients susceptible to air degradation, investing in an airless pump not only ensures superior product integrity, but it also helps ensure consumer satisfaction.

Airless Pump

An airless pump system operates without the need for a traditional dip tube and instead relies on a vacuum mechanism to dispense the product. Below is an overview of how it works:

1. Vacuum Creation: When the pump is pressed, it creates a vacuum that pulls the product up from the bottom of the container.
2. Rising Piston: Inside the container, there is typically a disc or piston that moves upward as the product is dispensed, reducing the air gap and keeping the product airtight.

3. Reduced Air Introduction: Since the system is sealed, no air enters the container, which prevents oxidation and contamination, preserving the product's integrity and extending shelf life.
4. Dispensing: As the piston moves up, the product is forced through the pump head and out of the nozzle.

Key Benefit Airless pumps are ideal for sensitive formulations like serums or creams because they keep the product free from air exposure, reducing the risk of spoilage.

Dip-Tube Pump

A dip-tube pump (standard pump dispenser) uses a straw-like tube that extends into the product inside the container. Here is how it works:

1. Air Pressure and Suction: When the pump is pressed, it creates suction, drawing the product up through the dip tube.
2. Air Replacement: As the product is dispensed, air enters the container to replace the product being used, keeping the pressure inside the bottle balanced.
3. Dispensing: The product is forced through the tube and exits via the pump head, with the amount dispensed controlled by the pump mechanism.

Key Challenge Over time, air introduced into the container may lead to oxidation or degradation of the product, making it less suitable for sensitive or preservative-free formulas [5].

Manufacturing Processes

Overview of Packaging Manufacturing Techniques

Common manufacturing methods include:

Extrusion Blow Molding (EBM)
Extrusion Blow Molding is used to produce hollow packaging with relatively simple shapes, such as bottles. In this process, plastic is melted and extruded into a hollow tube, called a parison, which is then inflated inside a mold to form the desired shape. Once the plastic cools and hardens, the mold is opened, and the finished product is ejected.

For example, cosmetic bottles are commonly created using EBM. The process allows for flexible packaging where needed, but it is typically used for rigid or semi-rigid containers.

Materials Common materials include HDPE (High-Density Polyethylene), PP (Polypropylene), and PET (Polyethylene Terephthalate).

Injection Blow Molding (IBM)

Injection Blow Molding is used for more complex shapes and rigid packaging. It involves creating a preform by injecting molten plastic into a preform mold. The preform is then transferred to a blow mold, where it is inflated to match the final packaging shape. After cooling, the finished product is removed from the mold.

This method is ideal for creating high-precision packaging with consistent wall thickness.

Materials Common materials include PP (Polypropylene), HDPE (High-Density Polyethylene), and PET (Polyethylene Terephthalate) [13].

Tube Manufacturing Process

Cosmetic tubes, commonly used for creams, gels, and lotions, are typically produced through extrusion or lamination processes, depending on the desired structure and material.

1. Extrusion Process (for Monolayer Tubes)

   - Plastic Granules (usually PE) are melted and extruded into a cylindrical shape to form a single continuous wall.
   - The tube body is then cut to the desired length, and the shoulder and neck of the tube are added in a secondary step through injection molding.
   - The tube is closed on one end and left open on the dispensing end, which will later be sealed after the product is filled [38].

2. Lamination Process (for Multilayer Tubes)

   - Laminated Tubes consist of multiple layers of materials (such as PE, aluminum, or EVOH), providing barrier properties for sensitive formulations.
   - In this process, sheets of laminated materials are cut and shaped into a tube, with the layers fused together through heat and pressure.
   - The shoulder and neck are molded and attached, and the tube is sealed at the bottom after filling.

Materials Tubes are typically made from PE (Polyethylene) for flexibility, but can also include aluminum for multilayer laminated structures that offer additional barrier properties [33].

Materials Selection and Production Methods

Material selection plays a huge role in package sustainability, cost, and product protection. Materials should be selected based on their compatibility with a product's formula and ingredients, their ability to protect the product's integrity over the product life cycle, their alignment with a brand's sustainability goals, and, often, the deciding factor in packaging selection, their alignment with a brand's budget and cost targets for a given product.

Generally, plastic packaging is the most cost-effective, with other popular materials like glass and metal costing more to produce, as well as transport, due to their increased weight. Virgin materials are less expensive than recycled; however, with growing sustainability concerns, many brands are willing to pay more for recycled materials in order to satisfy consumer demand for more sustainable packaging choices. When selecting material with recycled content, be sure to ask for certificates of analysis, which can help ensure the material supplier has verified the quality, compliance, and safety of the recycled material.

While there has been a surge in demand for recycled materials in cosmetic packaging, there has also been an increase in fraudulent recycled materials that lack a verified percentage, origin, or quality within the industry.

Efficiency and Cost Considerations in Manufacturing

Materials As stated in the previous section, plastic tends to be the most cost-effective material for cosmetic packaging production, followed by recycled plastic.

Recycled content for plastic (and other packaging materials) primarily falls into two categories: postconsumer recycled (PCR) and postindustrial recycled (PIR). PCR content is recycled after a consumer has used and discarded a package, whereas PIR content is recycled after a packaging manufacturer collects the production scrap or defective packaging for use in future production. Most recycled content we see used in cosmetic packaging is PCR content [3].

Order Volume Higher-order quantities result in lower costs per unit of packaging, as manufacturers can take advantage of bulk pricing for raw materials and production efficiencies. This can make initial packaging costs high for small brands that do not have the budget or demand for large volumes; however, as they scale up their volumes, their unit costs should reduce.

Stock Versus Custom While stock packaging is more readily available and requires less of an upfront investment, custom packaging can be price competitive once the initial tooling investment is paid. However, tooling investments can range from a couple thousand dollars to tens of thousands (and potentially even hundreds of thousands if you're investing in multiple packaging SKUs). While some manufacturers require the tooling investment upfront prior to starting the custom mold tool, some will allow the tooling investment to be amortized into the unit cost of a final package.

When considering stock versus custom packaging, brands must weigh the upfront costs of custom molds, which can run from four to five figures depending on the complexity. Stock packaging, while cheaper and faster to produce, may lack the brand differentiation that custom packaging offers. Additionally, decoration costs (such as silk-screen printing or embossing) are recurring with each production run, increasing the overall cost of the packaging.

Domestic (US) Versus Overseas US packaging production is expensive, especially compared to countries with lower labor costs. The majority of cosmetic packaging is produced in China, as it is among the most cost-effective. Note: packaging produced in China and shipped to the USA is subject to the tariff the US government imposes on goods produced in China and shipped to the USA. The tariff is paid by the brand importing the packaging; however, even with the tariff, packaging produced in China continues to be some of the most cost-competitive in the world [8].

Conclusion

Recap of Key Points Covered in the Chapter

This chapter has explored the multifaceted relationship between packaging design and consumer perception, illustrating how design influences brand identity, marketing strategies, and ultimately consumer purchase decisions. We have also covered the functional role of packaging, which extends from protecting the product formula to enhancing usability and adhering to regulatory safety standards. Additionally, sustainability in packaging has been a central theme, highlighting the importance of material selection and sustainable design techniques. It is clear that sustainable packaging strategies must be tailored to the specific product format, packaging type, material choice, and consumer behavior to create an environmentally responsible approach without compromising functionality or aesthetic appeal.

Implications of Packaging Trends and Innovations for the Beauty Industry

The shift toward sustainable, minimalist, and user-friendly packaging solutions reflects broader cultural shifts toward conscious consumption and environmental stewardship. Brands that embrace these trends can strengthen their market position while meeting the evolving needs of modern beauty consumers.

Future Directions and Areas for Further Research

Ongoing research into innovative materials, such as biodegradable or fully compostable packaging, will drive the next wave of packaging advancements. Additionally, developments in refillable packaging models and closed-loop systems are expected to grow as brands strive to reduce environmental impact while maintaining aesthetics and functionality. Circularity continues to be the end goal; however, successful implementation is dependent on creating the proper infrastructure, which requires legislative support, consumer adoption, and brand/retailer implementation.

References

1. Alassali, A., Picuno, C., Chong, Z. K., Guo, J., Maletz, R., & Kuchta, K. (2021). Towards higher quality of recycled plastics: Limitations from the material's perspective. *Sustainability, 13*(23), 13266. https://doi.org/10.3390/su132313266
2. ASTM International. (n.d.). *ASTM F1927 – Standard test method for determination of oxygen gas transmission rate through plastic film and sheeting using a coulometric sensor*. https://www.document-center.com/standards/show/ASTM-F1927
3. ATS Tanner. (n.d.). *What is the difference between post-consumer recycled material (PCR) and post-industrial recycled materials (PIR)?* https://www.ats-tanner.com/en/news-and-applications/what-is-the-difference-between-post-consumer-recycled-material-pcr-and-post-industrial-recycled-materials-pir-183621
4. Bioplastics News. (n.d.). *Bioplastics*. Bioplastics News. https://bioplasticsnews.com/bioplastics/
5. BlueSky Solutions. (n.d.). *Harnessing the power of pump packaging: Airless vs. conventional lotion pumps*. BlueSky Solutions. https://blueskysolutionsuk.com/harnessing-the-power-of-pump-packaging-airless-vs-conventional-lotion-pumps/
6. Briasco, B., Capra, P., Cozzi, A. C., Mannucci, B., & Perugini, P. (2016). Packaging evaluation approach to improve cosmetic product safety. *MDPI, 3*(3), 32. https://www.mdpi.com/2079-9284/3/3/32
7. CE.way Regulatory Consultants. (n.d.). *Cosmetic stability and compatibility testing*. CE.way. https://news.ceway.eu/cosmetic-stability-compatibility-testing/
8. Congressional Research Service. (2020, June 16). *Trends in plastic waste*. https://crsreports.congress.gov/product/pdf/IF/IF11582
9. Cosmopacks. (n.d.). *The importance of package compatibility testing for cosmetic packaging*. Cosmopacks. https://www.cosmopacks.com/the-importance-of-package-compatibility-testing-for-cosmetic-packaging/
10. Electronic Code of Federal Regulations. (n.d.). *49 CFR § 178.603 – Drop test*. https://www.ecfr.gov/current/title-49/subtitle-B/chapter-I/subchapter-C/part-178/subpart-M/section-178.603
11. FEVE. (n.d.). *What is the history of glass packaging?* https://FEVE.org/about-glass/history/
12. FlexPak. (n.d.). *Seal strength test method*. FlexPak. https://flexpakinc.com/seal-strength-test-method/
13. Formlabs. (n.d.). *What is blow molding? Process, materials, and types of blow molding*. Formlabs. https://formlabs.com/blog/blow-molding/
14. Friedman, L. (2020, January 21). *Why bioplastics will not solve the world's plastics problem*. Yale Environment 360. https://e360.yale.edu/features/why-bioplastics-will-not-solve-the-worlds-plastics-problem
15. How2Recycle. (2024, February). *How2Recycle guidelines abbreviated*. How2Recycle. https://how2recycle.info/wp-content/uploads/2024/01/How2Recycle_Guidelines_Abbreviated_Feb2024.pdf
16. Hubbe, M. A., & Ayoub, A. (2021). Renewable fibers and bio-based materials for packaging applications: A review of recent developments. *BioResources, 16*(2), 3293–3377. https://bioresources.cnr.ncsu.edu/resources/renewable-fibers-and-bio-based-materials-for-packaging-applications-a-review-of-recent-developments/

17. Ibrahim, I., Ashour, A. G., Zeiada, W., Salem, N., & Abdallah, M. (2024). A systematic review on the technical performance and sustainability of 3D printing filaments using recycled plastic. *Sustainability, 16*(18), 8247. https://doi.org/10.3390/su16188247
18. Ingwersen, W., & Safaei, A. (2022, May 31). *Introduction to life cycle assessment methodology and standards*. Workshop on Carbon LCA of Vehicles, UNECE, Geneva. US EPA & WBCSD. https://unece.org/sites/default/files/2022-06/1_2_GHG_EPA.pdf
19. International Council for Harmonisation of Technical Requirements for Pharmaceuticals for Human Use (ICH). (2018). *Q1F: Stability data package for registration applications in climatic zones III and IV* (WHO Version). https://database.ich.org/sites/default/files/Q1F_Stability_Guideline_WHO_2018.pdf
20. Jahn, M. (2018). Leachables and extractables: From regulatory expectations to laboratory assessment. In N. Warne & H. C. Mahler (Eds.), *Challenges in protein product development* (Vol. 38, pp. 493–523). Springer. https://doi.org/10.1007/978-3-319-90603-4_16
21. Keystone Compliance. (n.d.). *Personal care package testing*. Keystone Compliance. https://keystonecompliance.com/personal-care-package-testing/
22. Kumar, A. (2021). *Globalization and beauty: A historical and interdisciplinary perspective.* Retrieved from https://d1wqtxts1xzle7.cloudfront.net/69837777/Globalization_and_Beauty_A_Historical_an20210918-24175-9m2jre.pdf
23. Labthink. (n.d.). *Test items and methods for cosmetic packaging*. https://en.labthink.com/en-us/literatures/test-items-and-methods-for-cosmetic-packaging.html
24. Making Cosmetics. (n.d.). *Stability testing of cosmetics*. https://www.makingcosmetics.com/Stability-Testing-of-Cosmetics_ep_59.html
25. Manufacturing Engineering Technology (MET). (n.d.). *Sealed packs leak test*. MET. https://met.uk.com/uploads/images/pdfs/sealed_packs_leak_test.pdf
26. McGovern, G. (2023, April 12). *How bauxite mining destroys nature and communities*. CMSWire. https://www.cmswire.com/digital-experience/how-bauxite-mining-destroys-nature-and-communities/
27. Murat, P., Puttaswamy, S. H., Ferret, P.-J., Coslédan, S., & Simon, V. (2020). Identification of potential extractables and leachables in cosmetic plastic packaging by microchambers-thermal extraction and pyrolysis-gas chromatography-mass spectrometry. *Molecules, 25*(9), 2115. https://doi.org/10.3390/molecules25092115
28. Nguyen, L. H., Allemann, F. P., & Verones, F. (2020). Environmental life cycle assessment of bioplastics: A review of their potential environmental impacts compared to conventional plastics. *Environmental Science and Pollution Research, 27*(23), 28820–28839. https://doi.org/10.1007/s11356-020-09284-0
29. Novak, M., Amin, V., & Liang, H. (2024). Biodegradable biobased polymers: A review of the state of the art. *MDPI Polymers*. https://www.mdpi.com/2073-4360/16/16/2262
30. Obadă, D.-R., Grădinaru, I.-A., & Dabija, D.-C. (2022). Enhancing sustainable cosmetics brand purchase: A comprehensive approach based on the SOR model and the triple bottom line. *Sustainability, 14*(21), 14118. https://doi.org/10.3390/su142114118
31. Olam, M. (2023). *Mechanical and thermal properties of HDPE/PET microplastics, applications, and impact on environment and life*. IntechOpen. https://www.intechopen.com/chapters/86412
32. Packaging Europe. (2020). *Doing more, with less: The future of lightweight packaging*. https://www.packagingeurope.com
33. PBS Plastics. (n.d.). *What is the difference between plastics extrusion and injection molding?* https://www.pbsplastics.com/what-is-the-difference-between-plastics-extrusion-and-injection-molding/
34. Plastic Pollution Coalition. (2023). *The ugly side of beauty: The cosmetics industry's plastic packaging problem*. https://www.plasticpollutioncoalition.org
35. Plastics Recycling. (2018). *Life cycle inventory of postconsumer recycled resins (PCR) used in packaging applications*. Plastics Recycling. https://plasticsrecycling.org/images/library/2018-APR-LCI-report.pdf

36. Stoica, D., & Stuparu-Creţu, M. (2024). Review of bio-based biodegradable polymers: Smart solutions for sustainable food packaging. *MDPI*. https://www.mdpi.com/2304-8158/13/19/3027
37. The Aluminum Association. (2020). *Aluminum: A lightweight, recyclable, and corrosion-resistant material*. https://www.aluminum.org
38. Tube.org. (n.d.). *How plastic tubes are made*. Tube.org. https://tube.org/learning-center/how-plastic-tubes-are-made/
39. Turan, D. (2021). Water vapor transport properties of polyurethane films for packaging of respiring foods. *Food Engineering Reviews, 13*(1), 54–65. https://doi.org/10.1007/s12393-019-09205-z
40. U.S. Environmental Protection Agency. (2020a). *Aluminum: Material-specific data*. https://www.epa.gov/facts-and-figures-about-materials-waste-and-recycling/aluminum-material-specific-data
41. U.S. Environmental Protection Agency. (2020b). *Glass: Material-specific data*. https://www.epa.gov/facts-and-figures-about-materials-waste-and-recycling/glass-material-specific-data
42. U.S. Environmental Protection Agency. (2020c). *Plastics: Material-specific data*. https://www.epa.gov/facts-and-figures-about-materials-waste-and-recycling/plastics-material-specific-data
43. Vassallo, N., & Refalo, P. (2024). Reducing the environmental impacts of plastic cosmetic packaging: A multi-attribute life cycle assessment. *Cosmetics, 11*(2), 34. https://doi.org/10.3390/cosmetics11020034
44. Visonpack. (n.d.). *Plastic, metal, or glass: Which is the better choice for cosmetic packaging?* https://www.visonpack.com/blog/plastic-metal-or-glass-which-is-the-better-choice-for-cosmetic-packaging/#ib-toc-anchor-1
45. Vollmer, I., Jenks, M. J. F., Roelands, M. C. P., White, R. J., van Harmelen, T., Wild, P., Lammertsma, K., & Rabnawaz, M. (2021). Beyond mechanical recycling: Giving new life to plastic waste. *Circular Economy and Sustainability, 1*(2), 566–582. https://doi.org/10.1007/s43615-021-00145-7
46. Lagarde, E. (2022, March 21). Ancient times - packaging history. Re-Sources. https://re-sources.co/ancient-history-packaging/

Index

K. Alexander, P. Romanowski (eds.), *Cosmetic Science*, AAPS Introductions in the Pharmaceutical Sciences 15, https://doi.org/10.1007/978-3-032-03596-7

L

M

N

O

P

Q

R

MIX
Papier aus verantwortungsvollen Quellen
Paper from responsible sources
FSC® C105338

If you have any concerns about our products,
you can contact us on
ProductSafety@springernature.com

In case Publisher is established outside the EU,
the EU authorized representative is:
Springer Nature Customer Service Center GmbH
Europaplatz 3, 69115 Heidelberg, Germany

Printed by Libri Plureos GmbH
in Hamburg, Germany